Multistory Rigid Frame =

- Minimum: 5 borings or
  1 boring / 40,000 ft^2

usual: 100-200' grid spacing

Minimum: 3 z below foundation
50z 100' or $\Delta \sigma_v = 5\%$ to $10\% \sigma'_v$

Residential development: min: $\dfrac{1 \text{ boring}}{90,000 \text{ ft}^2}$

300' grid spacing

2φ's

$$\sigma_3 = \sigma_3' = \dfrac{\Delta \sigma_d}{\tan^2\left(45 + \frac{\varphi}{2}\right)_1}$$

$$\gamma_d = \dfrac{\gamma}{1+w}$$

$$e = \dfrac{G_s}{G_m} - 1$$

Pressure head b = (11 - headloss) + Elevation head

Fundamentals of
Geotechnical Engineering

Fundamentals of Geotechnical Engineering

THIRD EDITION

Braja M. Das

CENGAGE
Learning™

Australia • Brazil • Japan • Korea • Mexico • Singapore • Spain • United Kingdom • United States

CENGAGE
Learning

Fundamentals of Geotechnical Engineering, Third Edition
Braja M. Das

Director, Global Engineering Program:
   Chris Carson

Senior Developmental Editor: Hilda Gowans

Production Manager: Renate McCloy

Production Service: RPK Editorial Services

Copyeditor: Shelly Gerger-Knechtl

Proofreader: Martha McMaster

Indexer: Braja Das

Compositor: Integra

Creative Director: Angela Cluer

Internal Designer: Carmela Pereira

Cover Designer: Andrew Adams

Cover Image: Courtesy of Geopier
   Foundation Company, Inc.,
   www.geopier.com

Permissions Coordinator: Kristiina Bowering

For product information and technology assistance, contact us at **Cengage Learning Customer & Sales Support, 1-800-354-9706**.

For permission to use material from this text or product, submit all requests online at **www.cengage.com/permissions**. Further permissions questions can be emailed to **permissionrequest@cengage.com**.

Library of Congress Control Number: 2007939898

U.S. Student Edition:

ISBN-13: 978-0-495-29572-3
ISBN-10: 0-495-29572-8

Cengage Learning
200 First Stamford Place, Suite 400
Stamford, CT 06902
USA

Cengage Learning is a leading provider of customized learning solutions with office locations around the globe, including Singapore, the United Kingdom, Australia, Mexico, Brazil, and Japan. Locate your local office at: **international.cengage.com/region**.

Cengage Learning products are represented in Canada by Nelson Education Ltd.

For your course and learning solutions, visit **academic.cengage.com/engineering**.

Purchase any of our products at your local college store or at our preferred online store **www.ichapters.com**.

Printed in the United States of America
3 4 5 6 7 10 09

To our granddaughter, Elizabeth Madison

$$\gamma = \frac{W}{\forall} = \frac{1}{\forall}\left(u_s + w_w\right) \cdot \frac{w_s}{W_s}$$

$$= \frac{w_s}{\forall}\left(\frac{u_s}{W_s} + \frac{w_w}{w_s}\right)$$

$$\gamma = \gamma_c\left(1 + w\right)$$

Preface

Principles of Foundation Engineering and *Principles of Geotechnical Engineering* were originally published in 1984 and 1985, respectively. These texts were well received by instructors, students, and practitioners alike. Depending on the needs of the users, the texts were revised and are presently in their sixth editions.

Toward the latter part of 1998, there were several requests to prepare a single volume that was concise in nature but combined the essential components of *Principles of Foundation Engineering* and *Principles of Geotechnical Engineering*. In response to those requests, the first edition of *Fundamentals of Geotechnical Engineering* was published in 2000, followed by the second edition in 2004 with a 2005 copyright. These editions include the fundamental concepts of soil mechanics as well as foundation engineering, including bearing capacity and settlement of shallow foundations (spread footings and mats), retaining walls, braced cuts, piles, and drilled shafts.

This third edition has been revised and prepared based on comments received from the users. As in the previous editions, SI units are used throughout the text. This edition consists of 14 chapters. The major changes from the second edition include the following:

- The majority of example problems and homework problems are new.
- Chapter 2 on "Soil Deposits and Grain-Size Analysis" has an expanded discussion on residual soil, alluvial soil, lacustrine deposits, glacial deposits, aeolian deposits, and organic soil.
- Chapter 3 on "Weight-Volume Relationships, Plasticity, and Soil Classification" includes recently published relationships for maximum and minimum void ratios as they relate to the estimation of relative density of granular soils. The fall cone method to determine liquid and plastic limits has been added.
- Recently published empirical relationships to estimate the maximum unit weight and optimum moisture content of granular and cohesive soils are included in Chapter 4 on "Soil Compaction."
- Procedures to estimate the hydraulic conductivity of granular soil using the results of grain-size analysis via the Kozeny-Carman equation are provided in Chapter 5, "Hydraulic Conductivity and Seepage."

- Chapter 6 on "Stresses in a Soil Mass" has new sections on Westergaard's solution for vertical stress due to point load, line load of finite length, and rectangularly loaded area.
- Additional correlations for the degree of consolidation, time factor, and coefficient of secondary consolidation are provided in Chapter 7 on "Consolidation."
- Chapter 8 on "Shear Strength of Soil" has extended discussions on sensitivity, thixotropy, and anisotropy of clays.
- Spencer's solution for stability of simple slopes with steady-state seepage has been added in Chapter 9 on "Slope Stability."
- Recently developed correlations between relative density and corrected standard penetration number, as well as angle of friction and cone penetration resistance have been included in Chapter 10 on "Subsurface Exploration."
- Chapter 11 on "Lateral Earth Pressure" now has graphs and tables required to estimate passive earth pressure using the solution of Caquot and Kerisel.
- Elastic settlement calculation for shallow foundations on granular soil using the strain-influence factor has been incorporated into Chapter 12 on "Shallow Foundations—Bearing Capacity and Settlement."
- Design procedures for mechanically stabilized earth retaining walls is included in Chapter 12 on "Retaining Walls and Braced Cuts."

It is important to emphasize the difference between soil mechanics and foundation engineering in the classroom. Soil mechanics is the branch of engineering that involves the study of the properties of soils and their behavior under stresses and strains under idealized conditions. Foundation engineering applies the principles of soil mechanics and geology in the plan, design, and construction of foundations for buildings, highways, dams, and so forth. Approximations and deviations from idealized conditions of soil mechanics become necessary for proper foundation design because, in most cases, natural soil deposits are not homogeneous. However, if a structure is to function properly, these approximations can be made only by an engineer who has a good background in soil mechanics. This book provides that background.

Fundamentals of Geotechnical Engineering is abundantly illustrated to help students understand the material. Several examples are included in each chapter. At the end of each chapter, problems are provided for homework assignment, and they are all in SI units.

My wife, Janice, has been a constant source of inspiration and help in completing the project. I would also like to thank Christopher Carson, General Manager, and Hilda Gowans, Senior Development Editor, of Cengage Learning for their encouragement, help, and understanding throughout the preparation and publication of the manuscript.

BRAJA M. DAS
Henderson, Nevada

Contents

1   Geotechnical Engineering—A Historical Perspective   1

1.1   Geotechnical Engineering Prior to the 18th Century   1
1.2   Preclassical Period of Soil Mechanics (1700–1776)   4
1.3   Classical Soil Mechanics — Phase I (1776–1856)   5
1.4   Classical Soil Mechanics — Phase II (1856–1910)   5
1.5   Modern Soil Mechanics (1910–1927)   6
1.6   Geotechnical Engineering after 1927   7
     References   11

2   Soil Deposits and Grain-Size Analysis   13

2.1   Natural Soil Deposits-General   13
2.2   Residual Soil   14
2.3   Gravity Transported Soil   14
2.4   Alluvial Deposits   14
2.5   Lacustrine Deposits   16
2.6   Glacial Deposits   17
2.7   Aeolian Soil Deposits   17
2.8   Organic Soil   18
2.9   Soil-Particle Size   19
2.10   Clay Minerals   20
2.11   Specific Gravity (G_s)   23
2.12   Mechanical Analysis of Soil   24
2.13   Effective Size, Uniformity Coefficient, and Coefficient of Gradation   32
     Problems   35
     References   37

3 Weight–Volume Relationships, Plasticity, and Soil Classification 38

3.1 Weight–Volume Relationships 38
3.2 Relationships among Unit Weight, Void Ratio, Moisture Content, and Specific Gravity 41
3.3 Relationships among Unit Weight, Porosity, and Moisture Content 44
3.4 Relative Density 51
3.5 Consistency of Soil 53
3.6 Activity 60
3.7 Liquidity Index 62
3.8 Plasticity Chart 62
3.9 Soil Classification 63
      Problems 75
      References 77

4 Soil Compaction 78

4.1 Compaction — General Principles 78
4.2 Standard Proctor Test 79
4.3 Factors Affecting Compaction 83
4.4 Modified Proctor Test 86
4.5 Empirical Relationships 90
4.6 Field Compaction 91
4.7 Specifications for Field Compaction 94
4.8 Determination of Field Unit Weight after Compaction 96
4.9 Special Compaction Techniques 99
4.10 Effect of Compaction on Cohesive Soil Properties 104
      Problems 107
      References 109

5 Hydraulic Conductivity and Seepage 111

      Hydraulic Conductivity 111
5.1 Bernoulli's Equation 111
5.2 Darcy's Law 113
5.3 Hydraulic Conductivity 115
5.4 Laboratory Determination of Hydraulic Conductivity 116
5.5 Empirical Relations for Hydraulic Conductivity 122
5.6 Equivalent Hydraulic Conductivity in Stratified Soil 129
5.7 Permeability Test in the Field by Pumping from Wells 131
      Seepage 134
5.8 Laplace's Equation of Continuity 134
5.9 Flow Nets 136
      Problems 142
      References 146

6   Stresses in a Soil Mass   147

Effective Stress Concept   147
6.1   Stresses in Saturated Soil without Seepage   147
6.2   Stresses in Saturated Soil with Seepage   151
6.3   Effective Stress in Partially Saturated Soil   156
6.4   Seepage Force   157
6.5   Heaving in Soil Due to Flow Around Sheet Piles   159
Vertical Stress Increase Due to Various Types of Loading   161
6.6   Stress Caused by a Point Load   161
6.7   Westergaard's Solution for Vertical Stress Due to a Point Load   163
6.8   Vertical Stress Caused by a Line Load   165
6.9   Vertical Stress Caused by a Line Load of Finite Length   166
6.10   Vertical Stress Caused by a Strip Load (Finite Width and Infinite Length)   170
6.11   Vertical Stress Below a Uniformly Loaded Circular Area   172
6.12   Vertical Stress Caused by a Rectangularly Loaded Area   174
6.13   Solutions for Westergaard Material   179
Problems   180
References   185

7   Consolidation   186

7.1   Fundamentals of Consolidation   186
7.2   One-Dimensional Laboratory Consolidation Test   188
7.3   Void Ratio–Pressure Plots   190
7.4   Normally Consolidated and Overconsolidated Clays   192
7.5   Effect of Disturbance on Void Ratio–Pressure Relationship   194
7.6   Calculation of Settlement from One-Dimensional Primary Consolidation   196
7.7   Compression Index (C_c) and Swell Index (C_s)   198
7.8   Settlement from Secondary Consolidation   203
7.9   Time Rate of Consolidation   206
7.10   Coefficient of Consolidation   212
7.11   Calculation of Primary Consolidation Settlement under a Foundation   220
7.12   Skempton-Bjerrum Modification for Consolidation Settlement   223
7.13   Precompression — General Considerations   227
7.14   Sand Drains   231
Problems   237
References   241

8   Shear Strength of Soil   243

8.1   Mohr-Coulomb Failure Criteria   243
8.2   Inclination of the Plane of Failure Caused by Shear   245
Laboratory Determination of Shear Strength Parameters   247
8.3   Direct Shear Test   247

8.4   Triaxial Shear Test   255
8.5   Consolidated-Drained Test   256
8.6   Consolidated-Undrained Test   265
8.7   Unconsolidated-Undrained Test   270
8.8   Unconfined Compression Test on Saturated Clay   272
8.9   Sensitivity and Thixotropy of Clay   274
8.10   Anisotropy in Undrained Shear Strength   276
           Problems   278
           References   280

9   Slope Stability   282

9.1   Factor of Safety   283
9.2   Stability of Infinite Slopes   284
9.3   Finite Slopes   287
9.4   Analysis of Finite Slope with Circularly Cylindrical Failure Surface — General   290
9.5   Mass Procedure of Stability Analysis (Circularly Cylindrical Failure Surface)   292
9.6   Method of Slices   310
9.7   Bishop's Simplified Method of Slices   314
9.8   Analysis of Simple Slopes with Steady–State Seepage   318
9.9   Mass Procedure for Stability of Clay Slope with Earthquake Forces   322
           Problems   326
           References   329

10   Subsurface Exploration   330

10.1   Subsurface Exploration Program   330
10.2   Exploratory Borings in the Field   333
10.3   Procedures for Sampling Soil   337
10.4   Observation of Water Levels   343
10.5   Vane Shear Test   345
10.6   Cone Penetration Test   351
10.7   Pressuremeter Test (PMT)   358
10.8   Dilatometer Test   360
10.9   Coring of Rocks   363
10.10   Preparation of Boring Logs   365
10.11   Soil Exploration Report   367
            Problems   367
            References   371

11   Lateral Earth Pressure   373

11.1   Earth Pressure at Rest   373
11.2   Rankine's Theory of Active and Passive Earth Pressures   377
11.3   Diagrams for Lateral Earth Pressure Distribution against Retaining Walls   386

11.4   Rankine's Active and Passive Pressure with Sloping Backfill   400
11.5   Retaining Walls with Friction   405
11.6   Coulomb's Earth Pressure Theory   407
11.7   Passive Pressure Assuming Curved Failure Surface in Soil   415
    Problems   418
    References   420

12 Shallow Foundations—Bearing Capacity and Settlement   422

    Ultimate Bearing Capacity of Shallow Foundations   423
12.1   General Concepts   423
12.2   Ultimate Bearing Capacity Theory   425
12.3   Modification of Bearing Capacity Equations for Water Table   430
12.4   The Factor of Safety   431
12.5   Eccentrically Loaded Foundations   436
    Settlement of Shallow Foundations   447
12.6   Types of Foundation Settlement   447
12.7   Elastic Settlement   448
12.8   Range of Material Parameters for Computing Elastic Settlement   457
12.9   Settlement of Sandy Soil: Use of Strain Influence Factor   458
12.10   Allowable Bearing Pressure in Sand Based on Settlement Consideration   462
12.11   Common Types of Mat Foundations   463
12.12   Bearing Capacity of Mat Foundations   464
12.13   Compensated Foundations   467
    Problems   469
    References   473

13 Retaining Walls and Braced Cuts   475

    Retaining Walls   475
13.1   Retaining Walls — General   475
13.2   Proportioning Retaining Walls   477
13.3   Application of Lateral Earth Pressure Theories to Design   478
13.4   Check for Overturning   480
13.5   Check for Sliding along the Base   482
13.6   Check for Bearing Capacity Failure   484
    Mechanically Stabilized Retaining Walls   493
13.7   Soil Reinforcement   493
13.8   Considerations in Soil Reinforcement   493
13.9   General Design Considerations   496
13.10   Retaining Walls with Metallic Strip Reinforcement   496
13.11   Step-by-Step-Design Procedure Using Metallic Strip Reinforcement   499
13.12   Retaining Walls with Geotextile Reinforcement   505
13.13   Retaining Walls with Geogrid Reinforcement   508

Braced Cuts   510

13.14   Braced Cuts — General   510
13.15   Lateral Earth Pressure in Braced Cuts   514
13.16   Soil Parameters for Cuts in Layered Soil   516
13.17   Design of Various Components of a Braced Cut   517
13.18   Heave of the Bottom of a Cut in Clay   523
13.19   Lateral Yielding of Sheet Piles and Ground Settlement   526
Problems   527
References   531

14   Deep Foundations—Piles and Drilled Shafts   532

Pile Foundations   532

14.1   Need for Pile Foundations   532
14.2   Types of Piles and Their Structural Characteristics   534
14.3   Estimation of Pile Length   542
14.4   Installation of Piles   543
14.5   Load Transfer Mechanism   545
14.6   Equations for Estimation of Pile Capacity   546
14.7   Calculation of q_p—Meyerhof's Method   548
14.8   Frictional Resistance, Q_s 550
14.9   Allowable Pile Capacity   556
14.10   Load-Carrying Capacity of Pile Point Resting on Rock   557
14.11   Elastic Settlement of Piles   563
14.12   Pile-Driving Formulas   566
14.13   Negative Skin Friction   569
14.14   Group Piles — Efficiency   574
14.15   Elastic Settlement of Group Piles   579
14.16   Consolidation Settlement of Group Piles   580
Drilled Shafts   584
14.17   Types of Drilled Shafts   584
14.18   Construction Procedures   585
14.19   Estimation of Load-Bearing Capacity   589
14.20   Settlement of Drilled Shafts at Working Load   595
14.21   Load-Bearing Capacity Based on Settlement   595
Problems   603
References   609

Answers to Selected Problems   611

Index   615

1

Geotechnical Engineering— A Historical Perspective

For engineering purposes, *soil* is defined as the uncemented aggregate of mineral grains and decayed organic matter (solid particles) with liquid and gas in the empty spaces between the solid particles. Soil is used as a construction material in various civil engineering projects, and it supports structural foundations. Thus, civil engineers must study the properties of soil, such as its origin, grain-size distribution, ability to drain water, compressibility, shear strength, and load-bearing capacity. *Soil mechanics* is the branch of science that deals with the study of the physical properties of soil and the behavior of soil masses subjected to various types of forces. *Soil engineering* is the application of the principles of soil mechanics to practical problems. *Geotechnical engineering* is the subdiscipline of civil engineering that involves natural materials found close to the surface of the earth. It includes the application of the principles of soil mechanics and rock mechanics to the design of foundations, retaining structures, and earth structures.

1.1 Geotechnical Engineering Prior to the 18th Century

The record of a person's first use of soil as a construction material is lost in antiquity. In true engineering terms, the understanding of geotechnical engineering as it is known today began early in the 18th century (Skempton, 1985). For years the art of geotechnical engineering was based on only past experiences through a succession of experimentation without any real scientific character. Based on those experimentations, many structures were built—some of which have crumbled, while others are still standing.

Recorded history tells us that ancient civilizations flourished along the banks of rivers, such as the Nile (Egypt), the Tigris and Euphrates (Mesopotamia), the Huang Ho (Yellow River, China), and the Indus (India). Dykes dating back to about 2000 B.C. were built in the basin of the Indus to protect the town of Mohenjo Dara (in what became Pakistan after 1947). During the Chan dynasty in China (1120 B.C. to 249 B.C.), many dykes were built for irrigation purposes. There is no evidence that measures were taken to stabilize the foundations or check erosion caused by floods (Kerisel,

1985). Ancient Greek civilization used isolated pad footings and strip-and-raft foundations for building structures. Beginning around 2750 B.C., the five most important pyramids were built in Egypt in a period of less than a century (Saqqarah, Meidum, Dahshur South and North, and Cheops). This posed formidable challenges regarding foundations, stability of slopes, and construction of underground chambers. With the arrival of Buddhism in China during the Eastern Han dynasty in 68 A.D., thousands of pagodas were built. Many of these structures were constructed on silt and soft clay layers. In some cases the foundation pressure exceeded the load-bearing capacity of the soil and thereby caused extensive structural damage.

One of the most famous examples of problems related to soil-bearing capacity in the construction of structures prior to the 18th century is the Leaning Tower of Pisa in Italy. (Figure 1.1.) Construction of the tower began in 1173 A.D. when the Republic of Pisa was flourishing and continued in various stages for over 200 years.

Figure 1.1 Leaning Tower of Pisa, Italy (Courtesy of Braja Das)

The structure weighs about 15,700 metric tons and is supported by a circular base having a diameter of 20 m. The tower has tilted in the past to the east, north, west and, finally, to the south. Recent investigations showed that a weak clay layer exists at a depth of about 11 m below the ground surface, compression of which caused the tower to tilt. By 1990 it was more than 5 m out of plumb with the 54 m height. The tower was closed in 1990 because it was feared that it would either fall over or collapse. It has recently been stabilized by excavating soil from under the north side of the tower. About 70 metric tons of earth were removed in 41 separate extractions that spanned the width of the tower. As the ground gradually settled to fill the resulting space, the tilt of the tower eased. The tower now leans 5 degrees. The half-degree change is not noticeable, but it makes the structure considerably more stable. Figure 1.2 is an example of a similar problem. The towers shown in Figure 1.2 are located in Bologna, Italy, and they were built in the 12th century. The tower on the left is the Garisenda Tower. It is 48 m high and weighs about 4210 metric tons. It has

Figure 1.2 Tilting of Garisenda Tower (left) and Asinelli Tower (right) in Bologna, Italy (Courtesy of Braja Das)

tilted about 4 degree. The tower on the right is the Asinelli Tower, which is 97 m high and weighs 7300 metric tons. It has tilted about 1.3 degree.

After encountering several foundation-related problems during construction over centuries past, engineers and scientists began to address the properties and behavior of soils in a more methodical manner starting in the early part of the 18th century. Based on the emphasis and the nature of study in the area of geotechnical engineering, the time span extending from 1700 to 1927 can be divided into four major periods (Skempton, 1985):

1. Pre-classical (1700 to 1776 A.D.)
2. Classical soil mechanics — Phase I (1776 to 1856 A.D.)
3. Classical soil mechanics — Phase II (1856 to 1910 A.D.)
4. Modern soil mechanics (1910 to 1927 A.D.)

Brief descriptions of some significant developments during each of these four periods are discussed below.

1.2   *Preclassical Period of Soil Mechanics (1700–1776)*

This period concentrated on studies relating to natural slope and unit weights of various types of soils as well as the semiempirical earth pressure theories. In 1717 a French royal engineer, Henri Gautier (1660–1737), studied the natural slopes of soils when tipped in a heap for formulating the design procedures of retaining walls. The *natural slope* is what we now refer to as the *angle of repose*. According to this study, the natural slopes (see Chapter 8) of *clean dry sand* and *ordinary earth* were 31° and 45°, respectively. Also, the unit weights of clean dry sand (see Chapter 3) and ordinary earth were recommended to be 18.1 kN/m³ and 13.4 kN/m³, respectively. No test results on clay were reported. In 1729, Bernard Forest de Belidor (1694–1761) published a textbook for military and civil engineers in France. In the book, he proposed a theory for lateral earth pressure on retaining walls (see Chapter 13) that was a follow-up to Gautier's (1717) original study. He also specified a soil classification system in the manner shown in the following table. (See Chapter 3.)

Classification	Unit weight kN/m³
Rock	—
Firm or hard sand	16.7 to
Compressible sand	18.4
Ordinary earth (as found in dry locations)	13.4
Soft earth (primarily silt)	16.0
Clay	18.9
Peat	—

The first laboratory model test results on a 76-mm-high retaining wall built with sand backfill were reported in 1746 by a French engineer, Francois Gadroy

(1705–1759), who observed the existence of slip planes in the soil at failure. (See Chapter 11.) Gadroy's study was later summarized by J. J. Mayniel in 1808. Another notable contribution during this period is that by the French engineer Jean Rodolphe Perronet (1708–1794), who studied slope stability (Chapter 9) around 1769 and distinguished between intact ground and fills.

1.3   *Classical Soil Mechanics—Phase I (1776–1856)*

During this period, most of the developments in the area of geotechnical engineering came from engineers and scientists in France. In the preclassical period, practically all theoretical considerations used in calculating lateral earth pressure on retaining walls were based on an arbitrarily based failure surface in soil. In his famous paper presented in 1776, French scientist Charles Augustin Coulomb (1736–1806) used the principles of calculus for maxima and minima to determine the true position of the sliding surface in soil behind a retaining wall. (See Chapter 11.) In this analysis, Coulomb used the laws of friction and cohesion for solid bodies. In 1790, the distinguished French civil engineer, Gaspard Claire Marie Riche de Brony (1755–1839) included Coulomb's theory in his leading textbook, *Nouvelle Architecture Hydraulique* (Vol. 1). In 1820, special cases of Coulomb's work were studied by French engineer Jacques Frederic Francais (1775–1833) and by French applied-mechanics professor Claude Louis Marie Henri Navier (1785–1836). These special cases related to inclined backfills and backfills supporting surcharge. In 1840, Jean Victor Poncelet (1788–1867), an army engineer and professor of mechanics, extended Coulomb's theory by providing a graphical method for determining the magnitude of lateral earth pressure on vertical and inclined retaining walls with arbitrarily broken polygonal ground surfaces. Poncelet was also the first to use the symbol ϕ for soil friction angle. (See Chapter 8.) He also provided the first ultimate bearing-capacity theory for shallow foundations. (See Chapter 12.) In 1846, Alexandre Collin (1808–1890), an engineer, provided the details for deep slips in clay slopes, cutting, and embankments. (See Chapter 9.) Collin theorized that, in all cases, the failure takes place when the mobilized cohesion exceeds the existing cohesion of the soil. He also observed that the actual failure surfaces could be approximated as arcs of cycloids.

    The end of Phase I of the classical soil mechanics period is generally marked by the year (1857) of the first publication by William John Macquorn Rankine (1820–1872), a professor of civil engineering at the University of Glasgow. This study provided a notable theory on earth pressure and equilibrium of earth masses. (See Chapter 11.) Rankine's theory is a simplification of Coulomb's theory.

1.4   *Classical Soil Mechanics—Phase II (1856–1910)*

Several experimental results from laboratory tests on sand appeared in the literature in this phase. One of the earliest and most important publications is by French engineer Henri Philibert Gaspard Darcy (1803–1858). In 1856, he published a study on the permeability of sand filters. (See Chapter 5.) Based on those tests, Darcy defined the term *coefficient of permeability* (or hydraulic conductivity) of soil, a very useful parameter in geotechnical engineering to this day.

Sir George Howard Darwin (1845–1912), a professor of astronomy, conducted laboratory tests to determine the overturning moment on a hinged wall retaining sand in loose and dense states of compaction. Another noteworthy contribution, which was published in 1885 by Joseph Valentin Boussinesq (1842–1929), was the development of the theory of stress distribution under loaded bearing areas in a homogeneous, semiinfinite, elastic, and isotropic medium. (See Chapter 6.) In 1887, Osborne Reynolds (1842–1912) demonstrated the phenomenon of dilatency in sand. Other notable studies during this period are those by John Clibborn (1847–1938) and John Stuart Beresford (1845–1925) relating to the flow of water through sand bed and uplift pressure (Chapter 6). Clibborn's study was published in the *Treatise on Civil Engineering, Vol. 2: Irrigation Work in India*, Roorkee, 1901 and also in *Technical Paper No. 97*, Government of India, 1902. Beresford's 1898 study on uplift pressure on the Narora Weir on the Ganges River has been documented in *Technical Paper No. 97*, Government of India, 1902.

1.5 Modern Soil Mechanics (1910–1927)

In this period, results of research conducted on clays were published in which the fundamental properties and parameters of clay were established. The most notable publications are given in Table 1.1.

Table 1.1 Important Studies on Clays (1910–1927)

Investigator	Year	Topic
Albert Mauritz Atterberg (1846–1916), Sweden	1911	Consistency of soil, that is, liquid, plastic, and shrinkage properties (Chapter 3)
Jean Frontard (1884–1962), France	1914	Double shear tests (undrained) in clay under constant vertical load (Chapter 8)
Arthur Langtry Bell (1874–1956), England	1915	Lateral pressure and resistance of clay (Chapter 11); bearing capacity of clay (Chapter 12); and shear-box tests for measuring undrained shear strength using undisturbed specimens (Chapter 8)
Wolmar Fellenius (1876–1957), Sweden	1918 1926	Slip-circle analysis of saturated clay slopes (Chapter 9)
Karl Terzaghi (1883–1963), Austria	1925	Theory of consolidation for clays (Chapter 7)

1.6   *Geotechnical Engineering after 1927*

The publication of *Erdbaumechanik auf Bodenphysikalisher Grundlage* by Karl Terzaghi in 1925 gave birth to a new era in the development of soil mechanics. Karl Terzaghi is known as the father of modern soil mechanics, and rightfully so. Terzaghi (Figure 1.3) was born on October 2, 1883 in Prague, which was then the capital of the Austrian province of Bohemia. In 1904, he graduated from the Technische Hochschule in Graz, Austria, with an undergraduate degree in mechanical engineering. After graduation he served one year in the Austrian army. Following his army service, Terzaghi studied one more year, concentrating on geological subjects. In January 1912, he received the degree of Doctor of Technical Sciences from his alma mater in Graz. In 1916, he accepted a teaching position at the Imperial

Figure 1.3 Karl Terzaghi (1883–1963) (Photo courtesy of Ralph B. Peck)

School of Engineers in Istanbul. After the end of World War I, he accepted a lectureship at the American Robert College in Istanbul (1918–1925). There he began his research work on the behavior of soils and settlement of clays (see Chapter 7) and on the failure due to piping in sand under dams. The publication *Erdbaumechanik* is primarily the result of this research.

In 1925, Terzaghi accepted a visiting lectureship at Massachusetts Institute of Technology, where he worked until 1929. During that time, he became recognized as the leader of the new branch of civil engineering called soil mechanics. In October 1929, he returned to Europe to accept a professorship at the Technical University of Vienna, which soon became the nucleus for civil engineers interested in soil mechanics. In 1939, he returned to the United States to become a professor at Harvard University.

The first conference of the International Society of Soil Mechanics and Foundation Engineering (ISSMFE) was held at Harvard University in 1936 with Karl Terzaghi presiding. It was through the inspiration and guidance of Terzaghi over the preceding quarter-century that papers were brought to that conference covering a wide range of topics, such as shear strength (Chapter 8), effective stress (Chapter 6), *in situ* testing (Chapter 10), Dutch cone penetrometer (Chapter 10), centrifuge testing, consolidation settlement (Chapter 7), elastic stress distribution (Chapter 6), preloading for soil improvement, frost action, expansive clays, arching theory of earth pressure, and soil dynamics and earthquakes. For the next quarter-century, Terzaghi was the guiding spirit in the development of soil mechanics and geotechnical engineering throughout the world. To that effect, in 1985, Ralph Peck (Figure 1.4) wrote that "few people during Terzaghi's lifetime would have disagreed that he was not only the guiding spirit in soil mechanics, but that he was the clearing house for research and application throughout the world. Within the next few years he would be engaged on projects on every continent save Australia and Antarctica." Peck continued with, "Hence, even today, one can hardly improve on his contemporary assessments of the state of soil mechanics as expressed in his summary papers and presidential addresses." In 1939, Terzaghi delivered the 45th James Forrest Lecture at the Institution of Civil Engineers, London. His lecture was entitled "Soil Mechanics—A New Chapter in Engineering Science." In it he proclaimed that most of the foundation failures that occurred were no longer "acts of God."

Following are some highlights in the development of soil mechanics and geotechnical engineering that evolved after the first conference of the ISSMFE in 1936:

- Publication of the book *Theoretical Soil Mechanics* by Karl Terzaghi in 1943 (Wiley, New York);
- Publication of the book *Soil Mechanics in Engineering Practice* by Karl Terzaghi and Ralph Peck in 1948 (Wiley, New York);
- Publication of the book *Fundamentals of Soil Mechanics* by Donald W. Taylor in 1948 (Wiley, New York);
- Start of the publication of *Geotechnique*, the international journal of soil mechanics in 1948 in England;
- Presentation of the paper on $\phi = 0$ concept for clays by A. W. Skempton in 1948 (see Chapter 8);

Figure 1.4 Ralph B. Peck (Photo courtesy of Ralph B. Peck)

- Publication of A. W. Skempton's paper on *A* and *B* pore water pressure parameters in 1954 (see Chapter 8);
- Publication of the book *The Measurement of Soil Properties in the Triaxial Test* by A. W. Bishop and B. J. Henkel in 1957 (Arnold, London);
- ASCE's Research Conference on Shear Strength of Cohesive Soils held in Boulder, Colorado in 1960.

Since the early days, the profession of geotechnical engineering has come a long way and has matured. It is now an established branch of civil engineering, and thousands of civil engineers declare geotechnical engineering to be their preferred area of specialty.

Since the first conference in 1936, except for a brief interruption during World War II, the ISSMFE conferences have been held at four-year intervals. In 1997, the ISSMFE was changed to ISSMGE (International Society of Soil Mechanics and Geotechnical Engineering) to reflect its true scope. These international conferences

Table 1.2 Details of ISSMFE (1936–1997) and ISSMGE (1997–present) Conferences

Conference	Location	Year
I	Harvard University, Boston, U.S.A.	1936
II	Rotterdam, the Netherlands	1948
III	Zurich, Switzerland	1953
IV	London, England	1957
V	Paris, France	1961
VI	Montreal, Canada	1965
VII	Mexico City, Mexico	1969
VIII	Moscow, U.S.S.R.	1973
IX	Tokyo, Japan	1977
X	Stockholm, Sweden	1981
XI	San Francisco, U.S.A.	1985
XII	Rio de Janeiro, Brazil	1989
XIII	New Delhi, India	1994
XIV	Hamburg, Germany	1997
XV	Istanbul, Turkey	2001
XVI	Osaka, Japan	2005
XVII	Alexandria, Egypt	2009 (scheduled)

have been instrumental for exchange of information regarding new developments and ongoing research activities in geotechnical engineering. Table 1.2 gives the location and year in which each conference of ISSMFE/ISSMGE was held, and Table 1.3 gives a list of all of the presidents of the society. In 1997, a total of 34 technical committees of ISSMGE was in place. The names of most of these technical committees are given in Table 1.4.

Table 1.3 Presidents of ISSMFE (1936–1997) and ISSMGE (1997–present) Conferences

Year	President
1936–1957	K. Terzaghi (U.S.A.)
1957–1961	A. W. Skempton (U.K.)
1961–1965	A. Casagrande (U.S.A.)
1965–1969	L. Bjerrum (Norway)
1969–1973	R. B. Peck (U.S.A.)
1973–1977	J. Kerisel (France)
1977–1981	M. Fukuoka (Japan)
1981–1985	V. F. B. deMello (Brazil)
1985–1989	B. B. Broms (Singapore)
1989–1994	N. R. Morgenstern (Canada)
1994–1997	M. Jamiolkowski (Italy)
1997–2001	K. Ishihara (Japan)
2001–2005	W. F. Van Impe (Belgium)
2005–2009	P. S. Sêco e Pinto (Portugal)

Table 1.4 ISSMGE Technical Committees

Committee number	Committee name
TC-1	Instrumentation for Geotechnical Monitoring
TC-2	Centrifuge Testing
TC-3	Geotechnics of Pavements and Rail Tracks
TC-4	Earthquake Geotechnical Engineering
TC-5	Environmental Geotechnics
TC-6	Unsaturated Soils
TC-7	Tailing Dams
TC-8	Frost
TC-9	Geosynthetics and Earth Reinforcement
TC-10	Geophysical Site Characterization
TC-11	Landslides
TC-12	Validation of Computer Simulation
TC-14	Offshore Geotechnical Engineering
TC-15	Peat and Organic Soils
TC-16	Ground Property Characterization from In-situ Testing
TC-17	Ground Improvement
TC-18	Pile Foundations
TC-19	Preservation of Historic Sites
TC-20	Professional Practice
TC-22	Indurated Soils and Soft Rocks
TC-23	Limit State Design Geotechnical Engineering
TC-24	Soil Sampling, Evaluation and Interpretation
TC-25	Tropical and Residual Soils
TC-26	Calcareous Sediments
TC-28	Underground Construction in Soft Ground
TC-29	Stress-Strain Testing of Geomaterials in the Laboratory
TC-30	Coastal Geotechnical Engineering
TC-31	Education in Geotechnical Engineering
TC-32	Risk Assessment and Management
TC-33	Scour of Foundations
TC-34	Deformation of Earth Materials

References

ATTERBERG, A. M. (1911). "Über die physikalische Bodenuntersuchung, und über die Plastizität de Tone," International Mitteilungen für Bodenkunde, *Verlag für Fachliteratur.* G.m.b.H. Berlin, Vol. 1, 10–43.

BELIDOR, B. F. (1729). *La Science des Ingenieurs dans la Conduite des Travaux de Fortification et D'Architecture Civil*, Jombert, Paris.

BELL, A. L. (1915). "The Lateral Pressure and Resistance of Clay, and Supporting Power of Clay Foundations," *Min. Proceeding of Institute of Civil Engineers*, Vol. 199, 233–272.

BISHOP, A. W. and HENKEL, B. J. (1957). *The Measurement of Soil Properties in the Triaxial Test*, Arnold, London.

BOUSSINESQ, J. V. (1883). *Application des Potentiels â L'Etude de L'Équilibre et du Mouvement des Solides Élastiques*, Gauthier-Villars, Paris.

COLLIN, A. (1846). *Recherches Expérimentales sur les Glissements Spontanés des Terrains Argileux Accompagnées de Considérations sur Quelques Principes de la Mécanique Terrestre*, Carilian-Goeury, Paris.

COULOMB, C. A. (1776). "Essai sur une Application des Règles de Maximis et Minimis à Quelques Problèmes de Statique Relatifs à L'Architecture," *Mèmoires de la Mathèmatique et de Phisique*, présentés à l'Académie Royale des Sciences, par divers savans, et lûs dans sés Assemblées, De L'Imprimerie Royale, Paris, Vol. 7, Annee 1793, 343–382.

DARCY, H. P. G. (1856). *Les Fontaines Publiques de la Ville de Dijon*, Dalmont, Paris.

DARWIN, G. H. (1883). "On the Horizontal Thrust of a Mass of Sand," *Proceedings*, Institute of Civil Engineers, London, Vol. 71, 350–378.

FELLENIUS, W. (1918). "Kaj-och Jordrasen I Göteborg," *Teknisk Tidskrift*. Vol. 48, 17–19.

FRANCAIS, J. F. (1820). "Recherches sur la Poussée de Terres sur la Forme et Dimensions des Revêtments et sur la Talus D'Excavation," *Mémorial de L'Officier du Génie*, Paris, Vol. IV, 157–206.

FRONTARD, J. (1914). "Notice sur L'Accident de la Digue de Charmes," *Anns. Ponts et Chaussées 9th Ser.*, Vol. 23, 173–292.

GADROY, F. (1746). *Mémoire sur la Poussée des Terres*, summarized by Mayniel, 1808.

GAUTIER, H. (1717). *Dissertation sur L'Epaisseur des Culées des Ponts . . . sur L'Effort et al Pesanteur des Arches . . . et sur les Profiles de Maconnerie qui Doivent Supporter des Chaussées, des Terrasses, et des Remparts*. Cailleau, Paris.

KERISEL, J. (1985). "The History of Geotechnical Engineering up until 1700," *Proceedings*, XI International Conference on Soil Mechanics and Foundation Engineering, San Francisco, Golden Jubilee Volume, A. A. Balkema, 3–93.

MAYNIEL, J. J. (1808). *Traité Experimentale, Analytique et Pratique de la Poussé des Terres*. Colas, Paris.

NAVIER, C. L. M. (1839). *Leçons sur L'Application de la Mécanique à L'Establissement des Constructions et des Machines*, 2nd ed., Paris.

PECK, R. B. (1985). "The Last Sixty Years," *Proceedings*, XI International Conference on Soil Mechanics and Foundation Engineering, San Francisco, Golden Jubilee Volume, A. A. Balkema, 123–133.

PONCELET, J. V. (1840). *Mémoire sur la Stabilité des Revêtments et de seurs Fondations*, Bachelier, Paris.

RANKINE, W. J. M. (1857). "On the Stability of Loose Earth," *Philosophical Transactions*, Royal Society, Vol. 147, London.

REYNOLDS, O. (1887). "Experiments Showing Dilatency, a Property of Granular Material Possibly Connected to Gravitation," *Proceedings*, Royal Society, London, Vol. 11, 354–363.

SKEMPTON, A. W. (1948). "The $\phi = 0$ Analysis of Stability and Its Theoretical Basis," *Proceedings*, II International Conference on Soil Mechanics and Foundation Engineering, Rotterdam, Vol. 1, 72–78.

SKEMPTON, A. W. (1954). "The Pore Pressure Coefficients A and B," *Geotechnique*, Vol. 4, 143–147.

SKEMPTON, A. W. (1985). "A History of Soil Properties, 1717–1927," *Proceedings*, XI International Conference on Soil Mechanics and Foundation Engineering, San Francisco, Golden Jubilee Volume, A. A. Balkema, 95–121.

TAYLOR, D. W. (1948). *Fundamentals of Soil Mechanics*, John Wiley, New York.

TERZAGHI, K. (1925). *Erdbaumechanik auf Bodenphysikalisher Grundlage*, Deuticke, Vienna.

TERZAGHI, K. (1939). "Soil Mechanics — A New Chapter in Engineering Science," *Institute of Civil Engineers Journal*, London, Vol. 12, No. 7, 106–142.

TERZAGHI, K. (1943). *Theoretical Soil Mechanics*, John Wiley, New York.

TERZAGHI, K. and PECK, R. B. (1948). *Soil Mechanics in Engineering Practice*, John Wiley, New York.

2

Soil Deposits and Grain-Size Analysis

2.1   Natural Soil Deposits-General

During the planning, design, and construction of foundations, embankments, and earth-retaining structures, engineers find it helpful to know the origin of the soil deposit over which the foundation is to be constructed because each soil deposit has it own unique physical attributes.

Most of the soils that cover the earth are formed by the weathering of various rocks. There are two general types of weathering: (1) mechanical weathering and (2) chemical weathering.

Mechanical weathering is the process by which rocks are broken into smaller and smaller pieces by physical forces, including running water, wind, ocean waves, glacier ice, frost, and expansion and contraction caused by the gain and loss of heat.

Chemical weathering is the process of chemical decomposition of the original rock. In the case of mechanical weathering, the rock breaks into smaller pieces without a change in its chemical composition. However, in chemical weathering, the original material may be changed to something entirely different. For example, the chemical weathering of feldspar can produce clay minerals. Most rock weathering is a combination of mechanical and chemical weathering.

Soil produced by the weathering of rocks can be transported by physical processes to other places. The resulting soil deposits are called *transported soils*. In contrast, some soils stay where they were formed and cover the rock surface from which they derive. These soils are referred to as *residual soils*.

Transported soils can be subdivided into five major categories based on the *transporting agent*:

1. *Gravity transported* soil
2. *Lacustrine (lake) deposits*
3. *Alluvial* or *fluvial* soil deposited by running water
4. *Glacial* deposited by glaciers
5. *Aeolian* deposited by the wind

In addition to transported and residual soils, there are *peats* and *organic soils*, which derive from the decomposition of organic materials.

A general overview of various types of soils described above is given in Sections 2.2 through 2.8.

2.2   Residual Soil

Residual soils are found in areas where the rate of weathering is more than the rate at which the weathered materials are carried away by transporting agents. The rate of weathering is higher in warm and humid regions compared to cooler and drier regions and, depending on the climatic conditions, the effect of weathering may vary widely.

Residual soil deposits are common in the tropics. The nature of a residual soil deposit will generally depend on the parent rock. When hard rocks, such as granite and gneiss, undergo weathering, most of the materials are likely to remain in place. These soil deposits generally have a top layer of clayey or silty clay material, below which are silty or sandy soil layers. These layers in turn, are generally underlain by a partially weathered rock, and then sound bedrock. The depth of the sound bedrock may vary widely, even within a distance of a few meters.

In contrast to hard rocks, there are some chemical rocks, such as limestone, that are chiefly made up of calcite ($CaCo_3$) mineral. Chalk and dolomite have large concentrations of dolomite minerals [$Ca\,Mg(Co_3)_2$]. These rocks have large amounts of soluble materials, some of which are removed by groundwater, leaving behind the insoluble fraction of the rock. Residual soils that derive from chemical rocks do not possess a gradual transition zone to the bedrock. The residual soils derived from the weathering of limestone-like rocks are mostly red in color. Although uniform in kind, the depth of weathering may vary greatly. The residual soils immediately above the bedrock may be normally consolidated. Large foundations with heavy loads may be susceptible to large consolidation settlements on these soils.

2.3   Gravity Transported Soil

Residual soils on a steep natural slope can move slowly downward, and this is usually referred to as *creep*. When the downward soil movement is sudden and rapid, it is called a *landslide*. The soil deposits formed by landslides are *colluvium. Mud flows* are one type of gravity transported soil. In this case, highly saturated, loose sandy residual soils, on relatively flat slopes, move downward like a viscous liquid and come to rest in a more dense condition. The soil deposits derived from past mud flows are highly heterogeneous in composition.

2.4   Alluvial Deposits

Alluvial soil deposits derive from the action of streams and rivers and can be divided into two major categories: (1) *braided-stream deposits*, and (2) deposits caused by the *meandering belt of streams*.

Deposits from Braided Streams

Braided streams are high-gradient, rapidly flowing streams that are highly erosive and carry large amounts of sediment. Because of the high bed load, a minor change in the velocity of flow will cause sediments to deposit. By this process, these streams may build up a complex tangle of converging and diverging channels, separated by sandbars and islands.

The deposits formed from braided streams are highly irregular in stratification and have a wide range of grain sizes. Figure 2.1 shows a cross section of such a deposit. These deposits share several characteristics:

1. The grain sizes usually range from gravel to silt. Clay-sized particles are generally *not* found in deposits from braided streams.
2. Although grain size varies widely, the soil in a given pocket or lens is rather uniform.
3. At any given depth, the void ratio and unit weight may vary over a wide range within a lateral distance of only a few meters.

Meander Belt Deposits

The term *meander* is derived from the Greek work *maiandros*, after the Maiandros (now Menderes) River in Asia, famous for its winding course. Mature streams in a valley curve back and forth. The valley floor in which a river meanders is referred to as the *meander belt*. In a meandering river, the soil from the bank is continually eroded from the points where it is concave in shape and is deposited at points where the bank is convex in shape, as shown in Figure 2.2. These deposits are called *point bar deposits*, and they usually consist of sand and silt-sized particles. Sometimes, during the process of erosion and deposition, the river abandons a meander and cuts a shorter path. The abandoned meander, when filled with water, is called an *oxbow lake*. (See Figure 2.2.)

During floods, rivers overflow low-lying areas. The sand and silt-size particles carried by the river are deposited along the banks to form ridges known as *natural levees* (Figure 2.3). Finer soil particles consisting of silts and clays are carried by the water farther onto the floodplains. These particles settle at different rates to form *backswamp deposits* (Figure 2.3), often highly plastic clays.

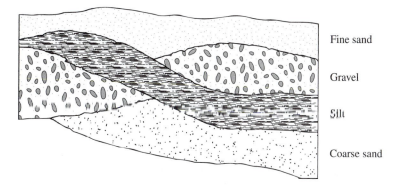

Fine sand

Gravel

Silt

Coarse sand

Figure 2.1 Cross section of a braided-stream deposit

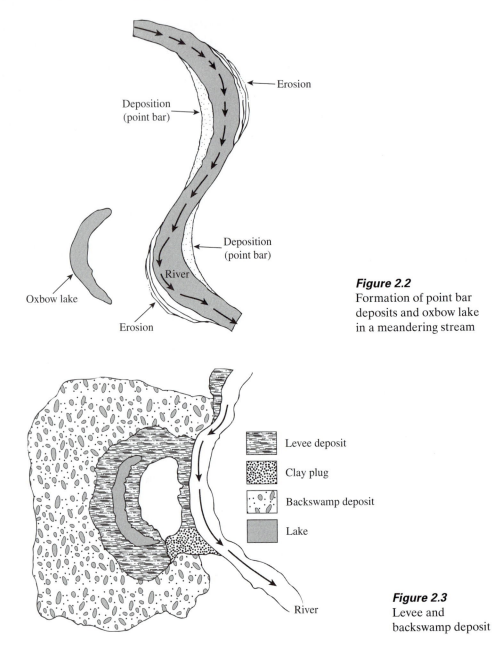

Figure 2.2
Formation of point bar deposits and oxbow lake in a meandering stream

Figure 2.3
Levee and backswamp deposit

2.5   *Lacustrine Deposits*

Water from rivers and springs flows into lakes. In arid regions, streams carry large amounts of suspended solids. Where the stream enters the lake, granular particles are deposited in the area forming a delta. Some coarser particles and the finer particles; that is, silt and clay, that are carried into the lake are deposited onto the lake bottom in alternate layers of coarse-grained and fine-grained particles. The deltas formed in humid regions usually have finer grained soil deposits compared to those in arid regions.

2.6   Glacial Deposits

During the Pleistocene Ice Age, glaciers covered large areas of the earth. The glaciers advanced and retreated with time. During their advance, the glaciers carried large amounts of sand, silt, clay, gravel, and boulders. *Drift* is a general term usually applied to the deposits laid down by glaciers. Unstratified deposits laid down by melting glaciers are referred to as *till*. The physical characteristics of till may vary from glacier to glacier.

The landforms that developed from the deposits of till are called *moraines*. A *terminal moraine* (Figure 2.4) is a ridge of till that marks the maximum limit of a glacier's advance. *Recessional moraines* are ridges of till developed behind the terminal moraine at varying distances apart. They are the result of temporary stabilization of the glacier during the recessional period. The till deposited by the glacier between the moraines is referred to as *ground moraine* (Figure 2.4). Ground moraines constitute large areas of the central United States and are called *till plains*.

The sand, silt, and gravel that are carried by the melting water from the front of a glacier are called *outwash*. In a pattern similar to that of braided-stream deposits, the melted water deposits the outwash, forming *outwash plains* (Figure 2.4), also called *glaciofluvial deposits*. The range of grain sizes present in a given till varies greatly.

2.7   Aeolian Soil Deposits

Wind is also a major transporting agent leading to the formation of soil deposits. When large areas of sand lie exposed, wind can blow the sand away and redeposit it elsewhere. Deposits of windblown sand generally take the shape of *dunes* (Figure 2.5). As dunes are formed, the sand is blown over the crest by the wind. Beyond the crest, the sand particles roll down the slope. The process tends to form a *compact sand*

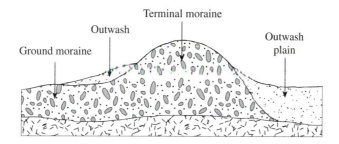

Figure 2.4 Terminal moraine, ground moraine, and outwash plain

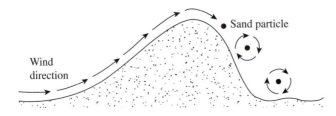

Figure 2.5 Sand dune

deposit on the *windward side*, and a rather *loose deposit* on the *leeward side*, of the dune. Following are some of the typical properties of *dune sand*:

1. The grain-size distribution of the sand at any particular location is surprisingly uniform. This uniformity can be attributed to the sorting action of the wind.
2. The general grain size decreases with distance from the source, because the wind carries the small particles farther than the large ones.
3. The relative density of sand deposited on the windward side of dunes may be as high as 50 to 65%, decreasing to about 0 to 15% on the leeward side.

Loess is an aeolian deposit consisting of silt and silt-sized particles. The grain-size distribution of loess is rather uniform. The cohesion of loess is generally derived from a clay coating over the silt-sized particles, which contributes to a stable soil structure in an unsaturated state. The cohesion may also be the result of the precipitation of chemicals leached by rainwater. Loess is a *collapsing* soil, because when the soil becomes saturated, it loses its binding strength between particles. Special precautions need to be taken for the construction of foundations over loessial deposits.

Volcanic ash (with grain sizes between 0.25 to 4 mm), and volcanic dust (with grain sizes less than 0.25 mm), may be classified as wind-transported soil. Volcanic ash is a lightweight sand or sandy gravel. Decomposition of volcanic ash results in highly plastic and compressible clays.

2.8 Organic Soil

Organic soils are usually found in low-lying areas where the water table is near or above the ground surface. The presence of a high water table helps in the growth of aquatic plants that, when decomposed, form organic soil. This type of soil deposit is usually encountered in coastal areas and in glaciated regions. Organic soils show the following characteristics:

1. Their natural moisture content may range from 200 to 300%.
2. They are highly compressible.
3. Laboratory tests have shown that, under loads, a large amount of settlement is derived from secondary consolidation.

2.9 Soil-Particle Size

Irrespective of the origin of soil, the sizes of particles in general, that make up soil, vary over a wide range. Soils are generally called *gravel, sand, silt,* or *clay,* depending on the predominant size of particles within the soil. To describe soils by their particle size, several organizations have developed *soil-separate-size limits.* Table 2.1 shows the soil-separate-size limits developed by the Massachusetts Institute of Technology, the U.S. Department of Agriculture, the American Association of State Highway and Transportation Officials, and the U.S. Army Corps of Engineers, and U.S. Bureau of Reclamation. In this table, the MIT system is presented for illustration purposes only, because it plays an important role in the history of the development of soil-separate-size limits. Presently, however, the Unified System is almost universally accepted. The Unified Soil Classification System has now been adopted by the American Society for Testing and Materials. (Also see Figure 2.6.)

Gravels are pieces of rocks with occasional particles of quartz, feldspar, and other minerals.

Sand particles are made of mostly quartz and feldspar. Other mineral grains may also be present at times.

Silts are the microscopic soil fractions that consist of very fine quartz grains and some flake-shaped particles that are fragments of micaceous minerals.

Clays are mostly flake-shaped microscopic and submicroscopic particles of mica, clay minerals, and other minerals. As shown in Table 2.1, clays are generally defined as particles smaller than 0.002 mm. In some cases, particles between 0.002 and 0.005 mm in size are also referred to as clay. Particles are classified as *clay* on the basis of their size; they may not necessarily contain clay minerals. Clays are defined as those particles "which develop plasticity when mixed with a limited amount of water" (Grim, 1953). (Plasticity is the puttylike property of clays when they contain a certain amount of water.) Nonclay soils can contain particles of quartz, feldspar, or mica that are small enough to be within the clay size classification. Hence, it is appropriate for soil particles smaller than 2 μ, or 5 μ as defined under different systems, to be called clay-sized particles rather than clay. Clay particles are mostly of colloidal size range (<1 μ), and 2 μ appears to be the upper limit.

Table 2.1 Soil-separate-size limits

Name of organization	Grain size (mm)			
	Gravel	Sand	Silt	Clay
Massachusetts Institute of Technology (MIT)	>2	2 to 0.06	0.06 to 0.002	<0.002
U.S. Department of Agriculture (USDA)	>2	2 to 0.05	0.05 to 0.002	<0.002
American Association of State Highway and Transportation Officials (AASHTO)	76.2 to 2	2 to 0.075	0.075 to 0.002	<0.002
Unified Soil Classification System (U.S. Army Corps of Engineers; U.S. Bureau of Reclamation; American Society for Testing and Materials)	76.2 to 4.75	4.75 to 0.075	Fines (i.e., silts and clays) <0.075	

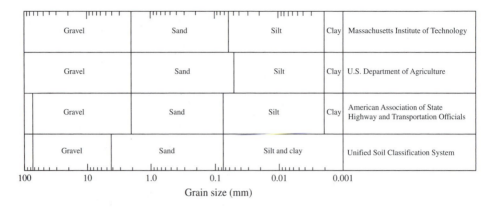

Figure 2.6 Soil-separate-size limits by various systems

2.10   *Clay Minerals*

Clay minerals are complex aluminum silicates composed of one of two basic units: (1) *silica tetrahedron* and (2) *alumina octahedron*. Each tetrahedron unit consists of four oxygen atoms surrounding a silicon atom (Figure 2.7a). The combination of tetrahedral silica units gives a *silica sheet* (Figure 2.7b). Three oxygen atoms at the base of each tetrahedron are shared by neighboring tetrahedra. The octahedral units consist of six hydroxyls surrounding an aluminum atom (Figure 2.7c), and the combination of the octahedral aluminum hydroxyl units gives an *octahedral sheet*. (This is also called a *gibbsite sheet*; Figure 2.7d.) Sometimes magnesium replaces the aluminum atoms in the octahedral units; in that case, the octahedral sheet is called a *brucite sheet*.

In a silica sheet, each silicon atom with a positive valence of four, is linked to four oxygen atoms, with a total negative valence of eight. But each oxygen atom at the base of the tetrahedron is linked to two silicon atoms. This means that the top oxygen atom of each tetrahedral unit has a negative valence charge of one to be counterbalanced. When the silica sheet is stacked over the octahedral sheet, as shown in Figure 2.7e, these oxygen atoms replace the hydroxyls to satisfy their valence bonds.

Kaolinite consists of repeating layers of elemental silica-gibbsite sheets, as shown in Figure 2.8a. Each layer is about 7.2 Å thick. The layers are held together by hydrogen bonding. Kaolinite occurs as platelets, each with a lateral dimension of 1000 to 20,000 Å and a thickness of 100 to 1000 Å. The surface area of the kaolinite particles per unit mass is about 15 m^2/g. The surface area per unit mass is defined as *specific surface*.

Illite consists of a gibbsite sheet bonded to two silica sheets—one at the top, and another at the bottom (Figure 2.8b). It is sometimes called *clay mica*. The illite layers are bonded together by potassium ions. The negative charge to balance the potassium ions comes from the substitution of aluminum for some silicon in the tetrahedral sheets. Substitution of one element for another with no change in the crystalline form is known as *isomorphous substitution*. Illite particles generally have lateral dimensions ranging from 1000 to 5000 Å, and thicknesses from 50 to 500 Å. The specific surface of the particles is about 80 m^2/g.

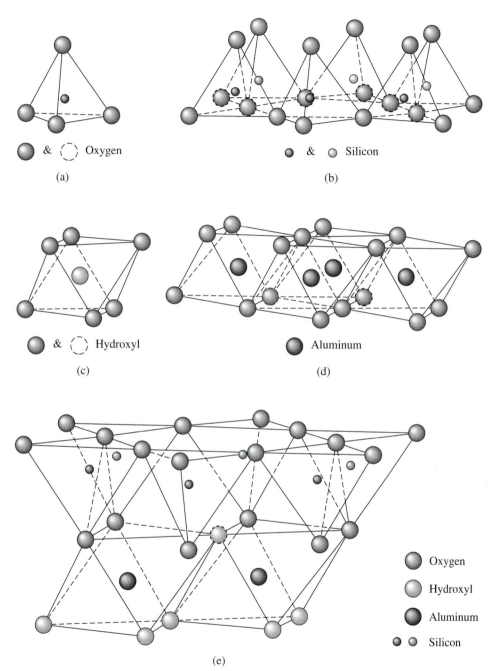

Figure 2.7 (a) Silica tetrahedron; (b) silica sheet; (c) alumina octahedron; (d) octahedral (gibbsite) sheet; (e) elemental silica-gibbsite sheet (After Grim, 1959)

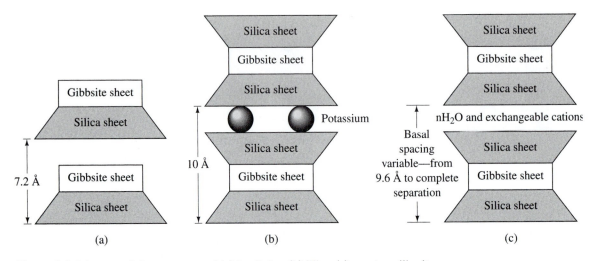

Figure 2.8 Diagram of the structures of (a) kaolinite; (b) illite; (c) montmorillonite

Montmorillonite has a similar structure to illite—that is, one gibbsite sheet sandwiched between two silica sheets (Figure 2.8c). In montmorillonite, there is isomorphous substitution of magnesium and iron for aluminum in the octahedral sheets. Potassium ions are not present here as in the case of illite, and a large amount of water is attracted into the space between the layers. Particles of montmorillonite have lateral dimensions of 1000 to 5000 Å and thicknesses of 10 to 50 Å. The specific surface is about 800 m^2/g.

Besides kaolinite, illite, and montmorillonite, other common clay minerals generally found are chlorite, halloysite, vermiculite, and attapulgite.

The clay particles carry a net negative charge on their surfaces. This is the result both of isomorphous substitution and of a break in continuity of the structure at its edges. Larger negative charges are derived from larger specific surfaces. Some positively charged sites also occur at the edges of the particles. A list for the reciprocal of the average surface density of the negative charge on the surface of some clay minerals (Yong and Warkentin, 1966) follows:

Clay mineral	Reciprocal of average surface density of charge ($Å^2$/electronic charge)
Kaolinite	25
Clay mica and chlorite	50
Montmorillonite	100
Vermiculite	75

In dry clay, the negative charge is balanced by exchangeable cations, like Ca^{++}, Mg^{++}, Na^+, and K^+, surrounding the particles being held by electrostatic attraction. When water is added to clay, these cations and a small number of anions float around the clay particles. This is referred to as *diffuse double layer* (Figure 2.9a). The cation concentration decreases with distance from the surface of the particle (Figure 2.9b).

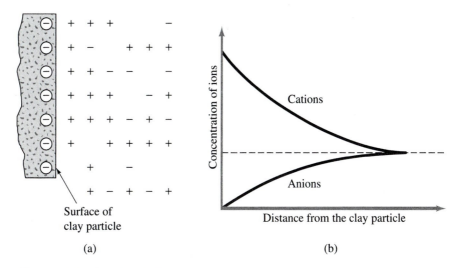

(a)   (b)

Figure 2.9 Diffuse double layer

Water molecules are polar. Hydrogen atoms are not arranged in a symmetric manner around an oxygen atom; instead, they occur at a bonded angle of 105°. As a result, a water molecule acts like a small rod with a positive charge at one end and a negative charge at the other end. It is known as a *dipole*.

The dipolar water is attracted both by the negatively charged surface of the clay particles, and by the cations in the double layer. The cations, in turn, are attracted to the soil particles. A third mechanism by which water is attracted to clay particles is *hydrogen bonding*, in which hydrogen atoms in the water molecules are shared with oxygen atoms on the surface of the clay. Some partially hydrated cations in the pore water are also attracted to the surface of clay particles. These cations attract dipolar water molecules. The force of attraction between water and clay decreases with distance from the surface of the particles. All of the water held to clay particles by force of attraction is known as *double-layer water*. The innermost layer of double-layer water, which is held very strongly by clay, is known as *adsorbed water*. This water is more viscous than is free water. The orientation of water around the clay particles gives clay soils their plastic properties.

2.11 Specific Gravity (G$_s$)

The specific gravity of the soil solids is used in various calculations in soil mechanics. The specific gravity can be determined accurately in the laboratory. Table 2.2 shows the specific gravity of some common minerals found in soils. Most of the minerals have a specific gravity that falls within a general range of 2.6 to 2.9. The specific gravity of solids of light-colored sand, which is made mostly of quartz, may be estimated to be about 2.65; for clayey and silty soils, it may vary from 2.6 to 2.9.

Table 2.2 Specific gravity of important minerals

Mineral	Specific gravity, G_s
Quartz	2.65
Kaolinite	2.6
Illite	2.8
Montmorillonite	2.65–2.80
Halloysite	2.0–2.55
Potassium feldspar	2.57
Sodium and calcium feldspar	2.62–2.76
Chlorite	2.6–2.9
Biotite	2.8–3.2
Muscovite	2.76–3.1
Hornblende	3.0–3.47
Limonite	3.6–4.0
Olivine	3.27–3.37

2.12 Mechanical Analysis of Soil

Mechanical analysis is the determination of the size range of particles present in a soil, expressed as a percentage of the total dry weight (or mass). Two methods are generally used to find the particle-size distribution of soil: (1) *sieve analysis*—for particle sizes larger than 0.075 mm in diameter, and (2) *hydrometer analysis*—for particle sizes smaller than 0.075 mm in diameter. The basic principles of sieve analysis and hydrometer analysis are described next.

Sieve Analysis

Sieve analysis consists of shaking the soil sample through a set of sieves that have progressively smaller openings. U.S. standard sieve numbers and the sizes of openings are given in Table 2.3.

The sieves used for soil analysis are generally 203 mm in diameter. To conduct a sieve analysis, one must first oven-dry the soil and then break all lumps into small particles. The soil is then shaken through a stack of sieves with openings of decreasing size from top to bottom (a pan is placed below the stack). Figure 2.10 shows a set of sieves in a shaker used for conducting the test in the laboratory. The smallest-size sieve that should be used for this type of test is the U.S. No. 200 sieve. After the soil is shaken, the mass of soil retained on each sieve is determined. When cohesive soils are analyzed, breaking the lumps into individual particles may be difficult. In this case, the soil may be mixed with water to make a slurry and then washed through the sieves. Portions retained on each sieve are collected separately and oven-dried before the mass retained on each sieve is measured.

Referring to Figure 2.11, we can step through the calculation procedure for a sieve analysis:

1. Determine the mass of soil retained on each sieve (i.e., M_1, M_2, $\cdots$ M_n) and in the pan (i.e., M_p) (Figures 2.11a and 2.11b).

Table 2.3 U.S. standard sieve sizes

Sieve no.	Opening (mm)
4	4.750
6	3.350
8	2.360
10	2.000
16	1.180
20	0.850
30	0.600
40	0.425
50	0.300
60	0.250
80	0.180
100	0.150
140	0.106
170	0.088
200	0.075
270	0.053

2. Determine the total mass of the soil: $M_1 + M_2 + \cdots + M_i + \cdots + M_n + M_p = \Sigma M$.
3. Determine the cumulative mass of soil retained above each sieve. For the ith sieve, it is $M_1 + M_2 + \cdots + M_i$ (Figure 2.11c).

Figure 2.10
A set of sieves
for a test in the
laboratory
(Courtesy of
Braja Das)

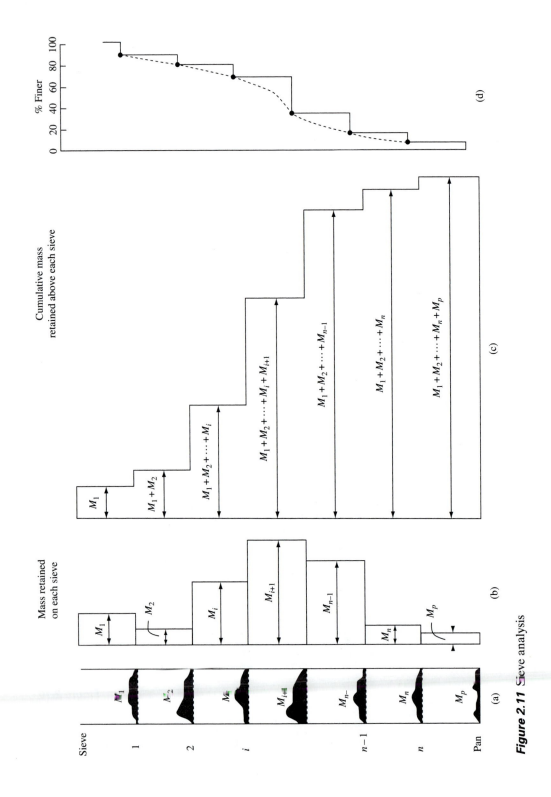

Figure 2.11 Sieve analysis

4. The mass of soil passing the ith sieve is $\Sigma M - (M_1 + M_2 + \cdots + M_i)$.
5. The percent of soil passing the ith sieve (or *percent finer*) (Figure 2.11d) is

$$F = \frac{\Sigma M - (M_1 + M_2 + \cdots + M_i)}{\Sigma M} \times 100$$

Once the percent finer for each sieve is calculated (step 5), the calculations are plotted on semilogarithmic graph paper (Figure 2.12) with percent finer as the ordinate (arithmetic scale) and sieve opening size as the abscissa (logarithmic scale). This plot is referred to as the *particle-size distribution curve*.

Hydrometer Analysis

Hydrometer analysis is based on the principle of sedimentation of soil grains in water. When a soil specimen is dispersed in water, the particles settle at different velocities, depending on their shape, size, and weight. For simplicity, it is assumed that all the soil particles are spheres, and the velocity of soil particles can be expressed by *Stokes' law*, according to which

$$v = \frac{\rho_s - \rho_w}{18\eta} D^2 \tag{2.1}$$

where
$\quad v$ = velocity
$\quad \rho_s$ = density of soil particles
$\quad \rho_w$ = density of water

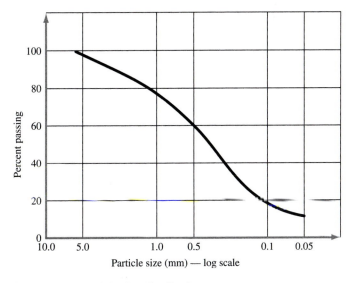

Figure 2.12 Particle-size distribution curve

η = viscosity of water
D = diameter of soil particles

Thus, from Eq. (2.1),

$$D = \sqrt{\frac{18\eta v}{\rho_s - \rho_w}} = \sqrt{\frac{18\eta}{\rho_s - \rho_w}}\sqrt{\frac{L}{t}} \tag{2.2}$$

where $v = \dfrac{\text{distance}}{\text{time}} = \dfrac{L}{t}$

Note that

$$\rho_s = G_s\rho_w \tag{2.3}$$

Thus, combining Eqs. (2.2) and (2.3) gives

$$D = \sqrt{\frac{18\eta}{(G_s - 1)\rho_w}}\sqrt{\frac{L}{t}} \tag{2.4}$$

If the units of η are $(g \cdot sec)/cm^2$, ρ_w is in g/cm^3, L is in cm, t is in min, and D is in mm, then

$$\frac{D\,(mm)}{10} = \sqrt{\frac{18\eta\,[(g \cdot sec)/cm^2]}{(G_s - 1)\rho_w\,(g/cm^3)}}\sqrt{\frac{L\,(cm)}{t\,(min) \times 60}}$$

or

$$D = \sqrt{\frac{30\eta}{(G_s - 1)\rho_w}}\sqrt{\frac{L}{t}}$$

Assuming ρ_w to be approximately equal to $1\ g/cm^3$, we have

$$D\,(mm) = K\sqrt{\frac{L\,(cm)}{t\,(min)}} \tag{2.5}$$

where $K = \sqrt{\dfrac{30\eta}{(G_s - 1)}}$ $\tag{2.6}$

Note that the value of K is a function of G_s and η, which are dependent on the temperature of the test. The variation of K with the temperature of the test and G_s is shown in Table 2.4.

In the laboratory, the hydrometer test is conducted in a sedimentation cylinder with 50 g of oven-dry sample. The sedimentation cylinder is 457 mm high and

Table 2.4 Variation of K with G_s

Temperature (°C)	G_s						
	2.50	2.55	2.60	2.65	2.70	2.75	2.80
17	0.0149	0.0146	0.0144	0.0142	0.0140	0.0138	0.0136
18	0.0147	0.0144	0.0142	0.0140	0.0138	0.0136	0.0134
19	0.0145	0.0143	0.0140	0.0138	0.0136	0.0134	0.0132
20	0.0143	0.0141	0.0139	0.0137	0.0134	0.0133	0.0131
21	0.0141	0.0139	0.0137	0.0135	0.0133	0.0131	0.0129
22	0.0140	0.0137	0.0135	0.0133	0.0131	0.0129	0.0128
23	0.0138	0.0136	0.0134	0.0132	0.0130	0.0128	0.0126
24	0.0137	0.0134	0.0132	0.0130	0.0128	0.0126	0.0125
25	0.0135	0.0133	0.0131	0.0129	0.0127	0.0125	0.0123
26	0.0133	0.0131	0.0129	0.0127	0.0125	0.0124	0.0122
27	0.0132	0.0130	0.0128	0.0126	0.0124	0.0122	0.0120
28	0.0130	0.0128	0.0126	0.0124	0.0123	0.0121	0.0119
29	0.0129	0.0127	0.0125	0.0123	0.0121	0.0120	0.0118
30	0.0128	0.0126	0.0124	0.0122	0.0120	0.0118	0.0117

63.5 mm in diameter. It is marked for a volume of 1000 ml. Sodium hexametaphosphate is generally used as the *dispersing agent*. The volume of the dispersed soil suspension is brought up to 1000 ml by adding distilled water.

When an ASTM 152H (ASTM, 2006) type of hydrometer is placed in the soil suspension (Figure 2.13) at a time t, measured from the start of sedimentation, it measures the specific gravity in the vicinity of its bulb at a depth L. The specific gravity is a function of the amount of soil particles present per unit volume of suspension at that depth. Also, at a time t, the soil particles in suspension at a depth L will have a diameter smaller than D as calculated in Eq. (2.5). The larger particles would have settled beyond the zone of measurement. Hydrometers are designed to give the amount of soil, in grams, that is still in suspension. Hydrometers are calibrated for soils that have a specific gravity (G_s) of 2.65; for soils of other specific gravity, it is necessary to make corrections.

By knowing the amount of soil in suspension, L, and t, we can calculate the percentage of soil by weight finer than a given diameter. Note that L is the depth measured from the surface of the water to the center of gravity of the hydrometer bulb at which the density of the suspension is measured. The value of L will change with time t; its variation with the hydrometer readings is given in Table 2.6. Hydrometer analysis is effective for separating soil fractions down to a size of about 0.5 μ.

In many instances, the results of sieve analysis and hydrometer analysis for finer fractions for a given soil are combined on one graph, such as the one shown in Figure 2.14. When these results are combined, a discontinuity generally occurs in the range where they overlap. This discontinuity occurs because soil particles are generally irregular in shape. Sieve analysis gives the intermediate dimensions of a particle; hydrometer analysis gives the diameter of an equivalent sphere that would settle at the same rate as the soil particle.

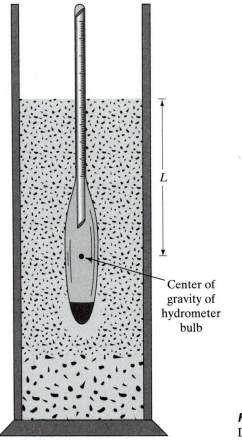

Figure 2.13
Definition of L in hydrometer test

Table 2.5 Variation of L with hydrometer reading—*ASTM 152-H* hydrometer

Hydrometer reading	L (cm)	Hydrometer reading	L (cm)
0	16.3	26	12.0
1	16.1	27	11.9
2	16.0	28	11.7
3	15.8	29	11.5
4	15.6	30	11.4
5	15.5	31	11.2
6	15.3	32	11.1
7	15.2	33	10.9
8	15.0	34	10.7
9	14.8	35	10.6
10	14.7	36	10.4
11	14.5	37	10.2
12	14.3	38	10.1
13	14.2	39	9.9
14	14.0	40	9.7

Table 2.5 (*continued*)

Hydrometer reading	L (cm)	Hydrometer reading	L (cm)
15	13.8	41	9.6
16	13.7	42	9.4
17	13.5	43	9.2
18	13.3	44	9.1
19	13.2	45	8.9
20	13.0	46	8.8
21	12.9	47	8.6
22	12.7	48	8.4
23	12.5	49	8.3
24	12.4	50	8.1
25	12.2	51	7.9

The percentages of gravel, sand, silt, and clay-size particles present in a soil can be obtained from the particle-size distribution curve. According to the Unified Soil Classification System, the soil in Figure 2.14 has these percentages:

Gravel (size limits—greater than 4.75 mm) = 0%
Sand (size limits—4.75 to 0.075 mm) = percent finer than 4.75 mm diameter
  − percent finer than 0.075 mm diameter = 100 − 62 = 38%
Silt and clay (size limits—less than 0.075 mm) = 62%

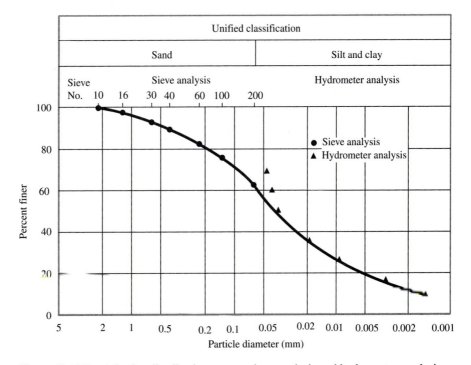

Figure 2.14 Particle-size distribution curve—sieve analysis and hydrometer analysis

Effective Size, Uniformity Coefficient, and Coefficient of Gradation

The particle-size distribution curve (Figure 2.15) can be used to compare different soils. Also, three basic soil parameters can be determined from these curves, and they can be used to classify granular soils. The three soil parameters are:

1. Effective size
2. Uniformity coefficient
3. Coefficient of gradation

The diameter in the particle-size distribution curve corresponding to 10% finer is defined as the *effective size*, or D_{10}. The *uniformity coefficient* is given by the relation

$$C_u = \frac{D_{60}}{D_{10}} \tag{2.7}$$

where
C_u = uniformity coefficient
D_{60} = the diameter corresponding to 60% finer in the particle-size distribution curve

The *coefficient of gradation* may be expressed as

$$C_c = \frac{D_{30}^2}{D_{60} \times D_{10}} \tag{2.8}$$

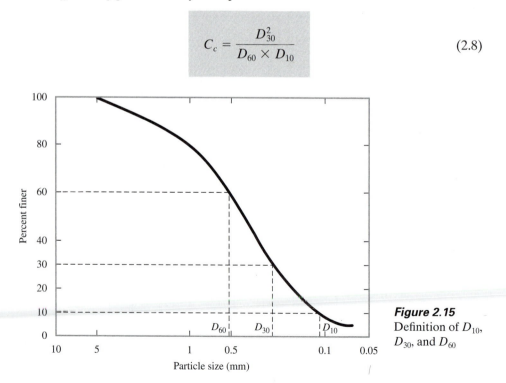

Figure 2.15
Definition of D_{10}, D_{30}, and D_{60}

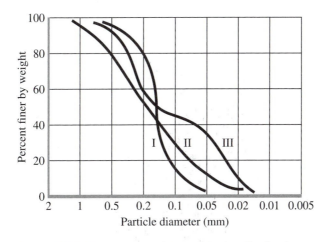

Figure 2.16 Different types of particle-size distribution curves

where

C_c = coefficient of gradation
D_{30} = diameter corresponding to 30% finer

The particle-size distribution curve shows not only the range of particle sizes present in a soil, but also the distribution of various size particles. Three curves are shown in Figure 2.16. Curve I represents a type of soil in which most of the soil grains are the same size. This is called *poorly graded* soil. Curve II represents a soil in which the particle sizes are distributed over a wide range and is termed *well graded*. A well-graded soil has a uniformity coefficient greater than about 4 for gravels, and 6 for sands, and a coefficient of gradation between 1 and 3 (for gravels and sands). A soil might have a combination of two or more uniformly graded fractions. Curve III represents such a soil, termed *gap graded*.

Example 2.1

Following are the results of a sieve analysis. Make the necessary calculations and draw a particle-size distribution curve.

U.S. sieve size	Mass of soil retained on each sieve (g)
4	0
10	40
20	60
40	89
60	140
80	122
100	210
200	56
Pan	12

Solution

The following table can now be prepared.

U.S. sieve (1)	Opening (mm) (2)	Mass retained on each sieve (g) (3)	Cumulative mass retained above each sieve (g) (4)	Percent finer[a] (5)
4	4.75	0	0	100
10	2.00	40	0 + 40 = 40	94.5
20	0.850	60	40 + 60 = 100	86.3
40	0.425	89	100 + 89 = 189	74.1
60	0.250	140	189 + 140 = 329	54.9
80	0.180	122	329 + 122 = 451	38.1
100	0.150	210	451 + 210 = 661	9.3
200	0.075	56	661 + 56 = 717	1.7
Pan	—	12	717 + 12 = 729 = ΣM	0

$$_a\frac{\Sigma M - \text{col.4}}{\Sigma M} \times 100 = \frac{729 - \text{col.4}}{729} \times 100$$

The particle-size distribution curve is shown in Figure 2.17.

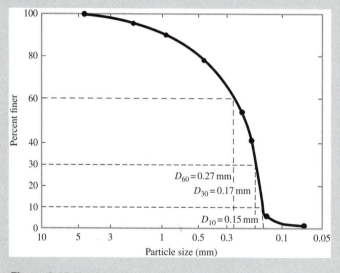

Figure 2.17 Particle-size distribution curve

Example 2.2

For the particle-size distribution curve shown in Figure 2.17, determine

a. D_{10}, D_{30}, and D_{60}
b. Uniformity coefficient, C_u
c. Coefficient of gradation, C_c

Solution

    a. From Figure 2.17,

$$D_{10} = \textbf{0.15 mm}$$

$$D_{30} = \textbf{0.17 mm}$$

$$D_{60} = \textbf{0.27 mm}$$

    b. $C_u = \dfrac{D_{60}}{D_{10}} = \dfrac{0.27}{0.15} = \textbf{1.8}$

    c. $C_c = \dfrac{D_{30}^2}{D_{60} \times D_{10}} = \dfrac{(0.17)^2}{(0.27)(0.15)} = \textbf{0.71}$ ■

Example 2.3

For the particle-size distribution curve shown in Figure 2.17, determine the percentages of gravel, sand, silt and clay-size particles present. Use the Unified Soil Classification System.

Solution
From Figure 2.17, we can prepare the following table.

Size (mm)	% finer	
76.2	100	
4.75	100	100 − 100 = 0% gravel
0.075	1.7	100 − 1.7 = 98.3% sand
—	0	1.7 − 0 = 1.7% silt and clay

    ■

Problems

2.1 Following are the results of a sieve analysis:

U.S. sieve No.	Mass of soil retained on each sieve (*g*)
4	0
10	21.6
20	49.5
40	102.6
60	89.1
100	95.6
200	60.4
pan	31.2

    a. Determine the percent finer than each sieve size and plot a grain-size distribution curve.

b. Determine D_{10}, D_{30}, and D_{60} from the grain-size distribution curve.

c. Calculate the uniformity coefficient, C_u.

d. Calculate the coefficient of gradation, C_c.

2.2 For a soil, given:

$D_{10} = 0.1$ mm

$D_{30} = 0.41$ mm

$D_{60} = 0.62$ mm

Calculate the uniformity coefficient and the coefficient of gradation of the soil.

2.3 Repeat Problem 2.2 for the following:

$D_{10} = 0.082$ mm

$D_{30} = 0.29$ mm

$D_{60} = 0.51$ mm

2.4 Repeat Problem 2.1 with the following results of a sieve analysis:

U.S. sieve No.	Mass of soil retained on each sieve (g)
4	0
6	30
10	48.7
20	127.3
40	96.8
60	76.6
100	55.2
200	43.4
pan	22

2.5 Repeat Problem 2.1 with the following results of a sieve analysis:

U.S. sieve No.	Mass of soil retained on each sieve (g)
4	0
6	0
10	0
20	9.1
40	249.4
60	179.8
100	22.7
200	15.5
pan	23.5

2.6 The particle characteristics of a soil are given below. Draw the particle-size distribution curve and find the percentages of gravel, sand, silt, and clay according to the MIT system (Table 2.1).

Size (mm)	Percent finer
0.850	100.0
0.425	92.1
0.250	85.8
0.150	77.3

Size (mm)	Percent finer
0.075	62.0
0.040	50.8
0.020	41.0
0.010	34.3
0.006	29.0
0.002	23.0

2.7 Redo Problem 2.6 according to the USDA system (Table 2.1).

2.8 Redo Problem 2.6 according to the AASHTO system (Table 2.1).

2.9 The particle-size characteristics of a soil are shown below. Find the percentages of gravel, sand, silt, and clay according to the MIT system (Table 2.1).

Size (mm)	Percent finer
0.850	100.0
0.425	100.0
0.250	94.1
0.150	79.3
0.075	34.1
0.040	28.0
0.020	25.2
0.010	21.8
0.006	18.9
0.002	14.0

2.10 Redo Problem 2.9 according to the USDA system (Table 2.1).

2.11 Redo Problem 2.9 according to the AASHTO system (Table 2.1).

2.12 In a hydrometer test, the results are as follows: $G_s = 2.60$, temperature of water = 24°, and hydrometer reading = 43 at 60 min after the start of sedimentation. What is the diameter, D, of the smallest-size particles that have settled beyond the zone of measurement at that time (that is, $t = 60$ min)?

2.13 Repeat Problem 2.8 with the following values: $G_s = 2.70$, temperature of water = 23°, and hydrometer reading = 25.

References

AMERICAN SOCIETY FOR TESTING AND MATERIALS (2006). *ASTM Book of Standards*, Vol. 04.08, West Conshohocken, PA.

GRIM, R F. (1953). *Clay Mineralogy*, McGraw-Hill, New York.

GRIM, R. L. (1959). "Physico Chemical Properties of Soils: Clay Minerals," *Journal of the Soil Mechanics and Foundations Division*, ASCE, Vol. 85, No. SM2, 1–17.

YONG, R. N., and WARKENTIN, B. P. (1966). *Introduction of Soil Behavior*, Macmillan, New York.

3

Weight–Volume Relationships, Plasticity, and Soil Classification

The preceding chapter presented the geological processes by which soils are formed, the description of the soil-particle size limits, and the mechanical analysis of soils. In natural occurrence, soils are three-phase systems consisting of soil solids, water, and air. This chapter discusses the weight–volume relationships of soil aggregates, their structures and plasticity, and their engineering classification.

3.1 ## Weight–Volume Relationships

Figure 3.1a shows an element of soil of volume V and weight W as it would exist in a natural state. To develop the weight–volume relationships, we separate the three phases; that is, solid, water, and air, as shown in Figure 3.1b. Thus, the total volume of a given soil sample can be expressed as

$$V = V_s + V_v = V_s + V_w + V_a \qquad (3.1)$$

where
V_s = volume of soil solids
V_v = volume of voids
V_w = volume of water in the voids
V_a = volume of air in the voids

Assuming the weight of the air to be negligible, we can give the total weight of the sample as

$$W = W_s + W_w \qquad (3.2)$$

where
W_s = weight of soil solids
W_w = weight of water

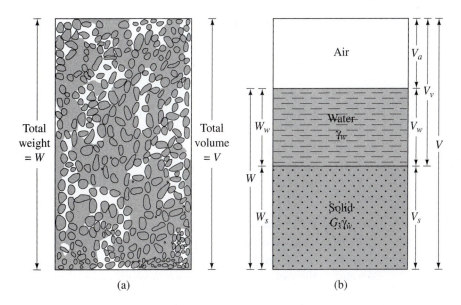

Figure 3.1 (a) Soil element in natural state; (b) three phases of the soil element

Volume Relationships

The volume relationships commonly used for the three phases in a soil element are *void ratio, porosity*, and *degree of saturation. Void ratio* (e) is defined as the ratio of the volume of voids to the volume of solids, or

$$e = \frac{V_v}{V_s} \tag{3.3}$$

Porosity (n) is defined as the ratio of the volume of voids to the total volume, or

$$n = \frac{V_v}{V} \tag{3.4}$$

Degree of saturation (S) is defined as the ratio of the volume of water to the volume of voids, or

$$S = \frac{V_w}{V_v} \tag{3.5}$$

The degree of saturation is commonly expressed as a percentage.

The relationship between void ratio and porosity can be derived from Eqs. (3.1), (3.3), and (3.4), as follows:

$$e = \frac{V_v}{V_s} = \frac{V_v}{V - V_v} = \frac{\left(\dfrac{V_v}{V}\right)}{1 - \left(\dfrac{V_v}{V}\right)} = \frac{n}{1 - n} \tag{3.6}$$

Also, from Eq. (3.6), we have

$$n = \frac{e}{1 + e} \tag{3.7}$$

Weight Relationships

The common weight relationships are *moisture content* and *unit weight*. *Moisture content* (*w*) is also referred to as *water content* and is defined as the ratio of the weight of water to the weight of solids in a given volume of soil, or

$$w = \frac{W_w}{W_s} \tag{3.8}$$

Unit weight (γ) is the weight of soil per unit volume:

$$\gamma = \frac{W}{V} \tag{3.9}$$

The unit weight can also be expressed in terms of weight of soil solids, moisture content, and total volume. From Eqs. (3.2), (3.8), and (3.9), we have

$$\gamma = \frac{W}{V} = \frac{W_s + W_w}{V} = \frac{W_s\left[1 + \left(\dfrac{W_w}{W_s}\right)\right]}{V} = \frac{W_s(1 + w)}{V} \tag{3.10}$$

Soils engineers sometimes refer to the unit weight defined by Eq. (3.9) as the *moist unit weight*.

It is sometimes necessary to know the weight per unit volume of soil excluding water. This is referred to as the *dry unit weight*, γ_d. Thus,

$$\gamma_d = \frac{W_s}{V} \tag{3.11}$$

From Eqs. (3.10) and (3.11), we can give the relationship among unit weight, dry unit weight, and moisture content as

$$\gamma_d = \frac{\gamma}{1 + w} \tag{3.12}$$

Unit weight is expressed in kilonewtons per cubic meter (kN/m³). Since the newton is a derived unit, it may sometimes be convenient to work with densities (ρ) of soil. The SI unit of density is kilograms per cubic meter (kg/m³). We can write the density equations [similar to Eqs. (3.9) and (3.11)] as

$$\rho = \frac{m}{V} \tag{3.13}$$

and

$$\rho_d = \frac{m_s}{V} \tag{3.14}$$

where

ρ = density of soil (kg/m³)
ρ_d = dry density of soil (kg/m³)
m = total mass of the soil sample (kg)
m_s = mass of soil solids in the sample (kg)

The unit of total volume, V, is m³.

The unit weights of soil in N/m³ can be obtained from densities in kg/m³ as

$$\gamma = \rho \cdot g = 9.81\rho \tag{3.15}$$

and

$$\gamma_d = \rho_d \cdot g = 9.81\rho_d \tag{3.16}$$

where g = acceleration due to gravity = 9.81 m/sec².

3.2   Relationships among Unit Weight, Void Ratio, Moisture Content, and Specific Gravity

To obtain a relationship among unit weight (or density), void ratio, and moisture content, consider a volume of soil in which the volume of the soil solids is 1, as shown in Figure 3.2. If the volume of the soil solids is 1, then the volume of voids is numerically equal to the void ratio, e [from Eq. (3.3)]. The weights of soil solids and water can be given as

$$W_s = G_s \gamma_w$$

$$W_w = wW_s = wG_s \gamma_w$$

where

G_s = specific gravity of soil solids
w = moisture content
γ_w = unit weight of water

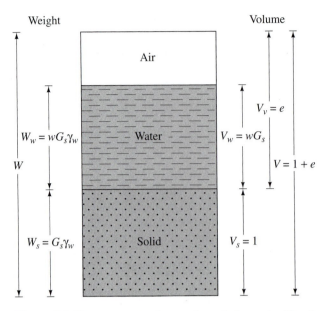

Figure 3.2 Three separate phases of a soil element with volume of soil solids equal to 1

The unit weight of water is 9.81 kN/m³. Now, using the definitions of unit weight and dry unit weight [Eqs. (3.9) and (3.11)], we can write

$$\gamma = \frac{W}{V} = \frac{W_s + W_w}{V} = \frac{G_s\gamma_w + wG_s\gamma_w}{1 + e} = \frac{(1 + w)G_s\gamma_w}{1 + e} \tag{3.17}$$

and

$$\gamma_d = \frac{W_s}{V} = \frac{G_s\gamma_w}{1 + e} \tag{3.18}$$

Since the weight of water in the soil element under consideration is $wG_s\gamma_w$, the volume occupied by it is

$$V_w = \frac{W_w}{\gamma_w} = \frac{wG_s\gamma_w}{\gamma_w} = wG_s$$

Hence, from the definition of degree of saturation [Eq. (3.5)], we have

$$S = \frac{V_w}{V_v} = \frac{wG_s}{e}$$

or

$$Se = wG_s \tag{3.19}$$

This is a very useful equation for solving problems involving three-phase relationships.

If the soil sample is *saturated*—that is, the void spaces are completely filled with water (Figure 3.3)—the relationship for saturated unit weight can be derived in a similar manner:

$$\gamma_{sat} = \frac{W}{V} = \frac{W_s + W_w}{V} = \frac{G_s\gamma_w + e\gamma_w}{1 + e} = \frac{(G_s + e)\gamma_w}{1 + e} \qquad (3.20)$$

where γ_{sat} = saturated unit weight of soil.

As mentioned before, because it is convenient to work with densities, the following equations [similar to the unit-weight relationships given in Eqs. (3.17), (3.18), and (3.20)] are useful:

$$\text{Density} = \rho = \frac{(1 + w)G_s\rho_w}{1 + e} \qquad (3.21)$$

bulk

↑ solve

$$\text{Dry density} = \rho_d = \frac{G_s\rho_w}{1 + e} \qquad (3.22)$$

$$\text{Saturated density} = \rho_{sat} = \frac{(G_s + e)\rho_w}{1 + e} \qquad (3.23)$$

where ρ_w = density of water = 1000 kg/m³.

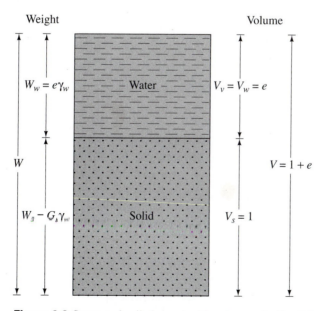

Figure 3.3 Saturated soil element with volume of soil solids equal to 1

Table 3.1 Void Ratio, Moisture Content, and Dry Unit Weight for Some Typical Soils in a Natural State

Type of soil	Void ratio, e	Natural moisture content in a saturated state (%)	Dry unit weight, γ_d (kN/m^3)
Loose uniform sand	0.8	30	14.5
Dense uniform sand	0.45	16	18
Loose angular-grained silty sand	0.65	25	16
Dense angular-grained silty sand	0.4	15	19
Stiff clay	0.6	21	17
Soft clay	0.9–1.4	30–50	11.5–14.5
Loess	0.9	25	13.5
Soft organic clay	2.5–3.2	90–120	6–8
Glacial till	0.3	10	21

Some typical values of void ratio, moisture content in a saturated condition, and dry unit weight for soils in a natural state are given in Table 3.1.

Relationships among Unit Weight, Porosity, and Moisture Content

The relationships among *unit weight, porosity,* and *moisture content* can be developed in a manner similar to that presented in the preceding section. Consider a soil that has a total volume equal to one, as shown in Figure 3.4. From Eq. (3.4),

$$n = \frac{V_v}{V}$$

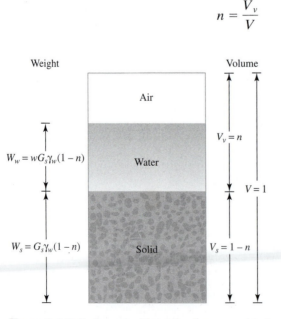

Figure 3.4 Soil element with total volume equal to 1

If V is equal to 1, then V_v is equal to n, so $V_s = 1 - n$. The weight of soil solids (W_s) and the weight of water (W_w) can then be expressed as follows:

$$W_s = G_s \gamma_w (1 - n) \tag{3.24}$$

$$W_w = w W_s = w G_s \gamma_w (1 - n) \tag{3.25}$$

So, the dry unit weight equals

$$\gamma_d = \frac{W_s}{V} = \frac{G_s \gamma_w (1 - n)}{1} = G_s \gamma_w (1 - n) \tag{3.26}$$

The moist unit weight equals

$$\gamma = \frac{W_s + W_w}{V} = G_s \gamma_w (1 - n)(1 + w) \tag{3.27}$$

Figure 3.5 shows a soil sample that is saturated and has $V = 1$. According to this figure,

$$\gamma_{\text{sat}} = \frac{W_s + W_w}{V} = \frac{(1 - n)G_s \gamma_w + n\gamma_w}{1} = [(1 - n)G_s + n]\gamma_w \tag{3.28}$$

The moisture content of a saturated soil sample can be expressed as

$$w = \frac{W_w}{W_s} = \frac{n\gamma_w}{(1 - n)\gamma_w G_s} = \frac{n}{(1 - n)G_s} \tag{3.29}$$

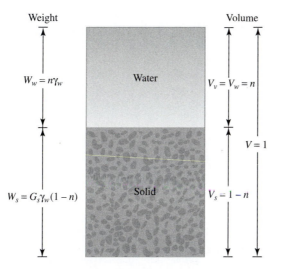

Figure 3.5 Saturated soil element with total volume equal to 1

Example 3.1

For a saturated soil, show that

$$\gamma_{\text{sat}} = \left(\frac{e}{w}\right)\left(\frac{1+w}{1+e}\right)\gamma_w$$

Solution

From Eqs. (3.19) and (3.20),

$$\gamma_{\text{sat}} = \frac{(G_s + e)\gamma_w}{1 + e} \tag{a}$$

and

$$e = wG_s$$

or

$$G_s = \frac{e}{w} \tag{b}$$

Combining Eqs. (a) and (b) gives

$$\gamma_{\text{sat}} = \frac{\left(\dfrac{e}{w} + e\right)\gamma_w}{1 + e} = \left(\frac{e}{w}\right)\left(\frac{1+w}{1+e}\right)\gamma_w$$

Example 3.2

A moist soil has these values: $V = 7.08 \times 10^{-3}\text{m}^3$, $m = 13.95$ kg, $w = 9.8\%$, and $G_s = 2.66$.

Determine the following:

a. ρ
b. ρ_d
c. e
d. n
e. $S(\%)$
f. Volume occupied by water

Solution

Part a.

From Eq. (3.13).

$$\rho = \frac{m}{V} = \frac{13.95}{7.08 \times 10^{-3}} = \mathbf{1970.3\ kg/m^3}$$

Part b.

From Eq. (3.12),

$$\rho_d = \frac{\rho}{1 + w} = \frac{1970.3}{1 + \left(\dfrac{9.8}{100}\right)} = \mathbf{1794.4\ kg/m^3}$$

Part c.

From Eq. (3.22),

$$e = \frac{G_s \rho_w}{\rho_d} - 1$$

$$e = \frac{(2.66)(1000)}{1794.4} - 1 = \mathbf{0.48}$$

Part d.

From Eq. (3.7),

$$n = \frac{e}{1 + e} = \frac{0.48}{1 + 0.48} = \mathbf{0.324}$$

Part e.

From Eq. (3.19)

$$S(\%) = \left(\frac{wG_s}{e}\right)(100) = \frac{(0.098)(2.66)}{0.48}(100) = \mathbf{54.3\%}$$

Part f.

Mass of soil solids is

$$m_s = \frac{m}{1 + w} = \frac{13.95}{1 + 0.098} = \mathbf{12.7\ kg}$$

Thus, mass of water is

$$m_w = m - m_s = 13.95 - 12.7 = \mathbf{1.25\ kg}$$

Volume of water is

$$V_w = \frac{m_w}{\rho_w} = \frac{1.25}{1000} = \mathbf{0.00125\ m^3}$$

Example 3.3

In the natural state, a moist soil has a volume of 0.0093 m^3 and weighs 177.6 N. The oven dry weight of the soil is 153.6 N. If $G_s = 2.71$, calculate the moisture content, moist unit weight, dry unit weight, void ratio, porosity, and degree of saturation.

Solution

Refer to Figure 3.6. The moisture content [Eq. (3.8)] is

$$w = \frac{W_w}{W_s} = \frac{W - W_s}{W_s} = \frac{177.6 - 153.6}{153.6} = \frac{24}{153.6} \times 100 = \mathbf{15.6\%}$$

The moist unit weight [Eq. (3.9)] is

$$\gamma = \frac{W}{V} = \frac{177.6}{0.0093} = 19{,}096 \text{ N/m}^3 \approx \mathbf{19.1 \text{ kN/m}^3}$$

For dry unit weight [Eq. (3.11)], we have

$$\gamma_d = \frac{W_s}{V} = \frac{153.6}{0.0093} = 16{,}516 \text{ N/m}^3 \approx \mathbf{16.52 \text{ kN/m}^3}$$

The void ratio [Eq. (3.3)] is found as follows:

$$e = \frac{V_v}{V_s}$$

$$V_s = \frac{W_s}{G_s \gamma_w} = \frac{0.1536}{2.71 \times 9.81} = 0.0058 \text{ m}^3$$

$$V_v = V - V_s = 0.0093 - 0.0058 = 0.0035 \text{ m}^3$$

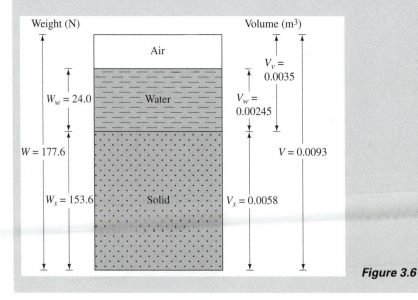

Figure 3.6

so

$$e = \frac{0.0035}{0.0058} \approx \mathbf{0.60}$$

For porosity [Eq. (3.7)], we have

$$n = \frac{e}{1 + e} = \frac{0.60}{1 + 0.60} = \mathbf{0.375}$$

We find the degree of saturation [Eq. (3.5)] as follows:

$$S = \frac{V_w}{V_v}$$

$$V_w = \frac{W_w}{\gamma_w} = \frac{0.024}{9.81} = 0.00245 \, \text{m}^3$$

so

$$S = \frac{0.00245}{0.0035} \times 100 = \mathbf{70\%}$$ ∎

Example 3.4

The dry density of a sand with a porosity of 0.387 is 1600 kg/m³. Find the void ratio of the soil and the specific gravity of the soil solids.

Solution
Void ratio
    From $n = 0.387$ and Eq. (3.6),

$$e = \frac{n}{1 - n} = \frac{0.387}{1 - 0.387} = \mathbf{0.631}$$

Specific gravity of soil solids From Eq. (3.22)

$$\rho_d = \frac{G\rho_w}{1 + e}$$

where
    ρ_d = dry density of soil
    ρ_w = density of water = 1000 kg/m³

Thus,

$$1600 = \frac{G_s(1000)}{1 + 0.631}$$

$$G_s = \mathbf{2.61}$$

Example 3.5

For a saturated soil, given $w = 40\%$ and $G_s = 2.71$, determine the saturated and dry unit weights.

Solution
For saturated soil, from Eq. (3.19)

$$e = wG_s = (0.4)(2.71) = 1.084$$

From Eq. (3.20),

$$\gamma_{sat} = \frac{(G_s + e)\gamma_w}{1 + e} = \frac{(2.71 + 1.084)9.81}{1 + 1.084} = \textbf{17.86 kN/m}^3$$

From Eq. (3.18),

$$\gamma_d = \frac{G_s\gamma_w}{1 + e} = \frac{(2.71)(9.81)}{1 + 1.084} = \textbf{12.76 kN/m}^3$$

Example 3.6

The mass of a moist soil sample collected from the field is 465 g, and its oven dry mass is 405.76 g. The specific gravity of the soil solids was determined in the laboratory to be 2.68. If the void ratio of the soil in the natural state is 0.83, find the following:

   a. The moist unit weight of the soil in the field (kN/m^3)
   b. The dry unit weight of the soil in the field (kN/m^3)
   c. The weight of water (in kN) to be added per cubic meter of soil in the field for saturation

Solution
Part a.
From Eq. (3.8),

$$w = \frac{m_w}{m_s} = \frac{465 - 405.76}{405.76} = \frac{59.24}{405.76} = 14.6\%$$

From Eq. (3.17),

$$\gamma = \frac{(1 + w)G_s\gamma_w}{1 + e} = \frac{(1 + 0.146)(2.68)(9.81)}{1 + 0.83} = \textbf{16.46 kN/m}^3$$

Part b.
From Eq. (3.18),

$$\gamma_d = \frac{G_s\gamma_w}{1 + e} = \frac{(2.68)(9.81)}{1 + 0.83} = \textbf{14.37 kN/m}^3$$

Part c.
From Eq. (3.20),

$$\gamma_{sat} = \frac{(G_s + e)\gamma_w}{1 + e} = \frac{(2.68 + 0.83)(9.81)}{1 + 0.83} = 18.82 \text{ kN/m}^3$$

So, the weight of water to be added is

$$\gamma_{sat} - \gamma = 18.82 - 16.46 = \mathbf{2.36 \text{ kN/m}^3}$$

3.4 Relative Density

The term *relative density* is commonly used to indicate the *in situ* denseness or looseness of granular soil. It is defined as

$$D_r = \frac{e_{max} - e}{e_{max} - e_{min}} \tag{3.30}$$

where

D_r = relative density, usually given as a percentage
e = *in situ* void ratio of the soil
e_{max} = void ratio of the soil in the loosest condition
e_{min} = void ratio of the soil in the densest condition

The values of D_r may vary from a minimum of 0 for very loose soil, to a maximum of 1 for very dense soil. Soils engineers qualitatively describe the granular soil deposits according to their relative densities, as shown in Table 3.2.

By using the definition of dry unit weight given in Eq. (3.18), we can also express relative density in terms of maximum and minimum possible dry unit weights. Thus,

$$D_r = \frac{\left[\dfrac{1}{\gamma_{d(min)}}\right] - \left[\dfrac{1}{\gamma_d}\right]}{\left[\dfrac{1}{\gamma_{d(min)}}\right] - \left[\dfrac{1}{\gamma_{d(max)}}\right]} = \left[\frac{\gamma_d - \gamma_{d(min)}}{\gamma_{d(max)} - \gamma_{d(min)}}\right]\left[\frac{\gamma_{d(max)}}{\gamma_d}\right] \tag{3.31}$$

Table 3.2 Qualitative description of granular soil deposits

Relative density (%)	Description of soil deposit
0–15	Very loose
15–50	Loose
50–70	Medium
70–85	Dense
85–100	Very dense

where

$\gamma_{d(min)}$ = dry unit weight in the loosest condition (at a void ratio of e_{max})

γ_d = *in situ* dry unit weight (at a void ratio of e)

$\gamma_{d(max)}$ = dry unit weight in the densest condition (at a void ratio of e_{min})

Cubrinovski and Ishihara (2002) studied the variation of e_{max} and e_{min} for a very large number of soils. Based on the best-fit linear-regression lines, they provided the following relationships.

- Clean sand (F_c = 0 to 5%)

$$e_{max} = 0.072 + 1.53\, e_{min} \tag{3.32}$$

- Sand with fines ($5 < F_c \le 15\%$)

$$e_{max} = 0.25 + 1.37\, e_{min} \tag{3.33}$$

- Sand with fines and clay ($15 < F_c \le 30\%; P_c = 5$ to 20%)

$$e_{max} = 0.44 + 1.21\, e_{min} \tag{3.34}$$

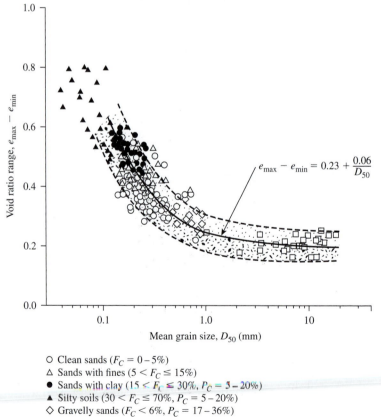

Clean sands ($F_C = 0-5\%$)
Sands with fines ($5 < F_C \le 15\%$)
Sands with clay ($15 < F_c \le 30\%, P_C = 5-20\%$)
Silty soils ($30 < F_C \le 70\%, P_C = 5-20\%$)
Gravelly sands ($F_C < 6\%, P_C = 17-36\%$)
Gravels

Figure 3.7 Plot of $e_{max} - e_{min}$ versus the mean grain size (from Cubrinovski and Ishihara, 2002)

- Silty soils ($30 < F_c \leqslant 70\%$; $P_c = 5$ to 20%)

$$e_{max} = 0.44 + 1.32\, e_{min} \qquad (3.35)$$

where

F_c = fine fraction for which grain size is smaller than 0.075 mm
P_c = clay-size fraction (< 0.005 mm)

Figure 3.7 shows a plot of $e_{max} - e_{min}$ versus the mean grain size (D_{50}) for a number of soils (Cubrinovski and Ishihara, 1999 and 2002). From this figure, the average plot for sandy and gravelly soils can be given by the relationship

$$e_{max} - e_{min} = 0.23 + \frac{0.06}{D_{50}\,(mm)} \qquad (3.36)$$

Example 3.7

For a given sandy soil, $e_{max} = 0.75$, $e_{min} = 0.46$, and $G_s = 2.68$. What is the moist unit weight of compaction (kN/m^3) in the field if $D_r = 78\%$ and $w = 9\%$?

Solution
From Eq. (3.30),

$$D_r = \frac{e_{max} - e}{e_{max} - e_{min}}$$

or

$$e = e_{max} - D_r(e_{max} - e_{min}) = 0.75 - 0.78(0.75 - 0.46) = 0.524$$

Again, from Eq. (3.17),

$$\gamma = \frac{G_s \gamma_w(1 + w)}{1 + e} = \frac{(2.68)(9.81)(1 + 0.09)}{1 + 0.524} = \textbf{18.8 kN/m}^3$$

3.5   *Consistency of Soil*

When clay minerals are present in fine-grained soil, that soil can be remolded in the presence of some moisture without crumbling. This cohesive nature is because of the adsorbed water surrounding the clay particles. In the early 1900s, a Swedish scientist named Albert Mauritz Atterberg developed a method to describe the consistency of fine-grained soils with varying moisture contents. At a very low moisture content, soil behaves more like a brittle solid. When the moisture content is very high, the soil and water may flow like a liquid. Hence, on an arbitrary basis, depending on the

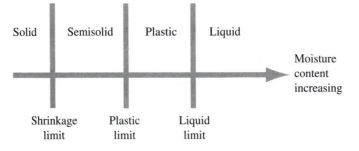

Figure 3.8 Atterberg limits

moisture content, the nature of soil behavior can be broken down into four basic states: *solid, semisolid, plastic*, and *liquid*, as shown in Figure 3.8.

The moisture content, in percent, at which the transition from solid to semisolid state takes place is defined as the *shrinkage limit*. The moisture content at the point of transition from semisolid to plastic state is the *plastic limit*, and from plastic to liquid state is the *liquid limit*. These limits are also known as *Atterberg limits*.

Liquid Limit (LL)

A schematic diagram (side view) of a liquid limit device is shown in Figure 3.9a. This device consists of a brass cup and hard rubber base. The brass cup can be dropped on the base by a cam operated by a crank. For the liquid limit test, a soil paste is placed in the cup. A groove is cut at the center of the soil pat, using the standard grooving tool (Figure 3.9b). Then, with the crank-operated cam, the cup is lifted and dropped from a height of 10 mm. The moisture content, in percent, required to close a distance of 12.7 mm along the bottom of the groove (see Figures 3.9c and 3.9d) after 25 blows is defined as the *liquid limit*.

The procedure for the liquid limit test is given in ASTM Test Designation D-4318. It is difficult to adjust the moisture content in the soil to meet the required 12.7 mm closure of the groove in the soil pat at 25 blows. Hence, at least four tests for the same soil are made at varying moisture content to determine the number of blows, *N*, required to achieve closure varying between 15 and 35. The moisture content of the soil, in percent, and the corresponding number of blows are plotted on semilogrithmic graph paper (Figure 3.10). The relation between moisture content and log *N* is approximated as a straight line. This is referred to as the *flow curve*. The moisture content corresponding to $N = 25$, determined from the flow curve, gives the liquid limit of the soil.

Another method of determining liquid limit that is popular in Europe and Asia is the *fall cone method* (British Standard—BS1377). In this test the liquid limit is defined as the moisture content at which a standard cone of apex angle 30° and weight of 0.78 N (80 gf) will penetrate a distance $d = 20$ mm in 5 seconds when allowed to drop from a position of point contact with the soil surface (Figure 3.11a). Due to the difficulty in achieving the liquid limit from a single test, four or more tests can be conducted at various moisture contents to determine the fall cone

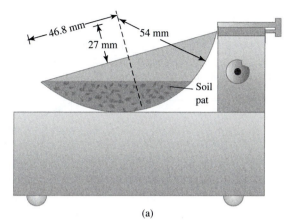

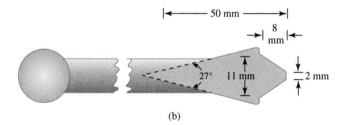

Section

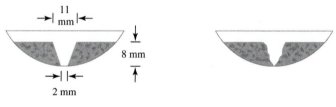

Plan

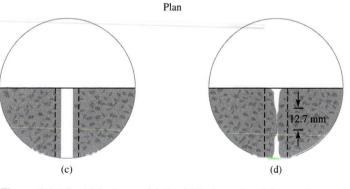

Figure 3.9 Liquid limit test: (a) liquid limit device; (b) grooving tool; (c) soil pat before test; (d) soil pat after test

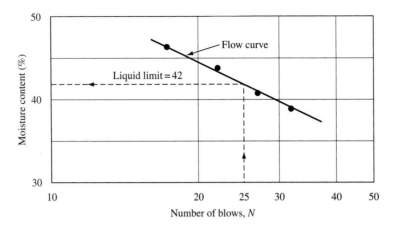

Figure 3.10 Flow curve for liquid limit determination of a silty clay

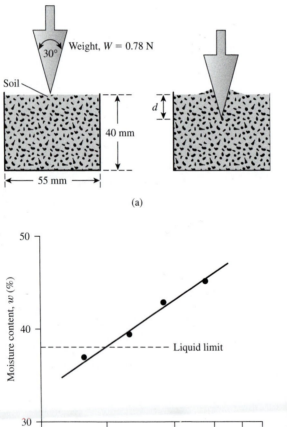

Figure 3.11
(a) Fall cone test (b) plot of moisture content vs. cone penetration for determination of liquid limit

penetration, *d*. A semilogarithmic graph can then be plotted with moisture content (*w*) versus cone penetration *d*. The plot results in a straight line. The moisture content corresponding to *d* = 20 mm is the liquid limit (Figure 3.11b).

Plastic Limit (PL)

The *plastic limit* is defined as the moisture content, in percent, at which the soil when rolled into threads of 3.2 mm in diameter, crumbles. The plastic limit is the lower limit of the plastic stage of soil. The test is simple and is performed by repeated rollings by hand of an ellipsoidal size soil mass on a ground glass plate (Figure 3.12).

The *plasticity index* (*PI*) is the difference between the liquid limit and plastic limit of a soil, or

$$PI = LL - PL \tag{3.37}$$

The procedure for the plastic limit test is given in ASTM Test Designation D-4318.

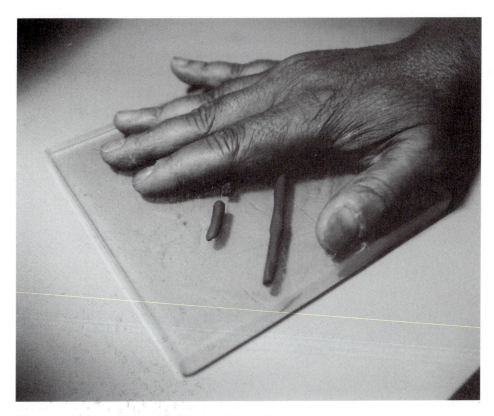

Figure 3.12 Plastic limit test (Courtesy of Braja Das)

As in the case of liquid limit determination, the fall cone method can be used to obtain the plastic limit. This can be achieved by using a cone of similar geometry, but with a mass of 2.35 N (240 gf). Three to four tests at varying moisture contents of soil are conducted, and the corresponding cone penetrations (d) are determined. The moisture content corresponding to a cone penetration of $d = 20$ mm is the plastic limit. Figure 3.13 shows the liquid and plastic limit determination of Cambridge Gault clay reported by Worth and Wood (1978).

Shrinkage Limit (SL)

Soil mass shrinks as moisture is gradually lost from the soil. With continuous loss of moisture, a stage of equilibrium is reached at which point more loss of moisture will result in no further volume change (Figure 3.14). The moisture content, in percent, at which the volume change of the soil mass ceases is defined as the *shrinkage limit*.

Shrinkage limit tests (ASTM Test Designation D-427) are performed in the laboratory with a porcelain dish about 44 mm in diameter and about 13 mm in height. The inside of the dish is coated with petroleum jelly and is then filled completely with wet soil. Excess soil standing above the edge of the dish is struck off with a straightedge. The mass of the wet soil inside the dish is recorded. The soil pat in the dish is then oven dried. The volume of the oven-dried soil pat is determined by the displacement of mercury. Figure 3.15 shows a photograph of the equipment needed for the shrinkage limit test. Because handling mercury can be hazardous, ASTM Test Designation D-4943 describes a method of dipping the oven-dried soil pat in a pot of melted wax. The wax-coated soil pat is then cooled. Its volume is determined by submerging it in water.

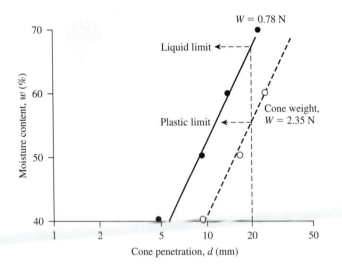

Figure 3.13 Liquid and plastic limits for Cambridge Gault clay determined by fall cone test

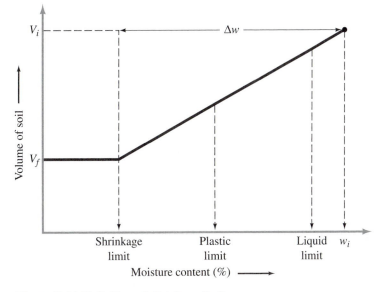

Figure 3.14 Definition of shrinkage limit

Referring to Figure 3.14, we can determine the shrinkage limit in the following manner:

$$SL = w_i \, (\%) - \Delta w \, (\%)$$
(3.38)

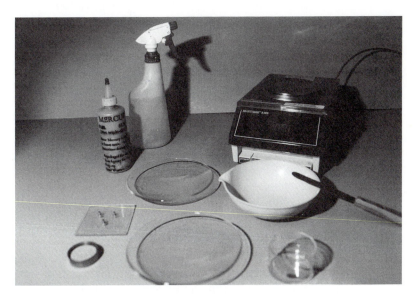

Figure 3.15 Equipment for shrinkage limit test (Courtesy of Braja Das)

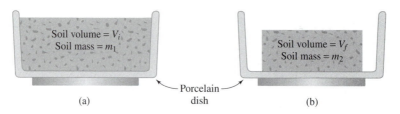

Figure 3.16 Shrinkage limit test: (a) soil pat before drying; (b) soil pat after drying

where
  w_i = initial moisture content when the soil is placed in the shrinkage limit dish
  Δw = change in moisture content (that is, between the initial moisture content and the moisture content at shrinkage limit)

However,

$$w_i(\%) = \frac{m_1 - m_2}{m_2} \times 100 \tag{3.39}$$

where
  m_1 = mass of the wet soil pat in the dish at the beginning of the test (g)
  m_2 = mass of the dry soil pat (g) (see Figure 3.16)
Also,

$$\Delta w(\%) = \frac{(V_i - V_f)\rho_w}{m_2} \times 100 \tag{3.40}$$

where
  V_i = initial volume of the wet soil pat (that is, inside volume of the dish, cm³)
  V_f = volume of the oven-dried soil pat (cm³)
  ρ_w = density of water (g/cm³)

Now, combining Eqs. (3.38), (3.39), and (3.40), we have

$$SL = \left(\frac{m_1 - m_2}{m_2}\right)(100) - \left[\frac{(V_i - V_f)\rho_w}{m_2}\right](100) \tag{3.41}$$

3.6   *Activity*

Since the plastic property of soil results from the adsorbed water that surrounds the clay particles, we can expect that the type of clay minerals and their proportional amounts in a soil will affect the liquid and plastic limits. Skempton (1953) observed that the plasticity index of a soil increases linearly with the percent of clay-size fraction (finer than 2 μ by weight) present in it. On the basis of these results, Skempton

Table 3.3 Activity of clay minerals

Mineral	Activity, A
Smectites	1–7
Illite	0.5–1
Kaolinite	0.5
Halloysite ($2H_2O$)	0.5
Holloysite ($4H_2O$)	0.1
Attapulgite	0.5–1.2
Allophane	0.5–1.2

defined a quantity called *activity*, which is the slope of the line correlating *PI* and percent finer than 2 μ. This activity may be expressed as

$$A = \frac{PI}{\text{percent of clay-size fraction, by weight}} \tag{3.42}$$

where A = activity. Activity is used as an index for identifying the swelling potential of clay soils. Typical values of activities for various clay minerals are listed in Table 3.4 (Mitchell, 1976).

Seed, Woodward, and Lundgren (1964) studied the plastic property of several artificially prepared mixtures of sand and clay. They concluded that although the relationship of the plasticity index to the percent of clay-size fraction is linear, as observed by Skempton, the line may not always pass through the origin. They showed that the relationship of the plasticity index to the percent of clay-size fraction present in a soil can be represented by two straight lines. This relationship is shown qualitatively in Figure 3.17. For clay-size fractions greater than 40%, the straight line passes through the origin when it is projected back.

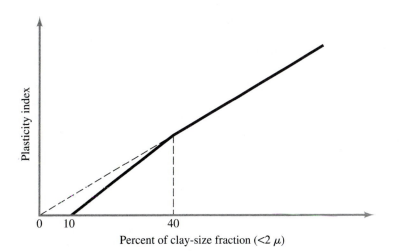

Figure 3.17 Simplified relationship between plasticity index and percent of clay-size fraction by weight

3.7	**Liquidity Index**

The relative consistency of a cohesive soil in the natural state can be defined by a ratio called the *liquidity index* (*LI*):

$$LI = \frac{w - PL}{LL - PL} \tag{3.43}$$

where w = *in situ* moisture content of soil.

The *in situ* moisture content of a sensitive clay may be greater than the liquid limit. In that case,

$$LI > 1$$

These soils, when remolded, can be transformed into a viscous form to flow like a liquid.

Soil deposits that are heavily overconsolidated may have a natural moisture content less than the plastic limit. In that case,

$$LI < 1$$

The values of the liquidity index for some of these soils may be negative.

3.8	**Plasticity Chart**

Liquid and plastic limits are determined by relatively simple laboratory tests that provide information about the nature of cohesive soils. The tests have been used extensively by engineers for the correlation of several physical soil parameters as well as for soil identification. Casagrande (1932) studied the relationship of the plasticity index to the liquid limit of a wide variety of natural soils. On the basis of the test results, he proposed a plasticity chart as shown in Figure 3.18. The important feature of this chart is the empirical *A*-line that is given by the equation $PI = 0.73(LL - 20)$. The *A*-line separates the inorganic clays from the inorganic silts. Plots of plasticity indexes against liquid limits for inorganic clays lie above the *A*-line, and those for inorganic silts lie below the *A*-line. Organic silts plot in the same region (below the *A*-line and with *LL* ranging from 30 to 50) as the inorganic silts of medium compressibility. Organic clays plot in the same region as the inorganic silts of high compressibility (below the *A*-line and *LL* greater than 50). The information provided in the plasticity chart is of great value and is the basis for the classification of fine-grained soils in the Unified Soil Classification System.

Note that a line called the *U*-line lies above the *A*-line. The *U*-line is approximately the upper limit of the relationship of the plasticity index to the liquid limit for any soil found so far. The equation for the *U*-line can be given as

$$PI = 0.9(LL - 8) \tag{3.44}$$

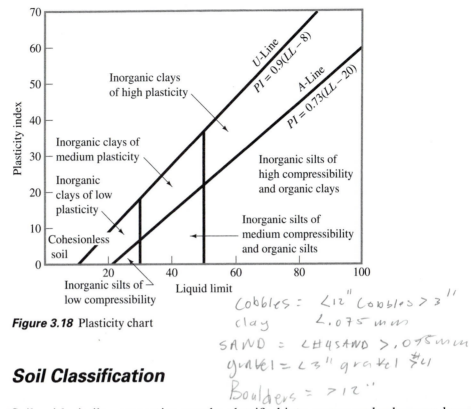

Figure 3.18 Plasticity chart

Cobbles = <12" Cobbles > 3"
clay   <.075 mm
SAND = <#4 SAND >.075 mm
gravel = <3" gravel #4
Boulders = >12"

3.9   *Soil Classification*

Soils with similar properties may be classified into groups and subgroups based on their engineering behavior. Classification systems provide a common language to express concisely the general characteristics of soils, which are infinitely varied, without a detailed description. At the present time, two elaborate classification systems that use the grain-size distribution and plasticity of soils are commonly used by soils engineers. They are the American Association of State Highway Officials (AASHTO) classification system and the Unified Soil Classification System. The AASHTO system is used mostly by state and county highway departments, whereas geotechnical engineers usually prefer to use the Unified System.

AASHTO Classification System

This system of soil classification was developed in 1929 as the Public Road Administration Classification System. It has undergone several revisions, with the present version proposed by the Committee on Classification of Materials for Subgrades and Granular Type Roads of the Highway Research Board in 1945 (ASTM Test Designation D-3282; AASHTO method M145).

The AASHTO classification system in present use is given in Table 3.4. According to this system, soil is classified into seven major groups: A-1 through A-7. Soils classified into groups A-1, A-2, and A-3 are granular materials, where 35% or less of the particles pass through the No. 200 sieve. Soils where more than

Table 3.4 Classification of highway subgrade materials

General classification	Granular materials (35% or less of total sample passing No. 200)						
	A-1			*A-2*			
Group classification	*A-1-a*	*A-1-b*	*A-3*	*A-2-4*	*A-2-5*	*A-2-6*	*A-2-7*
Sieve analysis (percent passing)							
No. 10	50 max.						
No. 40	30 max.	50 max.	51 min.				
No. 200	15 max.	25 max.	10 max.	35 max.	35 max.	35 max.	35 max.
Characteristics of fraction passing No. 40							
Liquid limit				40 max.	41 min.	40 max.	41 min.
Plasticity index	6 max.		NP	10 max.	10 max.	11 min.	11 min.
Usual types of significant constituent materials	Stone fragments, gravel, and sand		Fine sand	Silty or clayey gravel and sand			
General subgrade rating	Excellent to good						

General classification	Silt-clay materials (more than 35% of total sample passing No. 200)			
Group classification	*A-4*	*A-5*	*A-6*	*A-7* *A-7-5** *A-7-6[†]*
Sieve analysis (percent passing)				
No. 10				
No. 40				
No. 200	36 min.	36 min.	36 min.	36 min.
Characteristics of fraction passing No. 40				
Liquid limit	40 max.	41 min.	40 max.	41 min.
Plasticity index	10 max.	10 max.	11 min.	11 min.
Usual types of significant constituent materials	Silty soils		Clayey soils	
General subgrade rating	Fair to poor			

*For A-7-5, $PI \leq LL - 30$
[†]For A-7-6, $PI > LL - 30$

35% pass through the No. 200 sieve are classified into groups A-4, A-5, A-6, and A-7. These are mostly silt and clay-type materials. The classification system is based on the following criteria:

1. *Grain size*
   Gravel: fraction passing the 75 mm sieve and retained on the No. 10 (2 mm) U.S. sieve
   Sand: fraction passing the No. 10 (2 mm) U.S. sieve and retained on the No. 200 (0.075 mm) U.S. sieve
   Silt and clay: fraction passing the No. 200 U.S. sieve
2. *Plasticity*: The term *silty* is applied when the fine fractions of the soil have a plasticity index of 10 or less. The term *clayey* is applied when the fine fractions have a plasticity index of 11 or more.
3. If cobbles and *boulders* (size larger than 75 mm) are encountered, they are excluded from the portion of the soil sample on which classification is made. However, the percentage of such material is recorded.

To classify a soil according to Table 3.4, the test data are applied from left to right. By process of elimination, the first group from the left into which the test data will fit is the correct classification.

Figure 3.19 shows a plot of the range of the liquid limit and the plasticity index for soils which fall into groups A-2, A-4, A-5, A-6, and A-7.

For the evaluation of the quality of a soil as a highway subgrade material, a number called the *group index* (*GI*) is also incorporated with the groups and

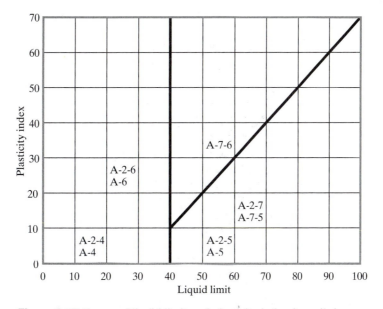

Figure 3.19 Range of liquid limit and plasticity index for soils in groups A-2, A-4, A-5, A-6, and A-7

subgroups of the soil. This number is written in parentheses after the group or sub-group designation. The group index is given by the equation

$$GI = (F - 35)[0.2 + 0.005(LL - 40)] + 0.01(F - 15)(PI - 10) \qquad (3.45)$$

where
F = percent passing the No. 200 sieve
LL = liquid limit
PI = plasticity index

The first term of Eq. (3.45)—that is, $(F - 35)[0.2 + 0.005(LL - 40)]$—is the partial group index determined from the liquid limit. The second term—that is, $0.01(F - 15)(PI - 10)$—is the partial group index determined from the plasticity index. Following are some rules for determining the group index:

1. If Eq. (3.45) yields a negative value for GI, it is taken as 0.
2. The group index calculated from Eq. (3.45) is rounded off to the nearest whole number (for example, $GI = 3.4$ is rounded off to 3; $GI = 3.5$ is rounded off to 4).
3. There is no upper limit for the group index.
4. The group index of soils belonging to groups A-1-a, A-1-b, A-2-4, A-2-5, and A-3 is always 0.
5. When calculating the group index for soils that belong to groups A-2-6 and A-2-7, use the partial group index for PI, or

$$GI = 0.01(F - 15)(PI - 10) \qquad (3.46)$$

In general, the quality of performance of a soil as a subgrade material is inversely proportional to the group index.

Example 3.8

Classify the following soils by the AASHTO classification system.

	Sieve analysis; % finer			Plasticity for the minus no. 40 fraction	
Soil	No. 10 sieve	No. 40 sieve	No. 200 sieve	Liquid limit	Plasticity index
A	83	48	20	20	5
B	100	92	86	70	32
C	48	28	6	—	Nonplastic
D	90	76	34	37	12

Solution

Soil *A*

According to Table 3.4, since 20% of the soil is passing through the No. 200 sieve, it falls under granular material classification—that is, A-1, A-3, or A-2. Proceeding from left to right, we see that it falls under A-1-b. The group index for A-1-b is zero. So, the classification is **A-1-b(0)**.

Soil *B*

Percent passing the No. 200 sieve is 86%. So, it is a silty clay material (that is, A-4, A-5, A-6, or A-7) as shown in Table 3.4. Proceeding from left to right, we see that it falls under A-7. For this case, $PI = 32 < LL - 30$. So, this is A-7-5. From Eq. (3.45)

$$GI = (F - 35)[0.2 + 0.005(LL - 40)] + 0.01(F - 15)(PI - 10)$$

Now, $F = 86$; $LL = 70$; $PI = 32$; so

$$GI = (86 - 35)[0.2 + 0.005(70 - 40)] + 0.01(86 - 15)(32 - 10)$$

$$= 33.47 \approx 33$$

Thus, the soil is **A-7-5(33)**.

Soil *C*

Percent passing the No. 200 sieve <35%. So, it is a granular material. Proceeding from left to right in Table 3.1, we find that it is A-1-a. The group index is zero. So the soil is **A-1-a(0)**.

Soil *D*

Percent passing the No. 200 sieve <35%. So, it is a granular material. From Table 3.4, it is A-2-6.

$$GI = 0.01(F - 15)(PI - 10)$$

Now, $F = 34$; $PI = 12$; so

$$GI = 0.01(34 - 15)(12 - 10) = 0.38 \approx 0$$

Thus, the soil is **A-2-6(0)**.

Unified Soil Classification System

The original form of this system was proposed by Casagrande in 1942 for use in the airfield construction works undertaken by the Army Corps of Engineers during World War II. In cooperation with the U.S. Bureau of Reclamation, this system was revised in 1952. At present, it is widely used by engineers (ASTM Test Designation D–2487). The Unified Classification System is presented in Table 3.5. This system classifies soils into two broad categories:

1. Coarse-grained soils that are gravelly and sandy in nature with less than 50% passing through the No. 200 sieve. The group symbols start with a

Table 3.5 Unified Soil Classification System (Based on Material Passing 75-mm Sieve)

Criteria for Assigning Group Symbols				Group Symbol
Coarse-Grained Soils More than 50% retained on No. 200 sieve	**Gravels** More than 50% of coarse fraction retained on No. 4 sieve	Clean Gravels Less than 5% fines[a]	$C_u \geq 4$ and $1 \leq C_c \leq 3^c$	GW
			$C_u < 4$ and/or $1 > C_c > 3^c$	GP
		Gravels with Fines More than 12% fines[a,d]	$PI < 4$ or plots below "A" line (Figure 3.20)	GM
			$PI > 7$ and plots on or above "A" line (Figure 3.20)	GC
	Sands 50% or more of coarse fraction passes No. 4 sieve	Clean Sands Less than 5% fines[b]	$C_u \geq 6$ and $1 \leq C_c \leq 3^c$	SW
			$C_u < 6$ and/or $1 > C_c > 3^c$	SP
		Sands with Fines More than 12% fines[b,d]	$PI < 4$ or plots below "A" line (Figure 3.20)	SM
			$PI > 7$ and plots on or above "A" line (Figure 3.20)	SC
Fine-Grained Soils 50% or more passes No. 200 sieve	**Silts and Clays** Liquid limit less than 50	Inorganic	$PI > 7$ and plots on or above "A" line (Figure 3.20)[e]	CL
			$PI < 4$ or plots below "A" line (Figure 3.20)[e]	ML
		Organic	$\dfrac{\text{Liquid limit–oven dried}}{\text{Liquid limit–not dried}} < 0.75$; see Figure 3.20; OL zone	OL
	Silts and Clays Liquid limit 50 or more	Inorganic	PI plots on or above "A" line (Figure 3.20)	CH
			PI plots below "A" line (Figure 3.20)	MH
		Organic	$\dfrac{\text{Liquid limit–oven dried}}{\text{Liquid limit–not dried}} < 0.75$; see Figure 3.20; OH zone	OH
Highly Organic Soils	Primarily organic matter, dark in color, and organic odor			Pt

[a] Gravels with 5 to 12% fine require dual symbols: GW-GM, GW-GC, GP-GM, GP-GC.

[b] Sands with 5 to 12% fines require dual symbols: SW-SM, SW-SC, SP-SM, SP-SC.

[c] $C_u = \dfrac{D_{60}}{D_{10}}$; $\quad C_c = \dfrac{(D_{30})^2}{D_{60} \times D_{10}}$

[d] If $4 \leq PI \leq 7$ and plots in the hatched area in Figure 3.16, use dual symbol GC-GM or SC-SM.

[e] If $4 \leq PI \leq 7$ and plots in the hatched area in Figure 3.16, use dual symbol CL-ML.

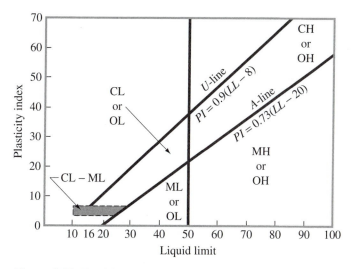

Figure 3.20 Plasticity chart

prefix of either G or S. G stands for gravel or gravelly soil, and S for sand or sandy soil.

2. Fine-grained soils with 50% or more passing through the No. 200 sieve. The group symbols start with a prefix of M, which stands for inorganic silt, C for inorganic clay, or O for organic silts and clays. The symbol Pt is used for peat, muck, and other highly organic soils.

Other symbols are also used for the classification:

- W—well graded
- P—poorly graded
- L—low plasticity (liquid limit less than 50)
- H—high plasticity (liquid limit more than 50)

For proper classification according to this system, some or all of the following information must be known:

1. Percent of gravel—that is, the fraction passing the 76.2-mm sieve and retained on the No. 4 sieve (4.75-mm opening)
2. Percent of sand—that is, the fraction passing the No. 4 sieve (4.75-mm opening) and retained on the No. 200 sieve (0.075-mm opening)
3. Percent of silt and clay—that is, the fraction finer than the No. 200 sieve (0.075-mm opening)
4. Uniformity coefficient (C_u) and the coefficient of gradation (C_c)
5. Liquid limit and plasticity index of the portion of soil passing the No. 40 sieve

The group symbols for coarse-grained gravelly soils are GW, GP, GM, GC, GC-GM, GW-GM, GW-GC, GP-GM, and GP-GC. Similarly, the group symbols for fine-grained soils are CL, ML, OL, CH, MH, OH, CL-ML, and Pt.

The group names of various soils classified under the Unified classification system can be determined using Figures 3.21, 3.22, and 3.23 . In using these figures, one needs to remember that in a given soil,

- Fine fraction = % passing No. 200 sieve
- Coarse fraction = % retained on No. 200 sieve
- Gravel fraction = % retained on No. 4 sieve
- Sand fraction = (% retained on No. 200 sieve) − (% retained on No. 4 sieve)

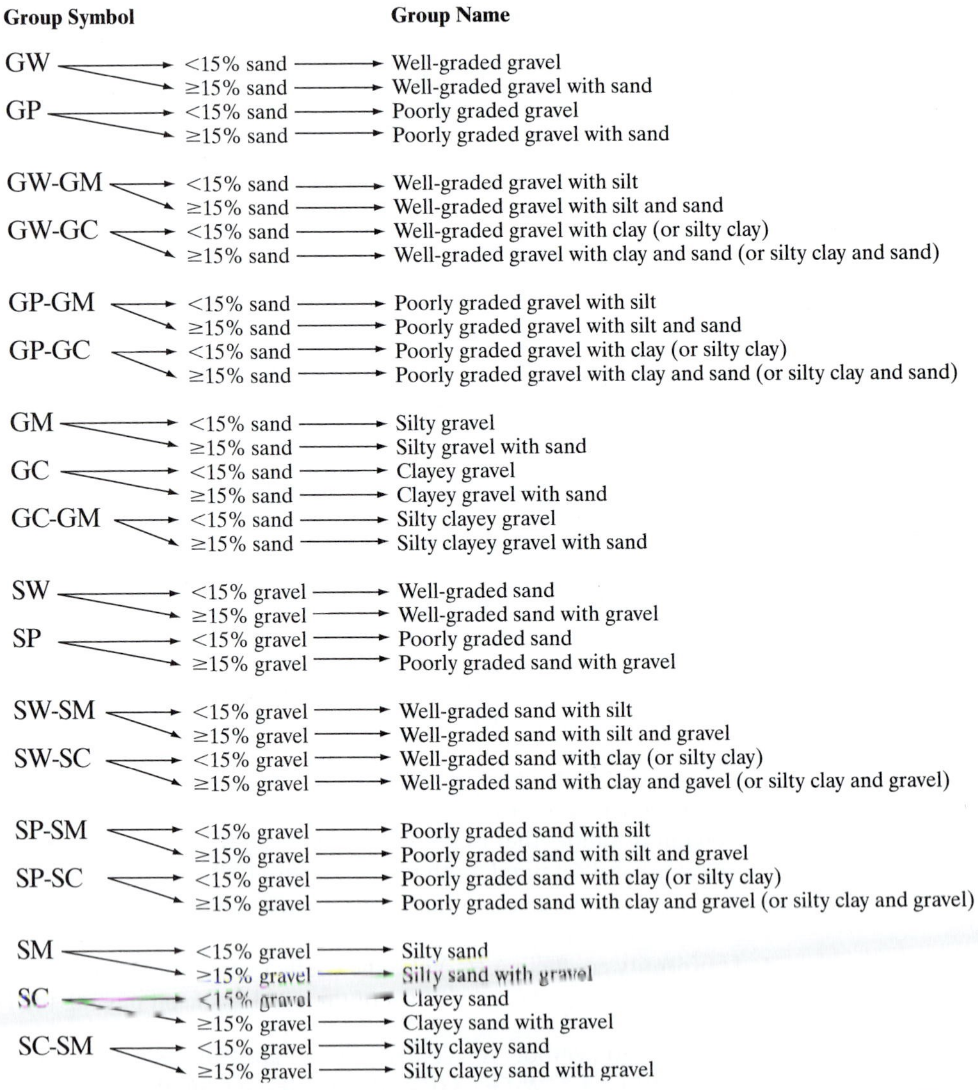

Group Symbol   **Group Name**

GW ——→ <15% sand ——→ Well-graded gravel
    ↘ ≥15% sand ——→ Well-graded gravel with sand
GP ——→ <15% sand ——→ Poorly graded gravel
    ↘ ≥15% sand ——→ Poorly graded gravel with sand

GW-GM ⟨ <15% sand ——→ Well-graded gravel with silt
       ≥15% sand ——→ Well-graded gravel with silt and sand
GW-GC ⟨ <15% sand ——→ Well-graded gravel with clay (or silty clay)
       ≥15% sand ——→ Well-graded gravel with clay and sand (or silty clay and sand)

GP-GM ⟨ <15% sand ——→ Poorly graded gravel with silt
       ≥15% sand ——→ Poorly graded gravel with silt and sand
GP-GC ⟨ <15% sand ——→ Poorly graded gravel with clay (or silty clay)
       ≥15% sand ——→ Poorly graded gravel with clay and sand (or silty clay and sand)

GM ——→ <15% sand ——→ Silty gravel
    ↘ ≥15% sand ——→ Silty gravel with sand
GC ——→ <15% sand ——→ Clayey gravel
    ↘ ≥15% sand ——→ Clayey gravel with sand
GC-GM ⟨ <15% sand ——→ Silty clayey gravel
       ≥15% sand ——→ Silty clayey gravel with sand

SW ——→ <15% gravel ——→ Well-graded sand
    ↘ ≥15% gravel ——→ Well-graded sand with gravel
SP ——→ <15% gravel ——→ Poorly graded sand
    ↘ ≥15% gravel ——→ Poorly graded sand with gravel

SW-SM ⟨ <15% gravel ——→ Well-graded sand with silt
       ≥15% gravel ——→ Well-graded sand with silt and gravel
SW-SC ⟨ <15% gravel ——→ Well-graded sand with clay (or silty clay)
       ≥15% gravel ——→ Well-graded sand with clay and gavel (or silty clay and gravel)

SP-SM ⟨ <15% gravel ——→ Poorly graded sand with silt
       ≥15% gravel ——→ Poorly graded sand with silt and gravel
SP-SC ⟨ <15% gravel ——→ Poorly graded sand with clay (or silty clay)
       ≥15% gravel ——→ Poorly graded sand with clay and gravel (or silty clay and gravel)

SM ——→ <15% gravel ——→ Silty sand
    ↘ >15% gravel ——→ Silty sand with gravel
SC ——→ <15% gravel ——→ Clayey sand
    ↘ ≥15% gravel ——→ Clayey sand with gravel
SC-SM ⟨ <15% gravel ——→ Silty clayey sand
       ≥15% gravel ——→ Silty clayey sand with gravel

Figure 3.21 Flowchart group names for gravelly and sandy soil (After ASTM, 2006)

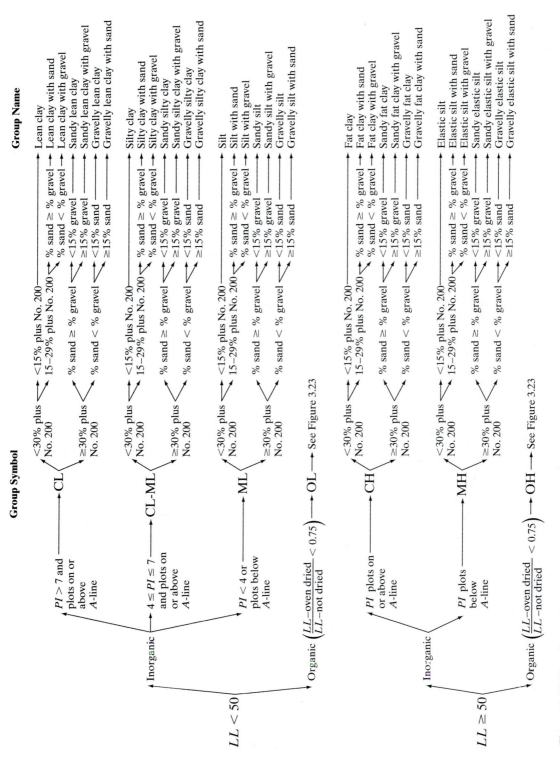

Figure 3.22 Flowchart group names for inorganic silty and clayey soils (After ASTM, 2006)

Figure 3.23 Flowchart group names for organic silty and clayey soils (After ASTM, 2006)

Example 3.9

The particle-size distributions of two soils are given in Figure 3.24. The liquid and plastic limits of the minus No. 40 fraction of the soils are as follows:

	Soil *A*	Soil *B*
Liquid limit	30	26
Plastic limit	22	20

Classify the soils by the Unified classification system.

Solution
Soil *A*

The particle-size distribution curve indicates that about 8% of the soil is finer than 0.075 mm in diameter (No. 200 sieve). Hence, this is a coarse-grained soil and, since this is within 5 to 12%, dual symbols need to be used.

Also, 100% of the total soil is finer than 4.75 mm (No. 4) sieve. Therefore, this is a sandy soil.

From Figure 3.24

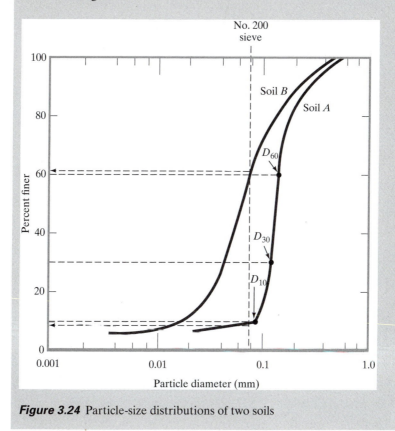

Figure 3.24 Particle-size distributions of two soils

$$D_{10} = 0.085 \text{ mm}$$

$$D_{30} = 0.12 \text{ mm}$$

$$D_{60} = 0.135 \text{ mm}$$

Thus

$$C_u = \frac{D_{60}}{D_{10}} = \frac{0.135}{0.085} = 1.59 < 6$$

$$C_c = \frac{(D_{30})^2}{D_{10} \times D_{60}} = \frac{(0.12)^2}{0.085 \times 0.135} = 1.25 > 1$$

With the liquid limit = 30 and the plasticity index = $30 - 22 = 8$ (which is greater than 7), it plots above the *A* line (Figure 3.20). So the classification is SP-SC.

Solution
Soil *B*
Sixty-one percent (that is, more than 50%) passes through the No. 200 sieve (0.075 mm diameter). Therefore, it is a fine-grained soil.

The liquid limit = 26 and the plasticity index = $26 - 20 = 6$. It falls inside the hatched area of the plasticity chart, so the classification is CL-ML. ∎

Example 3.10

The results of the particle-size analysis of a soil are as follows:

Percent passing through the No. 10 sieve = 100
Percent passing through the No. 40 sieve = 80
Percent passing through the No. 200 sieve = 58

The liquid limit and plasticity index of the minus No. 40 fraction of the soil are 30 and 10, respectively.

Classify the soil by the Unified classification system. Give the group symbol and group name.

Solution
Since 58% of the soil passes through the No. 200 sieve, it is a fine-grained soil. Referring to the plasticity chart in Figure 3.20, for $LL = 30$ and $PI = 10$, it can be classified (group symbol) as CL.

In order to determine the group name, we refer to Figure 3.22. The percent passing No. 200 sieve is more than 30%. Percent of gravel = 0; percent of sand = $(100 - 58) - (0) = 42$. Hence, percent sand > percent gravel. Also percent gravel is less than 15%. Hence, the group name is *sandy lean clay*. ∎

Problems

3.1   For a given soil, show that

  a. $\gamma_{sat} = \gamma_d + n\gamma_w$

  b. $\gamma_{sat} = n\left(\dfrac{1 + w_{sat}}{w_{sat}}\right)\gamma_w$

   where w_{sat} = moisture content at saturated state

  c. $\gamma_d = \dfrac{eS\gamma_w}{(1 + e)w}$

3.2   For a given soil, show that

$$w_{sat} = \dfrac{n\gamma_w}{\gamma_{sat} - n\gamma_w}$$

3.3   The moist mass of 2.8×10^{-3} m³ of soil is 5.53 kg. If the moisture content is 10% and the specific gravity of soil solids is 2.72, determine the following:

  a. Moist density
  b. Dry density
  c. Void ratio
  d. Porosity
  e. Degree of saturation
  f. Volume occupied by water

3.4   The moist unit weight of a soil is 18.7 kN/m³. Given that $G_s = 2.71$ and $w = 10.3\%$, determine

  a. Dry unit weight
  b. Void ratio
  c. Porosity
  d. Degree of saturation

3.5   For a given soil, the following are known: $G_s = 2.74$, moist unit weight $(\gamma) = 19.8$ kN/m³, and moisture content $(w) = 16.6\%$. Determine:

  a. Dry unit weight
  b. Void ratio
  c. Porosity
  d. Degree of saturation

3.6   Refer to Problem 3.5. Determine the weight of water, in kN, to be added per cubic meter (m³) of soil for

  a. 90% degree of saturation
  b. 100% degree of saturation

3.7   The dry density of a soil is 1750 kg/m³. Given that $G_s = 2.66$, what is the moisture content of the soil when it is saturated?

3.8   The porosity of a soil is 0.35. Given that $G_s = 2.72$, calculate

  a. Saturated unit weight (kN/m³)
  b. Moisture content when the moist unit weight (γ) is 18.3 kN/m³

3.9   For a saturated soil, the following are given: $w = 18\%$ and $G_s = 2.71$. Determine

  a. Saturated unit weight
  b. Dry unit weight
  c. Moist unit weight when the degree of saturation becomes 70%

3.10 The moisture content of a soil sample is 18.4%, and its dry unit weight is 15.7 kN/m³ Assuming that the specific gravity of solids is 2.65,
   a. Calculate the degree of saturation.
   b. What is the maximum dry unit weight to which this soil can be compacted without change in its moisture content?

3.11 A soil at a constant moisture content shows the following when compacted:

Degree of saturation (%)	Dry unit weight (kN/m³)
40	14.48
70	17.76

Determine the moisture content of the soil.

3.12 For a sandy soil, $e_{max} = 0.80$, $e_{min} = 0.46$, and $G_s = 2.71$. What is the void ratio at $D_r = 56\%$? Determine the moist unit weight of the soil when $w = 7\%$.

3.13 For a sandy soil, $e_{max} = 0.75$, $e_{min} = 0.52$, and $G_s = 2.7$. What are the void ratio and the dry unit weight at $D_r = 80\%$?

3.14 Following are the results from the liquid and plastic limit tests for a soil.
   Liquid limit test:

Number of blows, N	Moisture content (%)
15	42
20	40.8
28	39.1

Plastic limit test: PL = 18.7%
   a. Draw the flow curve and obtain the liquid limit.
   b. What is the plasticity index of the soil?

3.15 A saturated soil has the following characteristics: initial, volume $(V_i) = 24.6$ cm³, final volume $(V_f) = 15.9$ cm³, mass of wet soil $(m_1) = 44$ g, and mass of dry soil $(m_2) = 30.1$ g. Determine the shrinkage limit.

3.16 Classify the following soils by the AASHTO classification system and give the group indices.

	Percent finer than (sieve analysis)				Liquid limit*	Plasticity index*
Soil	No. 4	No. 10	No. 40	No. 200		
1	100	90	68	30	30	9
2	95	82	55	41	32	12
3	80	72	62	38	28	10
4	100	98	85	70	40	14
5	100	100	96	72	58	23
6	92	85	71	56	35	19
7	100	100	95	82	62	31
8	90	88	76	68	46	21
9	100	80	78	59	32	15
10	94	80	51	15	26	12

*Based on portion passing No. 40 sieve

3.17 Classify the following soils using the Unified Soil Classification System. Give the group symbols and the group names.

Soil	Sieve analysis, % finer No. 4	Sieve analysis, % finer No. 200	Liquid limit	Plastic limit	C_u	C_c
1	70	30	33	12		
2	48	20	41	19		
3	95	70	52	24		
4	100	82	30	11		
5	88	78	69	31		
6	71	4		NP	3.4	2.6
7	99	57	54	28		
8	71	11	32	16	4.8	2.9
9	100	2		NP	7.2	2.2
10	90	8	39	31	3.9	2.1

References

AMERICAN ASSOCIATION OF STATE HIGHWAY AND TRANSPORTATION OFFICIALS (1982). *AASHTO Materials, Part I, Specifications*, Washington, D.C.

AMERICAN SOCIETY FOR TESTING AND MATERIALS (2006). *ASTM Book of Standards*, Sec. 4, Vol. 04.08, West Conshohocken, PA.

BS:1377 (1990). *British Standard Methods of Tests for Soil for Engineering Purposes*, Part 2, BSI, London.

CASAGRANDE, A. (1948). "Classification and Identification of Soils," *Transactions*, ASCE, Vol. 113, 901–930.

CUBRINOVSKI, M., and ISHIHARA, K. (1999). "Empirical Correlation Between SPT N-Value and Relative Density for Sandy Soils," *Soils and Foundations*, Vol. 39, No. 5, 61–71.

CUBRINOVSKI, M., and ISHIHARA, K. (2002). "Maximum and Minimum Void Ratio Characteristics of Sands," *Soils and Foundations*, Vol. 42, No. 6, 65–78.

MITCHELL, J. K. (1976). *Fundamentals of Soil Behavior*, Wiley, New York.

SEED, H. B., WOODWARD, R. J., and LUNDGREN, R. (1964). "Fundamental Aspects of the Atterberg Limits," *Journal of the Soil Mechanics and Foundations Division*, ASCE, Vol. 90, No. SM6, 75–105.

SKEMPTON, A. W. (1953). "The Colloidal Activity of Clays," *Proceedings*, 3rd International Conference on Soil Mechanics and Foundation Engineering, London, Vol. 1, 57–61.

WORTH, C. P., and WOOD, D. M. (1978). "The Correlation of Index Properties with Some Basic Engineering Properties of Soils," *Canadian Geotechnical Journal*, Vol. 15, No. 2, 137–145.

4

Soil Compaction

In the construction of highway embankments, earth dams, and many other engineering structures, loose soils must be compacted to increase their unit weights. Compaction increases the strength characteristics of soils, thereby increasing the bearing capacity of foundations constructed over them. Compaction also decreases the amount of undesirable settlement of structures and increases the stability of slopes of embankments. Smooth-wheel rollers, sheepsfoot rollers, rubber-tired rollers, and vibratory rollers are generally used in the field for soil compaction. Vibratory rollers are used mostly for the densification of granular soils. This chapter discusses the principles of soil compaction in the laboratory and in the field.

4.1 Compaction—General Principles

Compaction, in general, is the densification of soil by removal of air, which requires mechanical energy. The degree of compaction of a soil is measured in terms of its dry unit weight. When water is added to the soil during compaction, it acts as a softening agent on the soil particles. The soil particles slip over each other and move into a densely packed position. The dry unit weight after compaction first increases as the moisture content increases (Figure 4.1). Note that at a moisture content $w = 0$, the moist unit weight (γ) is equal to the dry unit weight (γ_d), or

$$\gamma = \gamma_{d(w=0)} = \gamma_1$$

When the moisture content is gradually increased and the same compactive effort is used for compaction, the weight of the soil solids in a unit volume gradually increases. For example, at $w = w_1$, the moist unit weight is equal to

$$\gamma = \gamma_2$$

However, the dry unit weight at this moisture content is given by

$$\gamma_{d(w=w_1)} = \gamma_{d(w=0)} + \Delta\gamma_d$$

Beyond a certain moisture content $w = w_2$, (Figure 4.1), any increase in the moisture content tends to reduce the dry unit weight. This is because the water takes up the

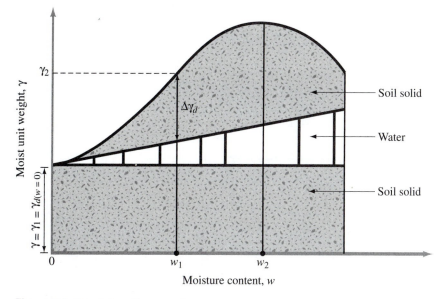

Figure 4.1 Principles of compaction

spaces that would have been occupied by the solid particles. The moisture content at which the maximum dry unit weight is attained is generally referred to as the *optimum moisture content*.

The laboratory test generally used to obtain the maximum dry unit weight of compaction and the optimum moisture content is called the *Proctor compaction test* (Proctor, 1933). The procedure for conducting this type of test is described in the following section.

4.2   Standard Proctor Test

In the Proctor test, the soil is compacted in a mold that has a volume of 943.3 cm³. The diameter of the mold is 101.6 mm. During the laboratory test, the mold is attached to a base plate at the bottom and to an extension at the top (Figure 4.2a). The soil is mixed with varying amounts of water and then compacted (Figure 4.3) in three equal layers by a hammer (Figure 4.2b) that delivers 25 blows to each layer. The hammer weighs 24.4 N (mass ≈ 2.5 kg), and has a drop of 304.8 mm. For each test, the moist unit weight of compaction γ can be calculated as

$$\gamma = \frac{W}{V_{(m)}} \tag{4.1}$$

where

W = weight of the compacted soil in the mold
$V_{(m)}$ = volume of the mold (= 943.3 cm³)

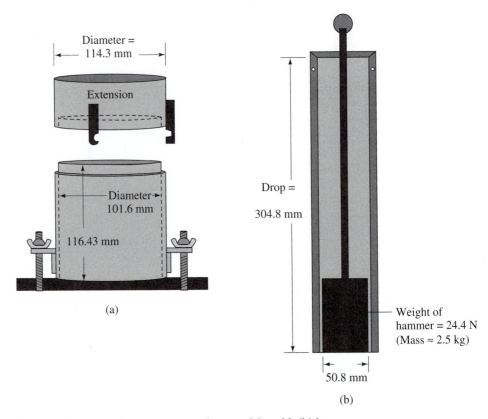

Figure 4.2 Standard Proctor test equipment: (a) mold; (b) hammer

For each test, the moisture content of the compacted soil is determined in the laboratory. With known moisture content, the dry unit weight γ_d can be calculated as

$$\gamma_d = \frac{\gamma}{1 + \dfrac{w(\%)}{100}} \qquad (4.2)$$

where $w(\%)$ = percentage of moisture content.

The values of γ_d determined from Eq. (4.2) can be plotted against the corresponding moisture contents to obtain the maximum dry unit weight and the optimum moisture content for the soil. Figure 4.4 shows such a compaction for a silty clay soil.

The procedure for the standard Proctor test is given in ASTM Test Designation D-698 and AASHTO Test Designation T-99.

For a given moisture content, the theoretical maximum dry unit weight is obtained when there is no air in the void spaces—that is, when the degree of saturation

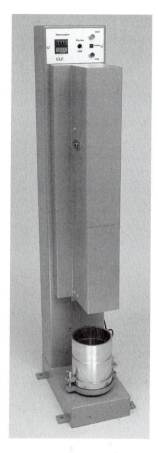

Figure 4.3
Standard Proctor test using a mechanical compactor
(Courtesy of ELE International)

equals 100%. Thus, the maximum dry unit weight at a given moisture content with zero air voids can be given by

$$\gamma_{zav} = \frac{G_s \gamma_w}{1 + e}$$

where

γ_{zav} = zero-air-void unit weight
γ_w = unit weight of water
e = void ratio
G_s = specific gravity of soil solids

For 100% saturation, $e = wG_s$, so

$$\gamma_{zav} = \frac{G_s \gamma_w}{1 + wG_s} = \frac{\gamma_w}{w + \dfrac{1}{G_s}} \qquad (4.3)$$

where w = moisture content.

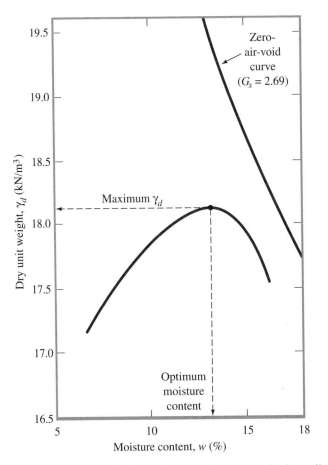

Figure 4.4 Standard Proctor compaction test results for a silty clay

To obtain the variation of γ_{zav} with moisture content, use the following procedure:

1. Determine the specific gravity of soil solids.
2. Know the unit weight of water (γ_w).
3. Assume several values of w, such as 5%, 10%, 15%, and so on.
4. Use Eq. (4.3) to calculate γ_{zav} for various values of w.

Figure 4.4 also shows the variation of γ_{zav} with moisture content and its relative location with respect to the compaction curve. Under no circumstances should any part of the compaction curve lie to the right of the zero-air-void curve.

Since Newton is a derived unit, in several instances it is more convenient to work with density (kg/m³) rather than unit weight. In that case, Eqs. (4.1), (4.2), and (4.3) can be rewritten as

$$\rho(kg/m^3) = \frac{m(kg)}{V_{(m)}(m^3)} \tag{4.4}$$

$$\rho_d(kg/m^3) = \frac{\rho(kg/m^3)}{1 + \frac{w(\%)}{100}} \tag{4.5}$$

$$\rho_{ZAV}(kg/m^3) = \frac{\rho_w(kg/m^3)}{w + \frac{1}{G_s}} \tag{4.6}$$

where

ρ, ρ_d, and ρ_{zav} = density, dry density, and zero-air-void density, respectively

m = mass of compacted soil in the mold

ρ_w = density of water (= 1000 kg/m³)

$V_{(m)}$ = volume of mold = 943.3 × 10⁻⁶ m³

4.3   *Factors Affecting Compaction*

The preceding section showed that moisture content has a great influence on the degree of compaction achieved by a given soil. Besides moisture content, other important factors that affect compaction are soil type and compaction effort (energy per unit volume). The importance of each of these two factors is described in more detail in this section.

Effect of Soil Type

The soil type—that is, grain-size distribution, shape of the soil grains, specific gravity of soil solids, and amount and type of clay minerals present—has a great influence on the maximum dry unit weight and optimum moisture content. Lee and Suedkamp (1972) studied compaction curves for 35 different soil samples. They observed four different types of compaction curves. These curves are shown in Figure 4.5. Type A compaction curves are the ones that have a single peak. This type of curve is generally found in soils that have a liquid limit between 30 and 70. Curve type B is a one and one-half peak curve, and curve type C is a double peak curve. Compaction curves of types B and C can be found in soils that have a liquid limit less than about 30. Compaction curves of type D are ones that do not have a definite peak. They are termed odd-shaped. Soils with a liquid limit greater than about 70 may exhibit compaction curves of type C or D. Soils that produce C- and D-type curves are not very common.

Effect of Compaction Effort

The compaction energy per unit volume, E, used for the standard Proctor test described in Section 4.2 can be given as

$$E = \frac{\left(\begin{array}{c}\text{number}\\ \text{of blows}\\ \text{per layer}\end{array}\right) \times \left(\begin{array}{c}\text{number}\\ \text{of}\\ \text{layers}\end{array}\right) \times \left(\begin{array}{c}\text{weight}\\ \text{of}\\ \text{hammer}\end{array}\right) \times \left(\begin{array}{c}\text{height of}\\ \text{drop of}\\ \text{hammer}\end{array}\right)}{\text{volume of mold}} \tag{4.7}$$

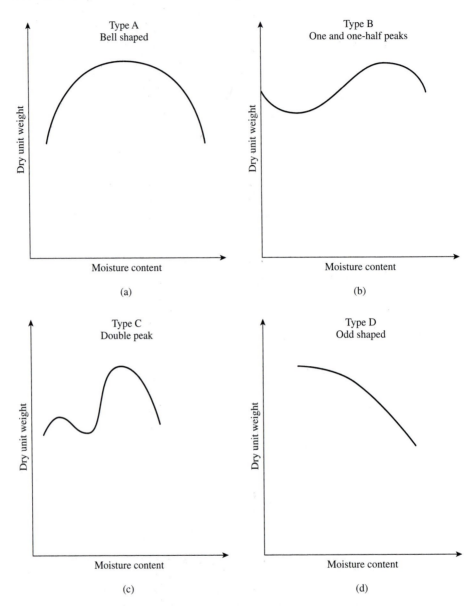

Figure 4.5 Various types of compaction curves encountered in soils

or

$$E = \frac{(25)(3)(24.4)(0.3048 \text{ m})}{943.3 \times 10^{-6} \text{ m}^3} = 591.3 \times 10^3 \text{ N-m/m}^3 = 591.3 \text{ kN-m/m}^3$$

If the compaction effort per unit volume of soil is changed, the moisture–unit weight curve will also change. This can be demonstrated with the aid of Figure 4.6,

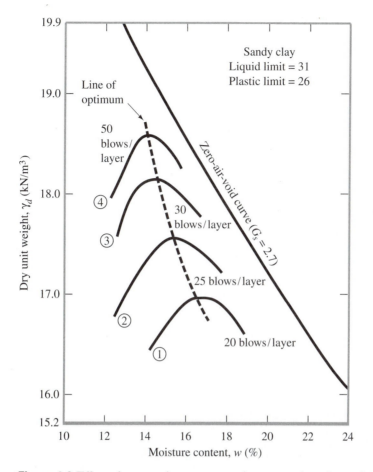

Figure 4.6 Effect of compaction energy on the compaction of a sandy clay

which shows four compaction curves for a sandy clay. The standard Proctor mold and hammer were used to obtain the compaction curves. The number of layers of soil used for compaction was kept at three for all cases. However, the number of hammer blows per each layer varied from 20 to 50. The compaction energy used per unit volume of soil for each curve can be calculated easily by using Eq. (4.7). These values are listed in Table 4.1.

Table 4.1 Compaction energy for tests shown in Figure 4.6

Curve number in Figure 4.6	Number of blows/layer	Compaction energy (kN-m/m³)
1	20	473.0
2	25	591.3
3	30	709.6
4	50	1182.6

From Table 4.1 and Figure 4.6, we can reach two conclusions:

1. As the compaction effort is increased, the maximum dry unit weight of compaction is also increased.
2. As the compaction effort is increased, the optimum moisture content is decreased to some extent.

The preceding statements are true for all soils. Note, however, that the degree of compaction is not directly proportional to the compaction effort.

4.4   *Modified Proctor Test*

With the development of heavy rollers and their use in field compaction, the standard Proctor test was modified to better represent field conditions. This is sometimes referred to as the *modified Proctor test* (ASTM Test Designation D-1557 and AASHTO Test Designation T-180). For conducting the modified Proctor test, the same mold is used, with a volume of 943.3 cm^3, as in the case of the standard Proctor test. However, the soil is compacted in five layers by a hammer that weighs 44.5 N (mass = 4.536 kg). The drop of the hammer is 457.2 mm. The number of hammer blows for each layer is kept at 25 as in the case of the standard Proctor test. Figure 4.7

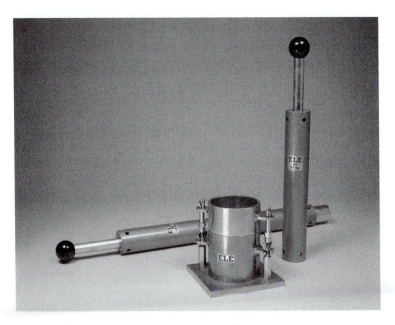

Figure 4.7 Hammer used for the modified Proctor test. (Courtesy of ELE International)

shows a hammer used for the modified Proctor test. The compaction energy for unit volume of soil in the modified test can be calculated as

$$E = \frac{(25 \text{ blows/layer})(5 \text{ layers})(44.5 \times 10^{-3} \text{ kN})(0.4572 \text{ m})}{943.3 \times 10^{-6} \text{ m}^3} = 2696 \text{ kN-m/m}^3$$

A comparison of the hammers used in the standard and modified Proctor tests is shown in Figure 4.8.

Because it increases the compactive effort, the modified Proctor test results in an increase of the maximum dry unit weight of the soil. The increase of the maximum dry unit weight is accompanied by a decrease of the optimum moisture content.

In the preceding discussions, the specifications given for Proctor tests adopted by ASTM and AASHTO regarding the volume of the mold (943.3 cm^3) and the number of blows (25 blows/layer) are generally the ones adopted for fine-grained soils that pass the U.S. No. 4 sieve. However, under each test designation, three different suggested methods reflect the size of the mold, the number of blows per layer, and the maximum particle size in a soil aggregate used for testing. A summary of the test methods is given in Tables 4.2 and 4.3.

Figure 4.8 Comparision of standard (left) and modified (right) Proctor hammers (Courtesy of Braja Das)

Table 4.2 Specifications for standard Proctor test (Based on ASTM Test Designation 698)

Item	Method A	Method B	Method C
Diameter of mold	101.6 mm	101.6 mm	152.4 mm
Volume of mold	943.3 cm³	943.3 cm³	2124 cm³
Weight of hammer	24.4 N	24.4 N	24.4 N
Height of hammer drop	304.8 mm	304.8 mm	304.8 mm
Number of hammer blows per layer of soil	25	25	56
Number of layers of compaction	3	3	3
Energy of compaction	591.3 kN-m/m³	591.3 kN-m/m³	591.3 kN-m/m³
Soil to be used	Portion passing No. 4 (4.57 mm) sieve. May be used if 20% *or less* by weight of material is retained on No. 4 sieve.	Portion passing 9.5-mm sieve. May be used if soil retained on No. 4 sieve *is more* than 20%, and 20% *or less* by weight is retained on 9.5-mm sieve.	Portion passing 19-mm sieve. May be used if *more than* 20% by weight of material is retained on 9.5-mm sieve, and *less than* 30% by weight is retained on 19-mm sieve.

Table 4.3 Specifications for modified Proctor test (Based on ASTM Test Designation 1557)

Item	Method A	Method B	Method C
Diameter of mold	101.6 mm	101.6 mm	152.4 mm
Volume of mold	943.3 cm³	943.3 cm³	2124 cm³
Weight of hammer	44.5 N	44.5 N	44.5 N
Height of hammer drop	457.2 mm	457.2 mm	457.2 mm
Number of hammer blows per layer of soil	25	25	56
Number of layers of compaction	5	5	5
Energy of compaction	2696 kN-m/m³	2696 kN-m/m³	2696 kN-m/m³
Soil to be used	Portion passing No. 4 (4.57 mm) sieve. May be used if 20% *or less* by weight of material is retained on No. 4 sieve.	Portion passing 9.5-mm sieve. May be used if soil retained on No. 4 sieve *is more* than 20%, and 20% *or less* by weight is retained on 9.5-mm sieve.	Portion passing 19-mm sieve. May be used if *more than* 20% by weight of material is retained on 9.5-mm sieve, and *less than* 30% by weight is retained on 19-mm sieve.

Example 4.1

The laboratory test data for a standard Proctor test are given in the table. Find the maximum dry unit weight and the optimum moisture content.

Volume of Proctor mold (cm³)	Mass of wet soil in the mold (kg)	Moisture content (%)
943.3	1.48	8.4
943.3	1.88	10.2
943.3	2.12	12.3
943.3	1.82	14.6
943.3	1.65	16.8

Solution

We can prepare the following table:

Volume (cm³)	Mass of wet soil (kg)	Moist density (kg/m³)	Moisture content, w (%)	Dry density, ρ_d (kg/m³)
943.3	1.48	1568.96	8.4	1447.38
943.3	1.88	1993.00	10.2	1808.53
943.3	2.12	2247.43	12.3	2001.27
943.3	1.82	1929.40	14.6	1683.60
943.3	1.68	1780.98	16.8	1524.81

The plot of ρ_d against w is shown in Figure 4.9. From the graph, we observe

Maximum dry density = **2020 kg/m³**

Optimum moisture content = **13%**

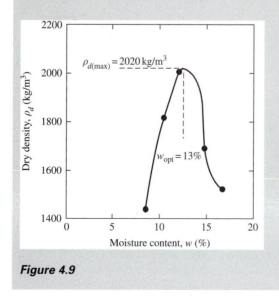

Figure 4.9

Example 4.2

Calculate the zero-air-void unit weights (in kN/m³) for a soil with $G_s = 2.68$ at moisture contents of 5, 10, 15, 20, and 25%. Plot a graph of γ_{zav} against moisture content.

Solution

From Eq. (4.3)

$$\gamma_{zav} = \frac{\gamma_w}{1 + \frac{1}{G_s}}$$

$$\gamma_w = 9.81 \text{ kN/m}^3; G_s = 2.68.$$

Refer to the following table:

w (%)	γ_{zav} (kN/m³)
5	23.18
10	20.73
15	18.75
20	17.12
25	15.74

The plot of γ_{zav} against w is shown in Figure 4.10

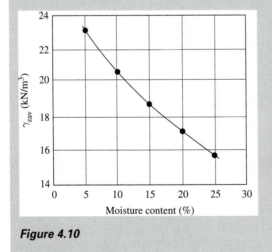

Figure 4.10

| 4.5 | ## Empirical Relationships |

Omar, *et al.* (2003) recently presented the results of modified Proctor compaction tests on 311 soil samples. Of these samples, 45 were gravelly soil (GP, GP-GM, GW, GW-GM, and GM), 264 were sandy soil (SP, SP-SM, SW-SM, SW, SC-SM, SC, and SM), and two were clay with low plasticity (CL). All compaction tests were conducted using ASTM 1557 method C. Based on the tests, the following correlations were developed.

$$\rho_{d(\max)}\ (\text{kg/m}^3) = [4,804,574G_s - 195.55(LL)^2 + 156,971(\text{R\#4})^{0.5} - 9,527,830]^{0.5} \tag{4.8}$$

$$\text{In}(w_{\text{opt}}) = 1.195 \times 10^{-4}(LL)^2 - 1.964G_s - 6.617 \times 10^{-5}(\text{R\#4}) + 7.651 \tag{4.9}$$

where

$\rho_{d(\max)}$ = maximum dry density
w_{opt} = optimum moisture content
G_s = specific gravity of soil solids
LL = liquid limit, in percent
R#4 = percent retained on No. 4 sieve

More recently, Gurtug and Sridharan (2004) proposed correlations for optimum moisture content and maximum dry unit weight with the plastic limit (*PL*) of cohesive soils. These correlations can be expressed as:

$$w_{\text{opt}}(\%) = [1.95 - 0.38(\log CE)]\ (PL) \tag{4.10}$$

$$\gamma_{d(\max)}\ (\text{kN/m}^3) = 22.68e^{-0.0183w_{\text{opt}}(\%)} \tag{4.11}$$

where

PL = plastic limit (%)
CE = compaction energy (kN-m/m^3)

For modified Proctor test, $CE = 2700$ kN/m^3. Hence,

$$w_{\text{opt}}(\%) \approx 0.65(PL)$$

and

$$\gamma_{d(\max)}\ (\text{kN/m}^3) \approx 22.68e^{-0.012(PL)}$$

4.6 Field Compaction

Most compaction in the field is done with rollers. There are four common types of rollers:

1. Smooth-wheel roller (or smooth-drum roller)
2. Pneumatic rubber-tired roller
3. Sheepsfoot roller
4. Vibratory roller

Smooth-wheel rollers (Figure 4.11) are suitable for proofrolling subgrades and for the finishing operation of fills with sandy and clayey soils. They provide 100% coverage under the wheels with ground contact pressures as high as 310–380 kN/m^2. They are not suitable for producing high unit weights of compaction when used on thicker layers.

Figure 4.11 Smooth-wheel roller (Courtesy of Ingram Compaction, LLC)

Figure 4.12 Pneumatic rubber-tired roller (Courtesy of Ingram Compaction, LLC)

Pneumatic rubber-tired rollers (Figure 4.12) are better in many respects than smooth-wheel rollers. The former are heavily loaded wagons with several rows of tires. These tires are closely spaced—four to six in a row. The contact pressure under the tires can range from 600 to 700 kN/m^2, and they produce 70% to 80% coverage. Pneumatic rollers can be used for sandy and clayey soil compaction. Compaction is achieved by a combination of pressure and kneading action.

Sheepsfoot rollers (Figure 4.13) are drums with a large number of projections. The area of each of these projections may range from 25 to 85 cm^2. Sheepsfoot rollers are most effective in compacting clayey soils. The contact pressure under the projections can range from 1380 to 6900 kN/m^2. During compaction in the field, the initial passes compact the lower portion of a lift. The top and middle portions of a lift are compacted at a later stage.

Vibratory rollers are very efficient in compacting granular soils. Vibrators can be attached to smooth-wheel, pneumatic rubber-tired, or sheepsfoot rollers to provide vibratory effects to the soil. Figure 4.14 demonstrates the principles of vibratory rollers. The vibration is produced by rotating off-center weights.

Hand-held vibrating plates can be used for effective compaction of granular soils over a limited area. Vibrating plates are also gang-mounted on machines, which can be used in less restricted areas.

In addition to soil type and moisture content, other factors must be considered to achieve the desired unit weight of compaction in the field. These factors include the thickness of lift, the intensity of pressure applied by the compacting equipment, and the area over which the pressure is applied. The pressure applied at the surface decreases with depth, resulting in a decrease in the degree of compaction of soil.

Figure 4.13 Sheepsfoot roller (Courtesy of David A. Carroll, Austin, Texas)

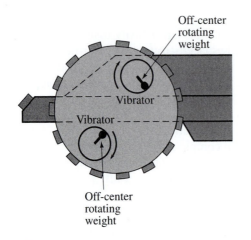

Figure 4.14 Principles of vibratory rollers

During compaction, the dry unit weight of soil is also affected by the number of roller passes. The dry unit weight of a soil at a given moisture content will increase up to a certain point with the number of passes of the roller. Beyond this point, it will remain approximately constant. In most cases, about 10 to 15 roller passes yield the maximum dry unit weight economically attainable.

4.7 *Specifications for Field Compaction*

In most specifications for earth work, one stipulation is that the contractor must achieve a compacted field dry unit weight of 90% to 95% of the maximum dry unit weight determined in the laboratory by either the standard or modified Proctor test. This specification is, in fact, for relative compaction R, which can be expressed as

$$R\,(\%) = \frac{\gamma_{d(\text{field})}}{\gamma_{d(\text{max}-\text{lab})}} \times 100 \tag{4.12}$$

In the compaction of granular soils, specifications are sometimes written in terms of the required relative density D_r or compaction. Relative density should not be confused with relative compaction. From Chapter 3, we can write

$$D_r = \left[\frac{\gamma_{d(\text{field})} - \gamma_{d(\text{min})}}{\gamma_{d(\text{max})} - \gamma_{d(\text{min})}}\right]\left[\frac{\gamma_{d(\text{max})}}{\gamma_{d(\text{field})}}\right] \tag{4.13}$$

Comparing Eqs. (4.12) and (4.13), we can see that

$$R = \frac{R_0}{1 - D_r(1 - R_0)} \tag{4.14}$$

where

$$R_0 = \frac{\gamma_{d(\min)}}{\gamma_{d(\max)}} \tag{4.15}$$

Based on the observation of 47 soil samples, Lee and Singh (1971) gave a correlation between R and D_r for granular soils:

$$R = 80 + 0.2 D_r \tag{4.16}$$

The specification for field compaction based on relative compaction or on relative density is an end-product specification. The contractor is expected to achieve a minimum dry unit weight regardless of the field procedure adopted. The most economical compaction condition can be explained with the aid of Figure 4.15. The compaction curves A, B, and C are for the same soil with varying compactive effort. Let curve A represent the conditions of maximum compactive effort that can be obtained from the existing equipment. Let it be required to achieve a minimum dry unit weight of $\gamma_{d(\text{field})} = R\gamma_{d(\max)}$. To achieve this, the moisture content w needs to be between w_1 and w_2. However, as can be seen from compaction curve C, the required $\gamma_{d(\text{field})}$ can be achieved with a lower compactive effort at a moisture content $w = w_3$. However, in practice, a compacted field unit weight of $\gamma_{d(\text{field})} = R\gamma_{d(\max)}$ cannot be achieved by the minimum compactive effort because it allows no margin for error considering the variability of field conditions. Hence, equipment with slightly more than the minimum compactive effort should be used. The compaction curve B

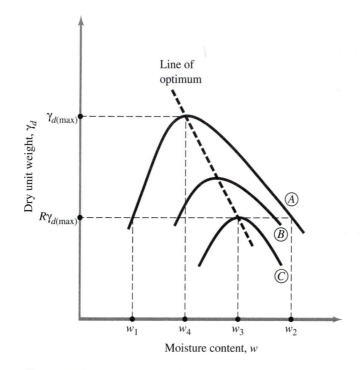

Figure 4.15 Most economical compaction condition

represents this condition. Now it can be seen from Figure 4.15 that the most economical moisture content is between w_3 and w_4. Note that $w = w_4$ is the optimum moisture content for curve A, which is for the maximum compactive effort.

The concept described in the preceding paragraph, along with Figure 4.15, is historically attributed to Seed (1964), who was a prominent figure in modern geotechnical engineering. The idea is elaborated on in more detail in Holtz and Kovacs (1981).

<table><tr><td>**4.8**</td></tr></table>

Determination of Field Unit Weight after Compaction

When the compaction work is progressing in the field, it is useful to know whether or not the unit weight specified is achieved. Three standard procedures are used for determining the field unit weight of compaction:

1. Sand cone method
2. Rubber balloon method
3. Nuclear method

Following is a brief description of each of these methods.

Sand Cone Method (ASTM Designation D-1556)

The sand cone device consists of a glass or plastic jar with a metal cone attached at its top (Figure 4.16). The jar is filled with very uniform dry Ottawa sand. The weight of the jar, the cone, and the sand filling the jar is determined (W_1). In the field, a small

Figure 4.16 Plastic jar and the metal cone for the sand cone device (*Note:* The jar is filled with Ottawa sand.) (Courtesy of Braja Das)

hole is excavated in the area where the soil has been compacted. If the weight of the moist soil excavated from the hole (W_2) is determined and the moisture content of the excavated soil is known, the dry weight of the soil (W_3) can be found as

$$W_3 = \frac{W_2}{1 + \dfrac{w\,(\%)}{100}} \qquad (4.17)$$

where w = moisture content.

After excavation of the hole, the cone with the sand-filled jar attached to it is inverted and placed over the hole (Figure 4.17). Sand is allowed to flow out of the jar into the hole and the cone. Once the hole and cone are filled, the weight of the jar the cone, and the remaining sand in the jar is determined (W_4), so

$$W_5 = W_1 - W_4 \qquad (4.18)$$

where W_5 = weight of sand to fill the hole and cone.

The volume of the hole excavated can now be determined as

$$V = \frac{W_5 - W_c}{\gamma_{d(sand)}} \qquad (4.19)$$

where

W_c = weight of sand to fill the cone only
$\gamma_{d(sand)}$ = dry unit weight of Ottawa sand used

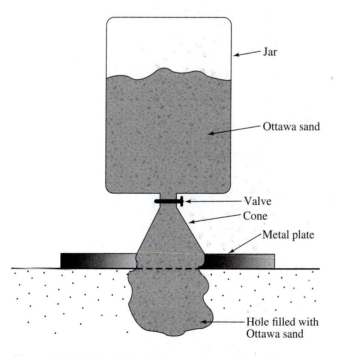

Figure 4.17 Field unit weight by sand cone method

The values of W_c and $\gamma_{d(\text{sand})}$ are determined from the calibration done in the laboratory. The dry unit weight of compaction made in the field can now be determined as

$$\gamma_d = \frac{\text{dry weight of the soil excavated from the hole}}{\text{volume of the hole}} = \frac{W_3}{V} \qquad (4.20)$$

Rubber Balloon Method (ASTM Designation D-2167)

The procedure for the rubber balloon method is similar to that for the sand cone method; a test hole is made, and the moist weight of the soil removed from the hole and its moisture content are determined. However, the volume of the hole is determined by introducing a rubber balloon filled with water from a calibrated vessel into the hole, from which the volume can be read directly. The dry unit weight of the compacted soil can be determined by using Eq. (4.17). Figure 4.18 shows a calibrated vessel used in this method.

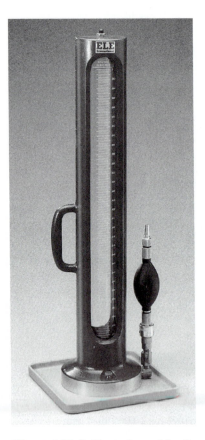

Figure 4.18 Calibrated vessel for the rubber balloon method for determination of field unit weight (Courtesy of ELE International)

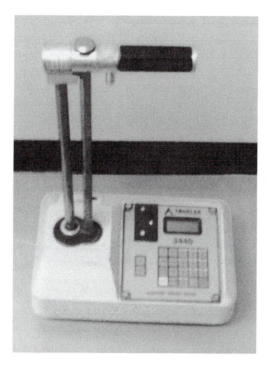

Figure 4.19 Nuclear density meter (Courtesy of Braja Das)

Nuclear Method

Nuclear density meters are now used often to determine the compacted dry unit weight of soil. The density meters operate either in drilled holes or from the ground surface. The instrument measures the weight of wet soil per unit volume and also the weight of water present in a unit volume of soil. The dry unit weight of compacted soil can be determined by subtracting the weight of water from the moist unit weight of soil. Figure 4.19 shows a photograph of a nuclear density meter.

4.9   *Special Compaction Techniques*

Several special types of compaction techniques have been developed for deep compaction of in-place soils, and these techniques are used in the field for large-scale compaction works. Among these, the popular methods are vibroflotation, dynamic compaction, and blasting. Details of these methods are provided in the following sections.

Vibroflotation

Vibroflotation is a technique for *in situ* densification of thick layers of loose granular soil deposits. It was developed in Germany in the 1930s. The first vibroflotation device was used in the United States about 10 years later. The process involves the use of a *Vibroflot* (also called the *vibrating unit*), which is about 2.1 m long (as shown in Figure 4.20.) This vibrating unit has an eccentric weight inside it and can develop a centrifugal force, which enables the vibrating unit to vibrate horizontally. There are openings at the bottom and top of the vibrating unit for water jets. The vibrating unit is attached to a follow-up pipe. Figure 4.20 shows the entire assembly of equipment necessary for conducting the field compaction.

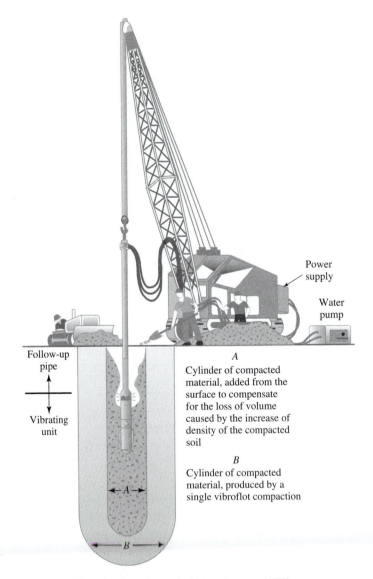

Power supply

Water pump

Follow-up pipe

Vibrating unit

A
Cylinder of compacted material, added from the surface to compensate for the loss of volume caused by the increase of density of the compacted soil

B
Cylinder of compacted material, produced by a single vibroflot compaction

Figure 4.20 Vibroflotation unit (After Brown, 1977)

The entire vibroflotation compaction process in the field can be divided into four stages:

Stage 1: The jet at the bottom of the Vibroflot is turned on and lowered into the ground.

Stage 2: The water jet creates a quick condition in the soil and it allows the vibrating unit to sink into the ground.

Stage 3: Granular material is poured from the top of the hole. The water from the lower jet is transferred to the jet at the top of the vibrating unit. This water carries the granular material down the hole.

Stage 4: The vibrating unit is gradually raised in about 0.3 m lifts and held vibrating for about 30 seconds at each lift. This process compacts the soil to the desired unit weight.

As for the vibrating units, 23-kW electric units have been used in the U.S. since the latter part of the 1940s. The 75-kW units were introduced in the early 1970s. The general description of the 75-kW electric and hydraulic Vibroflot units are as follows (Brown, 1977):

a. Vibrating tip

Length	2.1 m
Diameter	406 mm
Weight	17.8 kN
Maximum movement when full	12.5 mm
Centrifugal force	160 kN

b. Eccentric

Weight	1.2 kN
Offset	38 mm
Length	610 mm
Speed	1800 rpm

c. Pump

Operating flow rate	$0-1.6$ m^3/min
Pressure	$700-1050$ kN/m^2

The zone of compaction around a single probe varies with the type of Vibroflot used. The cylindrical zone of compaction has a radius of about 2 m for a 23-kW unit. This radius can extend to about 3 m for a 75-kW unit.

Compaction by vibroflotation is done in various probe spacings, depending on the zone of compaction. This spacing is shown in Figure 4.21. The capacity for successful densification of *in situ* soil depends on several factors, the most important of which is the grain-size distribution of the soil and the type of backfill used to fill the holes during the withdrawal period of the Vibroflot. The range of the grain-size distribution of *in situ* soil marked Zone 1 in Figure 4.22 is most suitable for compaction by vibroflotation. Soils that contain excessive amounts of fine sand and silt-size particles are difficult to compact, and considerable effort is needed to reach the proper relative density of compaction. Zone 2 in Figure 4.22 is the approximate lower limit of grain-size distribution for which compaction by vibroflotation is effective. Soil

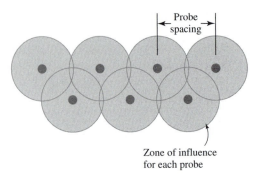

Figure 4.21 Probe spacing for vibroflotation

deposits whose grain-size distributions fall in Zone 3 contain appreciable amounts of gravel. For these soils, the rate of probe penetration may be slow and may prove uneconomical in the long run.

　　The grain-size distribution of the backfill material is an important factor that controls the rate of densification. Brown (1977) has defined a quantity called the *suitability number* for rating backfill as

$$S_N = 1.7\sqrt{\frac{3}{(D_{50})^2} + \frac{1}{(D_{20})^2} + \frac{1}{(D_{10})^2}} \qquad (4.21)$$

where D_{50}, D_{20}, and D_{10} are the diameters (in mm) through which, respectively, 50, 20, and 10% of the material passes.

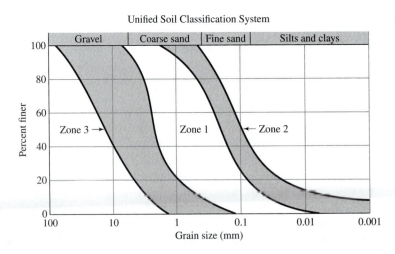

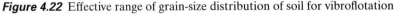

Figure 4.22 Effective range of grain-size distribution of soil for vibroflotation

The smaller the value of S_N, the more desirable the backfill material. Following is a backfill rating system proposed by Brown:

Range of S_N	Rating as backfill
0–10	Excellent
10–20	Good
20–30	Fair
30–50	Poor
>50	Unsuitable

Dynamic Compaction

Dynamic compaction is a technique that has gained popularity in the United States for the densification of granular soil deposits. This process consists primarily of dropping a heavy weight repeatedly on the ground at regular intervals. The weight of the hammer used varies over a range of 80 to 360 kN, and the height of the hammer drop varies between 7.5 and 30.5 m. The stress waves generated by the hammer drops aid in the densification. The degree of compaction achieved at a given site depends on the following three factors:

1. Weight of hammer
2. Height of hammer drop
3. Spacing of locations at which the hammer is dropped

Leonards, Cutter, and Holtz (1980) suggested that the significant depth of influence for compaction can be approximated by using the equation

$$D \simeq (\tfrac{1}{2})\sqrt{W_H h} \tag{4.22}$$

where
D = significant depth of densification (m)
W_H = dropping weight (metric ton)
h = height of drop (m)

Blasting

Blasting is a technique that has been used successfully in many projects (Mitchell, 1970) for the densification of granular soils. The general soil grain sizes suitable for compaction by blasting are the same as those for compaction by vibroflotation. The process involves the detonation of explosive charges such as 60% dynamite at a certain depth below the ground surface in saturated soil. The lateral spacing of the charges varies from about 3 to 10 m. Three to five successful detonations are usually necessary to achieve the desired compaction. Compaction up to a relative density of about 80% and up to a depth of about 20 m over a large area can easily be achieved by using this process. Usually, the explosive charges are placed at a depth of about two-thirds of the thickness of the soil layer desired to be compacted.

4.10   *Effect of Compaction on Cohesive Soil Properties*

Compaction induces variations in the structure of cohesive soils, which, in turn, affect the physical properties such as hydraulic conductivity and shear strength (Lambe, 1958). This can be explained by referring to Figure 4.23. Figure 4.23a shows a compaction curve (that is, variation of dry unit weight versus moisture content). If the clay is compacted with a moisture content on the dry side of the optimum, as represented by point *A*, it will possess a *flocculent structure* (that is, a loose random orientation of particles), as shown in Figure 4.23b. Each clay particle, at this time, has a thin layer of adsorbed water and a thicker layer of viscous double layer water. In this instance, the clay particles are being held together by electrostatic attraction of positively charged edges to negatively charged faces. At low moisture content, the diffuse double layer of ions surrounding the clay particles cannot be freely developed. When the moisture content of compaction is increased, as shown by point *B*, the diffuse double layers around the particles expand, thus increasing the repulsion between the clay particles and giving a lower degree of flocculation and a higher dry unit weight. A continued increase of moisture content from *B* to *C* will expand the double layers more,

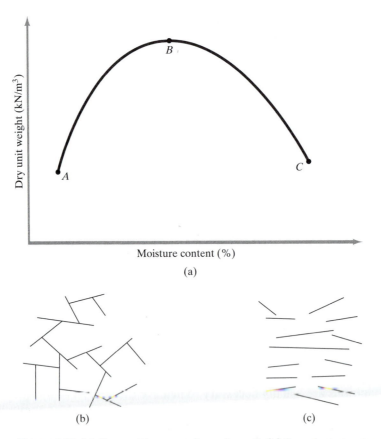

Figure 4.23 (a) Compaction curve for a clay soil, (b) flocculent structure of clay particles, (c) dispersed structure of clay particles

and this will result in a continued increase of repulsion between the particles. This will give a still greater degree of particle orientation and a more or less *dispersed* structure (Figure 4.23c). However, the dry unit weight will decrease because the added water will dilute the concentration of soil solids per unit volume. It is also important to point out that at a given moisture content, higher compactive effort tends to give a more parallel orientation to the clay particles, thereby giving a more dispersed structure. The particles are closer, and the soil has a higher unit weight of compaction. Figure 4.24 shows the degree of particle orientation with moisture content for compacted Boston Blue clay (Lambe, 1958) and kaolinite (Seed and Chan, 1959).

For a given soil and compaction energy, the hydraulic conductivity (Chapter 5) will change with the molding moisture content at which the compaction is conducted. Figure 4.25 shows the general nature of the variation of hydraulic conductivity with dry unit weight and molding moisture content. The hydraulic conductivity, which is a measure of how easily water flows through soil, decreases with the increase in moisture content. It reaches a minimum value at approximately the optimum moisture content. Beyond the optimum moisture content, the hydraulic conductivity increases slightly.

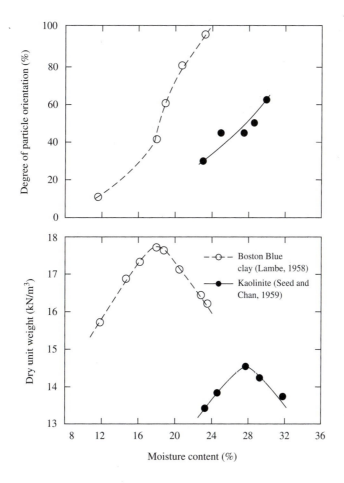

Figure 4.24
Variation of degree of particle orientation with molding moisture content (Compiled from Lambe, 1958; and Seed and Chan, 1959)

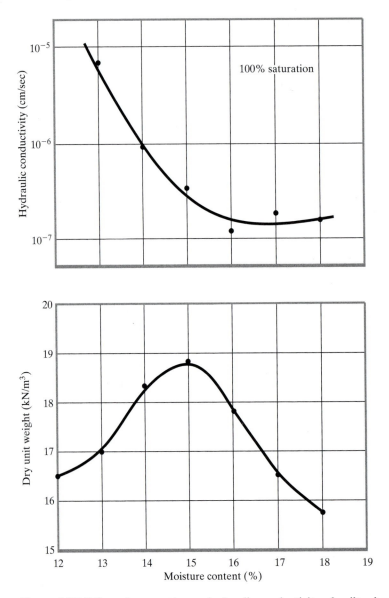

Figure 4.25 Effect of compaction on hydraulic conductivity of a silty clay

The strength of compacted clayey soils (see Chapter 8) generally decreases with the molding moisture content. This is shown in Figure 4.26. Note that at approximately optimum moisture content, there is a great loss of strength. This means that if two samples are compacted to the same dry unit weight, one of them on the dry side of the optimum, and the other on the wet side of the optimum, the specimen compacted on the dry side of the optimum (that is, with flocculent structure), will exhibit greater strength.

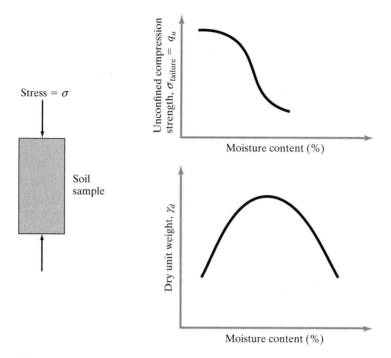

Figure 4.26 Effect of compaction on the strength of clayey soils

Problems

4.1   The laboratory test data for a standard Proctor test are given below. Find the maximum dry unit density and optimum moisture content.

Volume of Proctor mold (cm³)	Mass of wet soil in the mold (kg)	Moisture content (%)
943.3	1.78	5.0
943.3	1.87	7.5
943.3	1.95	10.0
943.3	1.98	12.5
943.3	2.02	15.0
943.3	1.97	17.5
943.3	1.91	20.0

4.2   For the soil described in Problem 4.1, given $G_s = 2.78$. Determine (at optimum moisture content):
   a. Void ratio
   b. Degree of saturation.

4.3   The results of a standard Proctor test are given in the following table. Determine the maximum dry unit weight of compaction and the optimum moisture content. Also determine the moisture content required to achieve 95% of $\gamma_{d(max)}$.

Volume of Proctor mold (cm³)	Mass of wet soil in the mold (kg)	Moisture content (%)
943.3	1.47	10.0
943.3	1.83	12.5
943.3	2.02	15.0
943.3	1.95	17.5
943.3	1.73	20.0
943.3	1.69	22.5

4.4 Calculate the zero-air-void density for a soil with $G_s = 2.70$ at moisture contents of 5, 10, 15, 20, and 25%. Plot a graph of ρ_{zav} versus moisture content.

4.5 A proposed embankment fill requires 3500 m³ of compacted soil. The void ratio of the compacted fill is specified as 0.65. Four borrow pits are available, as described in the following table, which lists the respective void ratios of the soil and the cost per cubic meter for moving the soil to the proposed construction site. Make the necessary calculations to select the pit from which the soil should be bought to minimize the cost. Assume G_s to be the same at all pits.

Borrow pit	Void ratio	Cost ($/m³)
A	0.85	9
B	1.2	6
C	0.95	7
D	0.75	10

4.6 The maximum and minimum dry unit weights of a sand were determined in the laboratory to be 17.5 kN/m³ and 14.8 kN/m³. What would be the relative compaction in the field if the relative density is 70%?

4.7 The relative compaction of a sand in the field is 92%. The maximum and minimum dry unit weights of the sand are 16.2 kN/m³ and 14.6 kN/m³, respectively. For the field condition, determine
a. Dry unit weight
b. Relative density of compaction
c. Moist unit weight at a moisture content of 10%

4.8 The *in situ* moisture content of a soil is 16% and the moist unit weight is 17.3 kN/m³. The specific gravity of soil solids is 2.72. This soil is to be excavated and transported to a construction site for use in a compacted fill. If the specifications call for the soil to be compacted to a minimum dry unit weight of 18.1 kN/m³ at the same moisture content of 16%, how many cubic meters of soil from the excavation site are needed to produce 2000 m³ of compacted fill?

4.9 Laboratory compaction test results on a clayey soil are listed in the table.

Moisture content (%)	Dry unit weight (kN/m³)
6	14.80
8	17.45
9	18.52
11	18.9
12	18.5
14	16.9

Following are the results of a field unit weight determination test on the same soil with the sand cone method:

- Calibrated dry density of Ottawa sand = 1570 kg/m^3
- Calibrated mass of Ottawa sand to fill the cone = 0.545 kg
- Mass of jar + cone + sand (before use) = 7.59 kg
- Mass of jar + cone + sand (after use) = 4.78 kg
- Mass of moist soil from hole = 3.007 kg
- Moisture content of moist soil = 10.2%

Determine

a. Dry unit weight of compaction in the field

b. Relative compaction in the field

4.10 Following are the results for the backfill material used in a vibroflotation project:

D_{10} = 0.11 mm

D_{20} = 0.19 mm

D_{50} = 1.3 mm

Determine the stability number, S_N. What would be its rating as a backfill?

4.11 For a dynamic compaction test, weight of hammer = 15 metric tons; height of drop = 12 m. Determine the significant depth of influence for compaction D in meters.

References

AMERICAN ASSOCIATION OF STATE HIGHWAY AND TRANSPORTATION OFFICIALS (1982). *AASHTO Materials, Part II*, Washington, D.C.

AMERICAN SOCIETY FOR TESTING AND MATERIALS (2006). *ASTM Standards*, Vol. 04.08, West Conshohocken, PA.

BROWN, E. (1977). "Vibroflotation Compaction of Cohesionless Soils," *Journal of the Geotechnical Engineering Division*, ASCE, Vol. 103, No. GT12, 1437–1451.

GURTUG, Y., and SRIDHARAN, A. (2004). "Compaction Behaviour and Prediction of Its Characteristics of Fine Grained Soils with Particular Reference to Compaction Energy," *Soils and Foundations*, Vol. 44, No. 5, 27–36.

HOLTZ, R. D., and KOVACS, W. D. (1981). *An Introduction to Geotechnical Engineering*, Prentice-Hall, Englewood Cliffs, NJ.

LAMBE, T. W. (1958). "The Structure of Compacted Clay," *Journal of the Soil Mechanics and Foundations Division*, ASCE, Vol. 84, No. SM2, 1654-1–1654-34.

LEE, K. W., and SINGH, A. (1971). "Relative Density and Relative Compaction," *Journal of the Soil Mechanics and Foundations Division*, ASCE, Vol. 97, No. SM7, 1049–1052.

LEE, P. Y., and SUEDKAMP, R. J. (1972). "Characteristics of Irregularly Shaped Compaction Curves of Soils," *Highway Research Record No. 381*, National Academy of Sciences, Washington, D.C., 1–9.

LEONARDS, G. A., CUTTER, W. A., and HOLTZ, R. D. (1980). "Dynamic Compaction of Granular Soils," *Journal of the Geotechnical Engineering Division*, ASCE, Vol. 106, No. GT1, 35–44.

MITCHELL, J. K. (1970). "In-Place Treatment of Foundation Soils," *Journal of the Soil Mechanics and Foundations Division*, ASCE, Vol. 96, No. SM1, 73–110.

OMAR, M., ABDALLAH, S., BASMA, A., and BARAKAT, S. (2003). "Compaction Characteristics of Granular Soils in United Arab Emirates," *Geotechnical and Geological Engineering*, Vol. 21, No. 3, 283–295.

PROCTOR, R. R. (1933). "Design and Construction of Rolled Earth Dams," *Engineering News Record*, Vol. 3, 245–248, 286–289, 348–351, 372–376.

SEED, H. B. (1964). Lecture Notes, CE 271, Seepage and Earth Dam Design, University of California, Berkeley.

SEED, H. B., and CHAN, C. K. (1959). "Structure and Strength Characteristics of Compacted Clays," *Journal of the Soil Mechanics and Foundations Division*, ASCE, Vol. 85, No. SM5, 87–128.

5

Hydraulic Conductivity and Seepage

Soils have interconnected voids through which water can flow from points of high energy to points of low energy. The study of the flow of water through porous soil media is important in soil mechanics. It is necessary for estimating the quantity of underground seepage under various hydraulic conditions, for investigating problems involving the pumping of water for underground construction, and for making stability analyses of earth dams and earth-retaining structures that are subject to seepage forces.

HYDRAULIC CONDUCTIVITY

5.1 Bernoulli's Equation

From fluid mechanics we know that, according to Bernoulli's equation, the total head at a point in water under motion can be given by the sum of the pressure, velocity, and elevation heads, or

$$h = \frac{u}{\gamma_w} + \frac{v^2}{2g} + Z \qquad (5.1)$$

$$\uparrow \qquad \uparrow \qquad \uparrow$$

Pressure   Velocity   Elevation
head   head   head

where
    h = total head
    u = pressure
    v = velocity
    g = acceleration due to gravity
    γ_w = unit weight of water

Note that the elevation head, Z, is the vertical distance of a given point above or below a datum plane. The pressure head is the water pressure, u, at that point divided by the unit weight of water, γ_w.

If Bernoulli's equation is applied to the flow of water through a porous soil medium, the term containing the velocity head can be neglected because the seepage velocity is small. Then the total head at any point can be adequately represented by

$$h = \frac{u}{\gamma_w} + Z \qquad (5.2)$$

Figure 5.1 shows the relationship among the pressure, elevation, and total heads for the flow of water through soil. Open standpipes called *piezometers* are installed at points A and B. The levels to which water rises in the piezometer tubes situated at points A and B are known as the *piezometric levels* of points A and B, respectively. The pressure head at a point is the height of the vertical column of water in the piezometer installed at that point.

The loss of head between two points, A and B, can be given by

$$\Delta h = h_A - h_B = \left(\frac{u_A}{\gamma_w} + Z_A \right) - \left(\frac{u_B}{\gamma_w} + Z_B \right) \qquad (5.3)$$

The head loss, Δh, can be expressed in a nondimensional form as

$$i = \frac{\Delta h}{L} \qquad (5.4)$$

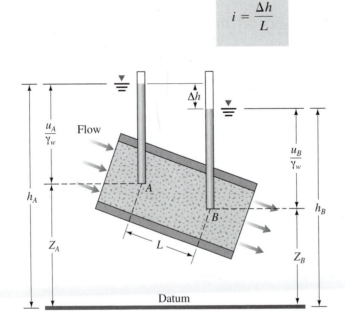

Figure 5.1 Pressure, elevation, and total heads for flow of water through soil

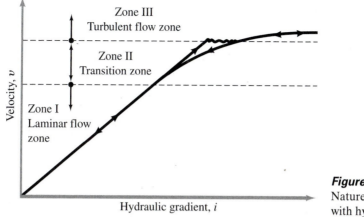

Figure 5.2
Nature of variation of v with hydraulic gradient, i

where

i = hydraulic gradient

L = distance between points A and B — that is, the length of flow over which the loss of head occurred

In general, the variation of the velocity, v, with the hydraulic gradient, i, is as shown in Figure 5.2. This figure is divided into three zones:

1. Laminar flow zone (Zone I)
2. Transition zone (Zone II)
3. Turbulent flow zone (Zone III)

When the hydraulic gradient is gradually increased, the flow remains laminar in Zones I and II, and the velocity, v, bears a linear relationship to the hydraulic gradient. At a higher hydraulic gradient, the flow becomes turbulent (Zone III). When the hydraulic gradient is decreased, laminar flow conditions exist only in Zone I.

In most soils, the flow of water through the void spaces can be considered laminar; thus,

$$v \propto i \tag{5.5}$$

In fractured rock, stones, gravels, and very coarse sands, turbulent flow conditions may exist, and Eq. (5.5) may not be valid.

5.2 Darcy's Law

In 1856, Henri Philibert Gaspard Darcy published a simple empirical equation for the discharge velocity of water through saturated soils. This equation was based primarily on Darcy's observations about the flow of water through clean sands and is given as

$$v = ki \tag{5.6}$$

where

v = *discharge velocity*, which is the quantity of water flowing in unit time through a unit gross cross-sectional area of soil at right angles to the direction of flow

k = hydraulic conductivity (otherwise known as the coefficient of permeability)

Hydraulic conductivity is expressed in cm/sec or m/sec, and discharge is in m³. It needs to be pointed out that the length is expressed in mm or m, so, in that sense, hydraulic conductivity should be expressed in mm/sec rather than cm/sec. However, geotechnical engineers continue to use cm/sec as the unit for hydraulic conductivity.

Note that Eq. (5.6) is similar to Eq. (5.5); both are valid for laminar flow conditions and applicable for a wide range of soils. In Eq. (5.6), v is the discharge velocity of water based on the gross cross-sectional area of the soil. However, the actual velocity of water (that is, the seepage velocity) through the void spaces is greater than v. A relationship between the discharge velocity and the seepage velocity can be derived by referring to Figure 5.3, which shows a soil of length L with a gross cross-sectional area A. If the quantity of water flowing through the soil in unit time is q, then

$$q = vA = A_v v_s \tag{5.7}$$

where

v_s = *seepage velocity*

A_v = area of void in the cross section of the specimen

However,

$$A = A_v + A_s \tag{5.8}$$

where A_s = area of soil solids in the cross section of the specimen. Combining Eqs. (5.7) and (5.8) gives

$$q = v(A_v + A_s) = A_v v_s$$

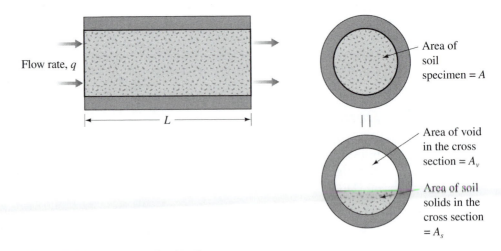

Figure 5.3 Derivation of Eq. (5.10)

or

$$v_s = \frac{v(A_v + A_s)}{A_v} = \frac{v(A_v + A_s)L}{A_v L} = \frac{v(V_v + V_s)}{V_v} \tag{5.9}$$

where

V_v = volume of voids in the specimen
V_s = volume of soil solids in the specimen

Equation (5.9) can be rewritten as

$$v_s = v \left[\frac{1 + \left(\dfrac{V_v}{V_s} \right)}{\dfrac{V_v}{V_s}} \right] = v \left(\frac{1 + e}{e} \right) = \frac{v}{n} \tag{5.10}$$

where

e = void ratio
n = porosity

Keep in mind that the terms *actual velocity* and *seepage velocity* are defined in an average sense. The actual and seepage velocities will vary with location within the pore volume of the soil.

5.3 Hydraulic Conductivity

The hydraulic conductivity of soils depends on several factors: fluid viscosity, pore-size distribution, grain-size distribution, void ratio, roughness of mineral particles, and degree of soil saturation. In clayey soils, structure plays an important role in hydraulic conductivity. Other major factors that affect the hydraulic conductivity of clays are the ionic concentration and the thickness of layers of water held to the clay particles.

The value of hydraulic conductivity, k, varies widely for different soils. Some typical values for saturated soils are given in Table 5.1. The hydraulic conductivity of unsaturated soils is lower and increases rapidly with the degree of saturation.

Table 5.1 Typical values of hydraulic conductivity for saturated soils

Soil type	k (cm /sec)
Clean gravel	100–1
Coarse sand	1.0–0.01
Fine sand	0.01–0.001
Silty clay	0.001–0.00001
Clay	<0.000001

The hydraulic conductivity of a soil is also related to the properties of the fluid flowing through it by the following equation:

$$k = \frac{\gamma_w}{\eta} \overline{K}$$

(5.11)

where

γ_w = unit weight of water
η = viscosity of water
$\overline{K}$ = absolute permeability

The *absolute permeability*, $\overline{K}$, is expressed in units of length squared (that is, cm^2).

5.4 *Laboratory Determination of Hydraulic Conductivity*

Two standard laboratory tests are used to determine the hydraulic conductivity of soil: the constant head test and the falling head test. The constant head test is used primarily for coarse-grained soils. For fine-grained soils, however, the flow rates through the soil are too small and, therefore, falling head tests are preferred. A brief description of each follows.

Constant Head Test

A typical arrangement of the constant head permeability test is shown in Figure 5.4. In this type of laboratory setup, the water supply at the inlet is adjusted in such a way that the difference of head between the inlet and the outlet remains constant during the test period. After a constant flow rate is established, water is collected in a graduated flask for a known duration.

The total volume of water, Q, collected may be expressed as

$$Q = Avt = A(ki)t$$

(5.12)

where

A = area of cross section of the soil specimen
t = duration of water collection

Also, because

$$i = \frac{h}{L}$$

(5.13)

where L = length of the specimen, Eq. (5.13) can be substituted into Eq. (5.12) to yield

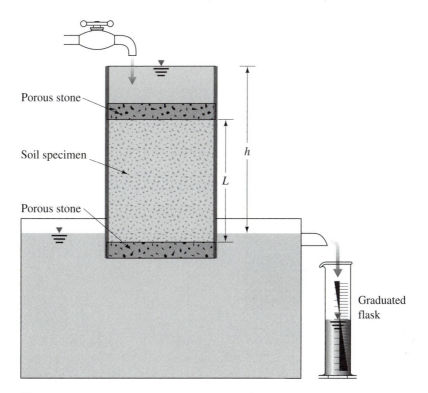

Porous stone

Soil specimen

Porous stone

Graduated flask

Figure 5.4 Constant head permeability test

$$Q = A\left(k\frac{h}{L}\right)t \tag{5.14}$$

or

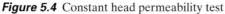

$$k = \frac{QL}{Aht} \tag{5.15}$$

Falling Head Test

A typical arrangement of the falling head permeability test is shown in Figure 5.5. Water from a standpipe flows through the soil. The initial head difference, h_1, at time $t = 0$ is recorded, and water is allowed to flow through the soil specimen such that the final head difference at time $t = t_2$ is h_2.

   The rate of flow of the water, q, through the specimen at any time t can be given by

$$q = k\frac{h}{L}A = -a\frac{dh}{dt} \tag{5.16}$$

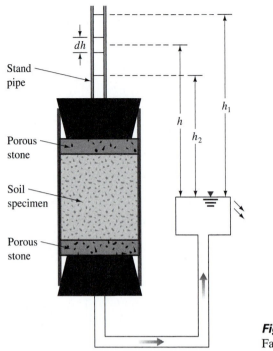

Figure 5.5
Falling head permeability test

where

a = cross-sectional area of the standpipe
A = cross-sectional area of the soil specimen

Rearranging Eq. (5.16) gives

$$dt = \frac{aL}{Ak}\left(-\frac{dh}{h}\right) \tag{5.17}$$

Integration of the left side of Eq. (5.17) with limits of time from 0 to t and the right side with limits of head difference from h_1 to h_2 gives

$$t = \frac{aL}{Ak}\log_e\frac{h_1}{h_2}$$

or

$$k = 2.303\frac{aL}{At}\log_{10}\frac{h_1}{h_2} \tag{5.18}$$

Example 5.1

For a constant head laboratory permeability test on a fine sand, the following values are given (refer to Figure 5.4):

► Length of specimen = 250 mm
► Diameter of specimen = 64 mm
► Head difference = 460 mm
► Water collected in 2 min = 0.51 cm^3

Determine

a. Hydraulic conductivity, k, of the soil (cm/min)
b. Discharge velocity (cm/min)
c. Seepage velocity (cm/min)

The void ratio of the soil specimen is 0.46.

Solution

a. From Eq. (5.15),

$$k = \frac{QL}{Aht} = \frac{(0.51)\,(25)}{\left(\frac{\pi}{4}\,6.4^2\right)(46)(2)} = 0.431 \times 10^{-2}\,\text{cm/min}$$

b. From Eq. (5.6),

$$v = ki = (0.431 \times 10^{-2})\left(\frac{46}{25}\right) = 0.793 \times 10^{-2}\,\text{cm/min}$$

c. From Eq. (5.10),

$$v_s = v\left(\frac{1+e}{e}\right) = (0.793 \times 10^{-2})\left(\frac{1+0.46}{0.46}\right) = 2.52 \times 10^{-2}\,\text{cm/min}$$

Example 5.2

For a constant head permeability test, the following values are given:

► $L = 300$ mm
► A = specimen area = 32 cm^2
► $k = 0.0244$ cm/sec

The head difference was slowly changed in steps to 800, 700, 600, 500, and 400 mm. Calculate and plot the rate of flow, q, through the specimen, in cm^3/sec, against the head difference.

Solution

From Eqs. (5.16), and given that $L = 300$ mm,

$$q = kiA = (0.0244)\,(32)\left(\frac{h}{L}\right) = 0.7808\left(\frac{h}{300}\right)$$

Now, the following table can be prepared:

h(mm)	q(cm³/sec)
800	2.08
700	1.82
600	1.56
500	1.30
400	1.04

The plot of q versus h is shown in Figure 5.6.

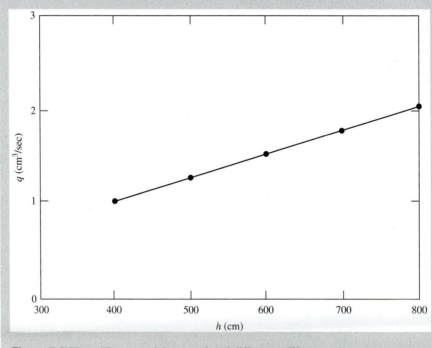

Figure 5.6 Plot of flow rate (q) versus head difference (h)

Example 5.3

A permeable soil layer is underlain by an impervious layer, as shown in Figure 5.7a. With $k = 4.8 \times 10^{-3}$ cm/sec for the permeable layer, calculate the rate of seepage through it in m³/hr/m width if $H = 3$ m and $\alpha = 5°$.

Solution
From Figure 5.7b and Eqs. (5.13),

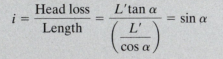

$$i = \frac{\text{Head loss}}{\text{Length}} = \frac{L' \tan \alpha}{\left(\dfrac{L'}{\cos \alpha} \right)} = \sin \alpha$$

$q = kiA = (k) (\sin \alpha) (3 \cos \alpha) (1); \ k = 4.8 \times 10^{-3}$ cm/sec $= 4.8 \times 10^{-5}$ m/sec; $q = (4.8 \times 10^{-5}) (\sin 5°) (3 \cos 5°) (3600) = 0.045$ m³/hr/m

↑
To change to
m/hr

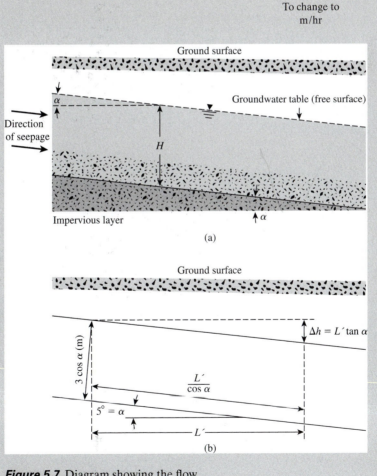

(a)

(b)

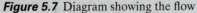

Figure 5.7 Diagram showing the flow

Example 5.4

For a falling head permeability test, the following values are given: length of specimen = 38 cm, area of specimen = 19 cm^2, and k = 0.175 cm/min. What should be the area of the standpipe for the head to drop from 64 cm to 30 cm in 8 min?

Solution
From Eq. (5.18),

$$k = 2.303 \frac{al}{At} \log_{10} \frac{h_1}{h_2}$$

$$0.175 = 2.303 \left(\frac{a \times 38}{19 \times 8} \right) \log_{10} \left(\frac{64}{30} \right)$$

$$a = \textbf{0.924 cm}^2$$

Example 5.5

The hydraulic conductivity of a clayey soil is 3×10^{-7} cm/sec. The viscosity of water at 25 °C is 0.0911×10^{-4} g·sec/cm^2. Calculate the absolute permeability, $\overline{K}$, of the soil.

Solution
From Eq. (5.11),

$$k = \frac{\gamma_w}{\eta} \overline{K} = 3 \times 10^{-7} \text{ cm/sec}$$

so

$$3 \times 10^{-7} = \left(\frac{1\text{g/cm}^3}{0.0911 \times 10^{-4}} \right) \overline{K}$$

$$\overline{K} = \textbf{0.2733} \times \textbf{10}^{-11} \textbf{ cm}^2$$

5.5 *Empirical Relations for Hydraulic Conductivity*

Several empirical equations for estimating hydraulic conductivity have been proposed over the years. Some of these are discussed briefly in this section.

Granular Soil

For fairly uniform sand (that is, a small uniformity coefficient), Hazen (1930) proposed an empirical relationship for hydraulic conductivity in the form

$$k \text{ (cm/sec)} = cD_{10}^2 \tag{5.19}$$

where

c = a constant that varies from 1.0 to 1.5

D_{10} = the effective size (mm)

Equation (5.19) is based primarily on Hazen's observations of loose, clean, filter sands. A small quantity of silts and clays, when present in a sandy soil, may change the hydraulic conductivity substantially. Over the last several years, experimental observations have shown that the magnitude of c for various types of granular soils may vary by three orders of magnitude (Carrier, 2003) and, hence, is not very reliable.

On the basis of laboratory experiments, the U.S. Department of Navy (1971) provided an empirical correlation between k (cm/min) and D_{10} (mm) for granular

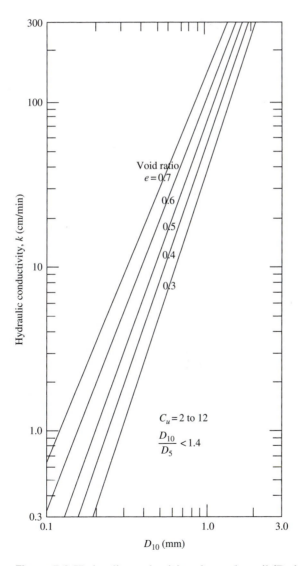

Figure 5.8 Hydraulic conductivity of granular soil (Redrawn from U. S. Department of Navy, 1971)

soils with the uniformity coefficient varying between 2 and 12 and $D_{10}/D_5 < 1.4$. This correlation is shown in Figure 5.8.

Another form of equation that gives fairly good results in estimating the hydraulic conductivity of sandy soils is based on the Kozeny-Carman equation (Kozeny, 1927; Carman, 1938, 1956). The derivation of this equation is not presented here. Interested readers are referred to any advanced soil mechanics book. According to the Kozeny-Carman equation

$$k = \frac{1}{C_s S^2 T^2} \frac{\gamma_w}{\eta} \frac{e^3}{1 + e} \tag{5.20}$$

where
  C_s = shape factor, which is a function of the shape of flow channels
  S_s = specific surface area per unit volume of particles
  T = tortuosity of flow channels
  γ_ω = unit weight of water
  η = viscosity of permeant
  e = void ratio

For practical use, Carrier (2003) has modified Eq. (5.20) in the following manner. At 20 °C, γ_ω/η for water is about $9.93 \times 10^4 \left(\frac{1}{\text{cm} \cdot \text{s}} \right)$. Also, $(C_s T^2)$ is approximately equal to 5. Substituting these values in Eq. (5.20), we obtain

$$k = 1.99 \times 10^4 \left(\frac{1}{S_s} \right)^2 \frac{e^3}{1 + e} \tag{5.21}$$

Again,

$$S_s = \frac{SF}{D_{\text{eff}}} \left(\frac{1}{\text{cm}} \right) \tag{5.22}$$

with

$$D_{\text{eff}} = \frac{100\%}{\Sigma \left(\dfrac{f_i}{D_{(av)i}} \right)} \tag{5.23}$$

where
  f_i = fraction of particles between two sieve sizes, in percent (*Note*: larger sieve, l; smaller sieve, s)
$$D_{(av)i} \,(\text{cm}) = [D_{li}\,(\text{cm})]^{0.5} \times [D_{si}\,(\text{cm})]^{0.5} \tag{5.24}$$
  SF = shape factor

Combining Eqs. (5.21), (5.22), (5.23), and (5.24),

$$k = 1.99 \times 10^4 \left[\frac{100\%}{\Sigma \dfrac{f_i}{D_{li}^{0.5} \times D_{si}^{0.5}}} \right]^2 \left(\frac{1}{SF} \right)^2 \left(\frac{e^3}{1 + e} \right) \tag{5.25}$$

The magnitude of SF may vary between 6 to 8, depending on the angularity of the soil particles.

Carrier (2003) further suggested a slight modification to Eq. (5.25), which can be written as

$$k = 1.99 \times 10^4 \left[\frac{100\%}{\Sigma \dfrac{f_i}{D_{li}^{0.404} \times D_{si}^{0.595}}} \right]^2 \left(\frac{1}{SF} \right)^2 \left(\frac{e^3}{1+e} \right) \qquad (5.26)$$

Equation (5.26) suggests that

$$k \propto \frac{e^3}{1+e} \qquad (5.27)$$

The author recommends the use of Eqs. (5.26) and (5.27).

Cohesive Soil

Tavenas, *et al.* (1983) also gave a correlation between the void ratio and the hydraulic conductivity of clayey soil for flow in vertical direction. This correlation is shown in Figure 5.9. An important point to note, however, is that in Figure 5.9, *PI*, the plasticity index, and *CF*, the clay-size fraction in the soil, are in *fraction* (decimal) form.

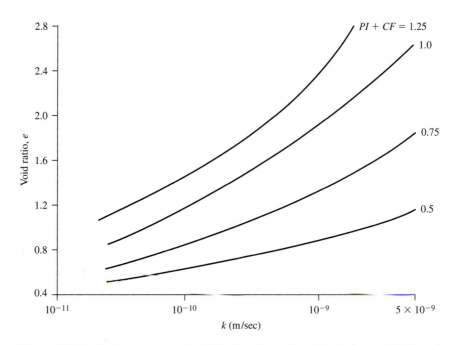

Figure 5.9 Variation of void ratio with hydraulic conductivity of clayey soils (Based on Tavenas, *et al.*, 1983)

According to their experimental observations, Samarasinghe, Huang, and Drnevich (1982) suggested that the hydraulic conductivity of normally consolidated clays (see Chapter 7 for definition) can be given by the following equation:

$$k = C\left(\frac{e^n}{1 + e}\right) \tag{5.28}$$

where C and n are constants to be determined experimentally.

Example 5.6

The hydraulic conductivity of a sand at a void ratio of 0.71 is 0.044 cm/sec. Estimate the hydraulic conductivity of this sand at a void ratio of 0.55. Use Eq. (5.27).

Solution
From Eq. (5.27)

$$k \propto \frac{e^3}{1 + e}$$

So

$$\frac{k_{0.71}}{k_{0.55}} = \frac{\left[\dfrac{0.71^3}{1 + 0.71}\right]}{\left[\dfrac{0.55^3}{1 + 0.55}\right]} = \frac{0.209}{0.107} = 1.95$$

Hence

$$k_{0.55} = \frac{k_{0.71}}{1.95} = \frac{0.044}{1.95} \simeq \mathbf{0.023\ cm/sec} \qquad \blacksquare$$

Example 5.7

The void ratio and hydraulic conductivity relation for a normally consolidated clay are given below.

Void ratio	k (cm/sec)
1.2	0.6×10^{-7}
1.52	1.519×10^{-7}

Estimate the value of k for the same clay with a void ratio of 1.4.

Solution
From Eq. (5.28)

$$\frac{k_1}{k_2} = \frac{\left[\dfrac{e_1^n}{1 + e_1}\right]}{\left[\dfrac{e_2^n}{1 + e_2}\right]}$$

Substitution of $e_1 = 1.2$, $k_1 = 0.6 \times 10^{-7}$ cm/sec, $e_2 = 1.52$, $k_2 = 1.159 \times 10^{-7}$ cm/sec in the preceding equation gives

$$\frac{0.6}{1.519} = \left(\frac{1.2}{1.52}\right)^n\left(\frac{2.52}{2.2}\right)$$

or

$$n = 4.5$$

Again, from Eq. (5.28)

$$k_1 = C\left(\frac{e_1^n}{1 + e_1}\right)$$

$$0.6 \times 10^{-7} = C\left(\frac{1.2^{4.5}}{1 + 1.2}\right)$$

or

$$C = 0.581 \times 10^{-7} \text{ cm/sec}$$

So

$$k = (0.581 \times 10^{-7})\left(\frac{e^{4.5}}{1 + e}\right) \text{ cm/sec}$$

Now, substituting $e = 1.4$ in the preceding equation

$$k = (0.581 \times 10^{-7})\left(\frac{1.4^{4.5}}{1 + 1.4}\right) = \mathbf{1.1 \times 10^{-7} \text{ cm/sec}} \qquad \blacksquare$$

Example 5.8

The grain-size distribution curve for a sand is shown in Figure 5.10. Estimate the hydraulic conductivity using Eq. (5.26). Given: the void ratio of the sand is 0.6. Use $SF = 7$.

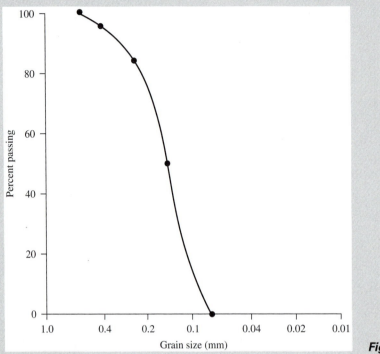

Figure 5.10

Solution

From Figure 5.10, the following table can be prepared.

Sieve no.	Sieve opening (cm)	Percent passing	Fraction of particles between two consecutive sieves (%)
30	0.06	100	
40	0.0425	96	4
60	0.02	84	12
100	0.015	50	34
200	0.0075	0	50

For fraction between Nos. 30 and 40 sieves:

$$\frac{f_i}{D_{li}^{0.404} \times D_{si}^{0.595}} = \frac{4}{(0.06)^{0.404} \times (0.0425)^{0.595}} = 81.62$$

For fraction between Nos. 40 and 60 sieves:

$$\frac{f_i}{D_{li}^{0.404} \times D_{si}^{0.595}} = \frac{12}{(0.0425)^{0.404} \times (0.02)^{0.595}} = 440.76$$

Similarly, for fraction between Nos. 60 and 100 sieves:

$$\frac{f_i}{D_{li}^{0.404} \times D_{si}^{0.595}} = \frac{34}{(0.02)^{0.404} \times (0.015)^{0.595}} = 2009.5$$

And, for between Nos. 100 and 200 sieves:

$$\frac{f_i}{D_{li}^{0.404} \times D_{si}^{0.595}} = \frac{50}{(0.015)^{0.404} \times (0.0075)^{0.595}} = 5013.8$$

$$\frac{100\%}{\Sigma \dfrac{f_i}{D_{li}^{0.404} \times D_{si}^{0.595}}} = \frac{100}{81.62 + 440.76 + 2009.5 + 5013.8} \approx 0.0133$$

From Eq. (5.26),

$$k = (1.99 \times 10^4)(0.0133)^2 \left(\frac{1}{7}\right)^2 \left(\frac{0.6^3}{1 + 0.6}\right) = \textbf{0.0097 cm/s}$$

5.6 *Equivalent Hydraulic Conductivity in Stratified Soil*

Depending on the nature of soil deposit, the hydraulic conductivity of a given layer of soil may vary with the direction of flow. In a stratified soil deposit where the hydraulic conductivity for flow in different directions changes from layer to layer, an equivalent hydraulic conductivity determination becomes necessary to simplify calculations. The following derivations relate to the equivalent hydraulic conductivity for flow in vertical and horizontal directions through multilayered soils with horizontal stratification.

Figure 5.11 shows n layers of soil with flow in the *horizontal direction*. Let us consider a cross-section of unit length passing through the n layer and perpendicular to the direction of flow. The total flow through the cross-section in unit time can be written as

$$
\begin{aligned}
q &= v \cdot 1 \cdot H \\
&= v_1 \cdot 1 \cdot H_1 + v_2 \cdot 1 \cdot H_2 + v_3 \cdot 1 \cdot H_3 + \cdots + v_n \cdot 1 \cdot H_n
\end{aligned}
\tag{5.29}
$$

where

v = average discharge velocity

$v_1, v_2, v_3, \ldots, v_n$ = discharge velocities of flow in layers denoted by the subscripts.

If $k_{H_1}, k_{H_2}, k_{H_3}, \ldots, k_{H_n}$ are the hydraulic conductivity of the individual layers in the horizontal direction, and $k_{H(eq)}$ is the equivalent hydraulic conductivity in the horizontal direction, then from Darcy's law

$$v = k_{H(eq)} i_{eq}; \; v_1 = k_{H_1} i_1; \; v_2 = k_{H_2} i_2; \; v_3 = k_{H_3} i_3; \ldots; \; v_n = k_{H_n} i_n$$

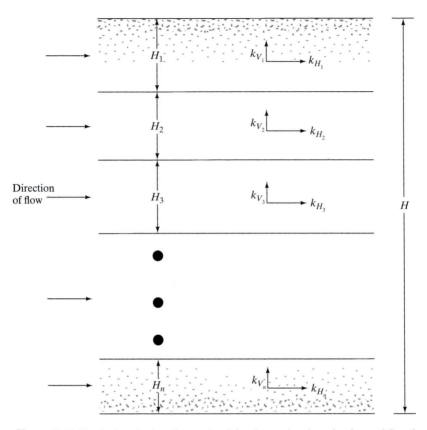

Figure 5.11 Equivalent hydraulic conductivity determination—horizontal flow in stratified soil

Substitution of the preceding relations for velocities in Eq. (5.29) and noting $i_{eq} = i_1 = i_2 = i_3 = \cdots i_n$ results in

$$k_{H(eq)} = \frac{1}{H}(k_{H_1}H_1 + k_{H_2}H_2 + k_{H_3}H_3 + \cdots + k_{H_n}H_n) \tag{5.30}$$

Figure 5.12 shows n layers of soil with flow in the vertical direction. In this case, the velocity of flow through all the layers is the same. However, the total head loss, h, is equal to the sum of the head loss in each layer. Thus

$$v = v_1 = v_2 = v_3 = \cdots = v_n \tag{5.31}$$

and

$$h = h_1 + h_2 + h_3 + \cdots + h_n \tag{5.32}$$

Using Darcy's law, Eq. (5.31) can be rewritten as

$$k_{V(eq)}\frac{h}{H} = k_{V_1}i_1 = k_{V_2}i_2 = k_{V_3}i_3 = \cdots = k_{V_n}i_n \tag{5.33}$$

where $k_{V1}, k_{V2}, k_{V3}, \ldots , k_{Vn}$ are the hydraulic conductivities of the individual layers in the vertical direction and $k_{V(eq)}$ is the equivalent hydraulic conductivity.

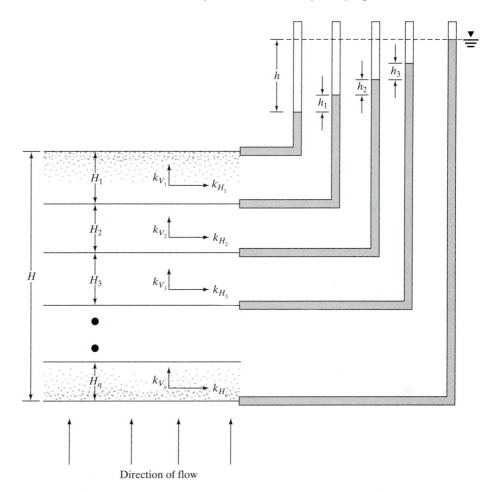

Figure 5.12 Equivalent hydraulic conductivity determination—vertical flow in stratified soil

Again, from Eq. (5.32)

$$h = H_1 i_1 + H_2 i_2 + H_3 i_3 + \cdots + H_n i_n \qquad (5.34)$$

Solution of Eqs. (5.33) and (5.34) gives

$$k_{V(eq)} = \frac{H}{\left(\dfrac{H_1}{k_{V_1}}\right) + \left(\dfrac{H_2}{k_{V_2}}\right) + \left(\dfrac{H_3}{k_{V_3}}\right) + \cdots + \left(\dfrac{H_n}{k_{V_n}}\right)} \qquad (5.35)$$

5.7 *Permeability Test in the Field by Pumping from Wells*

In the field, the average hydraulic conductivity of a soil deposit in the direction of flow can be determined by performing pumping tests from wells. Figure 5.13 shows a case where the top permeable layer, whose hydraulic conductivity has to be determined, is

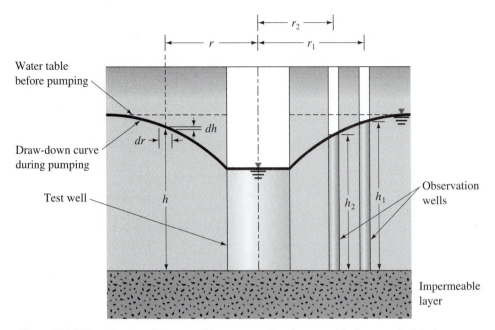

Figure 5.13 Pumping test from a well in an unconfined permeable layer underlain by an impermeable stratum

unconfined and underlain by an impermeable layer. During the test, water is pumped out at a constant rate from a test well that has a perforated casing. Several observation wells at various radial distances are made around the test well. Continuous observations of the water level in the test well and in the observation wells are made after the start of pumping, until a steady state is reached. The steady state is established when the water level in the test and observation wells becomes constant. The expression for the rate of flow of groundwater, q, into the well, which is equal to the rate of discharge from pumping, can be written as

$$q = k\left(\frac{dh}{dr}\right)2\pi rh \tag{5.36}$$

or

$$\int_{r_2}^{r_1} \frac{dr}{r} = \left(\frac{2\pi k}{q}\right)\int_{h_2}^{h_1} h\,dh$$

Thus,

$$k = \frac{2.303\,q\,\log_{10}\left(\dfrac{r_1}{r_2}\right)}{\pi(h_1^2 - h_2^2)} \tag{5.37}$$

From field measurements, if q, r_1, r_2, h_1, and h_2 are known, then the hydraulic conductivity can be calculated from the simple relationship presented in Eq. (5.37).

The average hydraulic conductivity for a confined aquifer can also be determined by conducting a pumping test from a well with a perforated casing that penetrates the full depth of the aquifer and by observing the piezometric level in a number of observation wells at various radial distances (Figure 5.14). Pumping is continued at a uniform rate q until a steady state is reached.

Because water can enter the test well only from the aquifer of thickness H, the steady state of discharge is

$$q = k\left(\frac{dh}{dr}\right)2\pi rH \tag{5.38}$$

or

$$\int_{r_2}^{r_1} \frac{dr}{r} = \int_{h_2}^{h_1} \frac{2\pi kH}{q} dh$$

This gives the hydraulic conductivity in the direction of flow as

$$k = \frac{q \log_{10}\left(\dfrac{r_1}{r_2}\right)}{2.727\, H(h_1 - h_2)} \tag{5.39}$$

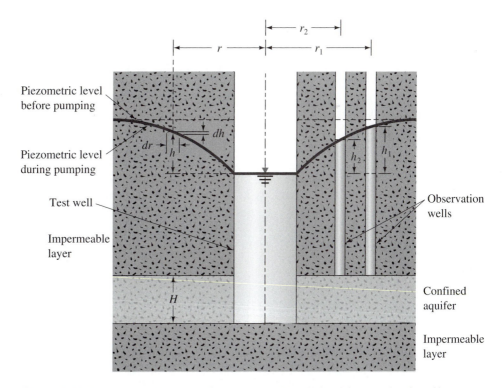

Figure 5.14 Pumping test from a well penetrating the full depth in a confined aquifer

SEEPAGE

In the preceding sections of this chapter, we considered some simple cases for which direct application of Darcy's law was required to calculate the flow of water through soil. In many instances, the flow of water through soil is not in one direction only, and it is not uniform over the entire area perpendicular to the flow. In such cases, the groundwater flow is generally calculated by the use of graphs referred to as *flow nets*. The concept of the flow net is based on *Laplace's equation of continuity*, which governs the steady flow condition for a given point in the soil mass. The following sections explain the derivation of Laplace's equation of continuity and its application to drawing flow nets.

5.8   *Laplace's Equation of Continuity*

To derive the Laplace differential equation of continuity, we consider a single row of sheet piles that have been driven into a permeable soil layer, as shown in Figure 5.15a. The row of sheet piles is assumed to be impervious. The steady-state flow of water from the upstream to the downstream side through the permeable layer is a two-dimensional flow. For flow at a point A, we consider an elemental soil block. The block has dimensions dx, dy, and dz (length dy is perpendicular to the plane of the paper); it is shown in an enlarged scale in Figure 5.15b. Let v_x and v_z be the components of the discharge velocity in the horizontal and vertical directions, respectively. The rate of flow of water into the elemental block in the horizontal direction is equal to $v_x\, dz\, dy$, and in the vertical direction it is $v_z\, dx\, dy$. The rates of outflow from the block in the horizontal and vertical directions are

$$\left(v_x + \frac{\partial v_x}{\partial x}dx\right)dz\, dy$$

and

$$\left(v_z + \frac{\partial v_z}{\partial z}dz\right)dx\, dy$$

respectively. Assuming that water is incompressible and that no volume change in the soil mass occurs, we know that the total rate of inflow should equal the total rate of outflow. Thus,

$$\left[\left(v_x + \frac{\partial v_x}{\partial x}dx\right)dz\, dy + \left(v_z + \frac{\partial v_z}{\partial z}dz\right)dx\, dy\right] - [v_x\, dz\, dy + v_z\, dx\, dy] = 0$$

or

$$\frac{\partial v_x}{\partial x} + \frac{\partial v_z}{\partial z} = 0 \tag{5.40}$$

With Darcy's law, the discharge velocities can be expressed as

$$v_x = k_x i_x = k_x\left(-\frac{\partial h}{\partial x}\right) \tag{5.41}$$

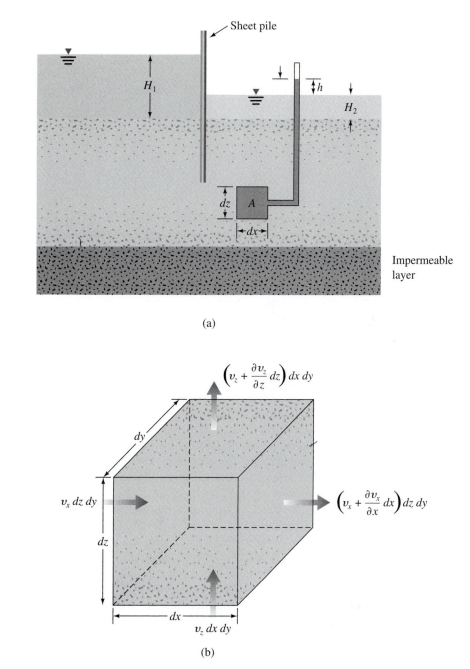

(a)

(b)

Figure 5.15 (a) Single-row sheet piles driven into a permeable layer, (b) flow at A

and

$$v_z = k_z i_z = k_z \left(-\frac{\partial h}{\partial z} \right) \tag{5.42}$$

where k_x and k_z are the hydraulic conductivities in the vertical and horizontal directions, respectively.

From Eqs. (5.40), (5.41), and (5.42), we can write

$$k_x \frac{\partial^2 h}{\partial x^2} + k_z \frac{\partial^2 h}{\partial z^2} = 0 \tag{5.43}$$

If the soil is isotropic with respect to the hydraulic conductivity—that is, $k_x = k_z$—the preceding continuity equation for two-dimensional flow simplifies to

$$\frac{\partial^2 h}{\partial x^2} + \frac{\partial^2 h}{\partial z^2} = 0 \tag{5.44}$$

5.9   *Flow Nets*

The continuity equation [Eq. (5.44)] in an isotropic medium represents two orthogonal families of curves: the flow lines and the equipotential lines. A *flow line* is a line along which a water particle will travel from the upstream to the downstream side in the permeable soil medium. An *equipotential line* is a line along which the potential head at all points is equal. Thus, if piezometers are placed at different points along an equipotential line, the water level will rise to the same elevation in all of them. Figure 5.16a demonstrates the definition of flow and equipotential lines for flow in the permeable soil layer around the row of sheet piles shown in Figure 5.15 (for $k_x = k_z = k$).

A combination of a number of flow lines and equipotential lines is called a *flow net*. Flow nets are constructed to calculate groundwater flow in the media. To complete the graphic construction of a flow net, one must draw the flow and equipotential lines in such a way that the equipotential lines intersect the flow lines at right angles and the flow elements formed are approximate squares.

Figure 5.16b shows an example of a completed flow net. Another example of a flow net in an isotropic permeable layer is shown in Figure 5.17. In these figures, N_f is the number of flow channels in the flow net, and N_d is the number of potential drops (defined later in this chapter).

Drawing a flow net takes several trials. While constructing the flow net, keep the boundary conditions in mind. For the flow net shown in Figure 5.16b, the following four boundary conditions apply:

1. The upstream and downstream surfaces of the permeable layer (lines *ab* and *de*) are equipotential lines.
2. Because *ab* and *de* are equipotential lines, all the flow lines intersect them at right angles.

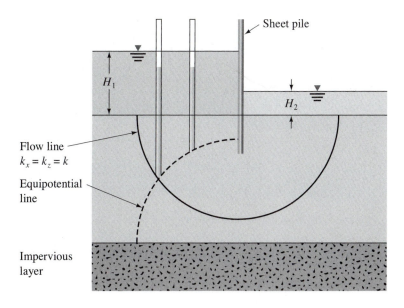

(a)

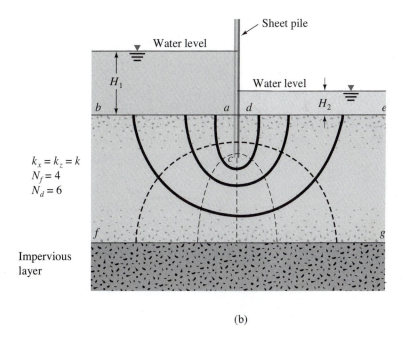

(b)

Figure 5.16 (a) Definition of flow lines and equipotential lines; (b) completed flow net

3. The boundary of the impervious layer—that is, line *fg*—is a flow line, and so is the surface of the impervious sheet pile, line *acd*.
4. The equipotential lines intersect *acd* and *fg* at right angles.

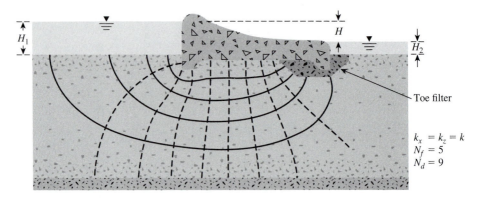

$$k_x = k_z = k$$
$$N_f = 5$$
$$N_d = 9$$

Figure 5.17 Flow net under a dam with toe filter

Seepage Calculation from a Flow Net

In any flow net, the strip between any two adjacent flow lines is called a *flow channel*. Figure 5.18 shows a flow channel with the equipotential lines forming square elements. Let $h_1, h_2, h_3, h_4, \ldots, h_n$ be the piezometric levels corresponding to the equipotential lines. The rate of seepage through the flow channel per unit length (perpendicular to the vertical section through the permeable layer) can be calculated as follows: Because there is no flow across the flow lines,

$$\Delta q_1 = \Delta q_2 = \Delta q_3 = \cdots = \Delta q \tag{5.45}$$

From Darcy's law, the flow rate is equal to kiA. Thus, Eq. (5.45) can be written as

$$\Delta q = k\left(\frac{h_1 - h_2}{l_1}\right)l_1 = k\left(\frac{h_2 - h_3}{l_2}\right)l_2 = k\left(\frac{h_3 - h_4}{l_3}\right)l_3 = \cdots \tag{5.46}$$

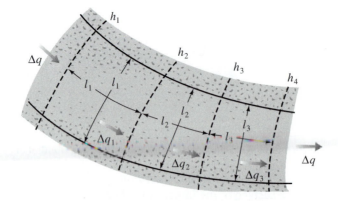

Figure 5.18 Seepage through a flow channel with square elements

Equation (5.46) shows that if the flow elements are drawn as approximate squares, then the drop in the piezometric level between any two adjacent equipotential lines is the same. This is called the *potential drop*. Thus,

$$h_1 - h_2 = h_2 - h_3 = h_3 - h_4 = \cdots = \frac{H}{N_d} \qquad (5.47)$$

and

$$\Delta q = k\frac{H}{N_d} \qquad (5.48)$$

where
H = head difference between the upstream and downstream sides
N_d = number of potential drops

In Figure 5.16b, for any flow channel, $H = H_1 - H_2$ and $N_d = 6$.
　　If the number of flow channels in a flow net is equal to N_f, the total rate of flow through all the channels per unit length can be given by

$$q = k\frac{HN_f}{N_d} \qquad (5.49)$$

　　Although drawing square elements for a flow net is convenient, it is not always necessary. Alternatively, one can draw a rectangular mesh for a flow channel, as shown in Figure 5.19, provided that the width-to-length ratios for all the rectangular elements in the flow net are the same. In this case, Eq. (5.46) for rate of flow through the channel can be modified to

$$\Delta q = k\left(\frac{h_1 - h_2}{l_1}\right)b_1 = k\left(\frac{h_2 - h_3}{l_2}\right)b_2 = k\left(\frac{h_3 - h_4}{l_3}\right)b_3 = \cdots \qquad (5.50)$$

If $b_1/l_1 = b_2/l_2 = b_3/l_3 = \cdots = n$ (i.e., the elements are not square), Eqs. (5.48) and (5.49) can be modified:

$$\Delta q = kH\left(\frac{n}{N_d}\right) \qquad (5.51)$$

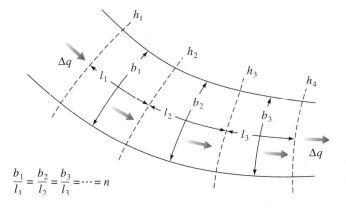

$$\frac{b_1}{l_1} = \frac{b_2}{l_2} = \frac{b_3}{l_3} = \cdots = n$$

Figure 5.19 Seepage through a flow channel with rectangular elements

or

$$q = kH\left(\frac{N_f}{N_d}\right)n \qquad (5.52)$$

Figure 5.20 shows a flow net for seepage around a single row of sheet piles. Note that flow channels 1 and 2 have square elements. Hence, the rate of flow through these two channels can be obtained from Eq. (5.48):

$$\Delta q_1 + \Delta q_2 = k\frac{H}{N_d} + k\frac{H}{N_d} = 2k\frac{H}{N_d}$$

However, flow channel 3 has rectangular elements. These elements have a width-to-length ratio of about 0.38; hence, from Eq. (5.51), we have

$$\Delta q_3 = kH\left(\frac{0.38}{N_d}\right)$$

So, the total rate of seepage can be given as

$$q = \Delta q_1 + \Delta q_2 + \Delta q_3 = 2.38\frac{kH}{N_d}$$

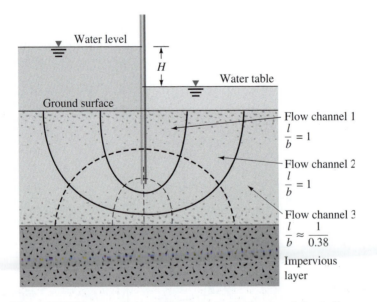

Figure 5.20 Flow net for seepage around a single row of sheet piles

Example 5.9

A flow net for flow around a single row of sheet piles in a permeable soil layer is shown in Figure 5.21. We are given that $k_x = k_z = k = 5 \times 10^{-3}$ cm/sec.

  a. How high (above the ground surface) will the water rise if piezometers are placed at points a, b, c, and d?
  b. What is the rate of seepage through flow channel II per unit length (perpendicular to the section shown)?

Solution

  a. From Figure 5.21, we see that $N_f = 3$ and $N_d = 6$. The head difference between the upstream and downstream sides is 3.33 m, so the head loss for each drop is 3.33/6 = 0.555 m. Point a is located on equipotential line 1, which means that the potential drop at a is 1×0.555 m. The water in the piezometer at a will rise to an elevation of $(5 - 0.555) = $ **4.445 m above the ground surface**. Similarly, we can calculate the other piezometric levels:

$$b = (5 - 2 \times 0.555) = \textbf{3.89 m above the ground surface}$$
$$c = (5 - 5 \times 0.555) = \textbf{2.225 m above the ground surface}$$
$$d = (5 - 5 \times 0.555) = \textbf{2.225 m above the ground surface}$$

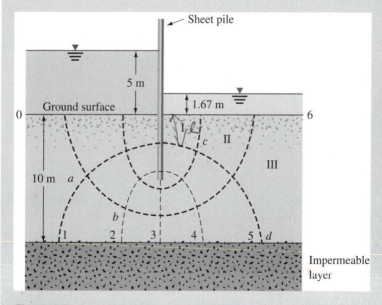

Figure 5.21 Flow net for flow around a single row of sheet piles in a permeable soil layer

b. From Eq. (5.48), we have

$$\Delta q = k \frac{H}{N_d}$$
$$k = 5 \times 10^{-3} \text{ cm/sec} = 5 \times 10^{-5} \text{ m/sec}$$
$$\Delta q = (5 \times 10^{-5})(0.555) = \textbf{2.775} \times \textbf{10}^{-5} \textbf{ m}^3\textbf{/sec/m}$$

Problems

5.1   A permeable soil layer is underlain by an impervious layer, as shown in Figure 5.22. With $k = 4.2 \times 10^{-3}$ cm/sec for the permeable layer, calculate the rate of seepage through it in m³/hr/m length. Given: $H = 4.8$ m and $\alpha = 6°$.

5.2   Refer to Figure 5.23. Find the flow rate in m³/sec/m length (at right angles to the cross-section shown) through the permeable soil layer given $H = 3$ m, $H_1 = 2.5$ m, $h = 2.8$ m, $L = 25$ m, $\alpha = 10°$, and $k = 0.04$ cm/sec.

5.3   Repeat Problem 5.2 with the following values: $H = 2.2$ m, $H_1 = 1.5$ m, $h = 2.7$ m, $L = 5$ m, $\alpha = 20°$, and $k = 1.12 \times 10^{-5}$ m/sec. The flow rate should be given in m³/hr/m length (at right angles to the cross-section shown).

5.4   Refer to the constant head permeability test shown in Figure 5.4. For a test, these values are given:
- $L = 460$ mm
- A = area of the specimen = 22.6 cm²
- Constant head difference = $h = 700$ mm
- Water collected in 3 min = 354 cm³

Calculate the hydraulic conductivity in cm/sec.

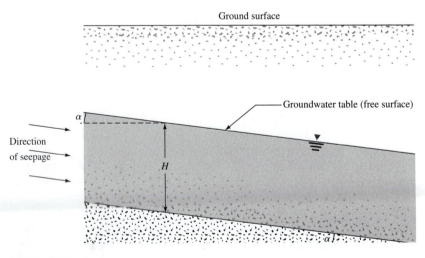

Figure 5.22

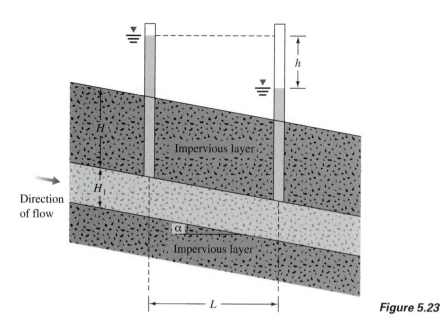

Figure 5.23

5.5 Refer to Figure 5.4. For a constant head permeability test in a sand, the following values are given:
- $L = 350$ mm
- $A = 125$ cm^2
- $h = 420$ mm
- Water collected in 3 min = 580 cm^3
- Void ratio of sand = 0.61

Determine:
a. Hydraulic conductivity, k (cm/sec)
b. Seepage velocity

5.6 In a constant head permeability test in the laboratory, the following values are given: $L = 250$ mm and $A = 105$ cm^2. If the value of $k = 0.014$ cm/sec and a flow rate of 120 cm^3/min must be maintained through the soil, what is the head difference, h, across the specimen? Also, determine the discharge velocity under the test conditions.

5.7 For a variable head permeability test, these values are given:
- Length of the soil specimen = 200 mm
- Area of the soil specimen = 1000 mm^2
- Area of the standpipe = 40 mm^2
- Head difference at time $t = 0$ is 500 mm
- Head difference at time $t = 3$ min is 300 mm

a. Determine the hydraulic conductivity of the soil in cm/sec.
b. What was the head difference at time $t = 100$ scc?

5.8 The hydraulic conductivity, k, of a soil is 0.832×10^{-5} cm/sec at a temperature of 20 °C. Determine its absolute permeability at 20 °C, given that at 20 °C, $\gamma_w = 9.789$ kN/m^3 and $\eta = 1.005 \times 10^{-3}$ N · sec/m^2 (Newton-second per square meter).

5.9   The hydraulic conductivity of a sand at a void ratio of 0.62 is 0.03 cm/sec. Estimate its hydraulic conductivity at a void ratio of 0.48. Use Eq. (5.27).

5.10   For a sand, we have porosity $(n) = 0.31$ and $k = 0.066$ cm/sec. Determine k when $n = 0.4$. Use Eq. (5.27).

5.11   The maximum dry unit weight determined in the laboratory for a quartz sand is 16.0 kN/m^3. In the field, if the relative compaction is 90%, determine the hydraulic conductivity of the sand in the field compaction condition (given that k for the sand at the maximum dry unit weight condition is 0.03 cm/sec and $G_s = 2.7$). Use Eq. (5.27).

5.12   For a sandy soil, we have $e_{max} = 0.66$, $e_{min} = 0.36$, and k at a relative density of 90% = 0.008 cm/sec. Determine k at a relative density of 50%. Use Eq. (5.27).

5.13   A normally consolidated clay has the values given in the table:

Void ratio, e	k (cm/sec)
0.8	1.2×10^{-6}
1.4	3.6×10^{-6}

Estimate the hydraulic conductivity of the clay at a void ratio (e) of 0.62. Use Eq. (5.28).

5.14   A normally consolidated clay has the following values:

Void ratio, e	k (cm/sec)
1.2	0.2×10^{-6}
1.9	0.91×10^{-6}

Estimate the magnitude of k of the clay at a void ratio (e) of 0.9. Use Eq. (5.28).

5.15   The sieve analysis for a sand is given in the following table. Estimate the hydraulic conductivity of the sand at a void ratio of 0.5. Use Eq. (5.26) and $SF = 6.5$.

U.S. Sieve no.	Percent passing
30	100
40	80
60	68
100	28
200	0

5.16   A layered soil is shown in Figure 5.24. Estimate the equivalent hydraulic conductivity for flow in the vertical direction.

5.17   Refer to Figure 5.24. Estimate the equivalent hydraulic conductivity (cm/sec) for flow in the horizontal direction. Also calculate the ratio of $k_{V(eq)}/k_{H(eq)}$.

5.18   Refer to Figure 5.13 for field pumping from a well. For a steady state condition, given;
$q = 0.68$ m^3/min
$h_1 = 5.6$ m at $r_1 = 60$ m
$h_2 = 5$ m at $r_2 = 30$ m
Calculate the hydraulic conductivity (cm/sec) of the permeable layer.

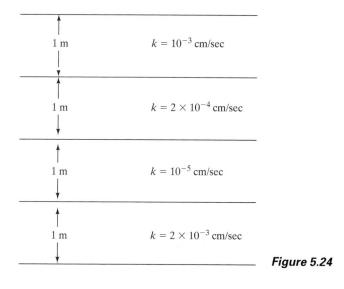

Figure 5.24

5.19 Refer to Figure 5.25 and use these values:

$H_1 = 7$ m $D = 3.5$ m
$H_2 = 1.75$ m $D_1 = 7$ m

Draw a flow net. Calculate the seepage loss per meter length of the sheet pile (at a right angle to the cross section shown).

5.20 Draw a flow net for a single row of sheet piles driven into a permeable layer as shown in Figure 5.25, given the following:

$H_1 = 5$ m $D = 4$ m
$H_2 = 0.7$ m $D_1 = 10$ m

Calculate the seepage loss per meter length of the sheet pile (at right angles to the cross section shown).

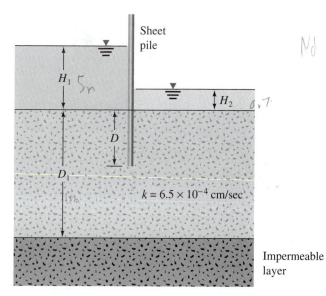

Figure 5.25

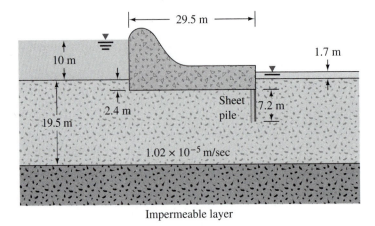

Figure 5.26

5.21 Draw a flow net for the weir shown in Figure 5.26. Calculate the rate of seepage under the weir.

References

CARMAN, P. C. (1938). "The Determination of the Specific Surface of Powders." *J. Soc. Chem. Ind. Trans.*, Vol. 57, 225.

CARMAN, P. C. (1956). *Flow of Gases through Porous Media*, Butterworths Scientific Publications, London.

CARRIER III, W. D. (2003). "Goodbye, Hazen; Hello, Kozeny-Carman," *Journal of Geotechnical and Geoenvironmental Engineering*, ASCE, Vol. 129, No. 11, 1054–1056.

DARCY, H. (1856). *Les Fontaines Publiques de la Ville de Dijon*, Dalmont, Paris.

HAZEN, A. (1930). "Water Supply," in *American Civil Engineers Handbook*, Wiley, New York.

KOZENY, J. (1927). "Ueber kapillare Leitung des Wassers in Boden." *Wien, Akad. Wiss.*, Vol. 136, No. 2a, 271.

SAMARASINGHE, A. M., HUANG, Y. H., and DRNEVICH, V. P. (1982). "Permeability and Consolidation of Normally Consolidated Soils," *Journal of the Geotechnical Engineering Division*, ASCE, Vol. 108, No. GT6, 835–850.

TAVENAS, F., JEAN, P., LEBLOND, F. T. P., and LEROUEIL, S. (1983). "The Permeability of Natural Soft Clays. Part II: Permeability Characteristics," *Canadian Geotechnical Journal*, Vol. 20, No. 4, 645–660.

U.S. DEPARTMENT OF NAVY (1971). "Design Manual — Soil Mechanics, Foundations, and Earth Structures," *NAVFAC DM-7*, U.S. Government Printing Office, Washington, D.C.

6

Stresses in a Soil Mass

As described in Chapter 3, soils are multiphase systems. In a given volume of soil, the solid particles are distributed randomly with void spaces in between. The void spaces are continuous and are occupied by water, air, or both. To analyze problems such as compressibility of soils, bearing capacity of foundations, stability of embankments, and lateral pressure on earth-retaining structures, engineers need to know the nature of the distribution of stress along a given cross section of the soil profile – that is, what fraction of the normal stress at a given depth in a soil mass is carried by water in the void spaces and what fraction is carried by the soil skeleton at the points of contact of the soil particles. This issue is referred to as the *effective stress concept*, and it is discussed in the first part of this chapter.

When a foundation is constructed, changes take place in the soil under the foundation. The net stress usually increases. This net stress increase in the soil depends on the load per unit area to which the foundation is subjected, the depth below the foundation at which the stress estimation is made, and other factors. It is necessary to estimate the net increase of vertical stress in soil that occurs as a result of the construction of a foundation so that settlement can be calculated. The second part of this chapter discusses the principles for estimating the *vertical stress increase* in soil caused by various types of loading, based on the theory of elasticity. Although natural soil deposits are not fully elastic, isotropic, or homogeneous materials, calculations for estimating increases in vertical stress yield fairly good results for practical work.

EFFECTIVE STRESS CONCEPT

6.1 Stresses in Saturated Soil without Seepage

Figure 6.1a shows a column of saturated soil mass with no seepage of water in any direction. The total stress at the elevation of point A, σ, can be obtained from the saturated unit weight of the soil and the unit weight of water above it. Thus,

$$\sigma = H\gamma_w + (H_A - H)\gamma_{sat} \tag{6.1}$$

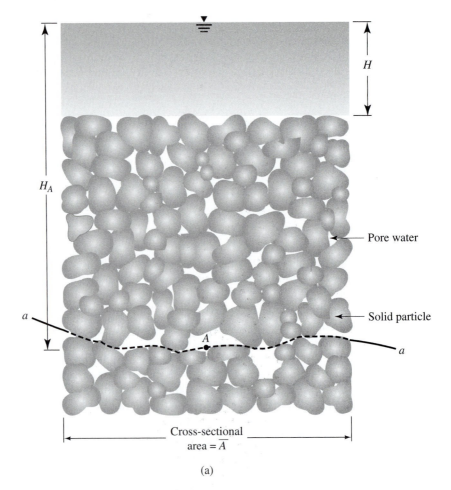

(a)

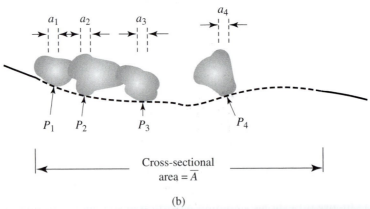

(b)

Figure 6.1 (a) Effective stress consideration for a saturated soil column without seepage; (b) forces acting at the points of contact of soil particles at the level of point A

where

γ_w = unit weight of water
γ_{sat} = saturated unit weight of the soil
H = height of water table from the top of the soil column
H_A = distance between point A and the water table

The total stress, σ, given by Eq. (6.1) can be divided into two parts:

1. A portion is carried by water in the continuous void spaces. This portion acts with equal intensity in all directions.

2. The rest of the total stress is carried by the soil solids at their points of contact. The sum of the vertical components of the forces developed at the points of contact of the solid particles per unit cross-sectional area of the soil mass is called the *effective stress*.

The concept of effective stress can be illustrated by drawing a wavy line, *a-a*, through the point A that passes through only the points of contacts of the solid particles. Let $P_1, P_2, P_3, \ldots, P_n$ be the forces that act at the points of contact of the soil particles (Figure 6.1b). The sum of the vertical components of all such forces over the unit cross-sectional area is equal to the effective stress, σ', or

$$\sigma' = \frac{P_{1(v)} + P_{2(v)} + P_{3(v)} + \cdots + P_{n(v)}}{\overline{A}} \tag{6.2}$$

where $P_{1(v)}, P_{2(v)}, P_{3(v)}, \ldots, P_{n(v)}$ are the vertical components of $P_1, P_2, P_3, \ldots, P_n$, respectively, and $\overline{A}$ is the cross-sectional area of the soil mass under consideration.

Again, if a_s is the cross-sectional area occupied by solid-to-solid contacts (that is, $a_s = a_1 + a_2 + a_3 + \cdots + a_n$), then the space occupied by water equals $(\overline{A} - a_s)$. So we can write

$$\sigma = \sigma' + \frac{u(\overline{A} - a_s)}{\overline{A}} = \sigma' + u(1 - a_s') \tag{6.3}$$

where

$u = H_A \gamma_w$ = pore water pressure (that is, the hydrostatic pressure at A)
$a_s' = a_s / \overline{A}$ = fraction of unit cross-sectional area of the soil mass occupied by solid-to-solid contacts

The value of a_s' is very small and can be neglected for the pressure ranges generally encountered in practical problems. Thus, Eq. (6.3) can be approximated by

$$\sigma = \sigma' + u \tag{6.4}$$

where u is also referred to as *neutral stress*. Substituting Eq. (6.1) for σ in Eq. (6.4) gives

$$\begin{aligned}
\sigma' &= [H\gamma_w + (H_A - H)\gamma_{sat}] - H_A\gamma_w \\
&= (H_A - H)(\gamma_{sat} - \gamma_w) \\
&= (\text{height of the soil column}) \times \gamma' \tag{6.5}
\end{aligned}$$

where $\gamma' = \gamma_{sat} - \gamma_w$ is the submerged unit weight of soil. Thus, it is clear that the effective stress at any point A is independent of the depth of water, H, above the submerged soil.

The principle of effective stress [Eq. (6.4)] was first developed by Terzaghi (1925, 1936). Skempton (1960) extended the work of Terzaghi and proposed the relationship between total and effective stress in the form of Eq. (6.3).

Example 6.1

A soil profile is shown in Figure 6.2. Calculate the total stress, pore water pressure, and effective stress at points A, B, C, and D.

Solution

For sand, $\gamma_d = \dfrac{G_s \gamma_w}{1 + e} = \dfrac{(2.68)(9.81)}{1 + 0.6} = 16.43 \text{ kN/m}^3$

For clay, $\gamma_{sat} = \dfrac{(G_s + e)\gamma_w}{1 + e} = \dfrac{(2.72 + 0.9)(9.81)}{1 + 0.9} = 18.69 \text{ kN/m}^3$

At A: Total stress: $\sigma_A = 0$
         Pore water pressure: $u_A = 0$
         Effective stress: $\sigma'_A = 0$

At B: $\sigma_B = 2\gamma_{dry(sand)} = 2 \times 16.43 = \mathbf{32.86 \text{ kN/m}^2}$

$u_B = \mathbf{0 \text{ kN/m}^2}$

$\sigma'_B = 32.86 - 0 = \mathbf{32.86 \text{ kN/m}^2}$

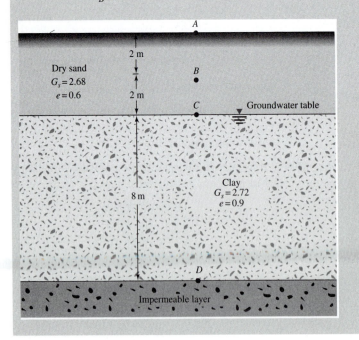

Figure 6.2

At C: $\sigma_C = 4\gamma_{\text{dry(sand)}} = 4 \times 16.43 = \mathbf{65.72\ kN/m^2}$

$u_C = \mathbf{0\ kN/m^2}$

$\sigma'_C = 65.72 - 0 = \mathbf{65.72\ kN/m^2}$

At D: $\sigma_D = 4\gamma_{\text{dry(sand)}} + 8\gamma_{\text{sat(clay)}}$

$= 4 \times 16.43 + 8 \times 18.69$

$= 65.72 + 149.52 = \mathbf{215.24\ kN/m^2}$

$u_D = 8\gamma_w = 8 \times 9.81 = \mathbf{78.48\ kN/m^2}$

$\sigma'_D = 215.24 - 78.48 = \mathbf{136.76\ kN/m^2}$ ∎

6.2 *Stresses in Saturated Soil with Seepage*

If water is seeping, the effective stress at any point in a soil mass will be different from the static case. It will increase or decrease, depending on the direction of seepage.

Upward Seepage

Figure 6.3a shows a layer of granular soil in a tank where upward seepage is caused by adding water through the valve at the bottom of the tank. The rate of water supply is kept constant. The loss of head caused by upward seepage between the levels of points A and B is h. Keeping in mind that the total stress at any point in the soil mass is determined solely by the weight of soil and the water above it, we find the effective stress calculations at points A and B:

At A
- Total stress: $\sigma_A = H_1\gamma_w$
- Pore water pressure: $u_A = H_1\gamma_w$
- Effective stress: $\sigma'_A = \sigma_A - u_A = 0$

At B
- Total stress: $\sigma_B = H_1\gamma_w + H_2\gamma_{\text{sat}}$
- Pore water pressure: $u_B = (H_1 + H_2 + h)\gamma_w$
- Effective stress: $\sigma'_B = \sigma_B - u_B$
  $= H_2(\gamma_{\text{sat}} - \gamma_w) - h\gamma_w$
  $= H_2\gamma' - h\gamma_w$

Similarly, we can calculate the effective stress at a point C located at a depth z below the top of the soil surface:

At C
- Total stress: $\sigma_C = H_1\gamma_w + z\gamma_{\text{sat}}$
- Pore water pressure: $u_C = \left(H_1 + z + \dfrac{h}{H_2}z \right)\gamma_w$

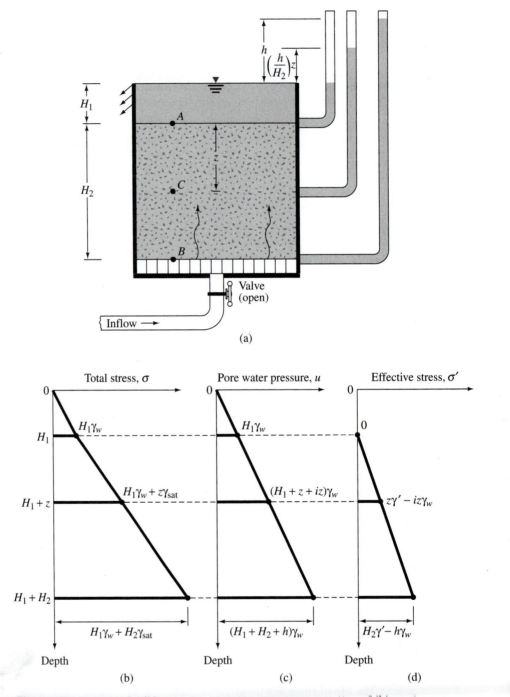

Figure 6.3 (a) Layer of soil in a tank with upward seepage; variation of (b) total stress; (c) pore water pressure; (d) effective stress with depth in a soil layer with upward seepage

- Effective stress: $\sigma'_C = \sigma_C - u_C$

$$= z(\gamma_{sat} - \gamma_w) - \frac{h}{H_2} z\gamma_w$$

$$= z\gamma' - \frac{h}{H_2} z\gamma_w$$

Note that h/H_2 is the hydraulic gradient i caused by the flow, and so

$$\sigma'_C = z\gamma' - iz\gamma_w \tag{6.6}$$

The variations of total stress, pore water pressure, and effective stress with depth are plotted in Figures 6.3b, c, and d, respectively. If the rate of seepage and thereby the hydraulic gradient are gradually increased, a limiting condition will be reached, at which point

$$\sigma'_C = z\gamma' - i_{cr}z\gamma_w = 0 \tag{6.7}$$

where i_{cr} = critical hydraulic gradient (for zero effective stress). In such a situation, the stability of the soil will be lost. This is generally referred to as *boiling*, or *quick condition*.

From Eq. (6.7), we have

$$i_{cr} = \frac{\gamma'}{\gamma_w} \tag{6.8}$$

For most soils, the value of i_{cr} varies from 0.9 to 1.1, with an average of 1.

Downward Seepage

The condition of downward seepage is shown in Figure 6.4a. The level of water in the soil tank is held constant by adjusting the supply from the top and the outflow at the bottom.

The hydraulic gradient caused by the downward seepage is $i = h/H_2$. The total stress, pore water pressure, and effective stress at any point C are, respectively,

$$\sigma_C = H_1\gamma_w + z\gamma_{sat}$$
$$u_C = (H_1 + z - iz)\gamma_w$$
$$\sigma'_C = (H_1\gamma_w + z\gamma_{sat}) - (H_1 + z - iz)\gamma_w$$
$$= z\gamma' + iz\gamma_w \tag{6.9}$$

The variations of total stress, pore water pressure, and effective stress with depth are also shown graphically in Figures 6.4b, c, and d.

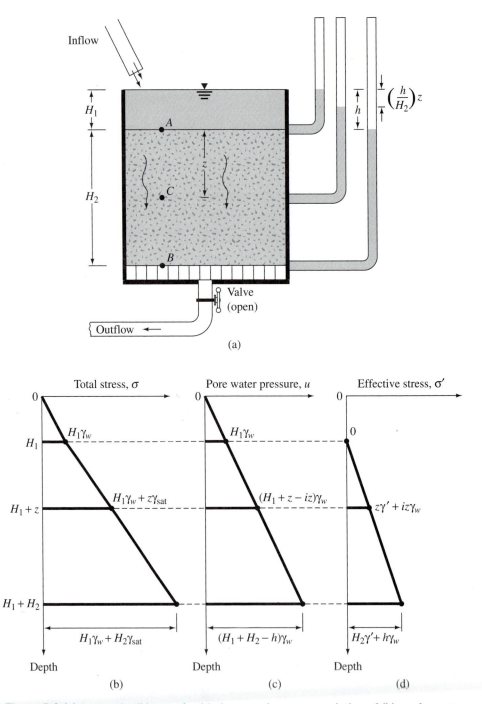

Figure 6.4 (a) Layer of soil in a tank with downward seepage; variation of (b) total stress; (c) pore water pressure; (d) effective stress with depth in a soil layer with downward seepage

Example 6.2

An exploratory drill hole was made in a saturated stiff clay (Figure 6.5). It was observed that the sand layer underlying the clay was under artesian pressure. Water in the drill hole rose to a height of H_1 above the top of the sand layer. If an open excavation is to be made in the clay, how deep can the excavation proceed before the bottom heaves? We are given $H = 8$ m, $H_1 = 4$ m, and $w = 32\%$.

Solution

Consider a point at the sand–clay interface. For heaving, $\sigma' = 0$, so

$$(H - H_{exc})\gamma_{sat(clay)} - H_1\gamma_w = 0$$

$$\gamma_{sat(clay)} = \frac{G_s\gamma_w + wG_s\gamma_w}{1 + e} = \frac{[2.70 + (0.32)(2.70)](9.81)}{1 + (0.32)(2.70)}$$

$$= 18.76 \text{ kN/m}^3$$

Thus,

$$(8 - H_{exc})(18.76) - (3)(9.81) = 0$$

$$H_{exc} = 8 - \frac{(3)(9.81)}{18.76} = \textbf{6.43 m}$$

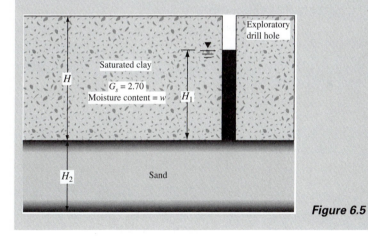

Figure 6.5

Example 6.3

A 10-m-thick layer of stiff saturated clay is underlain by a layer of sand (Figure 6.6). The sand is under artesian pressure. If $H = 7.2$ m, what would be the minimum height of water h in the cut so that the stability of the saturated clay is not lost?

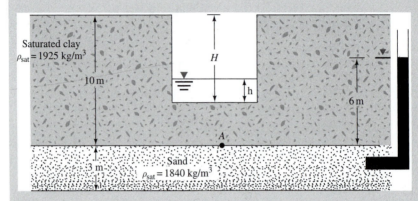

Figure 6.6 Layer of saturated clay underlain by layer of sand

Solution

Given: $\rho_{sat(clay)} = 1925$ kg/m^3. Thus

$$\gamma_{sat(clay)} = \frac{1925 \times 9.81}{1000} = 18.88 \text{ kN/m}^3$$

At point A

$$\sigma_A = (10 - 7.2)\gamma_{sat(clay)} + h\gamma_w = (2.8)(18.88) + h\gamma_w = 52.86 + h\gamma_w \text{ (kN/m}^2\text{)}$$

$$u_A = (6)(\gamma_w) = (6)(9.81) = 58.86 \text{ kN/m}^2$$

For heave

$$\sigma_A - u_A = 0$$

or

$$\sigma_A = u_A$$

$$52.86 + h\gamma_w = 58.86$$

$$h = \frac{58.86 - 52.86}{9.81} = \textbf{0.61 m}$$

6.3   Effective Stress in Partially Saturated Soil

In partially saturated soil, water in the void spaces is not continuous, and it is a three-phase system—that is, solid, pore water, and pore air (Figure 6.7). Hence, the total stress at any point in a soil profile is made up of intergranular, pore air, and pore

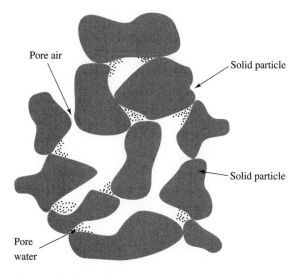

Figure 6.7 Partially saturated soil

water pressures. From laboratory test results, Bishop et al. (1960) gave the following equation for effective stress, σ', in partially saturated soils:

$$\sigma' = \sigma - u_a + \chi(u_a - u_w) \qquad (6.10)$$

where
σ = total stress
u_a = pore air pressure
u_w = pore water pressure

In Eq. (6.10), χ represents the fraction of a unit cross-sectional area of the soil occupied by water. For dry soil $\chi = 0$, and for saturated soil $\chi = 1$.

Bishop, *et al.*, pointed out that the intermediate values of χ depend primarily on the degree of saturation, S. However, these values are also influenced by factors such as soil structure.

6.4   *Seepage Force*

Section 6.2 showed that the effect of seepage is to increase or decrease the effective stress at a point in a layer of soil. It is often convenient to express the seepage force per unit volume of soil.

In Figure 6.1 it was shown that, with no seepage, the effective stress at a depth z measured from the surface of the soil layer is equal to $z\gamma'$. Thus the effective force on an area A is

$$P_1' = z\gamma'A$$

(the direction of the force P_1' is shown in Figure 6.8a.)

Again, if there is an upward seepage of water in the vertical direction through the same soil layer (Figure 6.3), the effective force on an area A at a depth z can be given by

$$P_2' = (z\gamma' - iz\gamma_w)A$$

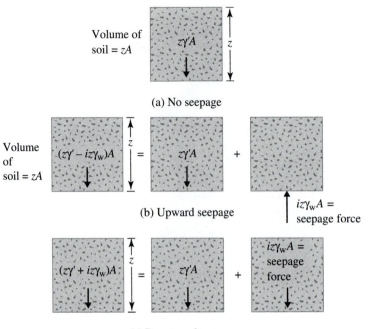

Volume of
soil = zA

$z\gamma'A$

z

(a) No seepage

Volume
of
soil = zA

$(z\gamma' - iz\gamma_w)A$

z

$=$

$z\gamma'A$

$+$

$iz\gamma_w A =$
seepage force

(b) Upward seepage

$(z\gamma' + iz\gamma_w)A$

z

$=$

$z\gamma'A$

$+$

$iz\gamma_w A =$
seepage
force

(c) Downward seepage

Figure 6.8 Force due to (a) no seepage; (b) upward seepage; (c) downward seepage
on a volume of soil

Hence, the decrease of total force because of seepage is

$$P_1' - P_2' = iz\gamma_w A \qquad (6.11)$$

The volume of the soil contributing to the effective force equals zA. So, the
seepage force per unit volume of soil is

$$\frac{P_1' - P_2'}{(\text{volume of soil})} = \frac{iz\gamma_w A}{zA} = i\gamma_w \qquad (6.12)$$

The force per unit volume, $i\gamma_w$, for this case acts in the upward direction — that
is, in the direction of flow. This is demonstrated in Figure 6.8b. Similarly, for downward seepage, it can be shown that the seepage force in the downward direction per
unit volume of soil is $i\gamma_w$ (Figure 6.8c).

From the preceding discussions, we can conclude that the seepage force per unit
volume of soil is equal to $i\gamma_w$, and in isotropic soils the force acts in the same direction
as the direction of flow. This statement is true for flow in any direction. Flow nets can
be used to find the hydraulic gradient at any point and, thus, the seepage force per
unit volume of soil.

This concept of seepage force can be effectively used to obtain the factor of safety against heave on the downstream side of a hydraulic structure. This is discussed in the following section.

6.5 *Heaving in Soil Due to Flow Around Sheet Piles*

Seepage force per unit volume of soil can be calculated for checking possible failure of sheet-pile structures where underground seepage may cause heaving of soil on the downstream side (Figure 6.9a). After conducting several model tests, Terzaghi (1922) concluded that heaving generally occurs within a distance of $D/2$ from the sheet piles (when D equals depth of embedment of sheet piles into the permeable layer). Therefore, we need to investigate the stability of soil in a zone measuring D by $D/2$ in cross-section as shown in Figure 6.9a.

The factor of safety against heaving can be given by (Figure 6.8b)

$$FS = \frac{W'}{U} \tag{6.13}$$

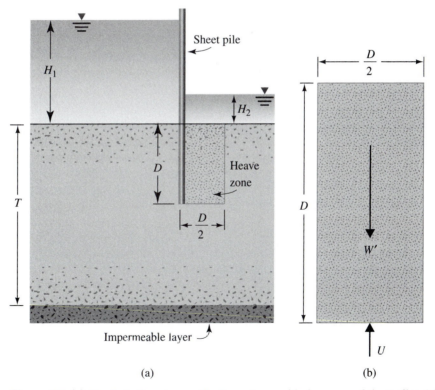

(a)   (b)

Figure 6.9 (a) Check for heaving on the downstream side for a row of sheet piles driven into a permeable layer; (b) enlargement of heave zone

where

FS = factor of safety

W' = effective weight of soil in the heave zone per unit length of sheet pile = $D(D/2)(\gamma_{sat} - \gamma_w) = (\frac{1}{2})D^2\gamma'$

U = uplifting force due to seepage on the same volume of soil

From Eq. (6.12)

$$U = (\text{soil volume}) \times (i_{av}\gamma_w) = \frac{1}{2}D^2 i_{av}\gamma_w$$

where i_{av} = average hydraulic gradient in the block of soil

Substituting the values of W' and U in Eq. (6.13), we can write

$$FS = \frac{\gamma'}{i_{av}\gamma_w} \tag{6.14}$$

For the case of *flow around a sheet pile* in a *homogeneous soil*, as shown in Figure 6.10, it can be demonstrated that

$$\frac{U}{0.5\gamma_w D(H_1 - H_2)} = C_o \tag{6.15}$$

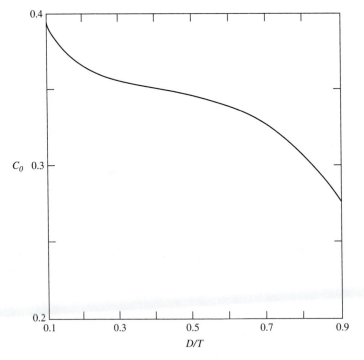

Figure 6.10 Variation of C_0 with D/T

where C_o is a function of D/T (see Figure 6.10). Hence, from Eq. (8.12),

$$FS = \frac{W'}{U} = \frac{0.5D^2\gamma'}{0.5C_o\gamma_w D(H_1 - H_2)} = \frac{D\gamma'}{C_o\gamma_w(H_1 - H_2)} \tag{6.16}$$

Example 6.4

Refer to Figure 6.9. Given $D = 3$ m, $T = 6$ m, $H_1 = 5$ m, $H_2 = 1$ m. For the permeable layer, $G_s = 2.68$ and $e = 0.7$. Calculate the factor of safety against downstream heave.

Solution
Eq. (6.16):

$$FS = \frac{D\gamma'}{C_o\gamma_w(H_1 - H_2)}$$

$$\gamma' = \frac{(G_s - 1)\gamma_w}{1 + e} = \frac{(2.68 - 1)(9.81)}{1 + 0.7} = 9.69 \text{ kN/m}^3$$

From Figure 6.10, for $D/T = 3/6 = 0.5$, the value of $C_o \approx 0.347$.

$$FS = \frac{(3)(9.69)}{(0.347)(9.81)(5 - 1)} = \textbf{2.13}$$

■

VERTICAL STRESS INCREASE DUE TO VARIOUS TYPES OF LOADING

6.6

Stress Caused by a Point Load

Boussinesq (1883) solved the problem of stresses produced at any point in a homogeneous, elastic, and isotropic medium as the result of a point load applied on the surface of an infinitely large half-space. According to Figure 6.11, Boussinesq's solution for normal stresses at a point A caused by the point load P is

$$\Delta\sigma_x = \frac{P}{2\pi}\left\{\frac{3x^2 z}{L^5} - (1 - 2\mu_s)\left[\frac{x^2 - y^2}{Lr^2(L + z)} + \frac{y^2 z}{L^3 r^2}\right]\right\} \tag{6.17}$$

$$\Delta\sigma_y = \frac{P}{2\pi}\left\{\frac{3y^2 z}{L^5} - (1 - 2\mu_s)\left[\frac{y^2 - x^2}{Lr^2(L + z)} + \frac{x^2 z}{L^3 r^2}\right]\right\} \tag{6.18}$$

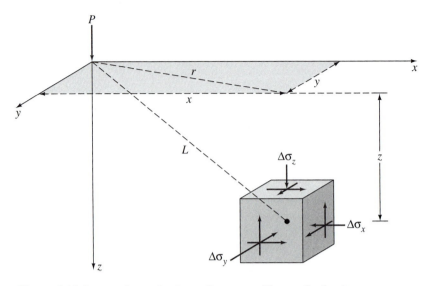

Figure 6.11 Stresses in an elastic medium caused by a point load

and

$$\Delta\sigma_z = \frac{3P}{2\pi}\frac{z^3}{L^5} = \frac{3P}{2\pi}\frac{z^3}{(r^2 + z^2)^{5/2}} \tag{6.19}$$

where

$r = \sqrt{x^2 + y^2}$
$L = \sqrt{x^2 + y^2 + z^2} = \sqrt{r^2 + z^2}$
μ_S = Poisson's ratio

Note that Eqs. (6.17) and (6.18), which are the expressions for horizontal normal stresses, are dependent on Poisson's ratio of the medium. However, the relationship for the vertical normal stress, $\Delta\sigma_z$, as given by Eq. (6.19), is independent of Poisson's ratio. The relationship for $\Delta\sigma_z$ can be rewritten in the following form:

$$\Delta\sigma_z = \frac{P}{z^2}\left\{\frac{3}{2\pi}\frac{1}{[(r/z)^2 + 1]^{5/2}}\right\} = \frac{P}{z^2}I_1 \tag{6.20}$$

where $I_1 = \dfrac{3}{2\pi}\dfrac{1}{[(r/z)^2 + 1]^{5/2}}$. $\tag{6.21}$

The variation of I_1 for various values of r/z is given in Table 6.1.

Typical values of Poisson's ratio for various soils are listed in Table 6.2.

Table 6.1 Variation of I_1 [Eq. (6.20)]

r/z	I_1	r/z	I_1
0	0.4775	0.9	0.1083
0.1	0.4657	1.0	0.0844
0.2	0.4329	1.5	0.0251
0.3	0.3849	1.75	0.0144
0.4	0.3295	2.0	0.0085
0.5	0.2733	2.5	0.0034
0.6	0.2214	3.0	0.0015
0.7	0.1762	4.0	0.0004
0.8	0.1386	5.0	0.00014

Table 6.2 Representative values of Poisson's ratio

Type of soil	Poisson's ratio, μ_S
Loose sand	0.2–0.4
Medium sand	0.25–0.4
Dense sand	0.3–0.45
Silty sand	0.2–0.4
Soft clay	0.15–0.25
Medium clay	0.2–0.5

6.7 *Westergaard's Solution for Vertical Stress Due to a Point Load*

Westergaard (1938) has proposed a solution for the determination of the vertical stress due to a point load P in an elastic solid medium in which there exist alternating layers with thin rigid reinforcements (Figure 6.12a). This type of assumption may be an idealization of a clay layer with thin seams of sand. For such an assumption the vertical stress increase at a point A (Figure 6.12) can be given as

$$\Delta\sigma_z = \frac{P\eta}{2\pi z^2}\left[\frac{1}{\eta^2 + (r/z)^2}\right]^{3/2} \tag{6.22}$$

where

$$\eta = \sqrt{\frac{1 - 2\mu_S}{2 - 2\mu_S}} \tag{6.23}$$

μ_S = Poisson's ratio of the solid between the rigid reinforcements

$r = \sqrt{x^2 + y^2}$

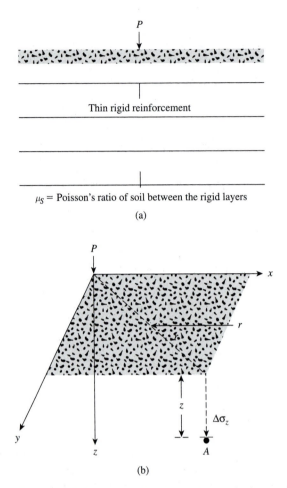

Figure 6.12 Westergaard's solution for vertical stress due to a point load

Equation (6.22) can be rewritten as

$$\Delta\sigma_z = \left(\frac{P}{z^2}\right)I_2 \qquad (6.24)$$

where

$$I_2 = \frac{1}{2\pi\eta^2}\left[\left(\frac{r}{\eta z}\right)^2 + 1\right]^{-3/2} \qquad (6.25)$$

Table 6.3 gives the variation of I_2 with μ_s.

In most practical problems of geotechnical engineering, Boussinesq's solution (Section 6.6) is preferred over Westergaard's solution. For that reason, *further development of stress calculation under various types of loading will use Boussinesq's solution in this chapter.*

Table 6.3 Variation of I_2 [Eq. (6.25)].

r/z	I_2		
	$\mu_s = 0$	$\mu_s = 0.2$	$\mu_s = 0.4$
0	0.3183	0.4244	0.9550
0.1	0.3090	0.4080	0.8750
0.2	0.2836	0.3646	0.6916
0.3	0.2483	0.3074	0.4997
0.4	0.2099	0.2491	0.3480
0.5	0.1733	0.1973	0.2416
0.6	0.1411	0.1547	0.1700
0.7	0.1143	0.1212	0.1221
0.8	0.0925	0.0953	0.0897
0.9	0.0751	0.0756	0.0673
1.0	0.0613	0.0605	0.0516
1.5	0.0247	0.0229	0.0173
2.0	0.0118	0.0107	0.0076
2.5	0.0064	0.0057	0.0040
3.0	0.0038	0.0034	0.0023
4.0	0.0017	0.0015	0.0010
5.0	0.0009	0.0008	0.0005

6.8   *Vertical Stress Caused by a Line Load*

Figure 6.13 shows a flexible line load of infinite length that has an intensity q per unit length on the surface of a semiinfinite soil mass. The vertical stress increase, $\Delta\sigma$, inside the soil mass can be determined by using the principles of the theory of elasticity, or

$$\Delta\sigma = \frac{2qz^3}{\pi(x^2 + z^2)^2} \tag{6.26}$$

The preceding equation can be rewritten as

$$\Delta\sigma = \frac{2q}{\pi z[(x/z)^2 + 1]^2}$$

or

$$\frac{\Delta\sigma}{(q/z)} = \frac{2}{\pi\left[\left(\dfrac{x}{z}\right)^2 + 1\right]^2} \tag{6.27}$$

Note that Eq. (6.27) is in a nondimensional form. Using this equation, we can calculate the variation of $\Delta\sigma/(q/z)$ with x/z. The variation is given in Table 6.4. The value of $\Delta\sigma$

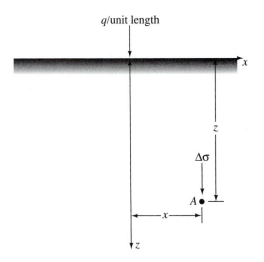

Figure 6.13 Line load over the surface of a semiinfinite soil mass

Table 6.4 Variation of $\Delta\sigma/(q/z)$ with x/z [Eq. (6.27)]

x/z	$\dfrac{\Delta\sigma}{q/z}$	x/z	$\dfrac{\Delta\sigma}{q/z}$
0	0.637	0.7	0.287
0.1	0.624	0.8	0.237
0.2	0.589	0.9	0.194
0.3	0.536	1.0	0.159
0.4	0.473	1.5	0.060
0.5	0.407	2.0	0.025
0.6	0.344	3.0	0.006

calculated by using Eq. (6.27) is the additional stress on soil caused by the line load. The value of $\Delta\sigma$ does not include the overburden pressure of the soil above the point A.

6.9   ## Vertical Stress Caused by a Line Load of Finite Length

Figure 6.14 shows a line load having a length L located on the surface of a semiinfinite soil mass. The intensity of the load per unit length is q. The vertical stress increase ($\Delta\sigma$) at a point $A(0, 0, z)$ can be obtained by integration of Boussinesq's solution [Eq. (6.19)] as

$$\Delta\sigma = \frac{q}{z}I_3 \tag{6.28}$$

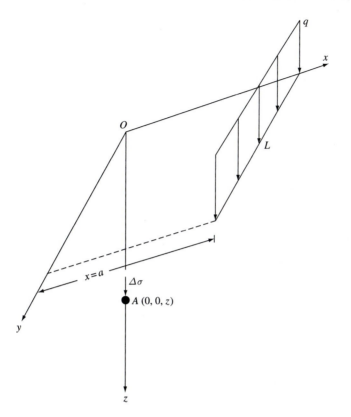

Figure 6.14 Line load of length L on the surface of a semiinfinite soil mass

where

$$I_3 = \frac{1}{2\pi(m^2 + 1)^2}\left[\frac{3n}{\sqrt{m^2 + n^2 + 1}} - \left(\frac{n}{\sqrt{(m^2 + n^2 + 1)}}\right)^3\right] \qquad (6.29)$$

$$m = \frac{a}{z} \qquad (6.30)$$

$$n = \frac{L}{z} \qquad (6.31)$$

Figure 6.15 shows the plot of I_3 for various values of m and n.

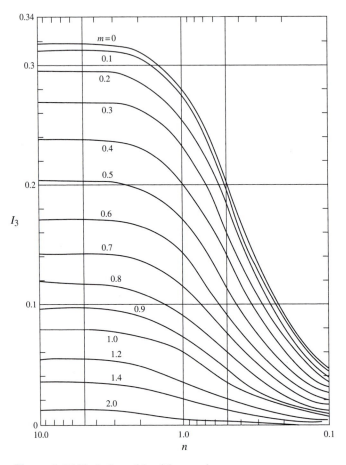

Figure 6.15 Variation of I_3 with m and n

Example 6.5

Refer to Figure 6.14. Given $a = 3$ m, $L = 4.8$ m, $q = 50$ kN/m. Determine the increase in stress, $\Delta\sigma$, due to the line load at

    a. Point with coordinates (0, 0, 6 m)
    b. Point with coordinates (0, 2.4 m, 6 m).

$$n' = \frac{b}{z} \qquad m' = \frac{B}{z}$$

Solution
Part a.

$$m = \frac{a}{z} = \frac{3}{6} = 0.5$$

$$n = \frac{L}{z} = \frac{4.8}{6} = 0.8$$

From Figure 6.15, for $m = 0.5$ and $n = 0.8$, the value of I_3 is about 0.158. So

$$\Delta\sigma = \frac{q}{z}(I_3) = \frac{50}{6}(0.158) = \textbf{1.32 kN/m}^2$$

Part b.
As shown in Figure 6.16. the method of superposition can be used. Referring to Figure 6.16

$$\Delta\sigma = \Delta\sigma_1 + \Delta\sigma_2$$

For obtaining $\Delta\sigma_1$ (Figure 6.16a)

$$m_1 = \frac{3}{6} = 0.5$$

$$n_1 = \frac{L_1}{z} = \frac{2.4}{6} = 0.4$$

From Figure 6.15, $I_{3(1)} \approx 0.1$. Similarly, for $\Delta\sigma_2$ (Figure 6.16b)

$$m_2 = 0.5$$

$$n_2 = \frac{L_2}{z} = \frac{2.4}{6} = 0.4$$

So, $I_{3(2)} \approx 0.1$. Hence

$$\Delta\sigma = \frac{q}{z}[I_{3(1)} + I_{3(2)}] = \frac{50}{6}(0.1 + 0.1) = \textbf{1.67 kN/m}^2$$ ∎

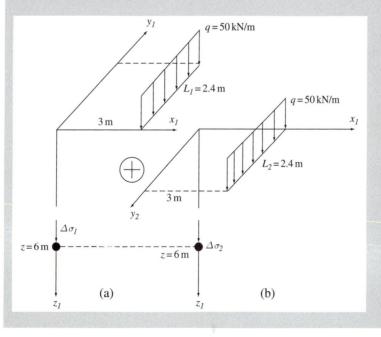

Figure 6.16

6.10   ## Vertical Stress Caused by a Strip Load (Finite Width and Infinite Length)

The fundamental equation for the vertical stress increase at a point in a soil mass as the result of a line load (see Section 6.8) can be used to determine the vertical stress at a point caused by a flexible strip load of width B (Figure 6.17). Let the load per unit area of the strip shown in Figure 6.17 be equal to q. If we consider an elemental strip of width dr, the load per unit length of this strip will be equal to $q\,dr$. This elemental strip can be treated as a line load. Equation (6.26) gives the vertical stress increase, $d\sigma$, at point A inside the soil mass caused by this elemental strip load. To calculate the vertical stress increase, we need to substitute $q\,dr$ for q and $(x - r)$ for x. So

$$d\sigma = \frac{2(q\,dr)z^3}{\pi[(x - r)^2 + z^2]^2} \tag{6.32}$$

The total increase in the vertical stress ($\Delta\sigma$) at point A caused by the entire strip load of width B can be determined by the integration of Eq. (6.32) with limits of r from $-B/2$ to $+B/2$, or

$$\Delta\sigma = \int d\sigma = \int_{-B/2}^{+B/2} \left(\frac{2q}{\pi}\right) \left\{ \frac{z^3}{[(x - r)^2 + z^2]^2} \right\} dr$$

$$= \frac{q}{\pi} \left\{ \tan^{-1}\left[\frac{z}{x - (B/2)}\right] - \tan^{-1}\left[\frac{z}{x + (B/2)}\right] \right.$$

$$\left. - \frac{Bz[x^2 - z^2 - (B^2/4)]}{[x^2 + z^2 - (B^2/4)]^2 + B^2 z^2} \right\} \tag{6.33}$$

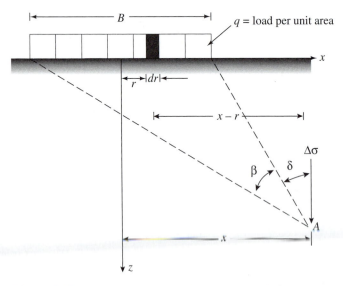

Figure 6.17 Vertical stress caused by a flexible strip load
(*Note:* Angles measured in counter-clockwise direction are taken as positive.)

Table 6.5 Variation of $\Delta\sigma/q$ with $2z/B$ and $2x/B$

2z/B	2x/B				
	0	0.5	1.0	1.5	2.0
0	1.000	1.000	0.500	—	—
0.5	0.959	0.903	0.497	0.089	0.019
1.0	0.818	0.735	0.480	0.249	0.078
1.5	0.668	0.607	0.448	0.270	0.146
2.0	0.550	0.510	0.409	0.288	0.185
2.5	0.462	0.437	0.370	0.285	0.205
3.0	0.396	0.379	0.334	0.273	0.211
3.5	0.345	0.334	0.302	0.258	0.216
4.0	0.306	0.298	0.275	0.242	0.205
4.5	0.274	0.268	0.251	0.226	0.197
5.0	0.248	0.244	0.231	0.212	0.188

Equation (6.33) can be simplified to the form

$$\Delta\sigma = \frac{q}{\pi}[\beta + \sin\beta\cos(\beta + 2\delta)] \tag{6.34}$$

The angles β and δ are defined in Figure 6.17.

Table 6.5 shows the variation of $\Delta\sigma/q$ with $2z/B$ for $2x/B$ equal to 0, 0.5, 1.0, 1.5, and 2.0. This table can be conveniently used to calculate the vertical stress at a point caused by a flexible strip load.

Example 6.6

With reference to Figure 6.17, we are given $q = 200$ kN/m², $B = 6$ m, and $z = 3$ m. Determine the vertical stress increase at $x = \pm 9$ m, ± 6 m, ± 3 m, and 0 m. Plot a graph of $\Delta\sigma$ against x.

Solution
We create the following table:

x (m)	2x/B	2z/B	$\Delta\sigma/q$*	$\Delta\sigma^\dagger$ kN/m²
±9	±3	1	0.0171	3.42
±6	±2	1	0.078	15.6
±3	±1	1	0.480	96.0
0	0	1	0.8183	163.66

* From Table 6.5
$^\dagger q = 200$ kN/m²

The plot of $\Delta\sigma$ versus x is given in Figure 6.18.

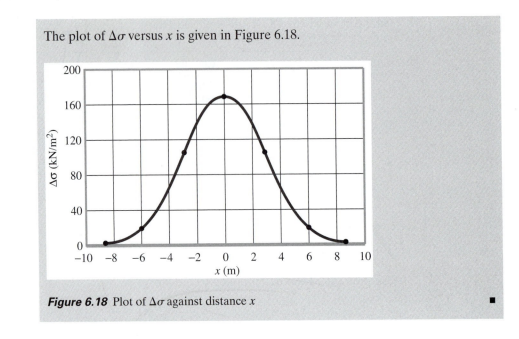

Figure 6.18 Plot of $\Delta\sigma$ against distance x ∎

6.11   *Vertical Stress Below a Uniformly Loaded Circular Area*

Using Boussinesq's solution for vertical stress $\Delta\sigma$ caused by a point load [Eq. (6.19)], we can also develop an expression for the vertical stress below the center of a uniformly loaded flexible circular area.

From Figure 6.19, let the intensity of pressure on the circular area of radius R be equal to q. The total load on the elemental area (shaded in the figure) $= qr\,dr\,d\alpha$. The vertical stress, $d\sigma$, at point A caused by the load on the elemental area (which may be assumed to be a concentrated load) can be obtained from Eq. (6.19):

$$d\sigma = \frac{3(qr\,dr\,d\alpha)}{2\pi}\frac{z^3}{(r^2 + z^2)^{5/2}} \tag{6.35}$$

The increase in the stress at point A caused by the entire loaded area can be found by integrating Eq. (6.35), or

$$\Delta\sigma = \int d\sigma = \int_{\alpha=0}^{\alpha=2\pi}\int_{r=0}^{r=R}\frac{3q}{2\pi}\frac{z^3 r}{(r^2 + z^2)^{5/2}}\,dr\,d\alpha$$

So

$$\Delta\sigma = q\left\{1 - \frac{1}{[(R/z)^2 + 1]^{3/2}}\right\} \tag{6.36}$$

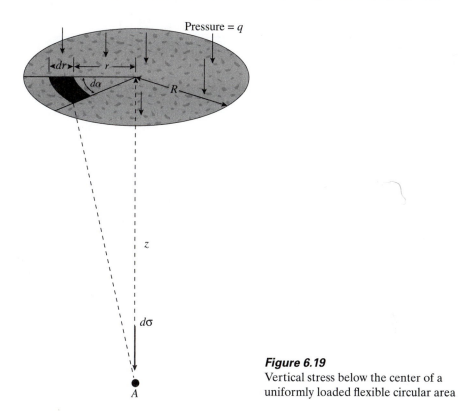

Figure 6.19
Vertical stress below the center of a uniformly loaded flexible circular area

The variation of $\Delta\sigma/q$ with z/R obtained from Eq. (6.36) is given in Table 6.6. Note that the value of $\Delta\sigma$ decreases rapidly with depth, and, at $z = 5R$, it is about 6% of q, which is the intensity of pressure at the ground surface.

Equation (6.36) is valid for determination of vertical stress increase ($\Delta\sigma$) at any depth z below the center of the flexible loaded circular area. Similarly, the stress increase at any depth z located at a radial distance r measured horizontally from the center of the loaded area can be given as

$$\Delta\sigma = f\left(q, \frac{r}{R}, \frac{z}{R}\right)$$

Table 6.6 Variation of $\Delta\sigma/q$ with z/R [Eq. (6.36)]

z/R	$\Delta\sigma/q$	z/R	$\Delta\sigma/q$
0	1	1.0	0.6465
0.02	0.9999	1.5	0.4240
0.05	0.9998	2.0	0.2845
0.10	0.9990	2.5	0.1996
0.2	0.9925	3.0	0.1436
0.4	0.9488	4.0	0.0869
0.5	0.9106	5.0	0.0571
0.8	0.7562		

Table 6.7 Variation of I_4 [Eq. (6.37)]

z/R	r/R					
	0	0.2	0.4	0.6	0.8	1.0
0	1.000	1.000	1.000	1.000	1.000	1.000
0.1	0.999	0.999	0.998	0.996	0.976	0.484
0.2	0.992	0.991	0.987	0.970	0.890	0.468
0.3	0.976	0.973	0.963	0.922	0.793	0.451
0.4	0.949	0.943	0.920	0.860	0.712	0.435
0.5	0.911	0.902	0.869	0.796	0.646	0.417
0.6	0.864	0.852	0.814	0.732	0.591	0.400
0.7	0.811	0.798	0.756	0.674	0.545	0.367
0.8	0.756	0.743	0.699	0.619	0.504	0.366
0.9	0.701	0.688	0.644	0.570	0.467	0.348
1.0	0.646	0.633	0.591	0.525	0.434	0.332
1.2	0.546	0.535	0.501	0.447	0.377	0.300
1.5	0.424	0.416	0.392	0.355	0.308	0.256
2.0	0.286	0.286	0.268	0.248	0.224	0.196
2.5	0.200	0.197	0.191	0.180	0.167	0.151
3.0	0.146	0.145	0.141	0.135	0.127	0.118
4.0	0.087	0.086	0.085	0.082	0.080	0.075

or

$$\frac{\Delta\sigma}{q} = I_4 \qquad (6.37)$$

The variation of I_4 with r/R and z/R is given in Table 6.7.

6.12 Vertical Stress Caused by a Rectangularly Loaded Area

Boussinesq's solution can also be used to calculate the vertical stress increase below a flexible rectangular loaded area, as shown in Figure 6.20. The loaded area is located at the ground surface and has length L and width B. The uniformly distributed load per unit area is equal to q. To determine the increase in the vertical stress $\Delta\sigma$ at point A located at depth z below the corner of the rectangular area, we need to consider a small elemental area $dx\,dy$ of the rectangle (Figure 6.20). The load on this elemental area can be given by

$$dq = q\,dx\,dy \qquad (6.38)$$

The increase in the stress $d\sigma$ at point A caused by the load dq can be determined by using Eq. (6.19). However, we need to replace P with $dq = q\,dx\,dy$ and r' with $x^2 + y^2$. Thus,

$$d\sigma = \frac{3q\,dx\,dy\,z^3}{2\pi(x^2 + y^2 + z^2)^{5/2}} \qquad (6.39)$$

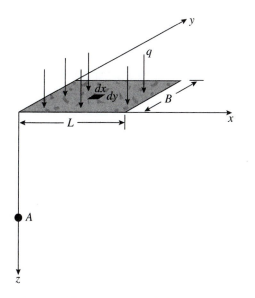

Figure 6.20 Vertical stress below the corner of a uniformly loaded flexible rectangular area

The increase in the stress $\Delta\sigma$ at point A caused by the entire loaded area can now be determined by integrating the preceding equation:

$$\Delta\sigma = \int d\sigma = \int_{y=0}^{B} \int_{x=0}^{L} \frac{3qz^3(dx\,dy)}{2\pi(x^2 + y^2 + z^2)^{5/2}} = qI_5 \tag{6.40}$$

where

$$I_5 = \frac{1}{4\pi}\left[\frac{2m'n'\sqrt{m^{2\prime} + n^{2\prime} + 1}}{m^{2\prime} + n^{2\prime} + m^{2\prime}n^{2\prime} + 1}\left(\frac{m^{2\prime} + n^{2\prime} + 2}{m^{2\prime} + n^{2\prime} + 1}\right)\right.$$

$$\left. + \tan^{-1}\left(\frac{2m'n'\sqrt{m^{2\prime} + n^{2\prime} + 1}}{m^{2\prime} + n^{2\prime} - m^{2\prime}n^{2\prime} + 1}\right)\right] \tag{6.41}$$

$$m' = \frac{B}{z} \tag{6.42}$$

$$n' = \frac{L}{z} \tag{6.43}$$

The variation of I_5 with m' and n' is shown in Figure 6.21.

The increase in the stress at any point below a rectangularly loaded area can be found by using Eq. (6.40) and Figure 6.21. This concept can further be explained by referring to Figure 6.22. Let us determine the stress at a point below point A' at depth z. The loaded area can be divided into four rectangles as shown. The point A' is the corner common to all four rectangles. The increase in the stress at depth z below point A'

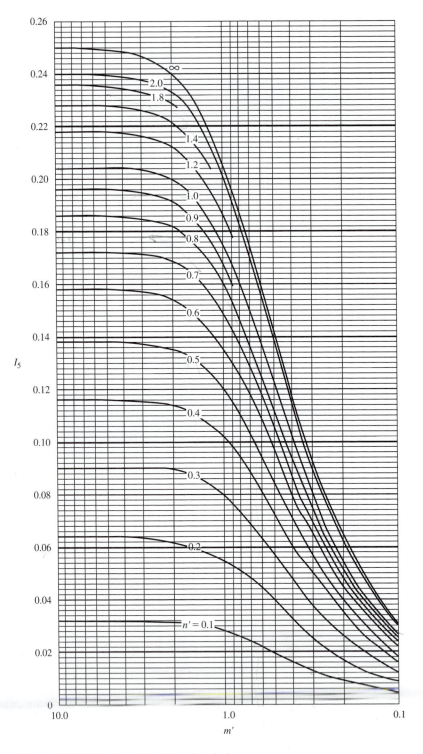

Figure 6.21 Variation of I_5 with m' and n'

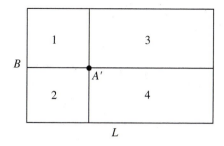

Figure 6.22 Increase of stress at any point below a rectangularly loaded flexible area

due to each rectangular area can now be calculated by using Eq. (6.40). The total stress increase caused by the entire loaded area can be given by

$$\Delta\sigma = q[I_{5(1)} + I_{5(2)} + I_{5(3)} + I_{5(4)}] \tag{6.44}$$

In many circumstances it may be necessary to calculate the stress increase below the center of a uniformly loaded rectangular. For convenience the stress increase may be expressed as

$$\Delta\sigma_c = qI_c \tag{6.45}$$

where

$$I_c = f(m_1, n_1) \tag{6.46}$$

$$m_1 = \frac{L}{B} \tag{6.47}$$

and

$$n_1 = \frac{z}{\dfrac{B}{2}} \tag{6.48}$$

Table 6.8 gives the variation of I_c with m_1 and n_1.

Table 6.8 Variation of I_c with m_1 and n_1 [Eq. (6.45)]

	m_1									
n_1	1	2	3	4	5	6	7	8	9	10
0.20	0.994	0.997	0.997	0.997	0.997	0.997	0.997	0.997	0.997	0.997
0.40	0.960	0.976	0.977	0.977	0.977	0.977	0.977	0.977	0.977	0.977
0.60	0.892	0.932	0.936	0.936	0.937	0.937	0.937	0.937	0.937	0.937
0.80	0.800	0.870	0.878	0.880	0.881	0.881	0.881	0.881	0.881	0.881
1.00	0.701	0.800	0.814	0.817	0.818	0.818	0.818	0.818	0.818	0.818
1.20	0.606	0.727	0.748	0.753	0.754	0.755	0.755	0.755	0.755	0.755
1.40	0.522	0.658	0.685	0.692	0.694	0.695	0.695	0.696	0.696	0.696
1.60	0.449	0.593	0.627	0.636	0.639	0.640	0.641	0.641	0.641	0.642

(*continued*)

Table 6.8 *(continued)*

	m_1									
n_1	1	2	3	4	5	6	7	8	9	10
1.80	0.388	0.534	0.573	0.585	0.590	0.591	0.592	0.592	0.593	0.593
2.00	0.336	0.481	0.525	0.540	0.545	0.547	0.548	0.549	0.549	0.549
3.00	0.179	0.293	0.348	0.373	0.384	0.389	0.392	0.393	0.394	0.395
4.00	0.108	0.190	0.241	0.269	0.285	0.293	0.298	0.301	0.302	0.303
5.00	0.072	0.131	0.174	0.202	0.219	0.229	0.236	0.240	0.242	0.244
6.00	0.051	0.095	0.130	0.155	0.172	0.184	0.192	0.197	0.200	0.202
7.00	0.038	0.072	0.100	0.122	0.139	0.150	0.158	0.164	0.168	0.171
8.00	0.029	0.056	0.079	0.098	0.113	0.125	0.133	0.139	0.144	0.147
9.00	0.023	0.045	0.064	0.081	0.094	0.105	0.113	0.119	0.124	0.128
10.00	0.019	0.037	0.053	0.067	0.079	0.089	0.097	0.103	0.108	0.112

Example 6.7

The plan of a uniformly loaded rectangular area is shown in Figure 6.23a. Determine the vertical stress increase, $\Delta\sigma$, below point A' at a depth $z = 4$ m.

Solution

The stress increase, $\Delta\sigma$ can be written as

$$\Delta\sigma = \Delta\sigma_1 - \Delta\sigma_2$$

where $\Delta\sigma_1$ = stress increase due to the loaded area shown in Figure 6.23b
$\Delta\sigma_2$ = stress increase due to the loaded area shown in Figure 6.23c

For the loaded area shown in Figure 6.23b:

$$m' = \frac{B}{z} = \frac{2}{4} = 0.5$$

$$n' = \frac{L}{z} = \frac{4}{4} = 1$$

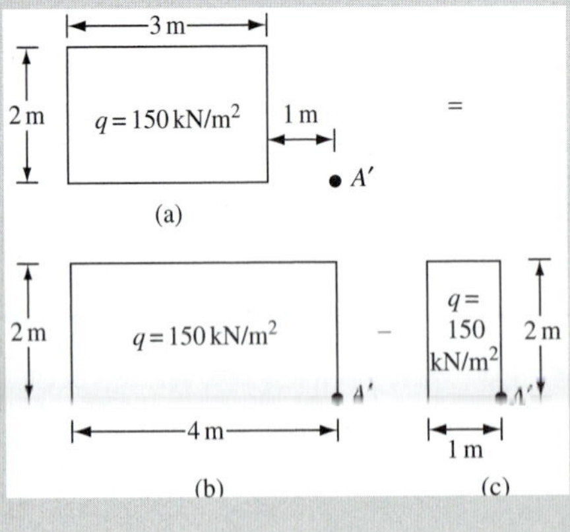

Figure 6.23

From Figure 6.21 for $m' = 0.5$ and $n' = 1$, the value of $I_5 = 0.1225$. So $\Delta\sigma_1 = qI_5 = (150)(0.1225) = 18.38$ kN/m²

Similarly, for the loaded area shown in Figure 6.23c:

$$m' = \frac{B}{z} = \frac{1}{4} = 0.25$$

$$n' = \frac{L}{z} = \frac{2}{4} = 0.5$$

Thus, $I_5 = 0.0473$., Hence

$$\Delta\sigma_2 = (150)(0.0473) = 7.1 \text{ kN/m}^2$$

So

$$\Delta\sigma = \Delta\sigma_1 - \Delta\sigma_2 = 18.38 - 7.1 = \mathbf{11.28 \text{ kN/m}^2}$$

6.13 *Solutions for Westergaard Material*

The Westergaard material was explained in Section 6.7, in which the semi-infinite mass is assumed to be homogeneous but reinforced internally so that no horizontal displacement can occur. Following are some solutions to obtain stress at a point due to surface loading on Westergaard material.

Vertical Stress ($\Delta\sigma$) Due to a Line Load of Finite Length

Referring to Figure 6.14, the stress at A

$$\Delta\sigma = \frac{q}{z}\frac{\eta}{2\pi}\left[\frac{n}{m^2 + \eta^2} \cdot \frac{1}{(m^2 + n^2 + \eta^2)^{0.5}}\right] \tag{6.49}$$

where

$$\eta = \sqrt{\frac{1 - 2\mu_s}{2 - 2\mu_s}}$$

$$m = \frac{a}{z}$$

$$n = \frac{L}{z}$$

Vertical Stress ($\Delta\sigma$) Due to a Rectangularly Loaded Area

Referring to Figure 6.20, the vertical stress at A

$$\Delta\sigma = \frac{q}{2\pi}\left\{\cot^{-1}\left[\eta^2\left(\frac{1}{m'^2} + \frac{1}{n'^2}\right) + \eta^4\left(\frac{1}{m'^2 n'^2}\right)\right]^{0.5}\right\} \tag{6.50}$$

where

$$m' = \frac{B}{z}$$

$$n' = \frac{L}{z}$$

Vertical Stress ($\Delta\sigma$) Due to a Circularly Loaded Area

Referring to Figure 6.19, the vertical stress at A

$$\Delta\sigma = q\left\{1 - \frac{\eta}{\left[\eta^2 + \left(\dfrac{R}{z}\right)^2\right]^{0.5}}\right\}$$ (6.51)

Problems

6.1   A soil profile is shown in Figure 6.24. Calculate the values of σ, u, and σ' at points A, B, C, and D. Plot the variation of σ, u, and σ' with depth. We are given the values in the table.

See Ex 6.1

Layer No.	Thickness (m)	Unit weight (kN/m³)
I	$H_1 = 2$	$\gamma_{dry} = 15$
II	$H_2 = 3$	$\gamma_{sat} = 17.8$
III	$H_3 = 7$	$\gamma_{sat} = 18.6$

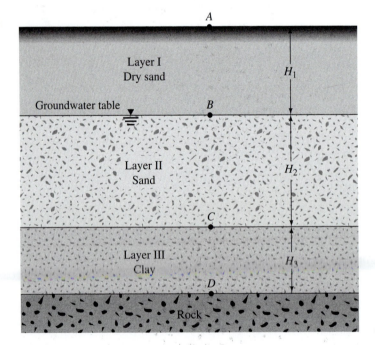

Figure 6.24

6.2   Repeat Problem 6.1 with the following:

Layer No.	Thickness (m)	Soil parameters
I	$H_1 = 4$	$e = 0.45$, $G_s = 2.68$
II	$H_2 = 3$	$e = 0.7$, $G_s = 2.5$
III	$H_3 = 5$	$e = 0.81$, $G_s = 2.75$

6.3   Refer to the soil profile shown in Figure 6.25. Given $H_1 = 4$ m and $H_2 = 3$ m. If the ground water table rises to 2 m below the ground surface, what will be the net change in effective stress at the bottom of the clay layer?

6.4   Refer to Figure 6.3a, in which there is an upward seepage of the water. Given: $H_1 = 0.5$ m, $H_2 = 2$ m, $h = 0.5$ m, void ratio $e = 0.55$, $G_s = 2.68$,
   a. Calculate the total stress, pore water pressure, and effective stress at C. (*Note*: $z = 0.7$ m.)
   b. What is the upward seepage force per unit volume of soil?

6.5   In Problem 6.4, what is the rate of upward seepage of water? Given: hydraulic conductivity of soil, $k = 0.13$ cm/sec, and area of tank $= 0.52$ m^2. Give your answer in m^3/min.

6.6   A sand has $G_s = 2.66$. Calculate the hydraulic gradient that will cause boiling for $e = 0.35, 0.45, 0.55, 0.7$, and 0.8. Plot a graph for i_{cr} versus e.

6.7   A 8 m-thick layer of stiff saturated clay is underlain by a layer of sand (Figure 6.26). The sand is under artesian pressure. Calculate the maximum depth of cut, H, that can be made in the clay.

6.8   Refer to Figure 6.11. Given $P = 30$ kN, determine the vertical stress increase at a point with $x = 5$ m, $y = 4$ m, and $z = 6$ m. Use Boussinesq's solution.

6.9   Solve Problem 6.8 using Westergaard solution. Given $\mu_s = 0.3$.

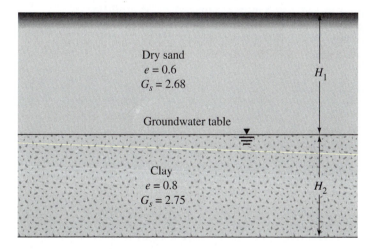

Dry sand
$e = 0.6$
$G_s = 2.68$

H_1

Groundwater table

Clay
$e = 0.8$
$G_s = 2.75$

H_2

Figure 6.25

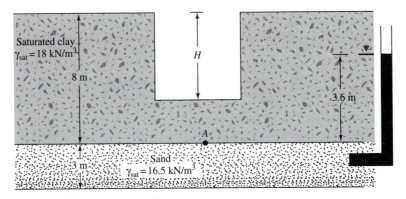

Figure 6.26 Layer of saturated clay underlain by layer of sand

6.10 Point loads of magnitude 9, 18, and 27 kN act at A, B, and C, respectively (Figure 6.27). Determine the increase in vertical stress at a depth of 3 m below point D. Use Boussinesq's equation.

6.11 Solve Problem 6.10 using Westergaard solution. Use $\mu_s = 0.4$

6.12 Refer to Figure 6.13. The magnitude of the line load q is 60 kN/m. Calculate and plot the variation of the vertical stress increase, $\Delta\sigma$, between the limits of $x = -10$ m and $x = +10$ m, given $z = 4$ m.

6.13 Refer to Figure 6.28. Determine the vertical stress increase, $\Delta\sigma$, at point A with the following values:

$q_1 = 100$ kN/m $x_1 = 3$ m $z = 2$ m
$q_2 = 200$ kN/m $x_2 = 2$ m

6.14 Repeat Problem 6.13 with the following values:

$q_1 = 100$ kN/m $x_1 = 3$ m $z = 2.5$ m
$q_2 = 260$ kN/m $x_2 = 2.5$ m

6.15 Figure 6.29 shows a line load of limited length. Given $q = 200$ kN/m, $L = 5$ m, $x = 4$ m. Determine the vertical stress increase at a point with coordinates $x = 1$ m, $y = 3$ m, $z = 5$ m.

6.16 Refer to Figure 6.17. Given $B = 5$ m, $q = 40$ kN/m^2, $x = 1.5$ m, and $z = 2$ m, determine the vertical stress increase, $\Delta\sigma$, at point A.

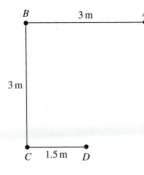

Figure 6.27

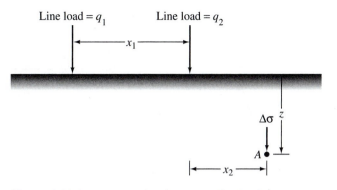

Line load = q_1 Line load = q_2

Figure 6.28 Stress at a point due to two line loads

6.17 Repeat Problem 6.16 for q = 700 kN/m², B = 2 m, x = 2 m, and z = 2.5 m.

6.18 Consider a circularly loaded flexible area on the ground surface. Given: radius of the circular area = R = 3 m; uniformly distributed load = q = 250 kN/m². Calculate the vertical stress increase $\Delta\sigma$ at a point located 5 m (z) below the ground surface (immediately below the center of the circular area).

6.19 Repeat Problem 6.18 with the following: R = 5 m, q = 300 kN/m², and z = 6 m.

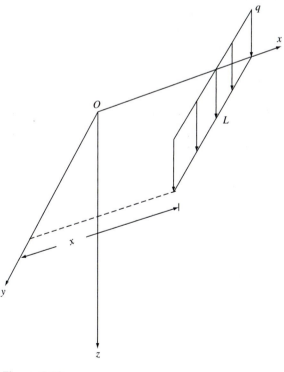

Figure 6.29

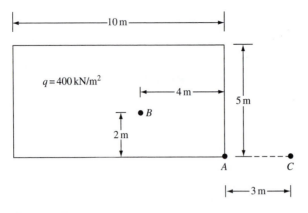

Figure 6.30

6.20 The plan of a flexible rectangular loaded area is shown in Figure 6.30. The uniformly distributed load on the flexible area (q) is 400 kN/m². Determine the increase in the vertical stress ($\Delta\sigma$) at a depth of $z = 5$ m below
  a. Point A
  b. Point B
  c. Point C
6.21 Refer to Figure 6.31. The circular flexible area is uniformly loaded. Given: $q = 320$ kN/m². Determine the vertical stress increase $\Delta\sigma$ at point A.
6.22 Refer to Figure 6.32. The flexible area is uniformly loaded. Given: $q = 300$ kN/m². Determine the vertical stress increase at point A.

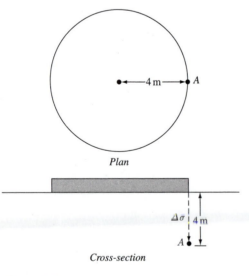

Figure 6.31

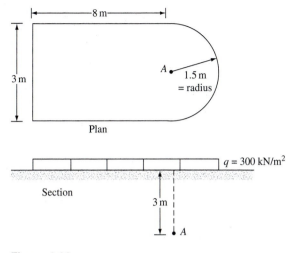

Figure 6.32

References

BISHOP, A. W., ALPEN, I., BLIGHT, G. C., and DONALD, I. B. (1960). "Factors Controlling the Strength of Partially Saturated Cohesive Soils," *Proceedings*, Research Conference on Shear Strength of Cohesive Soils, ASCE, 500–532.

BOUSSINESQ, J. (1883). *Application des Potentials à L'Etude de L'Equilibre et du Mouvement des Solides Elastiques*, Gauthier–Villars, Paris.

SKEMPTON, A. W. (1960). "Correspondence," *Geotechnique*, Vol. 10, No. 4, 186.

TERZAGHI, K. (1922). "Der Grundbruch an Stauwerken und seine Verhütung," *Die Wasserkraft*, Vol. 17, 445–449.

TERZAGHI, K. (1925). *Erdbaumechanik auf Bodenphysikalischer Grundlage*, Deuticke, Vienna.

TERZAGHI, K. (1936). "Relation between Soil Mechanics and Foundation Engineering: Presidential Address," *Proceedings*, First International Conference on Soil Mechanics and Foundation Engineering, Boston, Vol. 3, 13–18.

WESTERGAARD, H. M. (1938). "A Problem of Elasticity Suggested by a Problem in Soil Mechanics: Soft Material Reinforced by Numerous Strong Horizontal Sheets," in *Contribution to the Mechanics of Solids*, Stephen Timoshenko 60th Anniversary Vol., Macmillan, New York.

7

Consolidation

A stress increase caused by the construction of foundations or other loads compresses the soil layers. The compression is caused by (a) deformation of soil particles, (b) relocations of soil particles, and (c) expulsion of water or air from the void spaces. In general, the soil settlement caused by load may be divided into three broad categories:

1. *Immediate settlement*, which is caused by the elastic deformation of dry soil and of moist and saturated soils without any change in the moisture content. Immediate settlement calculations are generally based on equations derived from the theory of elasticity.
2. *Primary consolidation settlement*, which is the result of a volume change in saturated cohesive soils because of the expulsion of water that occupies the void spaces.
3. *Secondary consolidation settlement*, which is observed in saturated cohesive soils and is the result of the plastic adjustment of soil fabrics. It follows the primary consolidation settlement under a constant effective stress.

   This chapter presents the fundamental principles for estimating the consolidation settlement of soil layers under superimposed loadings.

7.1   Fundamentals of Consolidation

When a saturated soil layer is subjected to a stress increase, the pore water pressure suddenly increases. In sandy soils that are highly permeable, the drainage caused by the increase in the pore water pressure is completed immediately. Pore water drainage is accompanied by a reduction in the volume of the soil mass, resulting in settlement. Because of the rapid drainage of the pore water in sandy soils, immediate settlement and consolidation take place simultaneously. This is not the case, however, for clay soils, which have low hydraulic conductivity. The consolidation settlement is time dependent.

   Keeping this in mind, we can analyze the strain of a saturated clay layer subjected to a stress increase (Figure 7.1a). A layer of saturated clay of thickness H

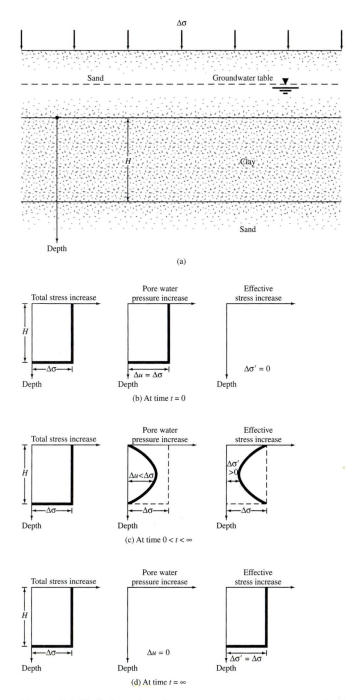

Figure 7.1 Variation of total stress, pore water pressure, and effective stress in a clay layer drained at top and bottom as the result of an added stress, $\Delta\sigma$

is confined between two layers of sand and is subjected to an instantaneous increase in *total stress* of $\Delta\sigma$. From Chapter 6, we know that

$$\Delta\sigma = \Delta\sigma' + \Delta u \qquad (7.1)$$

where

$\Delta\sigma'$ = increase in the effective stress
Δu = increase in the pore water pressure

Since clay has very low hydraulic conductivity and water is incompressible compared with the soil skeleton, at time $t = 0$, the entire incremental stress, $\Delta\sigma$, will be carried by water ($\Delta\sigma = \Delta u$) at all depths (Figure 7.1b). None will be carried by the soil skeleton (that is, incremental effective stress, $\Delta\sigma' = 0$).

After the application of incremental stress, $\Delta\sigma$, to the clay layer, the water in the void spaces will begin to be squeezed out and will drain in both directions into the sand layers. By this process, the excess pore water pressure at any depth on the clay layer will gradually decrease, and the stress carried by the soil solids (effective stress) will increase. Thus, at time $0 < t < \infty$,

$$\Delta\sigma = \Delta\sigma' + \Delta u \qquad (\Delta\sigma' > 0 \text{ and } \Delta u < \Delta\sigma)$$

However, the magnitudes of $\Delta\sigma'$ and Δu at various depths will change (Figure 7.1c), depending on the minimum distance of the drainage path to either the top or bottom sand layer.

Theoretically, at time $t = \infty$, the entire excess pore water pressure would dissipate by drainage from all points of the clay layer, thus giving $\Delta u = 0$. Then the total stress increase, $\Delta\sigma$, would be carried by the soil structure (Figure 7.1d), so

$$\Delta\sigma = \Delta\sigma'$$

This gradual process of drainage under the application of an additional load and the associated transfer of excess pore water pressure to effective stress causes the time-dependent settlement (consolidation) in the clay soil layer.

7.2 One-Dimensional Laboratory Consolidation Test

The one-dimensional consolidation testing procedure was first suggested by Terzaghi (1925). This test is performed in a consolidometer (sometimes referred to as an oedometer). Figure 7.2 is a schematic diagram of a consolidometer. The soil specimen is placed inside a metal ring with two porous stones, one at the top of the specimen and another at the bottom. The specimens are usually 63.5 mm in diameter and 25.4 mm thick. The load on the specimen is applied through a lever arm, and compression is measured by a micrometer dial gauge. The specimen is kept under water during the test. Each load is usually kept for 24 hours. After that, the load is usually doubled, thus doubling the pressure on the specimen, and the compression measurement is continued. At the end of the test, the dry weight of the test specimen is determined. Figure 7.3 shows a consolidation test in progress (right-hand side).

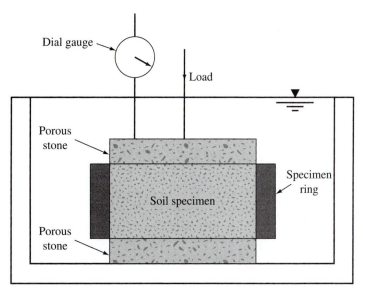

Figure 7.2 Consolidometer

The general shape of the plot of deformation of the specimen versus time for a given load increment is shown in Figure 7.4. From the plot, it can be observed that there are three distinct stages, which may be described as follows:

Stage I: Initial compression, which is mostly caused by preloading.

Stage II: Primary consolidation, during which excess pore water pressure is gradually transferred into effective stress by the expulsion of pore water.

Figure 7.3 Consolidation test in progress (right-hand side) (Courtesy of Braja Das)

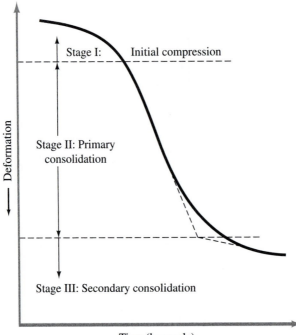

Deformation

Stage I: Initial compression

Stage II: Primary
consolidation

Stage III: Secondary consolidation

Time (log scale)

Figure 7.4 Time–deformation plot during consolidation for a given load increment

Stage III: Secondary consolidation, which occurs after complete dissipation of the excess pore water pressure, when some deformation of the specimen takes place because of the plastic readjustment of soil fabric.

7.3   *Void Ratio–Pressure Plots*

After the time–deformation plots for various loadings are obtained in the laboratory, it is necessary to study the change in the void ratio of the specimen with pressure. Following is a step-by-step procedure:

1. Calculate the height of solids, H_s, in the soil specimen (Figure 7.5):

$$H_S = \frac{W_s}{AG_s\gamma_w} \tag{7.2}$$

where
     W_s = dry weight of the specimen
     A = area of the specimen
     G_s = specific gravity of soil solids
     γ_w = unit weight of water

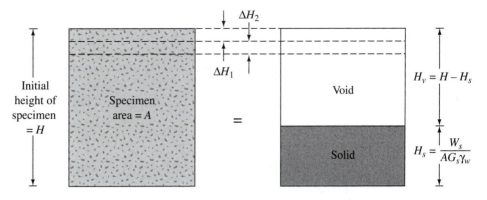

Figure 7.5 Change of height of specimen in one-dimensional consolidation test

2. Calculate the initial height of voids, H_v:

$$H_v = H - H_s \tag{7.3}$$

where H = initial height of the specimen.
3. Calculate the initial void ratio, e_0, of the specimen:

$$e_0 = \frac{V_v}{V_s} = \frac{H_v\, A}{H_s\, A} = \frac{H_v}{H_s} \tag{7.4}$$

4. For the first incremental loading σ_1 (total load/unit area of specimen), which causes deformation ΔH_1, calculate the change in the void ratio Δe_1:

$$\Delta e_1 = \frac{\Delta H_1}{H_s} \tag{7.5}$$

ΔH_1 is obtained from the initial and the final dial readings for the loading. At this time, the effective pressure on the specimen is $\sigma' = \sigma_1 = \sigma'_1$.
5. Calculate the new void ratio, e_1, after consolidation caused by the pressure increment σ_1:

$$e_1 = e_0 - \Delta e_1 \tag{7.6}$$

For the next loading, σ_2 (*note*: σ_2 equals the cumulative load per unit area of specimen), which causes additional deformation ΔH_2, the void ratio e_2 at the end of consolidation can be calculated as

$$e_2 = e_1 - \frac{\Delta H_2}{H_s} \tag{7.7}$$

Note that, at this time, the effective pressure on the specimen is $\sigma' = \sigma_2 = \sigma'_2$.

Proceeding in a similar manner, we can obtain the void ratios at the end of the consolidation for all load increments.

The effective pressures ($\sigma = \sigma'$) and the corresponding void ratios (e) at the end of consolidation are plotted on semilogarithmic graph paper. The typical shape of such a plot is shown in Figure 7.6.

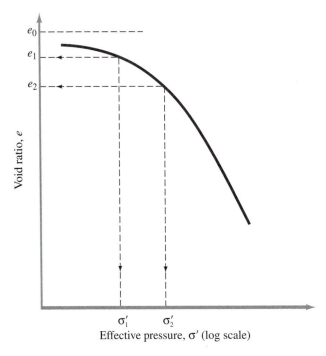

Figure 7.6 Typical plot of e versus $\log \sigma'$

7.4   *Normally Consolidated and Overconsolidated Clays*

Figure 7.6 showed that the upper part of the e–$\log \sigma'$ plot is somewhat curved with a flat slope, followed by a linear relationship for the void ratio, with $\log \sigma'$ having a steeper slope. This can be explained in the following manner.

A soil in the field at some depth has been subjected to a certain maximum effective past pressure in its geologic history. This maximum effective past pressure may be equal to or greater than the existing overburden pressure at the time of sampling. The reduction of pressure in the field may be caused by natural geologic processes or human processes. During the soil sampling, the existing effective overburden pressure is also released, resulting in some expansion. When this specimen is subjected to a consolidation test, a small amount of compression (that is, a small change in the void ratio) will occur when the total pressure applied is less than the maximum effective overburden pressure in the field to which the soil has been subjected in the past. When the total applied pressure on the specimen is greater than the maximum effective past pressure, the change in the void ratio is much larger, and the e–$\log \sigma'$ relationship is practically linear with a steeper slope.

This relationship can be verified in the laboratory by loading the specimen to exceed the maximum effective overburden pressure, and then unloading and reloading again. The e–$\log \sigma'$ plot for such cases is shown in Figure 7.7, in which *cd* represents unloading and *dfg* represents the reloading process.

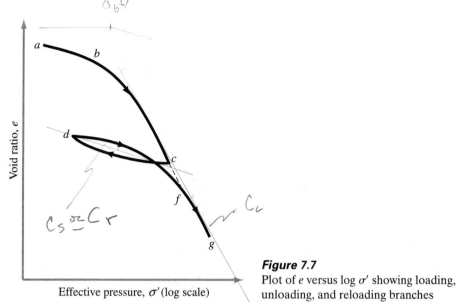

Figure 7.7
Plot of e versus log σ' showing loading, unloading, and reloading branches

This leads us to the two basic definitions of clay based on stress history:

1. *Normally consolidated*: The present effective overburden pressure is the maximum pressure to which the soil has been subjected in the past.
2. *Overconsolidated*: The present effective overburden pressure is less than that which the soil has experienced in the past. The maximum effective past pressure is called the *preconsolidation pressure*.

The past effective pressure cannot be determined explicitly because it is usually a function of geological processes and, consequently, it must be inferred from laboratory test results.

Casagrande (1936) suggested a simple graphic construction to determine the preconsolidation pressure, σ'_c, from the laboratory e–log σ' plot. The procedure follows (see Figure 7.8):

1. By visual observation, establish point a at which the e–log σ' plot has a minimum radius of curvature.
2. Draw a horizontal line ab.
3. Draw the line ac tangent at a.
4. Draw the line ad, which is the bisector of the angle bac.
5. Project the straight-line portion gh of the e–log σ' plot back to intersect ad at f. The abscissa of point f is the preconsolidation pressure, σ'_c.

The overconsolidation ratio (OCR) for a soil can now be defined as

$$OCR = \frac{\sigma'_c}{\sigma'}$$

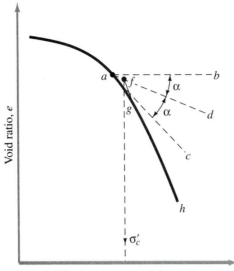

Figure 7.8 Graphic procedure for determining preconsolidation pressure

where

σ'_c = preconsolidation pressure of a specimen
σ' = present effective vertical pressure

7.5 Effect of Disturbance on Void Ratio–Pressure Relationship

A soil specimen will be remolded when it is subjected to some degree of disturbance. This will affect the void ratio–pressure relationship for the soil. For a normally consolidated clayey soil of low to medium sensitivity (Figure 7.9) under an effective overburden pressure of σ'_o and with a void ratio of e_0, the change in the void ratio with an increase of pressure in the field will be roughly as shown by curve 1. This is the *virgin compression curve*, which is approximately a straight line on a semilogarithmic plot. However, the laboratory consolidation curve for a fairly undisturbed specimen of the same soil (curve 2) will be located to the left of curve 1. If the soil is completely remolded and a consolidation test is conducted on it, the general position of the e–log σ' plot will be represented by curve 3. Curves 1, 2, and 3 will intersect approximately at a void ratio of $e = 0.4e_0$ (Terzaghi and Peck, 1967).

For an overconsolidated clayey soil of low to medium sensitivity that has been subjected to a preconsolidation pressure of σ'_c (Figure 7.10) and for which the present effective overburden pressure and the void ratio are σ'_o and e_0, respectively, the field consolidation curve will take a path represented approximately by *cbd*. Note that *bd* is a part of the virgin compression curve. The laboratory consolidation test results on a specimen subjected to moderate disturbance will be represented by

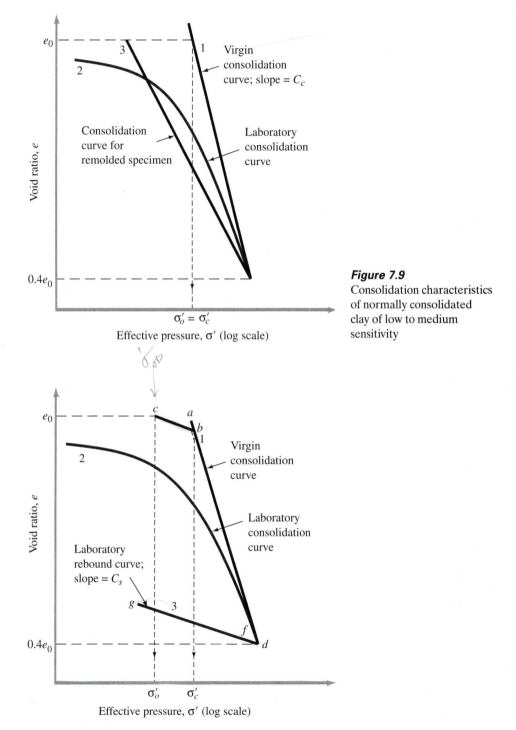

Figure 7.9
Consolidation characteristics of normally consolidated clay of low to medium sensitivity

Figure 7.10 Consolidation characteristics of overconsolidated clay of low to medium sensitivity

curve 2. Schmertmann (1953) concluded that the slope of line *cb*, which is the field recompression path, has approximately the same slope as the laboratory rebound curve *fg*.

Calculation of Settlement from One-Dimensional Primary Consolidation

With the knowledge gained from the analysis of consolidation test results, we can now proceed to calculate the probable settlement caused by primary consolidation in the field, assuming one-dimensional consolidation.

Let us consider a saturated clay layer of thickness H and cross-sectional area A under an existing average effective overburden pressure σ'_o. Because of an increase of pressure, $\Delta\sigma$, let the primary settlement be S_p. At the end of consolidation, $\Delta\sigma = \Delta\sigma'$. Thus, the change in volume (Figure 7.11) can be given by

$$\Delta V = V_0 - V_1 = HA - (H - S_p)A = S_p A \tag{7.8}$$

where V_0 and V_1 are the initial and final volumes, respectively. However, the change in the total volume is equal to the change in the volume of voids, ΔV_v. Thus,

$$\Delta V = S_p A = V_{v0} - V_{v1} = \Delta V_v \tag{7.9}$$

where V_{v0} and V_{v1} are the initial and final void volumes, respectively. From the definition of the void ratio, we have

$$\Delta V_v = \Delta e V_s \tag{7.10}$$

where Δe = change of void ratio. But

$$V_s = \frac{V_0}{1 + e_0} = \frac{AH}{1 + e_0} \tag{7.11}$$

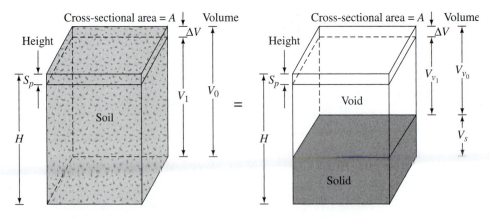

Figure 7.11 Settlement caused by one-dimensional consolidation

where e_0 = initial void ratio at volume V_0. Thus, from Eqs. (7.8), (7.9), (7.10), and (7.11), we get

$$\Delta V = S_p A = \Delta e V_s = \frac{AH}{1 + e_0} \Delta e$$

or

$$S_p = H \frac{\Delta e}{1 + e_0} \tag{7.12}$$

For normally consolidated clays that exhibit a linear e–log σ' relationship (Figure 7.9) (*note*: $\Delta\sigma = \Delta\sigma'$ at the end of consolidation),

$$\Delta e = C_c[\log(\sigma'_o + \Delta\sigma') - \log\sigma'_o] \tag{7.13}$$

where C_c = slope of the e–log σ'_o plot and is defined as the compression index. Substituting Eq. (7.13) into Eq. (7.12) gives

$$S_p = \frac{C_c H}{1 + e_0} \log\left(\frac{\sigma'_o + \Delta\sigma'}{\sigma'_o}\right) \tag{7.14}$$

For a thicker clay layer, a more accurate measurement of settlement can be made if the layer is divided into a number of sublayers and calculations are made for each sublayer. Thus, the total settlement for the entire layer can be given as

$$S_p = \sum\left[\frac{C_c H_i}{1 + e_0} \log\left(\frac{\sigma'_{o(i)} + \Delta\sigma'_{(i)}}{\sigma'_{o(i)}}\right)\right]$$

where
  H_i = thickness of sublayer i
  $\sigma'_{o(i)}$ = initial average effective overburden pressure for sublayer i
  $\Delta\sigma'_{(i)}$ = increase of vertical pressure for sublayer i

In overconsolidated clays (Figure 7.10), for $\sigma'_o + \Delta\sigma' \leq \sigma'_c$, field e–log σ' variation will be along the line cb, the slope of which will be approximately equal to the slope of the laboratory rebound curve. The slope of the rebound curve, C_s, is referred to as the *swell index*, so

$$\Delta e = C_s[\log(\sigma'_o + \Delta\sigma') - \log\sigma'_o] \tag{7.15}$$

From Eqs. (7.12) and (7.15), we have

$$S_p = \frac{C_s H}{1 + e_0} \log\left(\frac{\sigma'_o + \Delta\sigma'}{\sigma'_o}\right) \tag{7.16}$$

If $\sigma'_o + \Delta\sigma > \sigma'_c$, then

$$S_p = \frac{C_s H}{1 + e_0} \log \frac{\sigma'_c}{\sigma'_o} + \frac{C_c H}{1 + e_0} \log \left(\frac{\sigma'_o + \Delta\sigma'}{\sigma'_c} \right)$$

(7.17)

However, if the e–log σ' curve is given, it is possible simply to pick Δe off the plot for the appropriate range of pressures. This value may be substituted into Eq. (7.12) for calculating the settlement, S_p.

7.7 Compression Index (C_c) and Swell Index (C_s)

We can determine the compression index for field settlement caused by consolidation by graphic construction (as shown in Figure 7.8) after obtaining laboratory test results for the void ratio and pressure.

Skempton (1944) suggested empirical expressions for the compression index. For undisturbed clays:

$$C_c = 0.009(LL - 10)$$

(7.18)

For remolded clays:

$$C_c = 0.007(LL - 10)$$

(7.19)

where LL = liquid limit (%). In the absence of laboratory consolidation data, Eq. (7.18) is often used for an approximate calculation of primary consolidation in the field. Several other correlations for the compression index are also available now. Several of those correlations have been compiled by Rendon-Herrero (1980), and these are given in Table 7.1

Table 7.1 Correlations for Compression Index, C_c (compiled from Rendon-Herrero, 1980)

Equation	Region of applicability
$C_c = 0.01 w_N$	Chicago clays
$C_c = 1.15(e_O - 0.27)$	All clays
$C_c = 0.30(e_O - 0.27)$	Inorganic cohesive soil: silt, silty clay, clay
$C_c = 0.0115 w_N$	Organic soils, peats, organic silt, and clay
$C_c = 0.0046(LL - 9)$	Brazilian clays
$C_c = 0.75(e_O - 0.5)$	Soils with low plasticity
$C_c = 0.208 e_O + 0.0083$	Chicago clays
$C_c = 0.156 e_O + 0.0107$	All clays

Note: e_O = *in situ* void ratio; w_N = *in situ* water content.

Based on observations on several natural clays, Rendon-Herrero (1983) gave the relationship for the compression index in the form

$$C_c = 0.141G_s^{1.2}\left(\frac{1 + e_0}{G_s}\right)^{2.38} \tag{7.20}$$

More recently, Park and Koumoto (2004) expressed the compression index by the following relationship.

$$C_c = \frac{n_o}{371.747 - 4.275n_o} \tag{7.21}$$

where n_o = *in situ* porosity of the soil
Based on the modified Cam clay model, Wroth and Wood (1978) have shown that

$$C_c \approx 0.5G_s \frac{[PI(\%)]}{100} \tag{7.22}$$

where PI = plasticity index
If an average value of G_s is taken to be about 2.7 (Kulhawy and Mayne, 1990)

$$C_c \approx \frac{PI}{74} \tag{7.23}$$

The swell index is appreciably smaller in magnitude than the compression index and generally can be determined from laboratory tests. Typical values of the liquid limit, plastic limit, virgin compression index, and swell index for some natural soils are given in Table 7.2.

From Table 7.2, it can be seen that $C_s \approx 0.2$ to $0.3\ C_c$. Based on the modified Cam clay model, Kulhawy and Mayne (1990) have shown than

$$C_S \approx \frac{PI}{370} \tag{7.24}$$

Table 7.2 Compression and Swell of Natural Soils

Soil	Liquid limit	Plastic limit	Compression index, C_c	Swell index, C_s	C_s/C_c
Boston blue clay	41	20	0.35	0.07	0.2
Chicago clay	60	20	0.4	0.07	0.175
Ft. Gordon clay, Georgia	51	26	0.12	0.04	0.33
New Orleans clay	80	25	0.3	0.05	0.17
Montana clay	60	28	0.21	0.05	0.24

Example 7.1

Following are the results of a laboratory consolidation test on a soil specimen obtained from the field. Dry mass of specimen = 128 g, height of specimen at the beginning of the test = 2.54 cm, $G_s = 2.75$, and area of the specimen = 30.68 cm².

Pressure, σ' (kN/m²)	Final height of specimen at the end of consolidation (cm)
0	2.540
50	2.488
100	2.465
200	2.431
400	2.389
800	2.324
1600	2.225
3200	2.115

Make necessary calculations and draw an *e* vs. log σ' curve.

Solution

Calculation of H_s

From Eq. (7.2)

$$H_s = \frac{m_s}{AG_s\rho_w} = \frac{128 \text{ g}}{(30.68 \text{ cm}^2)(2.75)(1 \text{ g/cm}^3)} = 1.52 \text{ cm}$$

Now the following table can be prepared.

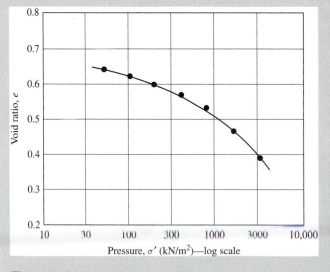

Figure 7.12 Variation of void ratio with pressure

Pressure, σ' kN/m^2	Height at the end of consolidation, H (cm)	$H_v = H - H_s$ (cm)	$e = H_v/H_s$
0	2.540	1.02	0.671
50	2.488	0.968	0.637
100	2.465	0.945	0.622
200	2.431	0.911	0.599
400	2.389	0.869	0.572
800	2.324	0.804	0.529
1600	2.225	0.705	0.464
3200	2.115	0.595	0.391

The e vs. log σ plot is shown in Figure 7.12.

Example 7.2

The laboratory consolidation data for an undisturbed clay sample are as follows:

$$e_1 = 1.1 \qquad \sigma'_1 = 95 \text{ kN/m}^2$$

$$e_2 = 0.9 \qquad \sigma'_2 = 475 \text{ kN/m}^2$$

What will be the void ratio for a pressure of 600 kN/m^2? (*Note:* $\sigma'_c < 95$ kN/m^2.)

Solution

From Figure 7.13

$$C_c = \frac{e_1 - e_2}{\log\sigma'_2 - \log\sigma'_1} = \frac{1.1 - 0.9}{\log 475 - \log 95} = 0.286$$

$$e_1 - e_3 = C_c(\log 600 - \log 95)$$

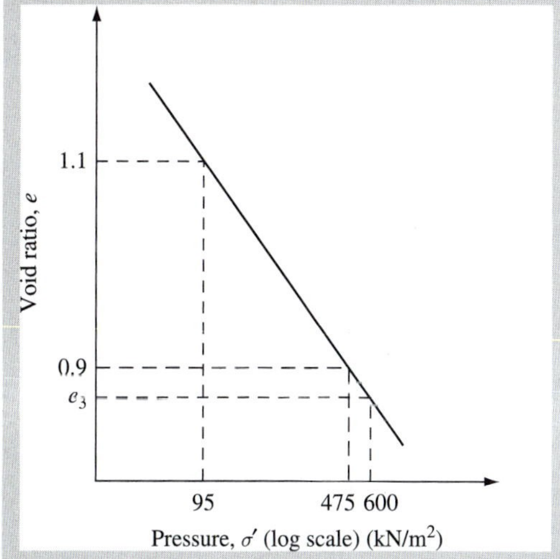

Figure 7.13

$$e_3 = e_1 - C_c \log \frac{600}{95}$$

$$= 1.1 - 0.286 \log \frac{600}{95} = 0.87$$

Example 7.3

A soil profile is shown in Figure 7.14. If a uniformly distributed load $\Delta\sigma$ is applied at the ground surface, what will be the settlement of the clay layer caused by primary consolidation? We are given that σ'_c for the clay is 125 kN/m² and $C_s = \frac{1}{6}C_c$.

Solution
The average effective stress at the middle of the clay layer is

$$\sigma'_o = 2.5\gamma_{\text{dry(sand)}} + (4.5)\left[\gamma_{\text{sat(sand)}} - \gamma_w\right] + \left(\frac{5}{2}\right)\left[\gamma_{\text{sat(clay)}} - \gamma_w\right]$$

or

$$\sigma'_o = (2.5)(16.5) + (4.5)(18.81 - 9.81) + (2.5)(19.24 - 9.81)$$

$$= 105.33 \text{ kN/m}^2$$

$$\sigma'_c = 125 \text{ kN/m}^2 > 105.33 \text{ kN/m}^2$$

$$\sigma'_o + \Delta\sigma' = 105.33 + 50 = 155.33 \ \text{kN/m}^2 > \sigma'_c$$

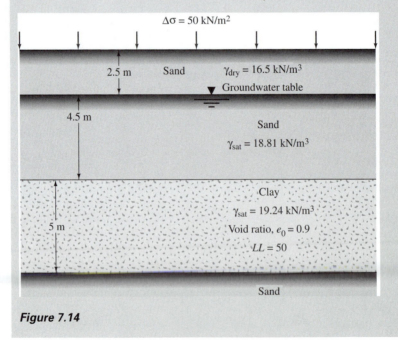

Figure 7.14

(*Note:* $\Delta\sigma = \Delta\sigma'$ at the end of consolidation.) So we need to use Eq. (7.17):

$$S_p = \frac{C_s H}{1 + e_0} \log\left(\frac{\sigma'_c}{\sigma'_o}\right) + \frac{C_c H}{1 + e_0} \log\left(\frac{\sigma'_o + \Delta\sigma'}{\sigma'_c}\right)$$

We have $H = 5$ m and $e_0 = 0.9$. From Eq. (7.18),

$$C_c = 0.009(LL - 10) = 0.009(50 - 10) = 0.36$$

$$C_s = \frac{1}{6}C_c = \frac{0.36}{6} = 0.06$$

Thus,

$$S_p = \frac{5}{1 + 0.9}\left[0.06 \log\left(\frac{125}{105.33}\right) + 0.36 \log\left(\frac{105.33 + 50}{125}\right)\right]$$

$$= 0.1011 \text{ m} \approx \textbf{101 mm} \qquad \blacksquare$$

7.8 Settlement from Secondary Consolidation

Section 7.2 showed that at the end of primary consolidation (that is, after complete dissipation of excess pore water pressure) some settlement is observed because of the plastic adjustment of soil fabrics, which is usually termed *creep*. This stage of consolidation is called *secondary consolidation*. During secondary consolidation, the plot of deformation versus the log of time is practically linear (Figure 7.4). The variation of the void ratio e with time t for a given load increment will be similar to that shown in Figure 7.4. This variation is illustrated in Figure 7.15.

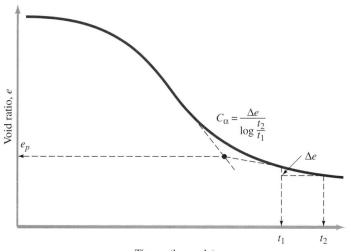

Figure 7.15 Variation of e with log t under a given load increment, and definition of secondary compression index

The secondary compression index can be defined from Figure 7.15 as

$$C_\alpha = \frac{\Delta e}{\log t_2 - \log t_1} = \frac{\Delta e}{\log(t_2/t_1)} \tag{7.25}$$

where

C_α = secondary compression index
Δe = change of void ratio
t_1, t_2 = time

The magnitude of the secondary consolidation can be calculated as

$$S_s = C'_\alpha H \log\left(\frac{t_2}{t_1}\right) \tag{7.26}$$

where

$$C'_\alpha = \frac{C_\alpha}{1 + e_p} \tag{7.27}$$

and

e_p = void ratio at the end of primary consolidation (Figure 7.15)
H = thickness of clay layer

Secondary consolidation settlement is more important than primary consolidation in organic and highly compressible inorganic soils. In overconsolidated inorganic clays, the secondary compression index is very small and has less practical significance. The variation of C'_α for various natural soil deposits is shown in Figure 7.16 (Mesri, 1973).

From Figure 7.16, it can be seen that C'_α for

- Overconsolidated clays ≈ 0.001 or less
- Normally consolidated clays ≈ 0.005 to 0.03
- Organic soils ≈ 0.04 or more

Mesri and Godlewski (1977) compiled the ratio of C'_α/C_c for a number of natural clays. From this study, it appears that C'_α/C_c for

- Inorganic clays and silts $\approx 0.04 \pm 0.01$
- Organic clays and silts $\approx 0.05 \pm 0.01$
- Peats $\approx 0.075 \pm 0.01$

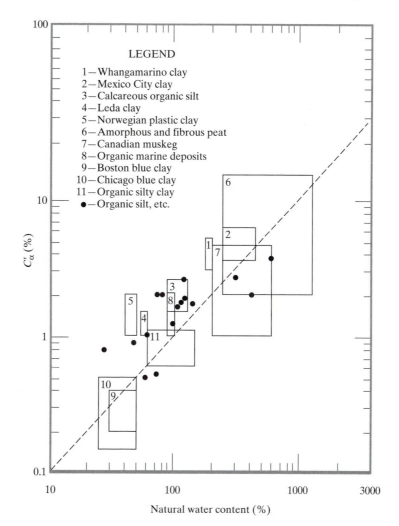

Figure 7.16 C'_α for natural soil deposits (After Mesri, 1973)

Example 7.4

For a normally consolidated clay layer in the field, the following values are given:

- Thickness of clay layer = 3 m
- Void ratio (e_o) = 0.8
- Compression index (C_c) = 0.28
- Average effective pressure on the clay layer (σ'_o) = 130 kN/m^2
- $\Delta\sigma'$ = 50 kN/m^2
- Secondary compression index (C_α) = 0.02

What is the total consolidation settlement of the clay layer five years after the completion of primary consolidation settlement? (*Note:* Time for completion of primary settlement = 1.5 years.)

Solution

From Eq. (7.27),

$$C_\alpha^j = \frac{C_\alpha}{1 + e_p}$$

The value of e_p can be calculated as

$$e_p = e_O - \Delta e_{primary}$$

Combining Eqs. (7.12) and (7.13), we find that

$$\Delta e = C_c \log\left(\frac{\sigma'_O + \Delta\sigma'}{\sigma'_O}\right) = 0.28 \log\left(\frac{130 + 50}{130}\right)$$
$$= 0.04$$

Primary consolidation, $S_p = \dfrac{\Delta e H}{1 + e_O} = \dfrac{(0.04)(3)}{1 + 0.8} = 0.067$ m

It is given that $e_O = 0.8$, and thus,

$$e_p = 0.8 - 0.04 = 0.76$$

Hence,

$$C'_\alpha = \frac{0.02}{1 + 0.76} = 0.011$$

From Eq. (7.26),

$$S_s = C'_\alpha H \log\left(\frac{t_2}{t_1}\right) = (0.011)(3) \log\left(\frac{5}{1.5}\right) \approx 0.017 \text{ m}$$

Total consolidation settlement = primary consolidation (S_p) + secondary settlement (S_s). So

Total consolidation settlement = $0.067 + 0.017 = 0.084$ m ≈ 84 mm ∎

7.9   ***Time Rate of Consolidation***

The total settlement caused by primary consolidation resulting from an increase in the stress on a soil layer can be calculated by using one of the three equations [(7.14), (7.16), or (7.17)] given in Section 7.6. However, the equations do not provide any information regarding the rate of primary consolidation. Terzaghi (1925) proposed the first theory to consider the rate of one-dimensional consolidation for saturated clay soils. The mathematical derivations are based on the following assumptions:

1. The clay–water system is homogeneous.
2. Saturation is complete.

3. Compressibility of water is negligible.
4. Compressibility of soil grains is negligible (but soil grains rearrange).
5. The flow of water is in one direction only (that is, in the direction of compression).
6. Darcy's law is valid.

Figure 7.17a shows a layer of clay of thickness $2H_{dr}$ located between two highly permeable sand layers. If the clay layer is subjected to an increased pressure of $\Delta\sigma$, the pore water pressure at any point A in the clay layer will increase. For one-dimensional consolidation, water will be squeezed out in the vertical direction toward the sand layers.

Figure 7.17b shows the flow of water through a prismatic element at A. For the soil element shown,

$$\begin{pmatrix} \text{rate of outflow} \\ \text{of water} \end{pmatrix} - \begin{pmatrix} \text{rate of inflow} \\ \text{of water} \end{pmatrix} = \begin{pmatrix} \text{rate of} \\ \text{volume changes} \end{pmatrix}$$

Thus,

$$\left(v_z + \frac{\partial v_z}{\partial z} dz \right) dx\, dy - v_z\, dx\, dy = \frac{\partial V}{\partial t}$$

where
   V = volume of the soil element
   v_z = velocity of flow in the z direction

or

$$\frac{\partial v_z}{\partial z} dx\, dy\, dz = \frac{\partial V}{\partial t} \tag{7.28}$$

Using Darcy's law, we have

$$v_z = ki = -k\frac{\partial h}{\partial z} = -\frac{k}{\gamma_w}\frac{\partial u}{\partial z} \tag{7.29}$$

where u = excess pore water pressure caused by the increase of stress. From Eqs. (7.28) and (7.29), we get

$$-\frac{k}{\gamma_w}\frac{\partial^2 u}{\partial z^2} = \frac{1}{dx\, dy\, dz}\frac{\partial V}{\partial t} \tag{7.30}$$

During consolidation, the rate of change in the volume of the soil element is equal to the rate of change in the volume of voids. So

$$\frac{\partial V}{\partial t} = \frac{\partial V_v}{\partial t} = \frac{\partial(V_s + eV_s)}{\partial t} = \frac{\partial V_s}{\partial t} + V_s\frac{\partial e}{\partial t} + e\frac{\partial V_s}{\partial t} \tag{7.31}$$

where
   V_s = volume of soil solids
   V_v = volume of voids

But (assuming that soil solids are incompressible),

$$\frac{\partial V_s}{\partial t} = 0$$

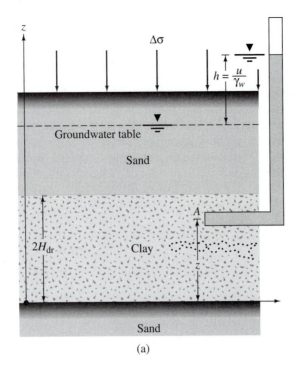

(a)

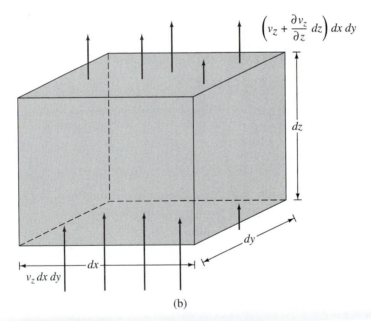

(b)

Figure 7.17 (a) Clay layer undergoing consolidation; (b) flow of water at A during consolidation

and

$$V_s = \frac{V}{1 + e_0} = \frac{dx\,dy\,dz}{1 + e_0}$$

Substituting for $\partial V_s/\partial t$ and V_s in Eq. (7.31) yields

$$\frac{\partial V}{\partial t} = \frac{dx\,dy\,dz}{1 + e_0}\frac{\partial e}{\partial t} \tag{7.32}$$

where e_0 = initial void ratio. Combining Eqs. (7.30) and (7.31) gives

$$-\frac{k}{\gamma_w}\frac{\partial^2 u}{\partial z^2} = \frac{1}{1 + e_0}\frac{\partial e}{\partial t} \tag{7.33}$$

The change in the void ratio is caused by the increase in the effective stress (that is, decrease of excess pore water pressure). Assuming that those values are linearly related, we have

$$\partial e = a_v\,\partial(\Delta\sigma') = -a_v\,\partial u \tag{7.34}$$

where
$\partial(\Delta\sigma')$ = change in effective pressure
$\quad a_v$ = coefficient of compressibility (a_v can be considered to be constant for a narrow range of pressure increases)

Combining Eqs. (7.33) and (7.34) gives

$$-\frac{k}{\gamma_w}\frac{\partial^2 u}{\partial z^2} = -\frac{a_v}{1 + e_0}\frac{\partial u}{\partial t} = -m_v\frac{\partial u}{\partial t}$$

where m_v = coefficient of volume compressibility = $a_v/(1 + e_0)$, or

$$\frac{\partial u}{\partial t} = c_v\frac{\partial^2 u}{\partial z^2} \tag{7.35}$$

where c_v = coefficient of consolidation = $k/(\gamma_w m_v)$.

Equation (7.35) is the basic differential equation of Terzaghi's consolidation theory and can be solved with the following boundary conditions:

$$z = 0,\quad u = 0$$

$$z = 2H_{dr},\quad u = 0$$

$$t = 0,\quad u = u_0$$

The solution yields

$$u = \sum_{m=0}^{m=\infty}\left[\frac{2u_0}{M}\sin\left(\frac{Mz}{H_{dr}}\right)\right]e^{-M^2 T_v} \tag{7.36}$$

where m is an integer

$$M = \frac{\pi}{2}(2m + 1)$$

u_0 = initial excess pore water pressure

and

$$T_v = \frac{c_v t}{H_{dr}^2} = \text{time factor}$$

The time factor is a nondimensional number.

Because consolidation progresses by dissipation of excess pore water pressure, the degree of consolidation at a distance z at any time t is

$$U_z = \frac{u_0 - u_z}{u_0} = 1 - \frac{u_z}{u_0} \tag{7.37}$$

where u_z = excess pore water pressure at time t. Equations (7.36) and (7.37) can be combined to obtain the degree of consolidation at any depth z. This is shown in Figure 7.18.

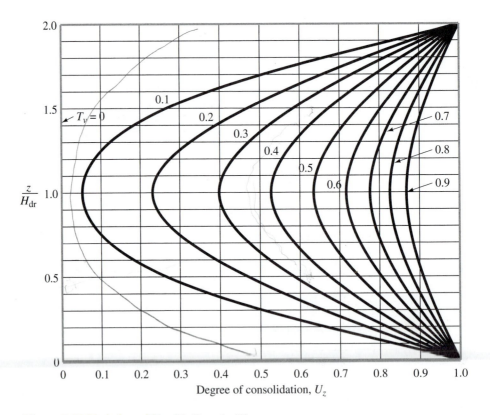

Figure 7.18 Variation of U_z with T_v and z/H_{dr}

The average degree of consolidation for the entire depth of the clay layer at any time t can be written from Eq. (7.37) as

$$U = \frac{S_t}{S_p} = 1 - \frac{\left(\dfrac{1}{2H_{dr}}\right) \displaystyle\int_0^{2H_{dr}} u_z\, dz}{u_0} \qquad (7.38)$$

where

U = average degree of consolidation
S_t = settlement of the layer at time t
S_p = ultimate settlement of the layer from primary consolidation

Substituting the expression for excess pore water pressure, u_z, given in Eq. (7.36) into Eq. (7.38) gives

$$U = 1 - \sum_{m=0}^{m=\infty} \frac{2}{M^2} e^{-M^2 T_v} \qquad (7.39)$$

The variation in the average degree of consolidation with the nondimensional time factor, T_v, is given in Table 7.3, which represents the case where u_0 is the same for the entire depth of the consolidating layer. The values of the time factor and their corresponding average degrees of consolidation may also be approximated by the following simple relationships:

$$\text{For } U = 0 \text{ to } 60\%, \quad T_v = \frac{\pi}{4}\left(\frac{U\%}{100}\right)^2 \qquad (7.40)$$

$$\text{For } U > 60\%, \quad T_v = 1.781 - 0.933 \log(100 - U\%) \qquad (7.41)$$

Sivaram and Swamee (1977) gave the following empirical relationships for U and T_v (for U varying from 0% to 100%):

$$\frac{U\%}{100} = \frac{\left(\dfrac{4T_v}{\pi}\right)^{0.5}}{\left[1 + \left(\dfrac{4T_v}{\pi}\right)^{2.8}\right]^{0.179}} \qquad (7.42)$$

and

$$T_v = \frac{\left(\dfrac{\pi}{4}\right)\left(\dfrac{U\%}{100}\right)^2}{\left[1 - \left(\dfrac{U\%}{100}\right)^{5.6}\right]^{0.357}} \qquad (7.43)$$

Table 7.3 Variation of time factor with degree of consolidation*

U (%)	T_v	U (%)	T_v	U (%)	T_v		
0	0	34	0.0907	68	0.377		
1	0.00008	35	0.0962	69	0.390		
2	0.0003	36	0.102	70	0.403		
3	0.00071	37	0.107	71	0.417		
4	0.00126	38	0.113	72	0.431		
5	0.00196	39	0.119	73	0.446		
6	0.00283	40	0.126	74	0.461		
7	0.00385	41	0.132	75	0.477		
8	0.00502	42	0.138	76	0.493		
9	0.00636	43	0.145	77	0.511		
10	0.00785	44	0.152	78	0.529		
11	0.0095	45	0.159	79	0.547		
12	0.0113	46	0.166	80	0.567		
13	0.0133	47	0.173	81	0.588		
14	0.0154	48	0.181	82	0.610		
15	0.0177	49	0.188	83	0.633		
16	0.0201	50	0.197	84	0.658		
17	0.0227	51	0.204	85	0.684		
18	0.0254	52	0.212	86	0.712		
19	0.0283	53	0.221	87	0.742		
20	0.0314	54	0.230	88	0.774		
21	0.0346	55	0.239	89	0.809		
22	0.0380	56	0.248	90	0.848		
23	0.0415	57	0.257	91	0.891		
24	0.0452	58	0.267	92	0.938		
25	0.0491	59	0.276	93	0.993		
26	0.0531	60	0.286	94	1.055		
27	0.0572	61	0.297	95	1.129		
28	0.0615	62	0.307	96	1.219		
29	0.0660	63	0.318	97	1.336		
30	0.0707	64	0.329	98	1.500		
31	0.0754	65	0.304	99	1.781		
32	0.0803	66	0.352	100	∞		
33	0.0855	67	0.364				

Different types of drainage with u_0 constant

*u_0 constant with depth.

7.10 Coefficient of Consolidation

The coefficient of consolidation, c_v, generally decreases as the liquid limit of soil increases. The range of variation of c_v for a given liquid limit of soil is rather wide.

For a given load increment on a specimen, there are two commonly used graphic methods for determining c_v from laboratory one-dimensional consolidation tests. One of them is the *logarithm-of-time method* proposed by Casagrande and Fadum (1940), and the other is the *square-root-of-time method* suggested by Taylor (1942). The general procedures for obtaining c_v by the two methods are described next.

Logarithm-of-Time Method

For a given incremental loading of the laboratory test, the specimen deformation versus log-of-time plot is shown in Figure 7.19. The following constructions are needed to determine c_v:

1. Extend the straight-line portions of primary and secondary consolidations to intersect at A. The ordinate of A is represented by d_{100}—that is, the deformation at the end of 100% primary consolidation.
2. The initial curved portion of the plot of deformation versus log t is approximated to be a parabola on the natural scale. Select times t_1 and t_2 on the curved portion such that $t_2 = 4t_1$. Let the difference of the specimen deformation during time $(t_2 - t_1)$ be equal to x.
3. Draw a horizontal line DE such that the vertical distance BD is equal to x. The deformation corresponding to the line DE is d_0 (that is, deformation at 0% consolidation).
4. The ordinate of point F on the consolidation curve represents the deformation at 50% primary consolidation, and its abscissa represents the corresponding time (t_{50}).
5. For 50% average degree of consolidation, $T_v = 0.197$ (Table 7.3);

$$T_{50} = \frac{c_v t_{50}}{H_{dr}^2}$$

or

$$C_v = \frac{0.197 H_{dr}^2}{t_{50}} \tag{7.44}$$

where H_{dr} = average longest drainage path during consolidation.

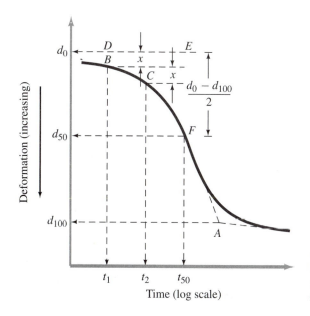

Figure 7.19
Logarithm-of-time method for determining coefficient of consolidation

For specimens drained at both top and bottom, H_{dr} equals one-half of the average height of the specimen during consolidation. For specimens drained on only one side, H_{dr} equals the average height of the specimen during consolidation.

Square-Root-of-Time Method

In this method, a plot of deformation versus the square root of time is drawn for the incremental loading (Figure 7.20). Other graphic constructions required are as follows:

1. Draw a line AB through the early portion of the curve.
2. Draw a line AC such that $\overline{OC} = 1.15\,\overline{OB}$. The abscissa of point D, which is the intersection of AC and the consolidation curve, gives the square root of time for 90% consolidation ($\sqrt{t_{90}}$).
3. For 90% consolidation, $T_{90} = 0.848$ (Table 7.3), so

$$T_{90} = 0.848 = \frac{c_v t_{90}}{H_{dr}^2}$$

or

$$c_v = \frac{0.848 H_{dr}^2}{t_{90}} \tag{7.45}$$

H_{dr} in Eq. (7.45) is determined in a manner similar to the logarithm-of-time method.

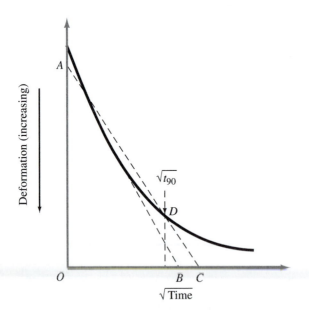

Figure 7.20 Square-root-of-time fitting method

Example 7.5

During a laboratory consolidation test, the time and dial gauge readings obtained from an increase of pressure on the specimen from 50 kN/m² to 100 kN/m² are given here.

Time (min)	Dial gauge reading (cm × 10⁴)	Time (min)	Dial gauge reading (cm × 10⁴)
0	3975	16.0	4572
0.1	4082	30.0	4737
0.25	4102	60.0	4923
0.5	4128	120.0	5080
1.0	4166	240.0	5207
2.0	4224	480.0	5283
4.0	4298	960.0	5334
8.0	4420	1440.0	5364

Using the logarithm-of-time method, determine c_v. The average height of the specimen during consolidation was 2.24 cm, and it was drained at the top and bottom.

Solution

The semi-logarithmic plot of dial reading vs. time is shown in Figure 7.21. For this, $t_1 = 0.1$ min, $t_2 = 0.4$ min to determine d_0. Following the procedure outlined in Figure 7.19, $t_{50} \approx 19$ min. From Eq. (7.44)

$$C_v = \frac{0.197 H_{dr}^2}{t_{50}} = \frac{0.197\left(\dfrac{2.24}{2}\right)^2}{19} = 0.013 \text{ cm}^2/\text{min} = \mathbf{2.17 \times 10^{-4} \text{ cm}^2/\text{sec}} \quad \blacksquare$$

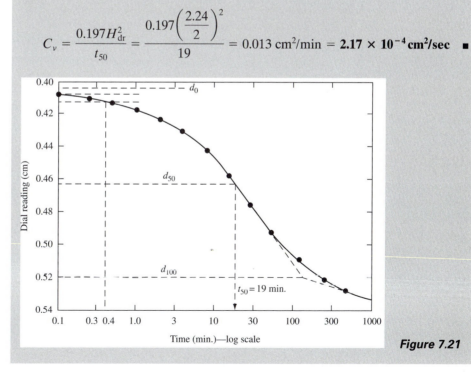

Figure 7.21

Example 7.6

A soil profile is shown in Figure 7.22. A surcharge load of 96 kN/m² is applied on the ground surface. Determine the following:

a. How high the water will rise in the piezometer immediately after the application of load.

b. After 104 days of the load application, $h = 4$ m. Determine the coefficient of consolidation (c_v) of the clay soil.

Solution
Part a.
Assuming uniform increase of initial excess pore water pressure through the 3 m depth of the clay layer

$$u_0 = \Delta\sigma = 96 \text{ kN/m}^2$$

$$h = \frac{96}{9.81} = \textbf{9.79 m}$$

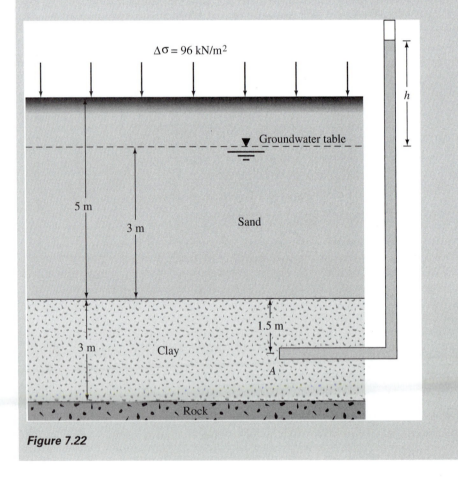

Figure 7.22

Part b.

$$U_A\% = \left(1 - \frac{u_A}{u_0}\right)100 = \left(1 - \frac{4 \times 9.81}{9.79 \times 9.81}\right)100 = \mathbf{59\%}$$

Since there is rock at the bottom of the clay layer, it is a one-way drainage case. For this type of drainage condition, the variation of U_z with z/H_{dr} for various values of T_v have been plotted in Figure 7.23. (Note: This has been taken from Figure 7.18, which is a two-way drainage case.). For this problem, $z/H_{dr} = 1.5/3 = 0.5$ and $U_z = 59\%$. Using these values of z/H_{dr} and U_z in Figure 7.23, we obtain $T_v \approx 0.3$.

$$T_v = \frac{c_v t}{H_{dr}^2}$$

$$0.3 = \frac{c_v(104 \times 24 \times 60 \times 60)}{(300 \text{ cm})^2}$$

$$c_v = \mathbf{0.003 \text{ cm}^2/\text{sec}}$$

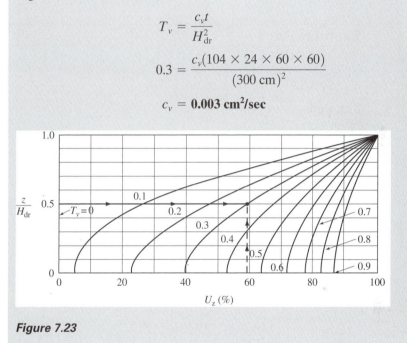

Figure 7.23

Example 7.7

The time required for 50% consolidation of a 25-mm-thick clay layer (drained at both top and bottom) in the laboratory is 2 min 20 sec. How long (in days) will it take for a 3-m-thick clay layer of the same clay in the field under the same pressure increment to reach 50% consolidation? In the field, there is a rock layer at the bottom of the clay.

Solution

$$T_{50} = \frac{c_v t_{tab}}{H_{dr(lab)}^2} = \frac{c_v t_{field}}{H_{dr(field)}^2}$$

or

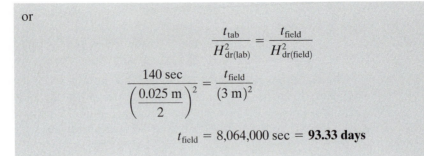

$$\frac{t_{tab}}{H^2_{dr(lab)}} = \frac{t_{field}}{H^2_{dr(field)}}$$

$$\frac{140 \text{ sec}}{\left(\dfrac{0.025 \text{ m}}{2}\right)^2} = \frac{t_{field}}{(3 \text{ m})^2}$$

$$t_{field} = 8{,}064{,}000 \text{ sec} = \textbf{93.33 days} \qquad \blacksquare$$

Example 7.8

Refer to Example 6.7. How long (in days) will it take in the field for 30% primary consolidation to occur? Use Eq. (7.40).

Solution
From Eq. (7.40), we have

$$\frac{c_v t_{field}}{H^2_{dr(field)}} = T_v \propto U^2$$

So

$$t \propto U^2$$

$$\frac{t_1}{t_2} = \frac{U_1^2}{U_2^2}$$

or

$$\frac{93.33 \text{ days}}{t_2} = \frac{50^2}{30^2}$$

$$t_2 = \textbf{33.6 days} \qquad \blacksquare$$

Example 7.9

For a normally consolidated clay,

$$\sigma'_o = 200 \text{ kN/m}^2 \qquad e = e_0 = 1.22$$

$$\sigma'_o + \Delta\sigma' = 400 \text{ kN/m}^2 \qquad e = 0.98$$

The hydraulic conductivity, k, of the clay for the loading range is 0.61×10^{-4} m/day

    a. How long (in days) will it take for a 4-m-thick clay layer (drained on one side) in the field to reach 60% consolidation?
    b. What is the settlement at that time (that is, at 60% consolidation)?

Solution

Part a.

The coefficient of volume compressibility is

$$m_v = \frac{a_v}{1 + e_{av}} = \frac{(\Delta e / \Delta \sigma')}{1 + e_{av}}$$

$$\Delta e = 1.22 - 0.98 = 0.24$$

$$\Delta \sigma' = 400 - 200 = 200 \text{ kN/m}^2$$

$$e_{av} = \frac{1.22 + 0.98}{2} = 1.1$$

So

$$m_v = \frac{0.24/200}{1 + 1.1} = 5.7 \times 10^{-4} \text{m}^2/\text{kN}$$

$$c_v = \frac{k}{m_v \gamma_w} = \frac{0.61 \times 10^{-4} \text{m/day}}{(5.7 \times 10^{-4} \text{m}^2/\text{kN})(9.81 \text{ kN/m}^2)} = 0.0109 \text{ m}^2/\text{day}$$

$$T_{60} = \frac{c_v t_{60}}{H_{dr}^2}$$

$$t_{60} = \frac{T_{60} H_{dr}^2}{c_v}$$

From Table 7.3, for $U = 60\%$, the value of T_{60} is 0.286, so

$$t_{60} = \frac{(0.286)(4)^2}{0.0109} = \textbf{419.8 days}$$

Part b.

$$C_c = \frac{e_1 - e_2}{\log(\sigma_2'/\sigma_1')} = \frac{1.22 - 0.98}{\log(400/200)} = 0.797$$

From Eq. (7.14), we have

$$S_p = \frac{C_c H}{1 + e_0} \log\left(\frac{\sigma_o' + \Delta \sigma'}{\sigma_o'}\right)$$

$$= \frac{(0.797)(4)}{1 + 1.22} \log\left(\frac{400}{200}\right) = 0.432 \text{ m}$$

$$S_p \text{ at } 60\% = (0.6)(0.432 \text{ m}) \approx \textbf{0.259 m}$$

Example 7.10

A laboratory consolidation test on a soil specimen (drained on both sides) determined the following results:

$$\text{thickness of the clay specimen} = 25 \text{ mm}$$

$$\sigma_1' = 50 \text{ kN/m}^2 \qquad e_1 = 0.92$$
$$\sigma_2' = 120 \text{ kN/m}^2 \qquad e_2 = 0.78$$

$$\text{time for 50\% consolidation} = 2.5 \text{ min}$$

Determine the hydraulic conductivity, k, of the clay for the loading range.

Solution

$$m_v = \frac{a_v}{1 + e_{av}} = \frac{(\Delta e/\Delta \sigma')}{1 + e_{av}}$$

$$= \frac{\dfrac{0.92 - 0.78}{120 - 50}}{1 + \dfrac{0.92 + 0.78}{2}} = 0.00108 \text{ m}^2/\text{kN}$$

$$c_v = \frac{T_{50}H_{dr}^2}{t_{50}}$$

From Table 7.3, for $U = 50\%$, the value of $T_v = 0.197$, so

$$c_v = \frac{(0.197)\left(\dfrac{0.025 \text{ m}}{2}\right)^2}{2.5 \text{ min}} = 1.23 \times 10^{-5} \text{ m}^2/\text{min}$$

$$k = c_v m_v \gamma_w = (1.23 \times 10^{-5})(0.00108)(9.81)$$

$$= \mathbf{1.303 \times 10^{-7} \, m/min} \qquad \blacksquare$$

7.11   *Calculation of Primary Consolidation Settlement under a Foundation*

Chapter 6 showed that the increase in the vertical stress in soil caused by a load applied over a limited area decreases with depth z measured from the ground surface downward. Hence, to estimate the one-dimensional settlement of a foundation, we can use Eq. (7.14), (7.16), or (7.17). However, the increase of effective stress $\Delta\sigma'$ in these equations should be the average increase below the center of the foundation,

Assuming the pressure increase varies parabolically, we can estimate the value of $\Delta\sigma_{av}'$ as (Simpson's rule)

$$\Delta\sigma_{av}' = \frac{\Delta\sigma_t + 4\Delta\sigma_m + \Delta\sigma_b}{6} \qquad (7.46)$$

Handwritten margin notes (left):

at o ft:
$\sigma_o = P/BL =$ load/area

at 5'
$\sigma_s = \dfrac{\sigma_o BL}{(B+2)}$

$P_{SAT} = \dfrac{M_w + M_s}{V_T}$

$M_{Avg} = \dfrac{SC}{SC}$

$H = M_S$

$P_w \; G_S \; A$

Handwritten margin notes (right/top):

$Z \; (B+Z)(L+Z) \; Area \; \Delta\sigma$

$0.42\,e_o$

$e = \dfrac{H}{H_S} - 1$

$e = \dfrac{H_v}{H_s}$

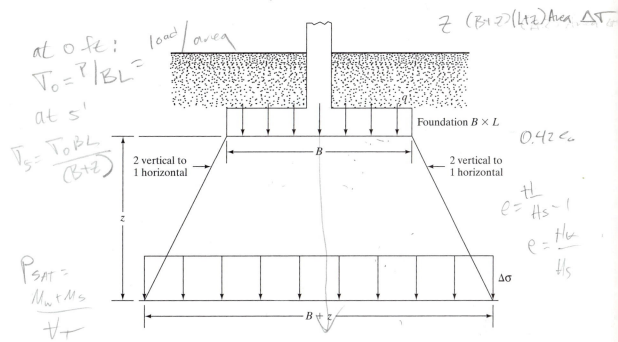

Figure 7.24 *2 : 1 method* of finding stress increase under a foundation

where $\Delta\sigma_t$, $\Delta\sigma_m$, and $\Delta\sigma_b$ represent the increase in the pressure at the top, middle, and bottom of the layer, respectively. The magnitudes of $\Delta\sigma_t$, $\Delta\sigma_m$, and $\Delta\sigma_b$ can be obtained using Eq. (6.45) and Table 6.8.

In several instances, foundation engineers use an approximate method to determine the increase of stress with depth caused by the construction of a foundation. This is referred to as the *2 : 1 method* (Figure 7.24). According to this method, the increase of stress at a depth z can be given as

$$\Delta\sigma = \frac{q \times B \times L}{(B + z)(L + z)} \qquad (7.47)$$

Note that Eq. (7.47) assumes that the stress from the foundation spreads out along lines with a *2 vertical* to *1 horizontal slope.*

Example 7.11

Calculate the primary consolidation settlement of the 3-m-thick clay layer (Figure 7.25) that will result from the load carried by a 1.5-m square footing. The clay is normally consolidated. Use *2 : 1 method* for calculation of $\Delta\sigma'$

Solution
For normally consolidated clay, from Eq. (7.14) we have

$$S_p = \frac{C_c H}{1 + e_0} \log\left(\frac{\sigma_o' + \Delta\sigma'}{\sigma_o'}\right)$$

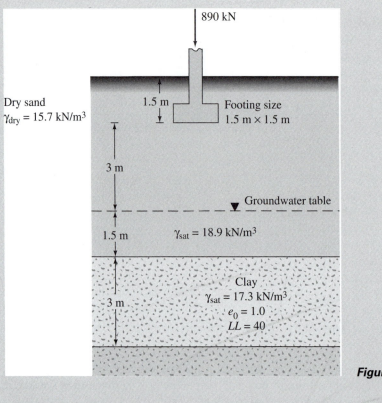

Figure 7.25

where

$$C_c = 0.009(LL - 10) = 0.009(40 - 10) = 0.27$$

$$H = 3000 \text{ mm}$$

$$e_0 = 1.0$$

$$\sigma_0' = 4.5 \times \gamma_{dry(sand)} + 1.5[\gamma_{sat(sand)} - 9.81] + \frac{3}{2}[\gamma_{sat(clay)} - 9.81]$$

$$= 4.5 \times 15.7 + 1.5(18.9 - 9.81) + 1.5(17.3 - 9.81) = 95.52 \text{ kN/m}^2$$

In order to calculate $\Delta\sigma'$, we can prepare the following table:

z (cm)	$B + z$ (m)	q^a (kN/m²)	$\Delta\sigma'$ [Eq. (7.47)]
4.5	6.0	395.6	$24.72 = \Delta\sigma_t'$
6.0	7.5	395.6	$15.82 = \Delta\sigma_m'$
7.5	9.0	395.6	$10.99 = \Delta\sigma_b'$

$$^a q = \frac{890}{1.5 \times 1.5} = 395.6 \text{ kN/m}^2$$

From Eq. (7.46)

$$\Delta\sigma'_{av} = \frac{24.72 + (4)(15.82) + 10.99}{6} \approx 16.5 \text{ kN/m}^2$$

$$S_p = \frac{(0.27)(3000)}{1 + 1} \log\left(\frac{95.52 + 16.5}{95.52}\right) = \textbf{28.0 mm}$$

Note: If we use Table 6.8 and Eq. (6.45) to estimate $\Delta\sigma'_{av}$, the value of S_p will be 21.3 mm.

7.12 Skempton-Bjerrum Modification for Consolidation Settlement

The consolidation settlement calculation presented in the preceding section is based on Eqs. (7.14), (7.16), and (7.17). These equations are based on one-dimensional laboratory consolidation tests. The underlying assumption for these equations is that the increase of pore water pressure (Δu) immediately after the load application is equal to the increase of stress ($\Delta\sigma$) at any depth. For this case

$$S_{p(oed)} = \int \frac{\Delta e}{1 + e_o} dz = \int m_v \, \Delta\sigma'_{(1)} \, dz \tag{7.48}$$

where

$S_{p(oed)}$ = primary consolidation settlement calculated by using Eqs. (7.14), (7.16), and (7.17)

$\Delta\sigma_{(1)}$ = vertical stress increase

m_v = volume coefficient of compressibility

In the field, however, when load is applied over a limited area on the ground surface, this assumption will not be correct. Consider the case of a circular foundation on a clay layer as shown in Figure 7.26. The vertical and the horizontal stress increases at a point in the clay layer immediately below the center of the foundation are $\Delta\sigma_{(1)}$ and $\Delta\sigma_{(3)}$, respectively. For a saturated clay, the pore water pressure increase at the depth can be given as (Chapter 8).

$$\Delta u = \Delta\sigma_{(3)} + A[\Delta\sigma_{(1)} - \Delta\sigma_{(3)}] \tag{7.49}$$

where A = pore water pressure parameter (see Chapter 8). For this case, one can write that

$$S_p = \int m_v \, \Delta u \, dz = \int (m_v)\{\Delta\sigma_{(3)} + A[\Delta\sigma_{(1)} - \Delta\sigma_{(3)}]\} \, dz \tag{7.50}$$

Combining Eqs. (7.48) and (7.50)

$$K_{cir} = \frac{S_p}{S_{p(oed)}} = \frac{\int_0^H m_v \, \Delta u \, dz}{\int_0^H m_v \, \Delta\sigma_{(1)} \, dz} = A + (1 - A)\left[\frac{\int_0^H \Delta\sigma_{(3)} \, dz}{\int_0^H \Delta\sigma_{(1)} \, dz}\right] \tag{7.51}$$

where K_{cir} = settlement ratio for circular foundations.

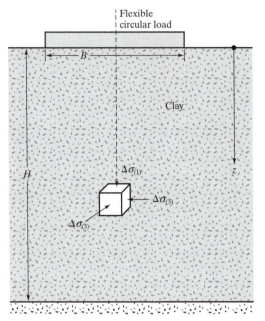

Figure 7.26
Circular foundation on a clay layer

The settlement ratio for a continuous foundation (K_{str}) can be determined in a manner similar to that for a circular foundation.) The variation of K_{cir} and K_{str} with A and H/B is given in Figure 7.27. (*Note: B* = diameter of a circular foundation, and *B* = width of a continuous foundation.)

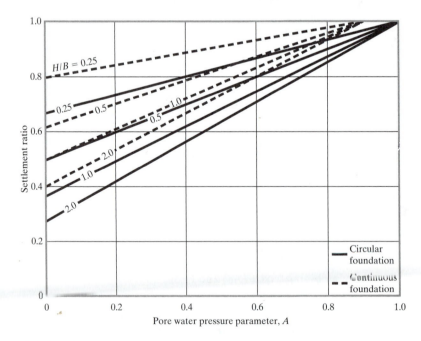

Figure 7.27 Settlement ratio for circular (K_{cir}) and continuous (K_{str}) foundations

Following is the procedure for determination of consolidation settlement according to Skempton and Bjerrum (1957).

1. Determine the primary consolidation settlement using the procedure outlined in Section 7.12. This is $S_{p(oed)}$. (Note the change of notation from S_p.)
2. Determine the pore water pressure parameter, A.
3. Determine H/B.
4. Obtain the settlement ratio—in this case, from Figure 7.27.
5. Calculate the actual consolidation settlement as

$$S_p = S_{p(oed)} \times \text{settlement ratio} \qquad (7.52)$$

$$\uparrow$$

$$\text{Step 1}$$

This technique is generally referred to as the *Skempton-Bjerrum modification* for consolidation settlement calculation.

Leonards (1976) considered the correction factor K_{cir} for three-dimensional consolidation effect in the field for a circular foundation located over *overconsolidated clay*. Referring to Figure 7.26,

$$S_p = K_{cir(OC)} S_{p(oed)} \qquad (7.53)$$

where

$$K_{cir(OC)} = f\left(OCR, \frac{B}{H}\right) \qquad (7.54)$$

$$OCR = \text{overconsolidation ratio} = \frac{\sigma_c'}{\sigma_o'} \qquad (7.55)$$

$$\sigma_c' = \text{preconsolidation pressure}$$

$$\sigma_o' = \text{present effective overburden pressure}$$

The interpolated values of $K_{cir(OC)}$ from the work of Leonards (1976) are given in Table 7.4. The procedure for using the afore-mentioned modification factors is demonstrated in Example 7.12.

Table 7.4 Variation of $K_{cir(OC)}$ with *OCR* and *B/H*

	$K_{cir(OC)}$		
OCR	**B/H = 4.0**	**B/H = 1.0**	**B/H = 0.2**
1	1	1	1
2	0.986	0.957	0.929
3	0.972	0.914	0.842
4	0.964	0.871	0.771
5	0.950	0.829	0.707
6	0.943	0.800	0.643
7	0.929	0.757	0.586
8	0.914	0.729	0.529
9	0.900	0.700	0.493
			(continued)

Table 7.4 (*continued*)

OCR	$K_{cir(OC)}$		
	$B/H = 4.0$	$B/H = 1.0$	$B/H = 0.2$
10	0.886	0.671	0.457
11	0.871	0.643	0.429
12	0.864	0.629	0.414
13	0.857	0.614	0.400
14	0.850	0.607	0.386
15	0.843	0.600	0.371
16	0.843	0.600	0.357

Example 7.12

Refer to Example 7.11. Assume that the clay is overconsolidated. Given $OCR = 3$, swell index $(C_s) \approx \frac{1}{4}C_c$.

    a. Calculate the primary consolidation settlement, S_p.

    b. Assuming the three dimensional effect, modify the settlement calculated in Part a.

Solution

Part a.

From Example 7.11, $\sigma'_o = 95.52 \text{ kN/m}^2$. Since $OCR = 3$, the preconsolidation pressure $\sigma'_c = (OCR)(\sigma'_o) = (3)(95.52) = 286.56 \text{ kN/m}^2$. For this case

$$\sigma'_o + \Delta\sigma'_{av} = 95.52 + 16.5 < \sigma'_c$$

So, Eq. (7.16) may be used

$$S_p = \frac{C_s H}{1 + e_0} \log\left(\frac{\sigma'_o + \Delta\sigma'_{av}}{\sigma'_o}\right) = \frac{\left(\frac{0.27}{4}\right)(3000)}{1 + 1} \log\left(\frac{95.52 + 16.5}{95.52}\right) = \textbf{7.0 mm}$$

Part b.

Assuming that the 2 : 1 method of stress increase holds good, the area of distribution of stress at the top of the clay layer will have dimensions of

$$B' = \text{width} = B + z = 1.5 + 4.5 = 6 \text{ m}$$

$$L' = \text{width} = L + z = 1.5 + 4.5 = 6 \text{ m}$$

The diameter of an equivalent circular area, B_{eq}, can be given as

$$\frac{\pi}{4}B_{eq}^2 = B'L'$$

$$B_{eq} = \sqrt{\frac{4B'L'}{\pi}} = \sqrt{\frac{(4)(6)(6)}{\pi}} = 6.77 \text{ m}$$

$$\frac{B_{eq}}{H} = \frac{6.77}{3} = 2.26$$

From Table 7.4, for $OCR = 3$ and $B_{eq}/H = 2.26$, $K_{cir(OC)} \approx 0.95$. Hence

$$S_p = K_{cir(OC)} S_{p(oed)} = (0.95)(7.0) = \mathbf{6.65 \ mm}$$ ∎

7.13 *Precompression—General Considerations*

When highly compressible, normally consolidated clayey soil layers lie at a limited depth and large consolidation settlements are expected as a result of the construction of large buildings, highway embankments, or earth dams, precompression of soil may be used to minimize postconstruction settlement. The principles of precompression are best explained by referring to Figure 7.28. Here, the proposed structural load per unit area is $\Delta\sigma_{(p)}$ and the thickness of the clay layer undergoing consolidation is H. The maximum primary consolidation settlement caused by the structural load, S_p, then is

$$S_p = \frac{C_c H}{1 + e_0} \log \frac{\sigma_o' + \Delta\sigma_{(p)}}{\sigma_o'} \tag{7.56}$$

Note that at the end of consolidation, $\Delta\sigma' = \Delta\sigma_{(p)}$.

The settlement–time relationship under the structural load will be like that shown in Figure 7.28b. However, if a surcharge of $\Delta\sigma_{(p)} + \Delta\sigma_{(f)}$ is placed on the ground, then the primary consolidation settlement, S_p', will be

$$S_{(p+f)} = \frac{C_c H}{1 + e_0} \log \frac{\sigma_o' + [\Delta\sigma_{(p)} + \Delta\sigma_{(f)}]}{\sigma_o'} \tag{7.57}$$

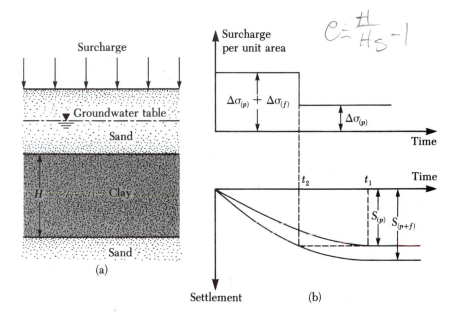

Figure 7.28 Principles of precompression

Note that at the end of consolidation,

$$\Delta\sigma' = \Delta\sigma_{(p)} + \Delta\sigma_{(f)}$$

The settlement–time relationship under a surcharge of $\Delta\sigma_{(p)} + \Delta\sigma_{(f)}$ is also shown in Figure 7.28b. Note that a total settlement of S_p would occur at a time t_2, which is much shorter than t_1. So, if a temporary total surcharge of $\Delta\sigma_{(f)} + \Delta\sigma_{(p)}$ is applied on the ground surface for time t_2, the settlement will equal S_p. At that time, if the surcharge is removed and a structure with a permanent load per unit area of $\Delta\sigma_{(p)}$ is built, no appreciable settlement will occur. The procedure just described is *precompression*. The total surcharge, $\Delta\sigma_{(p)} + \Delta\sigma_{(f)}$, can be applied by using temporary fills.

Derivation of Equations to Obtain $\Delta\sigma_{(f)}$ and t_2

Figure 7.28b shows that, under a surcharge of $\Delta\sigma_{(p)} + \Delta\sigma_{(f)}$, the degree of consolidation at time t_2 after load application is

$$U = \frac{S_p}{S_{(p+f)}} \tag{7.58}$$

Substitution of Eqs. (7.56) and (7.57) into Eq. (7.58) yields

$$U = \frac{\log\left[\dfrac{\sigma_o' + \Delta\sigma_{(p)}}{\sigma_o'}\right]}{\log\left[\dfrac{\sigma_o' + \Delta\sigma_{(p)} + \Delta\sigma_{(f)}}{\sigma_o'}\right]} = \frac{\log\left[1 + \dfrac{\Delta\sigma_{(p)}}{\sigma_o'}\right]}{\log\left\{1 + \dfrac{\Delta\sigma_{(p)}}{\sigma_o'}\left[1 + \dfrac{\Delta\sigma_{(f)}}{\Delta\sigma_{(p)}}\right]\right\}} \tag{7.59}$$

Figure 7.29 gives magnitudes of U for various combinations of $\Delta\sigma_{(p)}/\sigma_o'$ and $\Delta\sigma_{(f)}/\Delta\sigma_{(p)}$. The degree of consolidation referred to in Eq. (7.59) is actually the average degree of consolidation at time t_2, as shown in Figure 7.28. However, if the average degree of consolidation is used to determine time t_2, some construction problems might arise. The reason is that, after the removal of the surcharge and placement of the structural load, the portion of clay close to the drainage surface will continue to swell, and the soil close to the midplane will continue to settle (Figure 7.30). In some cases, net continuous settlement might result. A conservative approach may solve this problem; that is, assume that U in Eq. (7.59) is the midplane degree of consolidation (Johnson, 1970). Now, from Eq. (7.39), we have

$$U = f(T_v) \tag{7.60}$$

where

T_v = time factor = $c_v t_2/H^2_{dr}$
c_v = coefficient of consolidation
t_2 = time
H_{dr} = maximum drainage path ($H/2$ for two-way drainage and H for one-way drainage)

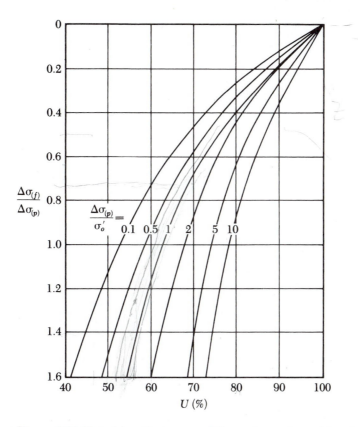

Figure 7.29 Plot of $\Delta\sigma_{(f)}/\Delta\sigma_{(p)}$ versus U for various values of $\Delta\sigma_{(p)}/\sigma'_o$—Eq. (7.59)

The variation of U (midplane degree of consolidation) with T_v is shown in Figure 7.31.

Procedure for Obtaining Precompression Parameters

Engineers may encounter two problems during precompression work in the field:

1. The value of $\Delta\sigma_{(f)}$ is known, but t_2 must be obtained. In such case, obtain σ'_o and $\Delta\sigma_{(p)}$ and solve for U using Eq. (7.59) or Figure 7.29. For this value of U, obtain T_v from Figure 7.31. Then

$$t_2 - \frac{T_v H_{dr}^2}{c_v} \qquad (7.61)$$

2. For a specified value of t_2, $\Delta\sigma_{(f)}$ must be obtained. In such case, calculate T_v. Then refer to Figure 7.31 to obtain the midplane degree of consolidation, U. With the estimated value of U, go to Figure 7.29 to find the required $\Delta\sigma_{(f)}/\Delta\sigma_{(p)}$ and then calculate $\Delta\sigma_{(f)}$.

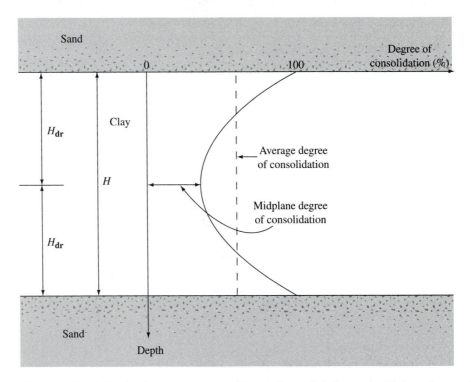

Figure 7.30 Distinction between average degree of consolidation and midplane degree of consolidation

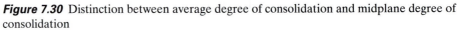

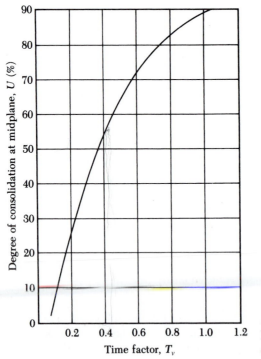

Figure 7.31
Plot of midplane degree of consolidation versus T_v

Sand Drains

The use of *sand drains* is another way to accelerate the consolidation settlement of soft, normally consolidated clay layers and achieve precompression before foundation construction. Sand drains are constructed by drilling holes through the clay layer(s) in the field at regular intervals. The holes are backfilled with highly permeable sand (see Figure 7.32a), and then a surcharge is applied at the ground surface. This surcharge will increase the pore water pressure in the clay. The excess pore water pressure in the clay will be dissipated by drainage—both vertically and radially to the sand drains—which accelerates settlement of the clay layer.

Note that the radius of the sand drains is r_w (Figure 7.32a). Figure 7.30b also shows the plan of the layout of the sand drains. The effective zone from which the radial drainage will be directed toward a given sand drain is approximately cylindrical, with a diameter of d_e.

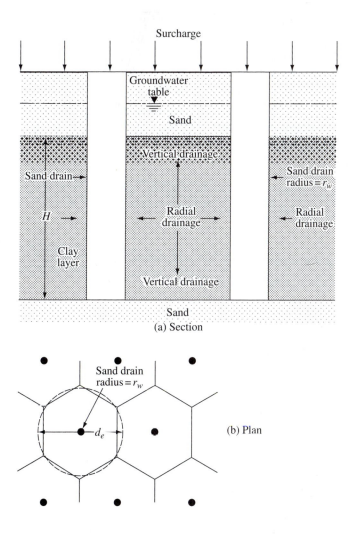

(a) Section

(b) Plan

Figure 7.32
Sand drains

To determine the surcharge that needs to be applied at the ground surface and the length of time that it has to be maintained, refer to Figure 7.28 and use the corresponding equation, Eq. (7.59):

$$U_{v,r} = \frac{\log\left[1 + \dfrac{\Delta\sigma_{(p)}}{\sigma'_o}\right]}{\log\left\{1 + \dfrac{\Delta\sigma_{(p)}}{\sigma'_o}\left[1 + \dfrac{\Delta\sigma_{(f)}}{\Delta\sigma_{(p)}}\right]\right\}} \tag{7.62}$$

The notations $\Delta\sigma_{(p)}$, σ'_o, and $\Delta\sigma_{(f)}$ are the same as those used in Eq. (7.59). However, unlike Eq. (7.59), the left-hand side of Eq. (7.62) is the *average degree* of consolidation instead of the degree of consolidation at midplane. Both *radial* and *vertical* drainage contribute to the average degree of consolidation. If $U_{v,r}$ can be determined for any time t_2 (see Figure 7.28b), then the total surcharge $\Delta\sigma_{(f)} + \Delta\sigma_{(p)}$ may be easily obtained from Figure 7.29. The procedure for determining the average degree of consolidation ($U_{v,r}$) is given in the following sections.

Average Degree of Consolidation Due to Radial Drainage Only

The theory for equal-strain consolidation due to radial drainage only (with no smear) was developed by Barron (1948). The theory is based on the assumption that there is *no drainage in the vertical direction*. According to this theory,

$$U_r = 1 - \exp\left(\frac{-8T_r}{m}\right) \tag{7.63}$$

where U_r = average degree of consolidation due to radial drainage only

$$m = \left(\frac{n^2}{n^2 - 1}\right)\ln(n) - \frac{3n^2 - 1}{4n^2} \tag{7.64}$$

$$n = \frac{d_e}{2r_w} \tag{7.65}$$

T_r = nondimensional time factor for radial drainage only

$$= \frac{c_{vr}t_2}{d_e^2} \tag{7.66}$$

c_{vr} = coefficient of consolidation for radial drainage

$$= \frac{k_h}{\left[\dfrac{\Delta e}{\Delta\sigma'(1 + e_{av})}\right]\gamma_w} \tag{7.67}$$

Note that Eq. (7.67) is similar to that defined in Eq. (7.35). In Eq. (7.35), k is the hydraulic conductivity in the vertical direction of the clay layer. In Eq. (7.67) k is replaced by k_h, the hydraulic conductivity for flow in the horizontal direction. In some cases, k_h may be assumed to equal k; however, for soils like varved clay, $k_h > k$. Table 7.5 gives the variation of U_r with T_r for various values of n.

Table 7.5 Solution for radial drainage

Degree of consolidation, U_r (%)	Time factor, T_r, for values of n				
	5	10	15	20	25
0	0	0	0	0	0
1	0.0012	0.0020	0.0025	0.0028	0.0031
2	0.0024	0.0040	0.0050	0.0057	0.0063
3	0.0036	0.0060	0.0075	0.0086	0.0094
4	0.0048	0.0081	0.0101	0.0115	0.0126
5	0.0060	0.0101	0.0126	0.0145	0.0159
6	0.0072	0.0122	0.0153	0.0174	0.0191
7	0.0085	0.0143	0.0179	0.0205	0.0225
8	0.0098	0.0165	0.0206	0.0235	0.0258
9	0.0110	0.0186	0.0232	0.0266	0.0292
10	0.0123	0.0208	0.0260	0.0297	0.0326
11	0.0136	0.0230	0.0287	0.0328	0.0360
12	0.0150	0.0252	0.0315	0.0360	0.0395
13	0.0163	0.0275	0.0343	0.0392	0.0431
14	0.0177	0.0298	0.0372	0.0425	0.0467
15	0.0190	0.0321	0.0401	0.0458	0.0503
16	0.0204	0.0344	0.0430	0.0491	0.0539
17	0.0218	0.0368	0.0459	0.0525	0.0576
18	0.0232	0.0392	0.0489	0.0559	0.0614
19	0.0247	0.0416	0.0519	0.0594	0.0652
20	0.0261	0.0440	0.0550	0.0629	0.0690
21	0.0276	0.0465	0.0581	0.0664	0.0729
22	0.0291	0.0490	0.0612	0.0700	0.0769
23	0.0306	0.0516	0.0644	0.0736	0.0808
24	0.0321	0.0541	0.0676	0.0773	0.0849
25	0.0337	0.0568	0.0709	0.0811	0.0890
26	0.0353	0.0594	0.0742	0.0848	0.0931
27	0.0368	0.0621	0.0776	0.0887	0.0973
28	0.0385	0.0648	0.0810	0.0926	0.1016
29	0.0401	0.0676	0.0844	0.0965	0.1059
30	0.0418	0.0704	0.0879	0.1005	0.1103
31	0.0434	0.0732	0.0914	0.1045	0.1148
32	0.0452	0.0761	0.0950	0.1087	0.1193
33	0.0469	0.0790	0.0987	0.1128	0.1239
34	0.0486	0.0820	0.1024	0.1171	0.1285
35	0.0504	0.0850	0.1062	0.1214	0.1332
36	0.0522	0.0881	0.1100	0.1257	0.1380

(*continued*)

Table 7.5 (*continued*)

Degree of consolidation, U_r (%)	Time factor, T_r, for values of n				
	5	10	15	20	25
37	0.0541	0.0912	0.1139	0.1302	0.1429
38	0.0560	0.0943	0.1178	0.1347	0.1479
39	0.0579	0.0975	0.1218	0.1393	0.1529
40	0.0598	0.1008	0.1259	0.1439	0.1580
41	0.0618	0.1041	0.1300	0.1487	0.1632
42	0.0638	0.1075	0.1342	0.1535	0.1685
43	0.0658	0.1109	0.1385	0.1584	0.1739
44	0.0679	0.1144	0.1429	0.1634	0.1793
45	0.0700	0.1180	0.1473	0.1684	0.1849
46	0.0721	0.1216	0.1518	0.1736	0.1906
47	0.0743	0.1253	0.1564	0.1789	0.1964
48	0.0766	0.1290	0.1611	0.1842	0.2023
49	0.0788	0.1329	0.1659	0.1897	0.2083
50	0.0811	0.1368	0.1708	0.1953	0.2144
51	0.0835	0.1407	0.1758	0.2020	0.2206
52	0.0859	0.1448	0.1809	0.2068	0.2270
53	0.0884	0.1490	0.1860	0.2127	0.2335
54	0.0909	0.1532	0.1913	0.2188	0.2402
55	0.0935	0.1575	0.1968	0.2250	0.2470
56	0.0961	0.1620	0.2023	0.2313	0.2539
57	0.0988	0.1665	0.2080	0.2378	0.2610
58	0.1016	0.1712	0.2138	0.2444	0.2683
59	0.1044	0.1759	0.2197	0.2512	0.2758
60	0.1073	0.1808	0.2258	0.2582	0.2834
61	0.1102	0.1858	0.2320	0.2653	0.2912
62	0.1133	0.1909	0.2384	0.2726	0.2993
63	0.1164	0.1962	0.2450	0.2801	0.3075
64	0.1196	0.2016	0.2517	0.2878	0.3160
65	0.1229	0.2071	0.2587	0.2958	0.3247
66	0.1263	0.2128	0.2658	0.3039	0.3337
67	0.1298	0.2187	0.2732	0.3124	0.3429
68	0.1334	0.2248	0.2808	0.3210	0.3524
69	0.1371	0.2311	0.2886	0.3300	0.3623
70	0.1409	0.2375	0.2967	0.3392	0.3724
71	0.1449	0.2442	0.3050	0.3488	0.3829
72	0.1490	0.2512	0.3134	0.3586	0.3937
73	0.1533	0.2583	0.3226	0.3689	0.4050
74	0.1577	0.2658	0.3319	0.3795	0.4167
75	0.1623	0.2735	0.3416	0.3906	0.4288
76	0.1671	0.2816	0.3517	0.4021	0.4414
77	0.1720	0.2900	0.3621	0.4141	0.4546
78	0.1773	0.2988	0.3731	0.4266	0.4683
79	0.1827	0.3079	0.3846	0.4397	0.4827
80	0.1884	0.3175	0.3966	0.4534	0.4978
81	0.1944	0.3277	0.4090	0.4679	0.5137
82	0.2007	0.3383	0.4225	0.4831	0.5304

Table 7.5 (*continued*)

Degree of consolidation, U_r (%)	Time factor, T_r, for values of n				
	5	10	15	20	25
83	0.2074	0.3496	0.4366	0.4922	0.5481
84	0.2146	0.3616	0.4516	0.5163	0.5668
85	0.2221	0.3743	0.4675	0.5345	0.5868
86	0.2302	0.3879	0.4845	0.5539	0.6081
87	0.2388	0.4025	0.5027	0.5748	0.6311
88	0.2482	0.4183	0.5225	0.5974	0.6558
89	0.2584	0.4355	0.5439	0.6219	0.6827
90	0.2696	0.4543	0.5674	0.6487	0.7122
91	0.2819	0.4751	0.5933	0.6784	0.7448
92	0.2957	0.4983	0.6224	0.7116	0.7812
93	0.3113	0.5247	0.6553	0.7492	0.8225
94	0.3293	0.5551	0.6932	0.7927	0.8702
95	0.3507	0.5910	0.7382	0.8440	0.9266
96	0.3768	0.6351	0.7932	0.9069	0.9956
97	0.4105	0.6918	0.8640	0.9879	1.0846
98	0.4580	0.7718	0.9640	1.1022	1.2100
99	0.5391	0.9086	1.1347	1.2974	1.4244

Average Degree of Consolidation Due to Vertical Drainage Only

The average degree of consolidation due to vertical drainage only may be obtained from Eqs. (7.40) and (7.41), (or Table 7.3):

$$T_v = \frac{\pi}{4}\left[\frac{U_v\%}{100}\right] \quad \text{for } U_v = 0\% \text{ to } 60\% \tag{7.68}$$

and

$$T_v = 1.781 - 0.933\log(100 - U_v\%) \quad \text{for } U_v > 60\% \tag{7.69}$$

where

U_v = average degree of consolidation due to vertical drainage only

$$T_v = \frac{c_v t_2}{H_{\mathrm{dr}}^2} \tag{7.70}$$

c_v = coefficient of consolidation for vertical drainage

Average Degree of Consolidation Due to Vertical and Radial Drainage

For a given surcharge and duration t_2, the average degree of consolidation due to drainage in the vertical and radial directions is

$$U_{v,r} = 1 - (1 - U_r)(1 - U_v) \tag{7.71}$$

Example 7.13

During the construction of a highway bridge, the average permanent load on the clay layer is expected to increase by about $115 \ kN/m^2$. The average effective overburden pressure at the middle of the clay layer is $210 \ kN/m^2$. Here, $H = 6 \ m$, $C_c = 0.28$, $e_0 = 0.9$, and $c_v = 0.36 \ m^2/mo$. The clay is normally consolidated.

    a. Determine the total primary consolidation settlement of the bridge without precompression.
    b. What is the surcharge, $\Delta\sigma_{(f)}$, needed to eliminate by precompression the entire primary consolidation settlement in 9 months?
    c. Redo part b with the addition of some sand drains. Assume that $r_w = 0.1$ m, $d_e = 3 \ m$, and $c_v = c_{vr}$.

Solution
Part a.
The total primary consolidation settlement will be calculated from Eq. (7.56):

$$S_p = \frac{C_c H}{1 + e_0} \log\left[\frac{\sigma_o' + \Delta\sigma_{(p)}}{\sigma_o'}\right] = \frac{(0.28)(6)}{1 + 0.9} \log\left[\frac{210 + 115}{210}\right]$$

$$= 0.1677 \ m = \textbf{167.7 mm}$$

Part b.

$$T_v = \frac{c_v t_2}{H_{dr}^2}$$

$$c_v = 0.36 \ m^2/mo.$$

$$H_{dr} = 3 \ m \ \text{(two-way drainage)}$$

$$t_2 = 9 \ mo.$$

Hence,

$$T_v = \frac{(0.36)(9)}{3^2} = 0.36$$

According to Figure 7.31, for $T_v = 0.36$, the value of U is 47%. Now

$$\Delta\sigma_{(p)} = 115 \ kN/m^2$$

$$\sigma_o' = 210 \ kN/m^2$$

So

$$\frac{\Delta\sigma_{(p)}}{\sigma_o'} = \frac{115}{210} - 0.548$$

According to Figure 7.29, for $U = 47\%$ and $\Delta\sigma_{(p)}/\sigma_o' = 0.548$, $\Delta\sigma_{(f)}/\Delta\sigma_{(p)} \approx 1.8$. So

$$\Delta\sigma_{(f)} = (1.8)(115) = \textbf{207 kN/m}^2 \qquad \blacksquare$$

Part c.
From Part b, $T_v = 0.36$. The value of U_v from Table 7.3 is about 67%. From Eq. (7.65), we have

$$n = \frac{d_e}{2r_w} = \frac{3}{2 \times 0.1} = 15$$

Again,

$$T_r = \frac{c_{vr}t_2}{d_e^2} = \frac{(0.36)(9)}{(3)^2} = 0.36$$

From Table 7.5 for $n = 15$ and $T_r = 0.36$, the value of U_r is about 77%. Hence,

$$U_{v,r} = 1 - (1 - U_v)(1 - U_r) = 1 - (1 - 0.67)(1 - 0.77)$$
$$= 0.924 = 92.4\%$$

Now, from Figure 7.29 for $\Delta\sigma_{(p)}/\sigma_o' = 0.548$ and $U_{v,r} = 92.4\%$, the value of $\Delta\sigma_{(f)}/\Delta\sigma_{(p)} \approx 0.12$. Hence, we have

$$\Delta\sigma_{(f)} = (115)(0.12) = \mathbf{13.8 \ kN/m^2} \qquad \blacksquare$$

Problems

7.1 The results of a laboratory consolidation test on a clay specimen are given in the table.

Pressure, σ' (kN/m²)	Total height of specimen at end of consolidation (mm)
25	17.65
50	17.40
100	17.03
200	16.56
400	16.15
800	15.88

Also, initial height of specimen = 19 mm, $G_s = 2.68$, mass of dry specimen = 95.2 g, and area of specimen = 31.68 cm².
a. Draw the e–$\log \sigma'$ graph.
b. Determine the preconsolidation pressure.
c. Determine the compression index, C_c.

7.2 Following are the results of a consolidation test:

e	Pressure, σ' (kN/m²)
1.1	25
1.085	50
1.055	100
1.01	200
0.94	400
0.79	800
0.63	1600

    a. Plot the e–log σ' curve.
    b. Using Casagrande's method, determine the preconsolidation pressure.
    c. Calculate the compression index, C_c.

7.3 The coordinates of two points on a virgin compression curve are given here:

$$\sigma'_1 = 190 \text{ kN/m}^2 \qquad e_1 = 1.75$$

$$\sigma'_2 = 385 \text{ kN/m}^2 \qquad e_2 = 1.49$$

Determine the void ratio that will correspond to an effective pressure of 500 kN/m^2.

7.4 Figure 7.33 shows a soil profile. The uniformly distributed load on the ground surface is $\Delta\sigma$. Estimate the primary consolidation settlement of the clay layer given these values:

$$H_1 = 1.5 \text{ m}, H_2 = 2 \text{ m}, H_3 = 2.5 \text{ m}$$

$$\text{Sand: } e = 0.62, G_s = 2.62$$

$$\text{Clay: } e = 0.98, G_s = 2.75, LL = 50$$

$$\Delta\sigma = 110 \text{ kN/m}^2$$

7.5 Repeat Problem 7.4 with the following values:

$$H_1 = 1.5 \text{ m}, H_2 = 2 \text{ m}, H_3 = 2 \text{ m}$$

$$\text{Sand: } e = 0.55, G_s = 2.67$$

$$\text{Clay: } e = 1.1, G_s = 2.73, LL = 45$$

$$\Delta\sigma = 120 \text{ kN/m}^2$$

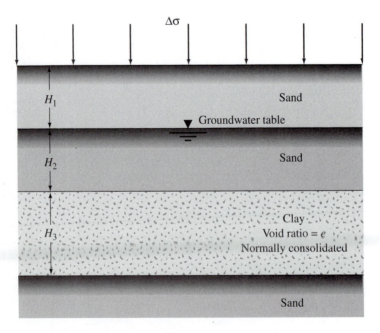

Figure 7.33

7.6 Repeat Problem 7.4 for these data:

$$\Delta\sigma = 87 \text{ kN/m}^2$$

$$H_1 = 1 \text{ m}, H_2 = 3 \text{ m}, H_3 = 3.2 \text{ m}$$

$$\text{Sand: } \gamma_{dry} = 14.6 \text{ kN/m}^3, \gamma_{sat} = 17.3 \text{ kN/m}^3$$

$$\text{Clay: } \gamma_{sat} = 19.3 \text{ kN/m}^3, LL = 38, e = 0.75$$

7.7 If the clay layer in Problem 7.6 is preconsolidated and the average preconsolidation pressure is 80 kN/m^2, what will be the expected primary consolidation settlement if $C_s = \frac{1}{5}C_c$.

7.8 Given are the relationships of e and σ' for a clay soil:

e	σ' (kN/m^2)
1.0	20
0.97	50
0.85	180
0.75	320

For this clay soil in the field, the following values are given: $H = 2.5$ m, $\sigma'_o = 60$ kN/m^2, and $\sigma'_o + \Delta\sigma' = 210$ kN/m^2. Calculate the expected settlement caused by primary consolidation.

7.9 Consider the virgin compression curve described in Problem 7.3.
   a. Find the coefficient of volume compressibility for the pressure range stated.
   b. If the coefficient of consolidation for the pressure range is 0.0023 cm^2/sec, find the hydraulic conductivity in (cm/sec) of the clay corresponding to the average void ratio.

7.10 Refer to Problem 7.4 Given $c_v = 0.003$ cm^2/sec, how long will it take for 50% primary consolidation to take place?

7.11 Laboratory tests on a 25-mm-thick clay specimen drained at the top and bottom show that 50% consolidation takes place in 8.5 min.
   a. How long will it take for a similar clay layer in the field, 3.2 m thick and drained at the top only, to undergo 50% consolidation?
   b. Find the time required for the clay layer in the field described in part (a) to reach 65% consolidation.

7.12 A 3-m-thick layer (two-way drainage) of saturated clay under a surcharge loading underwent 90% primary consolidation in 75 days. Find the coefficient of consolidation of clay for the pressure range.

7.13 For a 30-mm-thick undisturbed clay specimen described in Problem 7.12, how long will it take to undergo 90% consolidation in the laboratory for a similar consolidation pressure range? The laboratory test's specimen will have two-way drainage.

7.14 A normally consolidated clay layer is 5 m thick (one-way drainage). From the application of a given pressure, the total anticipated primary consolidation settlement will be 160 mm.
   a. What is the average degree of consolidation for the clay layer when the settlement is 50 mm?

b. If the average value of c_v for the pressure range is 0.003 cm²/sec, how long will it take for 50% settlement to occur?

c. How long will it take for 50% consolidation to occur if the clay layer is drained at both top and bottom?

7.15 In laboratory consolidation tests on a clay specimen (drained on both sides), the following results were obtained:
  - thickness of clay layer = 25 mm
  - $\sigma'_1 = 50$ kN/m² $e_1 = 0.75$
  - $\sigma'_2 = 100$ kN/m² $e_2 = 0.61$
  - time for 50% consolidation $(t_{50}) = 3.1$ min

Determine the hydraulic conductivity of the clay for the loading range.

7.16 A continuous foundation is shown in Figure 7.34. Find the vertical stresses at A, B, and C caused by the load carried by the foundation. Given width of foundation, $B = 1$ m.

7.17 Use Eqs. (7.14) and (7.46) to calculate the settlement of the footing described in Problem 7.16 from primary consolidation of the clay layer given

  Sand: $e = 0.6$, $G_s = 2.65$; degree of saturation of sand above groundwater table is 30%.

  Clay: $e = 0.85$, $G_s = 2.75$, $LL = 45$; the clay is normally consolidated.

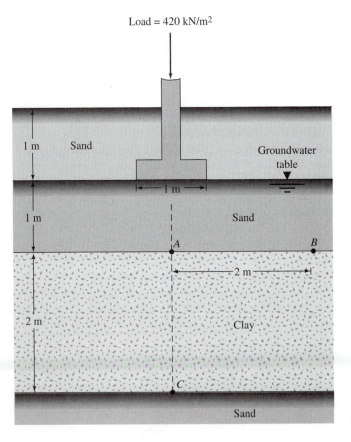

Figure 7.34

7.18 Refer to Problem 7.17. Using the results of Problem 7.17 and the Skempton-Bjeruum modification, calculate the primary consolidation settlement. Given pore water pressure parameter $A = 0.6$.

7.19 Refer to Figure 7.28. For the construction of an airport, a large fill operation is required. For the work, the average permanent load, $\Delta\sigma_{(p)}$, on the clay layer will increase by about 70 kN/m². The average effective overburden pressure on the clay layer before the fill operation is 95 kN/m². For the clay layer, which is normally consolidated and drained at top and bottom, $H = 5$ m, $C_c = 0.24$, $e_0 = 0.81$, and $c_v = 0.44$ m²/mo.

    a. Determine the primary consolidation settlement of the clay layer caused by the additional permanent load, $\Delta\sigma_{(p)}$.

    b. What is the time required for 90% of primary consolidation settlement under the additional permanent load only?

    c. What temporary surcharge, $\Delta\sigma_{(f)}$, will be required to eliminate the entire primary consolidation settlement in 6 months by the precompression technique?

[handwritten margin note: Table 7.3]

7.20 Redo part (c) of Problem 7.19 for a time of elimination of primary consolidation settlement of 7 months.

7.21 The diagram of a sand drain is shown in Figure 7.32. If $r_w = 0.3$ m, $d_e = 6$ m, $c_v = c_{vr} = 0.28$ m²/mo, and $H = 8.4$ m, determine the degree of consolidation caused only by the sand drain after 7 months of surcharge application.

7.22 Estimate the degree of consolidation for the clay layer described in Problem 7.21 that is caused by the combination of vertical drainage (drained on top and bottom) and radial drainage after 7 months of the application of surcharge.

7.23 A 4-m-thick clay layer is drained at top and bottom. Its characteristics are $c_{vr} = c_v$ (for vertical drainage) $= 0.0039$ m²/day, $r_w = 200$ mm, and $d_e = 2$ m. Estimate the degree of consolidation of the clay layer caused by the combination of vertical and radial drainage at $t = 0.2, 0.4, 0.8$, and 1 yr.

References

BARRON, R. A. (1948). "Consolidation of Fine-Grained Soils by Drain Wells," *Transactions*, American Society of Civil Engineers, Vol. 113, 718–754.

CASAGRANDE, A. (1936). "Determination of the Preconsolidation Load and Its Practical Significance," *Proceedings*, 1st International Conference on Soil Mechanics and Foundation Engineering, Cambridge, MA, Vol. 3, 60–64.

CASAGRANDE, A., and FADUM, R. E. (1940). "Notes on Soil Testing for Engineering Purposes," Harvard University Graduate School Engineering Publication No. 8.

JOHNSON, S. J. (1970). "Precompression for Improving Foundation Soils," *Journal of the Soil Mechanics and Foundations Division*, American Society of Civil Engineers, Vol. 96, No. SM1, 114–144.

KULHAWY, F. H., and MAYNE, P. W. (1990). *Manual on Estimating Soil Properties for Foundation Design*, Electric Power Research Institute, Palo Alto, California.

LEONARDS, G. A. (1976). "Estimating Consolidation Settlement of Shallow Foundations on Overconsolidated Clay," *Special Report No. 163*, Transportation Research Board, Washington, D.C., pp. 13–16.

MESRI, G. (1973). "Coefficient of Secondary Compression," *Journal of the Soil Mechanics and Foundations Division*, ASCE, Vol. 99, No. SM1, 122–137.

MESRI, G. and GODLEWSKI, P. M. (1977). "Time and Stress – Compressibility Interrelationship", *Journal of the Geotechnical Engineering Division*, ASCE, Vol. 103, No. GT5, 417–430.

RENDON-HERRERO, O. (1983). "Universal Compression Index Equation," *Discussion, Journal of Geotechnical Engineering*, ASCE, Vol. 109, No. 10, 1349.

RENDON-HERRERO, O. (1980). "Universal Compression Index Equation," *Journal of the Geotechnical Engineering Division*, ASCE, Vol. 106, No. GT11, 1179–1200.

SCHMERTMANN, J. H. (1953). "Undisturbed Consolidation Behavior of Clay," *Transactions*, ASCE, Vol. 120, 1201.

SIVARAM, B., and SWAMEE, A. (1977). "A Computational Method for Consolidation Coefficient," *Soils and Foundations*, Vol. 17, No. 2, 48–52.

SKEMPTON, A. W. (1944). "Notes on the Compressibility of Clays," *Quarterly Journal of the Geological Society of London*, Vol. 100, 119–135.

SKEMPTON, A. W., and BJERRUM, L. (1957). "A Contribution to Settlement Analysis of Foundations in Clay," *Geotechnique*, London, Vol. 7, 178.

TAYLOR, D. W. (1942). "Research on Consolidation of Clays," *Serial No. 82*, Department of Civil and Sanitary Engineering, Massachusetts Institute of Technology, Cambridge, MA.

TERZAGHI, K. (1925). *Erdbaumechanik auf Bodenphysikalischer Grundlage*, Deuticke, Vienna.

TERZAGHI, K., and PECK, R. B. (1967). *Soil Mechanics in Engineering Practice*, 2nd ed., Wiley, New York.

WROTH, C. P. and WOOD, D. M. (1978) "The Correlation of Index Properties with Some Basic Engineering Properties of Soils", *Canadian Geotechnical Journal*, Vol. 15, No. 2, 137–145.

8

Shear Strength of Soil

The *shear strength* of a soil mass is the internal resistance per unit area that the soil mass can offer to resist failure and sliding along any plane inside it. Engineers must understand the nature of shearing resistance in order to analyze soil stability problems such as bearing capacity, slope stability, and lateral pressure on earth-retaining structures.

8.1   Mohr-Coulomb Failure Criteria

Mohr (1900) presented a theory for rupture in materials. This theory contended that a material fails because of a critical combination of normal stress and shear stress, and not from either maximum normal or shear stress alone. Thus, the functional relationship between normal stress and shear stress on a failure plane can be expressed in the form

$$\tau_f = f(\sigma) \tag{8.1}$$

where
τ_f = shear stress on the failure plane
σ = normal stress on the failure plane

The failure envelope defined by Eq. (8.1) is a curved line. For most soil mechanics problems, it is sufficient to approximate the shear stress on the failure plane as a linear function of the normal stress (Coulomb, 1776). This relation can be written as

$$\tau_f = c + \sigma \tan \phi \tag{8.2}$$

where
c = cohesion
ϕ = angle of internal friction

The preceding equation is called the *Mohr-Coulomb failure criteria*.

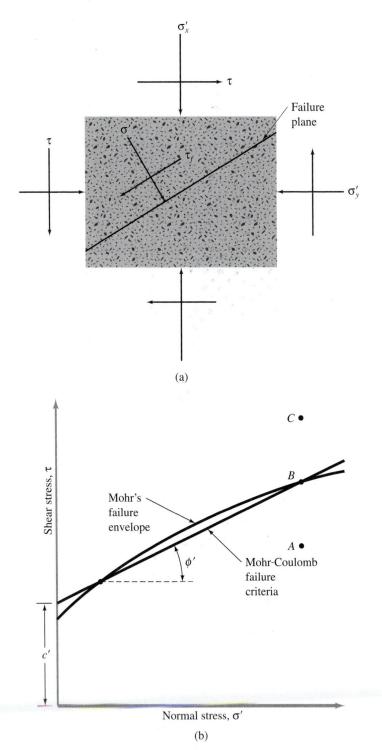

Figure 8.1 Mohr's failure envelope and the Mohr-Coulomb failure criteria

Table 8.1 Relationship between Relative Density and Angle of Friction of Cohesionless Soils

State of packing	Relative density (%)	Angle of friction, ϕ' (deg.)
Very loose	<20	<30
Loose	20–40	30–35
Compact	40–60	35–40
Dense	60–80	40–45
Very dense	>80	>45

In saturated soil, the total normal stress at a point is the sum of the effective stress and the pore water pressure, or

$$\sigma = \sigma' + u$$

The effective stress, σ', is carried by the soil solids. So, to apply Eq. (8.2) to soil mechanics, we need to rewrite it as

$$\tau_f = c' + (\sigma - u) \tan \phi' = c' + \sigma' \tan \phi' \qquad (8.3)$$

where
c' = effective stress cohesion
ϕ' = effective angle of friction

The significance of the failure envelope can be explained as follows: If the normal stress and the shear stress on a plane in a soil mass are such that they plot as point *A* in Figure 8.1b, then shear failure will not occur along that plane. If the normal stress and the shear stress on a plane plot as point *B* (which falls on the failure envelope), then shear failure will occur along that plane. A state of stress on a plane represented by point *C* cannot exist because it plots above the failure envelope, and shear failure in a soil would have occurred already.

The value of c' for sand and inorganic silt is 0. For normally consolidated clays, c' can be approximated at 0. Overconsolidated clays have values of c' that are greater than 0. The angle of friction, ϕ', is sometimes referred to as the *drained angle of friction*. Typical values of ϕ' for some granular soils are given in Table 8.1.

For normally consolidated clays, the friction angle ϕ' generally ranges from 20° to 30°. For overconsolidated clays, the magnitude of ϕ' decreases. For natural noncemented, overconsolidated clays with preconsolidation pressure less than about 1000 kN/m², the magnitude of c' is in the range of 5 to 15 kN/m².

8.2 *Inclination of the Plane of Failure Caused by Shear*

As stated by the Mohr-Coulomb failure criteria, failure from shear will occur when the shear stress on a plane reaches a value given by Eq. (8.3). To determine the inclination of the failure plane with the major principal plane, refer to Figure 8.2a, where

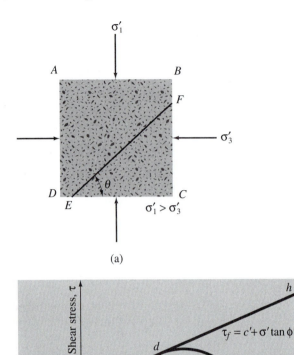

(a)

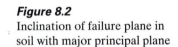

Normal stress, σ'

(b)

Figure 8.2
Inclination of failure plane in
soil with major principal plane

σ'_1 and σ'_3 are, respectively, the effective major and minor principal stresses. The failure plane EF makes an angle θ with the major principal plane. To determine the angle θ and the relationship between σ'_1 and σ'_3 refer to Figure 8.2b, which is a plot of the Mohr's circle for the state of stress shown in Figure 8.2a. In Figure 8.2b, *fgh* is the failure envelope defined by the relationship $\tau_f = c' + \sigma' \tan \phi'$. The radial line ab defines the major principal plane (CD in Figure 8.2a), and the radial line ad defines the failure plane (EF in Figure 8.2a). It can be shown that $<bad = 2\theta = 90 + \phi'$, or

$$\theta = 45 + \frac{\phi'}{2} \tag{8.4}$$

Again, from Figure 8.2b, we have

$$\frac{\overline{ad}}{\overline{fa}} = \sin \phi' \tag{8.5}$$

$$\overline{fa} = fO + Oa = c' \cot \phi' + \frac{\sigma'_1 + \sigma'_3}{2} \tag{8.6}$$

Also,

$$\overline{ad} = \frac{\sigma_1' - \sigma_3'}{2} \tag{8.7}$$

Substituting Eqs. (8.6) and (8.7) into Eq. (8.5), we obtain

$$\sin \phi' = \frac{\dfrac{\sigma_1' - \sigma_3'}{2}}{c' \cot \phi' + \dfrac{\sigma_1' + \sigma_3'}{2}}$$

or

$$\sigma_1' = \sigma_3' \left(\frac{1 + \sin \phi'}{1 - \sin \phi'} \right) + 2c' \left(\frac{\cos \phi'}{1 - \sin \phi'} \right) \tag{8.8}$$

However,

$$\frac{1 + \sin \phi'}{1 - \sin \phi'} = \tan^2 \left(45 + \frac{\phi'}{2} \right)$$

and

$$\frac{\cos \phi'}{1 - \sin \phi'} = \tan \left(45 + \frac{\phi'}{2} \right)$$

Thus,

$$\sigma_1' = \sigma_3' \tan^2 \left(45 + \frac{\phi'}{2} \right) + 2c' \tan \left(45 + \frac{\phi'}{2} \right) \tag{8.9}$$

The preceding relationship is Mohr-Coulomb's failure criteria restated in terms of failure stresses.

LABORATORY DETERMINATION OF SHEAR STRENGTH PARAMETERS

The shear strength parameters of a soil are determined in the laboratory primarily with two types of tests: direct shear test and triaxial test. The procedures for conducting each of these tests are explained in some detail in the following sections.

8.3   *Direct Shear Test*

This is the oldest and simplest form of shear test arrangement. A diagram of the direct shear test apparatus is shown in Figure 8.3. The test equipment consists of a metal shear box in which the soil specimen is placed. The soil specimens may be square or circular. The size of the specimens generally used is about 20 to 25 cm^2 across and 25 to 30 mm high. The box is split horizontally into halves. Normal force

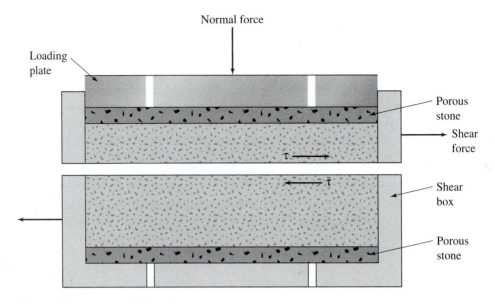

Figure 8.3 Diagram of direct shear test arrangement

on the specimen is applied from the top of the shear box. The normal stress on the specimens can be as great as 1000 kN/m². Shear force is applied by moving one half of the box relative to the other to cause failure in the soil specimen.

Depending on the equipment, the shear test can be either stress-controlled or strain-controlled. In stress-controlled tests, the shear force is applied in equal increments until the specimen fails. The failure takes place along the plane of split of the shear box. After the application of each incremental load, the shear displacement of the top half of the box is measured by a horizontal dial gauge. The change in the height of the specimen (and thus the volume change of the specimen) during the test can be obtained from the readings of a dial gauge that measures the vertical movement of the upper loading plate.

In strain-controlled tests, a constant rate of shear displacement is applied to one half of the box by a motor that acts through gears. The constant rate of shear displacement is measured by a horizontal dial gauge. The resisting shear force of the soil corresponding to any shear displacement can be measured by a horizontal proving ring or load cell. The volume change of the specimen during the test is obtained in a manner similar to the stress-controlled tests. Figure 8.4 is a photograph of strain-controlled direct shear test equipment.

The advantage of the strain-controlled tests is that, in the case of dense sand, peak shear resistance (that is, at failure) as well as lesser shear resistance (that is, at a point after failure called *ultimate strength*) can be observed and plotted. In stress-controlled tests, only peak shear resistance can be observed and plotted. Note that the peak shear resistance in stress-controlled tests can only be approximated. This is because failure occurs at a stress level somewhere between the prefailure load increment and the failure load increment. Nevertheless, stress-controlled tests probably simulate real field situations better than strain-controlled tests.

Figure 8.4 Direct shear test equipment (Courtesy of ELE International)

For a given test on dry soil, the normal stress can be calculated as

$$\sigma = \sigma' = \text{normal stress} = \frac{\text{normal force}}{\text{area of cross section of the specimen}} \qquad (8.10)$$

The resisting shear stress for any shear displacement can be calculated as

$$\tau = \text{shear stress} = \frac{\text{resisting shear force}}{\text{area of cross section of the specimen}} \qquad (8.11)$$

Figure 8.5 shows a typical plot of shear stress and change in the height of the specimen versus shear displacement for loose and dense sands. These observations

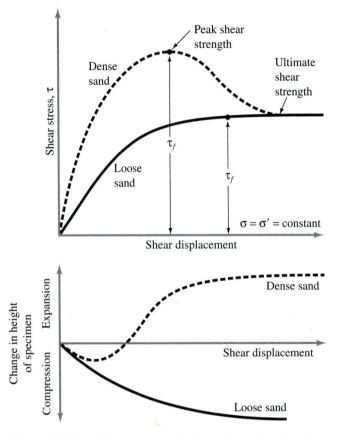

Figure 8.5 Plot of shear stress and change in height of specimen versus shear displacement for loose and dense dry sand (direct shear test)

were obtained from a strain-controlled test. The following generalizations can be made from Figure 8.5 regarding the variation of resisting shear stress with shear displacement:

1. In loose sand, the resisting shear stress increases with shear displacement until a failure shear stress of τ_f is reached. After that, the shear resistance remains approximately constant with any further increase in the shear displacement.

2. In dense sand, the resisting shear stress increases with shear displacement until it reaches a failure stress of τ_f. This τ_f is called the *peak shear strength*. After failure stress is attained, the resisting shear stress gradually decreases as shear displacement increases until it finally reaches a constant value called the *ultimate shear strength*.

Direct shear tests are repeated on similar specimens at various normal stresses. The normal stresses and the corresponding values of τ_f obtained from a number of tests are plotted on a graph, from which the shear strength parameters are determined.

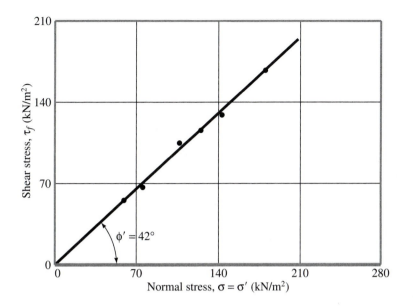

Figure 8.6 Determination of shear strength parameters for a dry sand using the results of direct shear tests

Figure 8.6 shows such a plot for tests on a dry sand. The equation for the average line obtained from experimental results is

$$\tau_f = \sigma' \tan \phi' \qquad (8.12)$$

(*Note:* $c' = 0$ for sand and $\sigma = \sigma'$ for dry conditions.) So the friction angle

$$\phi' = \tan^{-1}\left(\frac{\tau_f}{\sigma'}\right) \qquad (8.13)$$

Drained Direct Shear Test on Saturated Sand and Clay

The shear box that contains the soil specimen is generally kept inside a container that can be filled with water to saturate the specimen. A *drained test* is made on a saturated soil specimen by keeping the rate of loading slow enough so that the excess pore water pressure generated in the soil completely dissipates by drainage. Pore water from the specimen is drained through two porous stones (see Figure 8.3).

Since the hydraulic conductivity of sand is high, the excess pore water pressure generated because of loading (normal and shear) is dissipated quickly. Hence, for an ordinary loading rate, essentially full drainage conditions exist. The friction angle ϕ' obtained from a drained direct shear test of saturated sand will be the same as that for a similar specimen of dry sand.

The hydraulic conductivity of clay is very small compared with that of sand. When a normal load is applied to a clay soil specimen, a sufficient length of time must pass for full consolidation—that is, for dissipation of excess pore water pressure. For that reason, the shearing load has to be applied at a very slow rate. The test may last from 2 to 5 days.

General Comments on Direct Shear Test

The direct shear test is rather simple to perform, but it has some inherent shortcomings. The reliability of the results may be questioned. This is due to the fact that in this test the soil is not allowed to fail along the weakest plane but is forced to fail along the plane of split of the shear box. Also, the shear stress distribution over the shear surface of the specimen is not uniform. In spite of these shortcomings, the direct shear test is the simplest and most economical for a dry or saturated sandy soil.

In many foundation design problems, it will be necessary to determine the angle of friction between the soil and the material in which the foundation is constructed (Figure 8.7). The foundation material may be concrete, steel, or wood. The shear strength along the surface of contact of the soil and the foundation can be given as

$$\tau_f = c_a' + \sigma' \tan \delta' \tag{8.14}$$

where

c_a' = adhesion
δ' = effective angle of friction between the soil and the foundation material

Note that the preceding equation is similar in form to Eq. (8.3). The shear strength parameters between a soil and a foundation material can be conveniently determined by a direct shear test. This is a great advantage of the direct shear test. The foundation material can be placed in the bottom part of the direct shear test box and then the soil can be placed above it (that is, in the top part of the box), and the test can be conducted in the usual manner.

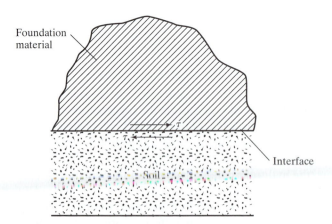

Figure 8.7 Interface of a foundation material and soil

Acar, Durgunoglu, and Tumay (1982) conducted several direct shear tests to determine the shear strength parameters between a quartz sand and foundation materials such as concrete, wood, and steel. Following are the details of the sand:

$$\text{Maximum void ratio, } e_{max} = 0.716$$

$$\text{Minimum void ratio, } e_{min} = 0.51$$

$$\text{Normal stress, } \sigma' = 100 \text{ kN/m}^2$$

The results of their tests can be expressed by the following relationships:

$$\text{Sand:} \quad \tan \phi' = 0.771\left(\frac{1}{e}\right) - 0.372$$

$$\text{Sand and concrete:} \quad \tan \delta' = 0.539\left(\frac{1}{e}\right) - 0.131$$

$$\text{Sand and wood:} \quad \tan \delta' = 0.386\left(\frac{1}{e}\right) - 0.022$$

$$\text{Sand and steel:} \quad \tan \delta' = 0.171\left(\frac{1}{e}\right) + 0.078$$

It is also important to realize that the relationships for ϕ' and δ' will vary depending on the magnitude of the effective normal stress, σ'. The reason for that can be explained by referring to Figure 8.8. It was mentioned in Section 8.1 that Mohr's failure envelope is actually curved, and Eq. (8.3) is only an approximation. If a direct shear test is conducted with $\sigma' = \sigma'_{(1)}$, the shear strength will be $\tau_{f(1)}$. So

$$\delta' = \delta'_1 = \tan^{-1}\left[\frac{\tau_{f(1)}}{\sigma'_{(1)}}\right]$$

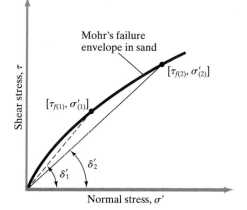

Figure 8.8 Curvilinear nature of Mohr's failure envelope in sand

This is shown in Figure 8.8. In a similar manner, if the test is conducted with $\sigma' = \sigma'_{(2)}$, then

$$\delta' = \delta'_2 = \tan^{-1}\left[\frac{\tau_{f(2)}}{\sigma'_{(2)}}\right]$$

As can be seen from Figure 8.8, $\delta'_2 < \delta'_1$ since $\sigma'_{(2)} > \sigma'_{(1)}$. Keeping this in mind, it must be realized that the values of ϕ' given in Table 8.1 are only the average values.

Example 8.1

Direct shear tests were performed on a dry, sandy soil. The size of the specimen was 50 mm $\times$ 50 mm $\times$ 20 mm. Tests results were as given in the table.

Test no.	Normal force (N)	Normal stress,* $\sigma = \sigma'$ (kN/m²)	Shear force at failure (N)	Shear stress at failure,[†] τ_f (kN/m²)
1	90	36	54	21.6
2	135	54	82.35	32.9
3	315	126	189.5	75.8
4	450	180	270.5	108.2

$$* \sigma = \frac{\text{normal force}}{\text{area of specimen}} = \frac{\text{normal force} \times 10^{-3}\text{kN}}{50 \times 50 \times 10^{-6}\text{m}^2}$$

$$^{†}\tau_f = \frac{\text{shear force}}{\text{area of specimen}} = \frac{\text{shear force} \times 10^{-3}\text{kN}}{50 \times 50 \times 10^{-6}\text{m}^2}$$

Find the shear stress parameters.

Solution
The shear stresses, τ_f, obtained from the tests are plotted against the normal stresses in Figure 8.9, from which we find $c' = 0$, $\phi' = 31°$. ■

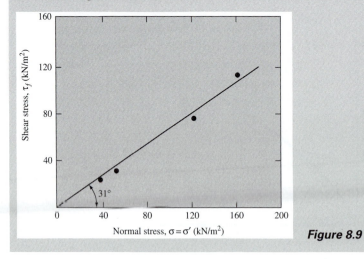

Figure 8.9

8.4 **Triaxial Shear Test**

The triaxial shear test is one of the most reliable methods available for determining the shear strength parameters. It is widely used for both research and conventional testing. The test is considered reliable for the following reasons:

1. It provides information on the stress–strain behavior of the soil that the direct shear test does not.
2. It provides more uniform stress conditions than the direct shear test does with its stress concentration along the failure plane.
3. It provides more flexibility in terms of loading path.

A diagram of the triaxial test layout is shown in Figure 8.10.

In the triaxial shear test, a soil specimen about 36 mm in diameter and 76 mm long is generally used. The specimen is encased by a thin rubber membrane and placed inside a plastic cylindrical chamber that is usually filled with water or glycerine. The specimen is subjected to a confining pressure by compression of the fluid in the chamber. (Note that air is sometimes used as a compression medium.) To cause shear failure in the specimen, axial stress is applied

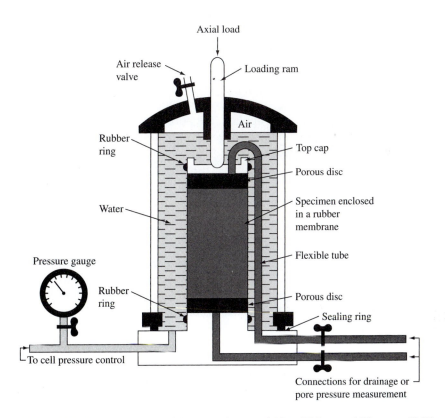

Figure 8.10 Diagram of triaxial test equipment (After Bishop and Bjerrum, 1960)

through a vertical loading ram (sometimes called *deviator stress*). Stress is added in one of two ways:

1. Application of dead weights or hydraulic pressure in equal increments until the specimen fails. (Axial deformation of the specimen resulting from the load applied through the ram is measured by a dial gauge.)
2. Application of axial deformation at a constant rate by a geared or hydraulic loading press. This is a strain-controlled test. The axial load applied by the loading ram corresponding to a given axial deformation is measured by a proving ring or load cell attached to the ram.

Connections to measure drainage into or out of the specimen, or to measure pressure in the pore water (as per the test conditions), are also provided. Three standard types of triaxial tests are generally conducted:

1. Consolidated-drained test or drained test (CD test)
2. Consolidated-undrained test (CU test)
3. Unconsolidated-undrained test or undrained test (UU test)

The general procedures and implications for each of the tests in *saturated soils* are described in the following sections.

8.5   *Consolidated-Drained Test*

In the consolidated-drained test, the specimen is first subjected to an all-around confining pressure, σ_3, by compression of the chamber fluid (Figure 8.11). As confining pressure is applied, the pore water pressure of the specimen increases by u_c. This increase in the pore water pressure can be expressed in the form of a nondimensional parameter:

$$B = \frac{u_c}{\sigma_3} \tag{8.15}$$

where B = Skempton's pore pressure parameter (Skempton, 1954).

For saturated soft soils, B is approximately equal to 1; however, for saturated stiff soils, the magnitude of B can be less than 1. Black and Lee (1973) gave the theoretical values of B for various soils at complete saturation. These values are listed in Table 8.2.

When the connection to drainage is kept open, dissipation of the excess pore water pressure, and thus consolidation, will occur. With time, u_c will become equal to 0. In saturated soil, the change in the volume of the specimen (ΔV_c) that takes place during consolidation can be obtained from the volume of pore water drained (Figure 8.12a). Then the deviator stress, $\Delta \sigma_d$, on the specimen is increased at a very slow rate (Figure 8.12b). The drainage connection is kept open, and the slow rate of

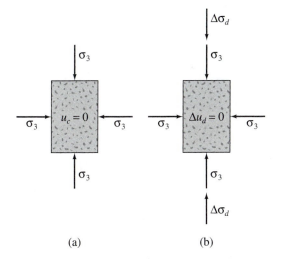

(a)   (b)

Figure 8.11 Consolidated-drained triaxial test: (a) specimen under chamber confining pressure; (b) deviator stress application

Table 8.2 Theoretical values of B at complete saturation

Type of soil	Theoretical value
Normally consolidated soft clay	0.9998
Lightly overconsolidated soft clays and silts	0.9988
Overconsolidated stiff clays and sands	0.9877
Very dense sands and very stiff clays at high confining pressures	0.9130

deviator stress application allows complete dissipation of any pore water pressure that developed as a result ($\Delta u_d = 0$).

A typical plot of the variation of deviator stress against strain in loose sand and normally consolidated clay is shown in Figure 8.12b. Figure 8.12c shows a similar plot for dense sand and overconsolidated clay. The volume change, ΔV_d, of specimens that occurs because of the application of deviator stress in various soils is also shown in Figures 8.12d and e.

Since the pore water pressure developed during the test is completely dissipated, we have

$$\text{total and effective confining stress} = \sigma_3 = \sigma_3'$$

and

$$\text{total and effective axial stress at failure} = \sigma_3 + (\Delta\sigma_d)_f = \sigma_1 = \sigma_1'$$

In a triaxial test, σ_1' is the major principal effective stress at failure and σ_3' is the minor principal effective stress at failure.

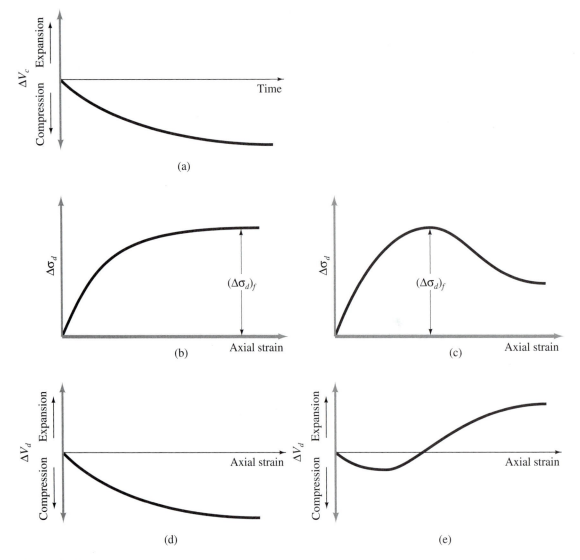

Figure 8.12 Consolidated-drained triaxial test: (a) volume change of specimen caused by chamber confining pressure; (b) plot of deviator stress against strain in the vertical direction for loose sand and normally consolidated clay; (c) plot of deviator stress against strain in the vertical direction for dense sand and overconsolidated clay; (d) volume change in loose sand and normally consolidated clay during deviator stress application; (e) volume change in dense sand and overconsolidated clay during deviator stress application

Figure 8.13 shows a soil specimen at failure during a consolidated-drained triaxial test.

Several tests on similar specimens can be conducted by varying the confining pressure. With the major and minor principal stresses at failure for each test, the Mohr's circles can be drawn and the failure envelopes can be obtained. Figure 8.14

Figure 8.13 Triaxial soil specimen at failure during consolidated-drained test (Courtesy of Braja Das)

shows the type of effective stress failure envelope obtained for tests in sand and normally consolidated clay. The coordinates of the point of tangency of the failure envelope with a Mohr's circle (that is, point A) give the stresses (normal and shear) on the failure plane of that test specimen.

Overconsolidation results when a clay is initially consolidated under an all-around chamber pressure of $\sigma_c \, (= \sigma'_c)$ and is allowed to swell as the chamber pressure is reduced to $\sigma_3 \, (= \sigma'_3)$ The failure envelope obtained from drained triaxial tests of such overconsolidated clay specimens shows two distinct branches (*ab* and *bc* in Figure 8.15). The portion *ab* has a flatter slope with a cohesion intercept, and the shear strength equation for this branch can be written as

$$\tau_f = c' + \sigma' \tan \phi'_1 \tag{8.16}$$

The portion *bc* of the failure envelope represents a normally consolidated stage of soil and follows the equation $\tau_f = \sigma' \tan \phi'$.

A consolidated-drained triaxial test on a clayey soil may take several days to complete. The time is needed to apply deviator stress at a very slow rate to ensure full drainage from the soil specimen. For that reason, the CD type of triaxial test is not commonly used.

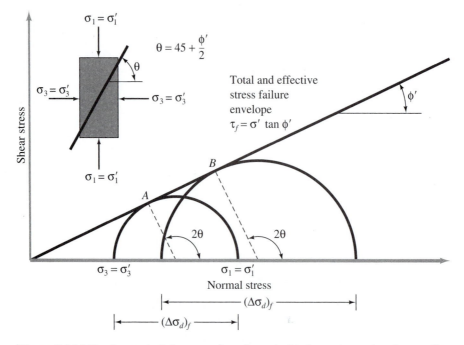

Figure 8.14 Effective stress failure envelope from drained tests in sand and normally consolidated clay

Effective Stress Friction Angle of Cohesive Soils

Figure 8.16 shows the variation of effective stress friction angle, ϕ', for several normally consolidated clays (Bejerrum and Simons, 1960; Kenney, 1959). It can be

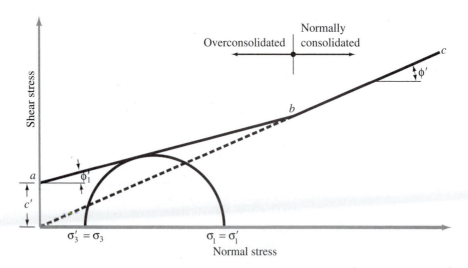

Figure 8.15 Effective stress failure envelope for overconsolidated clay

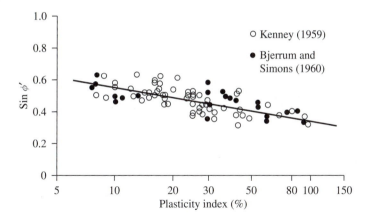

Figure 8.16 Variation of sin ϕ' with plasticity index (PI) for several normally consolidated clays

seen from the figure that, in general, the friction angle ϕ' decreases with the increase in plasticity index. The value of ϕ' generally decreases from about 37 to 38° with a plasticity index of about 10 to about 25° or less with a plasticity index of about 100.

Example 8.2

For a normally consolidated clay, these are the results of a drained triaxial test:

$$\text{chamber confining pressure} = 112 \text{ kN/m}^2$$

$$\text{deviator stress at failure} = 175 \text{ kN/m}^2$$

a. Find the angle of friction, ϕ'.
b. Determine the angle θ that the failure plane makes with the major principal plane.

Solution
For normally consolidated soil, the failure envelope equation is

$$\tau_f = \sigma' \tan \phi' \, (\text{since } c' = 0)$$

For the triaxial test, the effective major and minor principal stresses at failure are

$$\sigma'_1 = \sigma_1 = \sigma_3 + (\Delta\sigma_d)_f = 112 + 175 = 287 \text{ kN/m}^2$$

and

$$\sigma'_3 = \sigma_3 = 112 \text{ kN/m}^2$$

Part a.
The Mohr's circle and the failure envelope are shown in Figure 8.17, from which we get

$$\sin \phi' = \frac{AB}{OA} = \frac{\left(\dfrac{\sigma_1' - \sigma_3'}{2} \right)}{\left(\dfrac{\sigma_1' + \sigma_3'}{2} \right)}$$

or

$$\sin \phi' = \frac{\sigma_1' - \sigma_3'}{\sigma_1' + \sigma_3'} = \frac{287 - 112}{287 + 112} = 0.438$$

$$\phi' = \mathbf{26°}$$

Part b.

$$\theta = 45 + \frac{\phi'}{2} = 45° + \frac{26}{2} = \mathbf{58°}$$

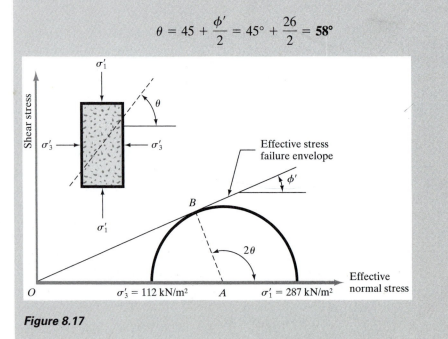

Figure 8.17

Example 8.3

Refer to Example 8.2.

a. Find the normal stress, σ', and the shear stress, τ_f, on the failure plane.
b. Determine the effective normal stress on the plane of maximum shear stress.

Solution
Part a.
From Figure 8.17, we can see that

$$\sigma' \text{ (on the failure plane)} = \frac{\sigma_1' + \sigma_3'}{2} + \frac{\sigma_1' - \sigma_3'}{2} \cos 2\theta \tag{a}$$

and

$$\tau_f = \frac{\sigma_1' - \sigma_3'}{2} \sin 2\theta \tag{b}$$

Substituting the values of $\sigma_1' = 287$ kN/m², $\sigma_3' = 112$ kN/m², and $\theta = 58°$ into the preceding equations, we get

$$\sigma' = \frac{287 + 112}{2} + \frac{287 - 112}{2} \cos(2 \times 58) = \textbf{161 kN/m}^2$$

and

$$\tau_f = \frac{287 - 112}{2} \sin(2 \times 58) = \textbf{78.6 kN/m}^2$$

Part b.
From Eq. (b), we can see that the maximum shear stress will occur on the plane with $\theta = 45°$. Substituting $\theta = 45°$ into Eq. (a) gives

$$\sigma' = \frac{287 + 112}{2} + \frac{287 - 112}{2} \cos 90 = \textbf{199.5 kN/m}^2 \qquad \blacksquare$$

Example 8.4

The equation of the effective stress failure envelope for normally consolidated clayey soil is $\tau_f = \sigma' \tan 30°$. A drained triaxial test was conducted with the same soil at a chamber confining pressure of 70 kN/m². Calculate the deviator stress at failure.

Solution
For normally consolidated clay, $c' = 0$. Thus, from Eq. (8.9), we have

$$\sigma_1' = \sigma_3' \tan^2 \left(45 + \frac{\phi'}{2} \right)$$

$$\phi' = 30°$$

$$\sigma_1' = 70 \tan^2 \left(45 + \frac{30}{2} \right) = 210 \text{ kN/m}^2$$

so

$$(\Delta\sigma_d)_f = \sigma_1' - \sigma_3' = 210 - 70 = \textbf{140 kN/m}^2 \qquad \blacksquare$$

Example 8.5

We have the results of two drained triaxial tests on a saturated clay:

$$Specimen\ I: \qquad \sigma_3 = 70\ kN/m^2$$

$$(\Delta\sigma_d)_f = 173\ kN/m^2$$

$$Specimen\ II: \qquad \sigma_3 = 105\ kN/m^2$$

$$(\Delta\sigma_d)_f = 235\ kN/m^2$$

Determine the shear strength parameters.

Solution

Refer to Figure 8.18. For specimen I, the principal stresses at failure are

$$\sigma'_3 = \sigma_3 = 70\ kN/m^2$$

and

$$\sigma'_1 = \sigma_1 = \sigma_3 + (\Delta\sigma_d)_f = 70 + 173 = 243\ kN/m^2$$

Similarly, the principal stresses at failure for specimen II are

$$\sigma'_3 = \sigma_3 = 105\ kN/m^2$$

and

$$\sigma'_1 = \sigma_1 = \sigma_3 + (\Delta\sigma_d)_f = 105 + 235 = 340\ kN/m^2$$

Using the relationship given by Eq. (8.9), we have

$$\sigma'_1 = \sigma'_3 \tan^2\left(45 + \frac{\phi'_1}{2}\right) + 2c' \tan\left(45 + \frac{\phi'_3}{2}\right)$$

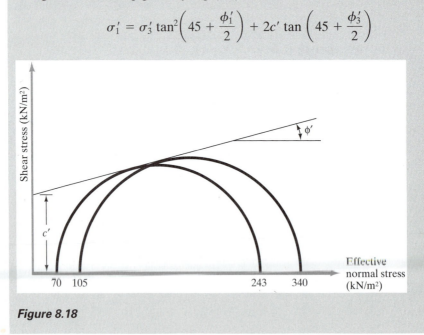

Figure 8.18

Thus, for specimen I,

$$243 = 70 \tan^2\left(45 + \frac{\phi'_1}{2}\right) + 2c'\tan\left(45 + \frac{\phi'_1}{2}\right)$$

and for specimen II,

$$340 = 105 \tan^2\left(45 + \frac{\phi'_1}{2}\right) + 2c'\tan\left(45 + \frac{\phi'_1}{2}\right)$$

Solving the two preceding equations, we obtain

$$\phi' = 28° \quad c' = 14.8 \text{ kN/m}^2 \qquad \blacksquare$$

8.6 *Consolidated-Undrained Test*

The consolidated-undrained test is the most common type of triaxial test. In this test, the saturated soil specimen is first consolidated by an all-round chamber fluid pressure, σ_3, that results in drainage. After the pore water pressure generated by the application of confining pressure is completely dissipated (that is, $u_c = B\sigma_3 = 0$), the deviator stress, $\Delta\sigma_d$, on the specimen is increased to cause shear failure. During this phase of the test, the drainage line from the specimen is kept closed. Since drainage is not permitted, the pore water pressure, Δu_d, will increase. During the test, measurements of $\Delta\sigma_d$ and Δu_d are made. The increase in the pore water pressure, Δu_d, can be expressed in a nondimensional form as

$$\overline{A} = \frac{\Delta u_d}{\Delta\sigma_d} \qquad (8.17)$$

where $\overline{A}$ = Skempton's pore pressure parameter (Skempton, 1954).

The general patterns of variation of $\Delta\sigma_d$ and Δu_d with axial strain for sand and clay soils are shown in Figures 8.19d, e, f, and g. In loose sand and normally consolidated clay, the pore water pressure increases with strain. In dense sand and overconsolidated clay, the pore water pressure increases with strain up to a certain limit, beyond which it decreases and becomes negative (with respect to the atmospheric pressure). This pattern is because the soil has a tendency to dilate.

Unlike in the consolidated-drained test, the total and effective principal stresses are not the same in the consolidated-undrained test. Since the pore water pressure at failure is measured in this test, the principal stresses may be analyzed as follows:

- Major principal stress at failure (total):

$$\sigma_3 + (\Delta\sigma_d)_f = \sigma_1$$

- Major principal stress at failure (effective):

$$\sigma_1 - (\Delta u_d)_f = \sigma'_1$$

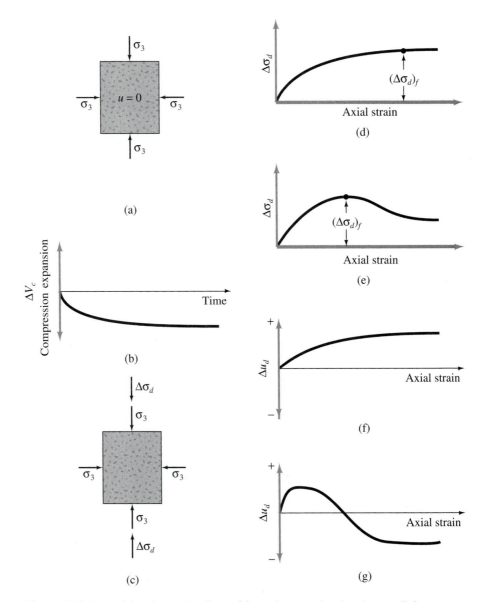

Figure 8.19 Consolidated-undrained test: (a) specimen under chamber confining pressure; (b) volume change in specimen caused by confining pressure; (c) deviator stress application; (d) deviator stress against axial strain for loose sand and normally consolidated clay; (e) deviator stress against axial strain for dense sand and overconsolidated clay; (f) variation of pore water pressure with axial strain for loose sand and normally consolidated clay; (g) variation of pore water pressure with axial strain for dense sand and overconsolidated clay

- Minor principal stress at failure (total):

$$\sigma_3$$

- Minor principal stress at failure (effective):

$$\sigma_3 - (\Delta u_d)_f = \sigma'_3$$

where $(\Delta u_d)_f$ = pore water pressure at failure. The preceding derivations show that

$$\sigma_1 - \sigma_3 = \sigma'_1 - \sigma'_3$$

Tests on several similar specimens with varying confining pressures may be done to determine the shear strength parameters. Figure 8.20 shows the total and effective stress Mohr's circles at failure obtained from consolidated-undrained triaxial tests in sand and normally consolidated clay. Note that A and B are two total stress Mohr's circles obtained from two tests. C and D are the effective stress Mohr's circles corresponding to total stress circles A and B, respectively. The diameters of circles A and C are the same; similarly, the diameters of circles B and D are the same.

In Figure 8.20, the total stress failure envelope can be obtained by drawing a line that touches all the total stress Mohr's circles. For sand and normally consolidated clays, this line will be approximately a straight line passing through the origin and may be expressed by the equation

$$\tau_f = \sigma \tan \phi \tag{8.18}$$

where

σ = total stress
ϕ = the angle that the total stress failure envelope makes with the normal stress axis, also known as the consolidated-undrained angle of shearing resistance

Equation (8.18) is seldom used for practical considerations.

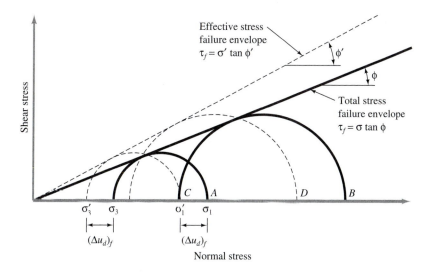

Figure 8.20 Total and effective stress failure envelopes for consolidated-undrained triaxial tests. (*Note:* The figure assumes that no back pressure is applied.)

Again referring to Figure 8.20, we see that the failure envelope that is tangent to all the effective stress Mohr's circles can be represented by the equation $\tau_f = \sigma'$ tan ϕ', which is the same as the failure envelope obtained from consolidated-drained tests (see Figure 8.14).

In overconsolidated clays, the total stress failure envelope obtained from consolidated-undrained tests takes the shape shown in Figure 8.21. The straight line $a'b'$ is represented by the equation

$$\tau_f = c + \sigma \tan \phi_1 \tag{8.19}$$

and the straight line $b'c'$ follows the relationship given by Eq. (8.18). The effective stress failure envelope drawn from the effective stress Mohr's circles is similar to that shown in Figure 8.21.

Consolidated-drained tests on clay soils take considerable time. For that reason, consolidated-undrained tests can be conducted on such soils with pore pressure measurements to obtain the drained shear strength parameters. Since drainage is not allowed in these tests during the application of deviator stress, the tests can be performed rather quickly.

Skempton's pore water pressure parameter $\overline{A}$ was defined in Eq. (8.17). At failure, the parameter $\overline{A}$ can be written as

$$\overline{A} = \overline{A}_f = \frac{(\Delta u_d)_f}{(\Delta \sigma_d)_f} \tag{8.20}$$

The general range of $\overline{A}_f$ values in most clay soils is as follows:

- Normally consolidated clays: 0.5 to 1
- Overconsolidated clays: −0.5 to 0

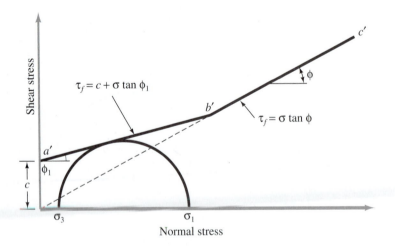

Figure 8.21 Total stress failure envelope obtained from consolidated-undrained tests in overconsolidated clay

Example 8.6

A consolidated-undrained test on a normally consolidated clay yielded the following results:

$$\sigma_3 = 84 \text{ kN/m}^2$$

$$\text{deviator stress, } (\Delta\sigma_d)_f = 63.7 \text{ kN/m}^2$$

$$\text{pore pressure, } (\Delta u_d)_f = 47.6 \text{ kN/m}^2$$

Calculate the consolidated-undrained friction angle and the drained friction angle.

Solution
Refer to Figure 8.22.

$$\sigma_3 = 84 \text{ kN/m}^2$$

$$\sigma_1 = \sigma_3 + (\Delta\sigma_d)_f = 84 + 63.7 = 147.7 \text{ kN/m}^2$$

$$\sigma_1 = \sigma_3 \tan^2\left(45 + \frac{\phi}{2}\right)$$

$$147.7 = 84 \tan^2\left(45 + \frac{\phi}{2}\right)$$

$$\phi = 2\left[\tan^{-1}\left(\frac{147.7}{84}\right)^{0.5} - 45\right] = \mathbf{16°}$$

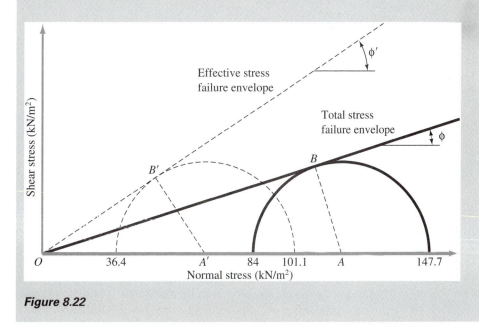

Figure 8.22

Again,

$$\sigma_3' = \sigma_3 - (\Delta u_d)_f = 84 - 47.6 = 36.4 \text{ kN/m}^2$$

$$\sigma_1' = \sigma_1 - (\Delta u_d)_f = 147.7 - 47.6 = 100.1 \text{ kN/m}^2$$

$$\sigma_1' = \sigma_3' \tan^2\left(45 + \frac{\phi'}{2}\right)$$

$$100.1 = 36.4 \tan^2\left(45 + \frac{\phi'}{2}\right)$$

$$\phi' = 2\left[\tan^{-1}\left(\frac{100.1}{36.4}\right)^{0.5} - 45\right] = \mathbf{27.8°} \qquad \blacksquare$$

8.7   *Unconsolidated-Undrained Test*

In unconsolidated-undrained tests, drainage from the soil specimen is not permitted during the application of chamber pressure, σ_3. The test specimen is sheared to failure by the application of deviator stress, $\Delta\sigma_d$, with no drainage allowed. Since drainage is not allowed at any stage, the test can be performed very quickly. Because of the application of chamber confining pressure, σ_3, the pore water pressure in the soil specimen will increase by u_c. There will be a further increase in the pore water pressure, Δu_d, because of the deviator stress application. Hence, the total pore water pressure, u, in the specimen at any stage of deviator stress application can be given as

$$u = u_c + \Delta u_d \qquad (8.21)$$

From Eqs. (8.15) and (8.17), we have $u_c = B\sigma_3$ and $\Delta u_d = \overline{A} \Delta\sigma_d$, so

$$u = B\sigma_3 + \overline{A} \Delta\sigma_d = B\sigma_3 + \overline{A}(\sigma_1 - \sigma_3) \qquad (8.22)$$

The unconsolidated-undrained test is usually conducted on clay specimens and depends on a very important strength concept for saturated cohesive soils. The added axial stress at failure $(\Delta\sigma_d)_f$ is practically the same regardless of the chamber confining pressure. This result is shown in Figure 8.23. The failure envelope for the total stress Mohr's circles becomes a horizontal line and hence is called a $\phi = 0$ condition, and

$$\tau_f = c_u \qquad (8.23)$$

where c_u is the undrained shear strength and is equal to the radius of the Mohr's circles.

The reason for obtaining the same added axial stress $(\Delta\sigma_d)_f$ regardless of the confining pressure is as follows: If a clay specimen (no. 1) is consolidated at a chamber pressure σ_3 and then sheared to failure with no drainage allowed, then the total stress conditions at failure can be represented by the Mohr's circle P in Figure 8.24. The pore pressure developed in the specimen at failure is equal to $(\Delta u_d)_f$. Thus, the major and minor principal effective stresses at failure are

$$\sigma_1' = [\sigma_3 + (\Delta\sigma_d)_f] - (\Delta u_d)_f = \sigma_1 - (\Delta u_d)_f$$

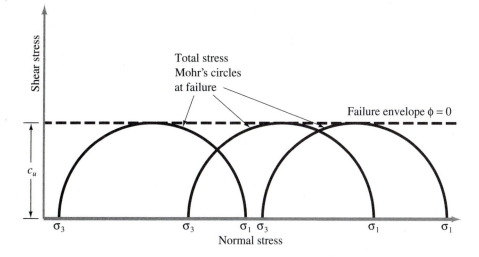

Figure 8.23 Total stress Mohr's circles and failure envelope ($\phi = 0$) obtained from unconsolidated-undrained triaxial tests

and

$$\sigma_3' = \sigma_3 - (\Delta u_d)_f$$

Q is the effective stress Mohr's circle drawn with the preceding principal stresses. Note that the diameters of circles P and Q are the same.

Now let us consider another similar clay specimen (no. 2) that is consolidated at a chamber pressure σ_3. If the chamber pressure is increased by $\Delta\sigma_3$ with no drainage allowed, then the pore water pressure increases by an amount Δu_c. For saturated

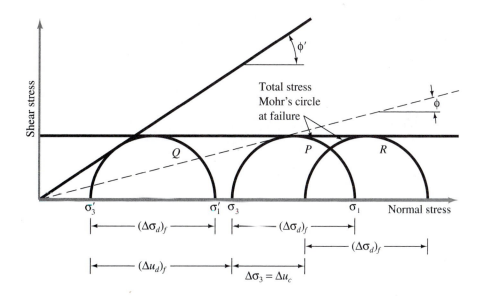

Figure 8.24 The $\phi = 0$ concept

soils under isotropic stresses, the pore water pressure increase is equal to the total stress increase, so $\Delta u_c = \Delta \sigma_3$. At this time, the effective confining pressure is equal to $\sigma_3 + \Delta \sigma_3 - \Delta u_c = \sigma_3 + \Delta \sigma_3 - \Delta \sigma_3 = \sigma_3$. This is the same as the effective confining pressure of specimen no. 1 before the application of deviator stress. Hence, if specimen no. 2 is sheared to failure by increasing the axial stress, it should fail at the same deviator stress $(\Delta \sigma_d)_f$ that was obtained for specimen no. 1. The total stress Mohr's circle at failure will be R (Figure 8.24). The added pore pressure increase caused by the application of $(\Delta \sigma_d)_f$ will be $(\Delta u_d)_f$.

At failure, the minor principal effective stress is

$$[\sigma_3 + \Delta \sigma_3] - [\Delta u_c + (\Delta u_d)_f] = \sigma_3 - (\Delta u_d)_f = \sigma'_3$$

and the major principal effective stress is

$$[\sigma_3 + \Delta \sigma_3 + (\Delta \sigma_d)_f] - [\Delta u_c + (\Delta u_d)_f] = [\sigma_3 + (\Delta \sigma_d)_f] - (\Delta u_d)_f$$
$$= \sigma_1 - (\Delta u_d)_f = \sigma'_1$$

Thus, the effective stress Mohr's circle will still be Q because strength is a function of effective stress. Note that the diameters of circles P, Q, and R are all the same.

Any value of $\Delta \sigma_3$ could have been chosen for testing specimen no. 2. In any case, the deviator stress $(\Delta \sigma_d)_f$ to cause failure would have been the same.

8.8 Unconfined Compression Test on Saturated Clay

The unconfined compression test is a special type of unconsolidated-undrained test that is commonly used for clay specimens. In this test, the confining pressure σ_3 is 0. An axial load is rapidly applied to the specimen to cause failure. At failure, the total minor principal stress is 0 and the total major principal stress is σ_1 (Figure 8.25). Since the undrained shear strength is independent of the confining pressure, we have

$$\tau_f = \frac{\sigma_1}{2} = \frac{q_u}{2} = c_u \tag{8.24}$$

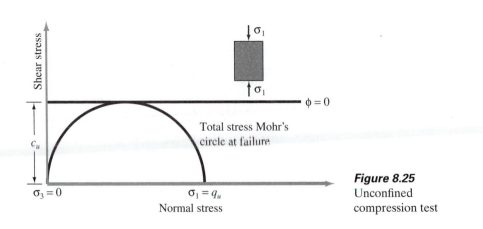

Figure 8.25
Unconfined
compression test

Table 8.3 General relationship of consistency and unconfined compression strength of clays

Consistency	q_u (kN/m²)
Very soft	0–25
Soft	25–50
Medium	50–100
Stiff	100–200
Very stiff	200–400
Hard	>400

where q_u is the *unconfined compression strength*. Table 8.3 gives the approximate consistencies of clays based on their unconfined compression strengths. A photograph of unconfined compression test equipment is shown in Figure 8.26.

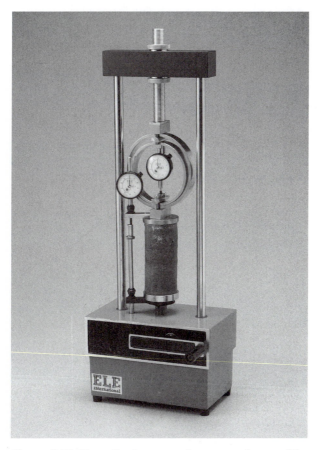

Figure 8.26 Unconfined compression test equipment (Courtesy of ELE International)

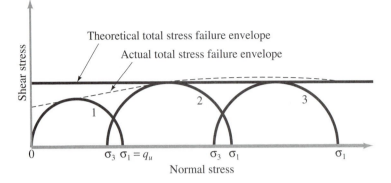

Figure 8.27 Comparison of results of unconfined compression tests and unconsolidated-undrained tests for a saturated clay soil. (*Note:* Mohr's circle no. 1 is for unconfined compression test; Mohr's circles no. 2 and 3 are for unconsolidated-undrained triaxial tests.)

Theoretically, for similar saturated clay specimens, the unconfined compression tests and the unconsolidated-undrained triaxial tests should yield the same values of c_u. In practice, however, unconfined compression tests on saturated clays yield slightly lower values of c_u than those obtained from unconsolidated-undrained tests. This fact is demonstrated in Figure 8.27.

8.9 *Sensitivity and Thixotropy of Clay*

For many naturally deposited clay soils, the unconfined compression strength is greatly reduced when the soils are tested after remolding without any change in the moisture content, as shown in Figure 8.28. This property of clay soils is called

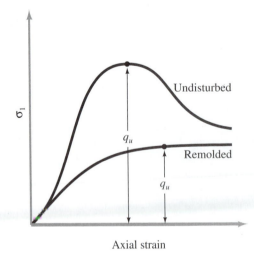

Figure 8.28 Unconfined compression strength for undisturbed and remolded clay

sensitivity. The degree of sensitivity may be defined as the ratio of the unconfined compression strength in an undisturbed state to that in a remolded state, or

$$S_t = \frac{q_{u(\text{undisturbed})}}{q_{u(\text{remolded})}}$$ (8.25)

The sensitivity ratio of most clays ranges from about 1 to 8; however, highly flocculent marine clay deposits may have sensitivity ratios ranging from about 10 to 80. There are also some clays that turn to viscous fluids upon remolding. These clays are found mostly in the previously glaciated areas of North America and Scandinavia and are referred to as "quick" clays. Rosenqvist (1953) classified clays on the basis of their sensitivity. This general classification is shown in Figure 8.29.

The loss of strength of clay soils from remolding is primarily caused by the destruction of the clay particle structure that was developed during the original process of sedimentation. If, however, after remolding, a soil specimen is kept in an undisturbed state (that is, without any change in the moisture content), it will continue to gain strength with time. This phenomenon is referred to as *thixotropy*. Thixotropy is a time-dependent reversible process in which materials under constant composition and volume soften when remolded. This loss of strength is gradually regained with time when the materials are allowed to rest.

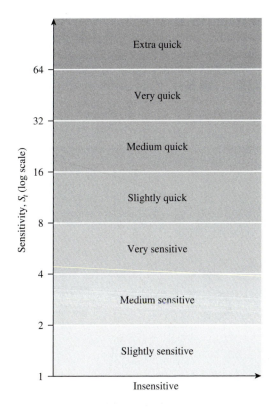

Figure 8.29
Classification of clays based on sensitivity

Most soils are partially thixotropic; part of the strength loss caused by remolding is never regained with time. For soils, the difference between the undisturbed strength and the strength after thixotropic hardening can be attributed to the destruction of the clay-particle structure that was developed during the original process of sedimentation.

8.10   Anisotropy in Undrained Shear Strength

Owing to the nature of the deposition of cohesive soils and subsequent consolidation, clay particles tend to become oriented perpendicular to the direction of the major principal stress. Parallel orientation of clay particles could cause the strength of the clay to vary with direction, or in other words, the clay could be anisotropic with respect to strength. This fact can be demonstrated with the aid of Figure 8.30, in which V and H are vertical and horizontal directions that coincide with lines perpendicular and parallel to the bedding planes of a soil deposit. If a soil specimen with its axis inclined at an angle i with the horizontal is collected and subjected to an undrained test, the undrained shear strength can be given by

$$c_{u(i)} = \frac{\sigma_1 - \sigma_3}{2} \qquad (8.26)$$

where $c_{u(i)}$ is the undrained shear strength when the major principal stress makes an angle i with the horizontal.

Let the undrained shear strength of a soil specimen with its axis vertical [i.e., $c_{u(i=90°)}$] be referred to as $c_{u(V)}$ (Figure 8.30a); similarly, let the undrained shear strength with its axis horizontal [i.e., $c_{u(i=0°)}$] be referred to as $c_{u(H)}$ (Figure 8.30c). If $c_{u(V)} = c_{u(i)} = c_{u(H)}$, the soil is isotropic with respect to strength, and the variation of undrained shear strength can be represented by a circle in a polar diagram, as shown by curve a in Figure 8.31. However, if the soil is anisotropic, $c_{u(i)}$ will change with

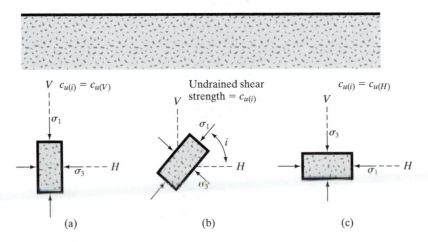

Figure 8.30 Strength anisotropy in clay

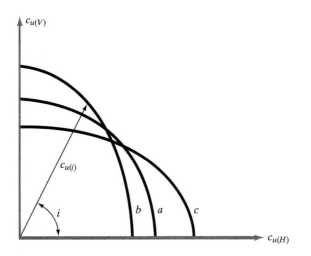

Figure 8.31 Directional variation of undrained strength of clay

direction. Casagrande and Carrillo (1944) proposed the following equation for the directional variation of the undrained shear strength:

$$c_{u(i)} = c_{u(H)} + [c_{u(V)} - c_{u(H)}] \sin^2 i \tag{8.27}$$

When $c_{u(V)} > c_{u(H)}$, the nature of variation of $c_{u(i)}$ can be represented by curve b in Figure 8.31. Again, if $c_{u(V)} < c_{u(H)}$, the variation of $c_{u(i)}$ is given by curve c. The coefficient of anisotropy can be defined as

$$K = \frac{c_{u(V)}}{c_{u(H)}} \tag{8.28}$$

In the case of natural soil deposits, the value of K can vary from 0.75 to 2.0. K is generally less than 1 in overconsolidated clays. Figure 8.32 shows the directional variation

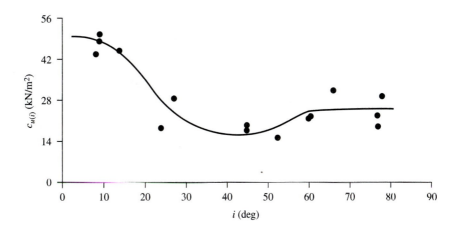

Figure 8.32 Directional variation of c_u for undisturbed Winnipeg Upper Brown clay (Based on Loh and Holt, 1974)

for $c_{u(\alpha)}$ based on Eq. (8.27). The anisotropy with respect to strength for clays can have an important effect on various stability calculations.

Problems

8.1 A direct shear test was conducted on a specimen of dry sand with a normal stress of 140 kN/m². Failure occurred at a shear stress of 94.5 kN/m². The size of the specimen tested was 50 mm × 50 mm × 25 mm (height). Determine the angle of friction, ϕ'. For a normal stress of 84 kN/m², what shear force would be required to cause failure of the specimen?

8.2 The size of a sand specimen in a direct shear test was 50 mm × 50 mm × 30 mm (height). It is known that, for the sand, tan $\phi' = 0.65/e$ (where e = void ratio) and the specific gravity of soil solids $G_s = 2.65$. During the test, a normal stress of 140 kN/m² was applied. Failure occurred at a shear stress of 105 kN/m². What was the mass of the sand specimen?

8.3 The angle of friction of a compacted dry sand is 38°. In a direct shear test on the sand, a normal stress of 84 kN/m² was applied. The size of the specimen was 50 mm × 50 mm × 30 mm (height). What shear force (in kN) will cause failure?

8.4 Repeat Problem 8.3 with the following changes:
friction angle = 37°
normal stress = 150 kN/m²

8.5 Following are the results of four drained direct shear tests on a normally consolidated clay:
diameter of specimen = 50 mm
height of specimen = 25 mm

Test no.	Normal force (N)	Shear force at failure (N)
1	271	120.6
2	406.25	170.64
3	474	204.1
4	541.65	244.3

Draw a graph for shear stress at failure against normal stress. Determine the drained angle of friction from the graph.

8.6 The relationship between the relative density, D_r, and the angle of friction, ϕ', of a sand can be given as $\phi'^\circ = 25 + 0.18D_r$ (D_r in %). A drained triaxial test on the same sand was conducted with a chamber confining pressure of 105 kN/m². The relative density of compaction was 45%. Calculate the major principal stress at failure.

8.7 Consider the triaxial test described in Problem 8.6.
   a. Estimate the angle that the failure plane makes with the major principal plane.
   b. Determine the normal and shear stresses (when the specimen failed) on a plane that makes an angle of 30° with the major principal plane.

8.8 The effective stress failure envelope of a sand can be given as $\tau_f = \sigma'$ tan 41°. A drained triaxial test was conducted on the same sand. The specimen failed when the deviator stress was 400.5 kN/m². What was the chamber confining pressure during the test?

8.9 Refer to Problem 8.8.

    a. Estimate the angle that the failure plane makes with the minor principal plane.

    b. Determine the normal stress and the shear stress on a plane that makes an angle of 35° with the minor principal plane.

8.10 For a normally consolidated clay, the results of a drained triaxial test are as follows:

- Chamber confining pressure = 150 kN/m²
- Deviator stress at failure = 275 kN/m²

Determine the soil friction angle, ϕ'.

8.11 For a normally consolidated clay, we are given $\phi' = 25°$. In a drained triaxial test, the specimen failed at a deviator stress of 154 kN/m². What was the chamber confining pressure, σ_3?

8.12 A consolidated-drained triaxial test was conducted on a normally consolidated clay. The results were as follows:

$$\sigma_3 = 276 \text{ kN/m}^2$$

$$(\Delta\sigma_d)_f = 276 \text{ kN/m}^2$$

    a. Find the angle of friction, ϕ'.

    b. What is the angle θ that the failure plane makes with the major principal stress?

    c. Determine the normal stress σ' and the shear stress τ_f on the failure plane.

8.13 Refer to Problem 8.12.

    a. Determine the effective normal stress on the plane of maximum shear stress.

    b. Explain why the shear failure took place along the plane as determined in part (b) and not along the plane of maximum shear stress.

8.14 The results of two drained triaxial tests on a saturated clay are given here:

- Specimen I: Chamber confining pressure = 69 kN/m²
                      Deviator stress at failure = 213 kN/m²
- Specimen II: Chamber confining pressure = 120 kN/m²
                      Deviator stress at failure = 258.7 kN/m²

Calculate the shear strength parameters of the soil.

8.15 A sandy soil has a drained angle of friction of 36°. In a drained triaxial test on the same soil, the deviator stress at failure is 268 kN/m². What is the chamber confining pressure?

8.16 A consolidated-undrained test was conducted on a normally consolidated specimen with a chamber confining pressure of 140 kN/m². The specimen failed while the deviator stress was 126 kN/m². The pore water pressure in the specimen at that time was 76.3 kN/m². Determine the consolidated-undrained and the drained friction angles.

8.17 Repeat Problem 8.16 with the following values:

$$\sigma_3 = 84 \text{ kN/m}^2$$

$$(\Delta\sigma_d)_f = 58.7 \text{ kN/m}^2$$

$$(\Delta u_d)_f = 39.2 \text{ kN/m}^2$$

8.18 The shear strength of a normally consolidated clay can be given by the equation $\tau_f = \sigma'\tan 28°$. A consolidated-undrained, triaxial test was conducted on the clay. Following are the results of the test:
- Chamber confining pressure = 105 kN/m²
- Deviator stress at failure = 97 kN/m²

a. Determine the consolidated-undrained friction angle, ϕ

b. What is the pore water pressure developed in the clay specimen at failure?

8.19 For the clay specimen described in Problem 8.18, what would have been the deviator stress at failure if a drained test had been conducted with the same chamber confining pressure (that is, $\sigma_3 = 105$ kN/m²)?

8.20 For a clay soil, we are given $\phi' = 28°$ and $\phi = 18°$. A consolidated-undrained triaxial test was conducted on this clay soil with a chamber confining pressure of 105 kN/m². Determine the deviator stress and the pore water pressure at failure.

8.21 During a consolidated-undrained triaxial test on a clayey soil specimen, the minor and major principal stresses at failure were 96 kN/m² and 187 kN/m², respectively. What will be the axial stress at failure if a similar specimen is subjected to an unconfined compression test?

8.22 The friction angle, ϕ', of a normally consolidated clay specimen collected during field exploration was determined from drained triaxial tests to be 22°. The unconfined compression strength, q_u, of a similar specimen was found to be 120 kN/m². Determine the pore water pressure at failure for the unconfined compression test.

8.23 Repeat Problem 8.22 with $\phi' = 25°$ and $q_u = 121.5$ kN/m².

References

ACAR, Y. B., DURGUNOGLU, H. T., and TUMAY, M. T. (1982). "Interface Properties of Sand," *Journal of the Geotechnical Engineering Division*, ASCE, Vol. 108, No. GT4, 648–654.

BISHOP, A. W., and BJERRUM, L. (1960). "The Relevance of the Triaxial Test to the Solution of Stability Problems," *Proceedings*, Research Conference on Shear Strength of Cohesive Soils, ASCE, 437–501.

BJERRUM, L., and SIMONS, N. E. (1960). "Compression of Shear Strength Characteristics of Normally Consolidated Clay," *Proceedings*, Research Conference on Shear Strength of Cohesive Soils, ASCE. 711–726.

BLACK, D. K., and LEE, K. L. (1973). "Saturating Laboratory Samples by Back Pressure," *Journal of the Soil Mechanics and Foundations Division*, ASCE, Vol. 99, No. SM1, 75–93.

CASAGRANDE, A., and CARRILLO, N. (1944). "Shear Failure of Anisotropic Materials," in *Contribution to Soil Mechanics 1941–1953*, Boston Society of Civil Engineers, Boston, MA.

COULOMB, C. A. (1776). "Essai sur une application des regles de Maximums et Minimis à quelques Problèmes de Statique, relatifs à l'Architecture," *Memoires de Mathematique et de Physique*, Présentés, à l'Academie Royale des Sciences, Paris, Vol. 3, 38.

KENNEY, T. C. (1959). "Discussion," *Proceedings*, ASCE, Vol. 85, No. SM3, 67–79.

LOH, A. K., and HOLT, R. T. (1974). "Directional Variation in Undrained Shear Strength and Fabric of Winnipeg Upper Brown Clay," *Canadian Geotechnical Journal*, Vol. 11, No. 3 430–437.

MOHR, O. (1900). "Welche Umstände Bedingen die Elastizitätsgrenze und den Bruch eines Materiales?" *Zeitschrift des Vereines Deutscher Ingenieure*, Vol. 44, 1524–1530, 1572–1577.

SKEMPTON, A. W. (1954). "The Pore Water Coefficients A and B," *Geotechnique*, Vol. 4, 143–147.

ROSENQVIST, I. TH. (1953). "Considerations on the Sensitivity of Norwegian Quick Clays, *Geotechnique*, Vol. 3, No. 5, 195–200.

9

Slope Stability

An exposed ground surface that stands at an angle with the horizontal is called an *unrestrained slope*. The slope can be natural or constructed. If the ground surface is not horizontal, a component of gravity will cause the soil to move downward, as shown in Figure 9.1. If the component of gravity is large enough, slope failure can occur; that is, the soil mass in zone *abcdea* can slide downward. The driving force overcomes the resistance from the shear strength of the soil along the rupture surface.

In many cases, civil engineers are expected to make calculations to check the safety of natural slopes, slopes of excavations, and compacted embankments. This process, called *slope stability analysis*, involves determining and comparing the shear stress developed along the most likely rupture surface with the shear strength of the soil.

The stability analysis of a slope is not an easy task. Evaluating variables such as the soil stratification and its in-place shear strength parameters may prove to be a formidable task. Seepage through the slope and the choice of a potential slip surface add to the complexity of the problem. This chapter explains the basic principles involved in slope stability analysis.

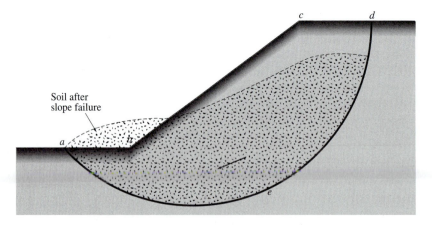

Figure 9.1 Slope failure

Factor of Safety

The task of the engineer charged with analyzing slope stability is to determine the factor of safety. Generally, the factor of safety is defined as

$$FS_s = \frac{\tau_f}{\tau_d} \qquad (9.1)$$

where

FS_s = factor of safety with respect to strength
τ_f = average shear strength of the soil
τ_d = average shear stress developed along the potential failure surface

The shear strength of a soil consists of two components, cohesion and friction, and may be expressed as

$$\tau_f = c' + \sigma' \tan \phi' \qquad (9.2)$$

where

c' = cohesion
ϕ' = drained angle of friction
σ' = effective normal stress on the potential failure surface

In a similar manner, we can also write

$$\tau_d = c'_d + \sigma' \tan \phi'_d \qquad (9.3)$$

where c'_d and ϕ'_d are, respectively, the effective cohesion and the angle of friction that develop along the potential failure surface. Substituting Eqs. (9.2) and (9.3) into Eq. (9.1), we get

$$FS_s = \frac{c' + \sigma' \tan \phi'}{c'_d + \sigma' \tan \phi'_d} \qquad (9.4)$$

Now we can introduce some other aspects of the factor of safety—that is, the factor of safety with respect to cohesion, $FS_{c'}$, and the factor of safety with respect to friction, $FS_{\phi'}$. They are defined as follows:

$$FS_{c'} = \frac{c'}{c'_d} \qquad (9.5)$$

and

$$FS_{\phi'} = \frac{\tan \phi'}{\tan \phi'_d} \qquad (9.6)$$

When Eqs. (9.4), (9.5), and (9.6) are compared, we see that when $FS_{c'}$ becomes equal to $FS_{\phi'}$, that is the factor of safety with respect to strength. Or, if

$$\frac{c'}{c'_d} = \frac{\tan \phi'}{\tan \phi'_d}$$

we can write

$$FS_s = FS_{c'} = FS_{\phi'} \tag{9.7}$$

When FS_s is equal to 1, the slope is in a state of impending failure. Generally, a value of 1.5 for the factor of safety with respect to strength is acceptable for the design of a stable slope.

9.2 Stability of Infinite Slopes

In considering the problem of slope stability, we may start with the case of an infinite slope, as shown in Figure 9.2. An infinite slope is one in which H is much greater than the slope height. The shear strength of the soil may be given by [Eq. (9.2)]

$$\tau_f = c' + \sigma' \tan \phi'$$

We will evaluate the factor of safety against a possible slope failure along a plane AB located at a depth H below the ground surface. The slope failure can occur by the movement of soil above the plane AB from right to left.

Let us consider a slope element, $abcd$, that has a unit length perpendicular to the plane of the section shown. The forces, F, that act on the faces ab and cd are

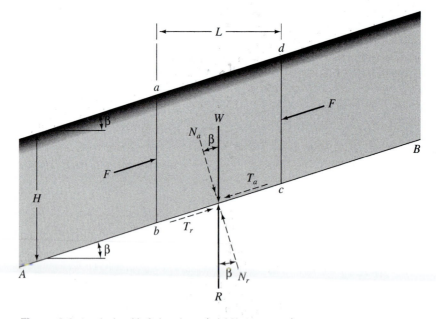

Figure 9.2 Analysis of infinite slope (without seepage)

equal and opposite and may be ignored. The effective weight of the soil element is (with pore water pressure equal to 0)

$$W = (\text{volume of the soil element}) \times (\text{unit weight of soil}) = \gamma LH \qquad (9.8)$$

The weight, W, can be resolved into two components:

1. Force perpendicular to the plane $AB = N_a = W \cos \beta = \gamma LH \cos \beta$.
2. Force parallel to the plane $AB = T_a = W \sin \beta = \gamma LH \sin \beta$. Note that this is the force that tends to cause the slip along the plane.

Thus, the effective normal stress σ' and the shear stress τ at the base of the slope element can be given as

$$\sigma' = \frac{N_a}{\text{area of the base}} = \frac{\gamma LH \cos \beta}{\left(\dfrac{L}{\cos \beta} \right)} = \gamma H \cos^2 \beta \qquad (9.9)$$

and

$$\tau = \frac{T_a}{\text{area of the base}} = \frac{\gamma LH \sin \beta}{\left(\dfrac{L}{\cos \beta} \right)} = \gamma H \cos \beta \sin \beta \qquad (9.10)$$

The reaction to the weight W is an equal and opposite force R. The normal and tangential components of R with respect to the plane AB are N_r and T_r:

$$N_r = R \cos \beta = W \cos \beta \qquad (9.11)$$

$$T_r = R \sin \beta = W \sin \beta \qquad (9.12)$$

For equilibrium, the resistive shear stress that develops at the base of the element is equal to $(T_r)/(\text{area of the base}) = \gamma H \sin \beta \cos \beta$. This may also be written in the form [Eq. (9.3)]

$$\tau_d = c'_d + \sigma' \tan \phi'_d$$

The value of the effective normal stress is given by Eq. (9.9). Substitution of Eq. (9.9) into Eq. (9.3) yields

$$\tau_d = c'_d + \gamma H \cos^2 \beta \tan \phi'_d \qquad (9.13)$$

Thus,

$$\gamma H \sin \beta \cos \beta = c'_d + \gamma H \cos^2 \beta \tan \phi'_d$$

or

$$\frac{c'_d}{\gamma H} = \sin \beta \cos \beta - \cos^2 \beta \tan \phi'_d$$

$$= \cos^2 \beta (\tan \beta - \tan \phi'_d) \qquad (9.14)$$

The factor of safety with respect to strength was defined in Eq. (9.7), from which

$$\tan \phi'_d = \frac{\tan \phi'}{FS_s} \quad \text{and} \quad c'_d = \frac{c'}{FS_s} \qquad (9.15)$$

Substituting the preceding relationships into Eq. (9.14), we obtain

$$FS_s = \frac{c'}{\gamma H \cos^2 \beta \tan \beta} + \frac{\tan \phi'}{\tan \beta} \tag{9.16}$$

For granular soils, $c' = 0$, and the factor of safety, FS_s, becomes equal to $(\tan \phi')/(\tan \beta)$. This indicates that, in an infinite slope in sand, the value of FS_s is independent of the height H, and the slope is stable as long as $\beta < \phi'$. The angle ϕ' for cohesionless soils is called the *angle of repose*.

If a soil possesses cohesion and friction, the depth of the plane along which critical equilibrium occurs may be determined by substituting $FS_s = 1$ and $H = H_{cr}$ into Eq. (9.16). Thus,

$$H_{cr} = \frac{c'}{\gamma \cos^2 \beta (\tan \beta - \tan \phi')} \tag{9.17}$$

If there is seepage through the soil and the ground water level coincides with the ground surface as shown in Figure 9.3, the factor of safety with respect to strength can be obtained as

$$FS_s = \frac{c'}{\gamma_{\text{sat}} H \cos^2 \beta \tan \beta} + \frac{\gamma'}{\gamma_{\text{sat}}} \frac{\tan \phi'}{\tan \beta} \tag{9.18}$$

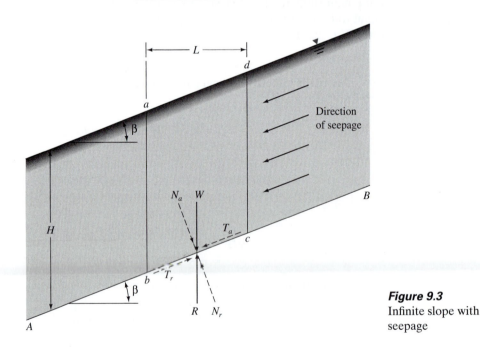

Figure 9.3
Infinite slope with seepage

where

γ_{sat} = saturated unit weight of the soil
γ' = effective unit weight of the soil

9.3 *Finite Slopes*

When the value of H_{cr} approaches the height of the slope, the slope is generally considered finite. When analyzing the stability of a finite slope in a homogeneous soil, for simplicity, we need to make an assumption about the general shape of the surface of potential failure. Although there is considerable evidence that slope failures usually occur on curved failure surfaces, Culmann (1875) approximated the surface of potential failure as a plane. The factor of safety, FS_s, calculated using Culmann's approximation gives fairly good results for near-vertical slopes only. After extensive investigation of slope failures in the 1920s, a Swedish geotechnical commission recommended that the actual surface of sliding may be approximated to be circularly cylindrical.

Since that time, most conventional stability analyses of slopes have been made by assuming that the curve of potential sliding is an arc of a circle. However, in many circumstances (for example, zoned dams and foundations on weak strata), stability analysis using plane failure of sliding is more appropriate and yields excellent results.

Analysis of Finite Slope with Plane Failure Surface (Culmann's Method)

This analysis is based on the assumption that the failure of a slope occurs along a plane when the average shearing stress that tends to cause the slip is greater than the shear strength of the soil. Also, the most critical plane is the one that has a minimum ratio of the average shearing stress that tends to cause failure to the shear strength of soil.

Figure 9.4 shows a slope of height H. The slope rises at an angle β with the horizontal. AC is a trial failure plane. If we consider a unit length perpendicular to the section of the slope, the weight of the wedge $ABC = W$:

$$W = \frac{1}{2}(H)(\overline{BC})(1)(\gamma)$$

$$= \frac{1}{2}H(H \cot \theta - H \cot \beta)\gamma$$

$$= \frac{1}{2}\gamma H^2 \left[\frac{\sin(\beta - \theta)}{\sin \beta \sin \theta} \right] \tag{9.19}$$

The normal and tangential components of W with respect to the plane AC are as follows:

$$N_a = \text{normal component} = W \cos \theta$$

$$= \frac{1}{2}\gamma H^2 \left[\frac{\sin(\beta - \theta)}{\sin \beta \sin \theta} \right] \cos \theta \tag{9.20}$$

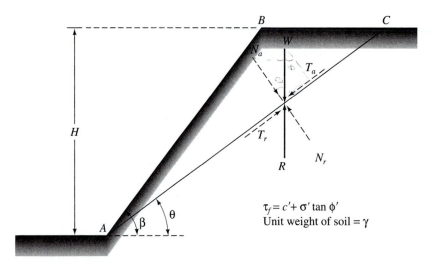

Figure 9.4 Finite slope analysis—Culmann's method

$$T_a = \text{tangential component} = W \sin \theta$$

$$= \frac{1}{2} \gamma H^2 \left[\frac{\sin(\beta - \theta)}{\sin \beta \sin \theta} \right] \sin \theta \tag{9.21}$$

The average effective normal stress and shear stress on the plane AC may be given by

$$\sigma' = \text{average effective normal stress}$$

$$= \frac{N_a}{(\overline{AC})(1)} = \frac{N_a}{\left(\dfrac{H}{\sin \theta} \right)}$$

$$= \frac{1}{2} \gamma H \left[\frac{\sin(\beta - \theta)}{\sin \beta \sin \theta} \right] \cos \theta \sin \theta \tag{9.22}$$

and

$$\tau = \text{average shear stress}$$

$$= \frac{T_a}{(\overline{AC})(1)} = \frac{T_a}{\left(\dfrac{H}{\sin \theta} \right)}$$

$$= \frac{1}{2} \gamma H \left[\frac{\sin(\beta - \theta)}{\sin \beta \sin \theta} \right] \sin^2 \theta \tag{9.23}$$

The average resistive shearing stress developed along the plane AC may also be expressed as

$$\tau_d = c_d' + \sigma' \tan \phi_d'$$

$$= c_d' + \frac{1}{2} \gamma H \left[\frac{\sin(\beta - \theta)}{\sin \beta \sin \theta} \right] \cos \theta \sin \theta \tan \phi_d' \tag{9.24}$$

Now, from Eqs. (9.23) and (9.24), we have

$$\frac{1}{2}\gamma H\left[\frac{\sin(\beta - \theta)}{\sin \beta \sin \theta}\right]\sin^2 \theta = c'_d + \frac{1}{2}\gamma H\left[\frac{\sin(\beta - \theta)}{\sin \beta \sin \theta}\right]\cos \theta \sin \theta \tan \phi'_d \quad (9.25)$$

or

$$c'_d = \frac{1}{2}\gamma H\left[\frac{\sin(\beta - \theta)(\sin \theta - \cos \theta \tan \phi'_d)}{\sin \beta}\right] \quad (9.26)$$

The expression in Eq. (9.26) is derived for the trial failure plane AC. In an effort to determine the critical failure plane, we use the principle of maxima and minima (for a given value of ϕ'_d) to find the angle θ at which the developed cohesion would be maximum. Thus, the first derivative of c'_d with respect to θ is set equal to 0, or

$$\frac{\partial c'_d}{\partial \theta} = 0 \quad (9.27)$$

Since γ, H, and β are constants in Eq. (9.26), we have

$$\frac{\partial}{\partial \theta}[\sin(\beta - \theta)(\sin \theta - \cos \theta \tan \phi'_d)] = 0 \quad (9.28)$$

Solving Eq. (9.28) gives the critical value of θ, or

$$\theta_{cr} = \frac{\beta + \phi'_d}{2} \quad (9.29)$$

Substitution of the value of $\theta = \theta_{cr}$ into Eq. (9.26) yields

$$c'_d = \frac{\gamma H}{4}\left[\frac{1 - \cos(\beta - \phi'_d)}{\sin \beta \cos \phi'_d}\right] \quad (9.30)$$

The maximum height of the slope for which critical equilibrium occurs can be obtained by substituting $c'_d = c'$ and $\phi'_d = \phi'$ into Eq. (9.30). Thus,

$$H_{cr} = \frac{4c'}{\gamma}\left[\frac{\sin \beta \cos \phi'}{1 - \cos(\beta - \phi')}\right] \quad (9.31)$$

Example 9.1

A cut is to be made in a soil that has $\gamma = 17$ kN/m^3, $c' = 40$ kN/m^2, and $\phi' = 15°$. The side of the cut slope will make an angle of 30° with the horizontal. What depth of the cut slope will have a factor of safety, FS_s, of 3?

Solution

We are given $\phi' = 15°$ and $c' = 40 \text{ kN/m}^2$. If $FS_s = 3$, then $FS_{c'}$ and $FS_{\phi'}$ should both be equal to 3. We have

$$FS_{c'} = \frac{c'}{c'_d}$$

or

$$c'_d = \frac{c'}{FS_{c'}} = \frac{c'}{FS_s} = \frac{40}{3} = 13.33 \text{ kN/m}^2$$

Similarly,

$$FS_{\phi'} = \frac{\tan \phi'}{\tan \phi'_d}$$

$$\tan \phi'_d = \frac{\tan \phi'}{FS_{\phi'}} = \frac{\tan \phi'}{FS_s} = \frac{\tan 15}{3}$$

or

$$\phi'_d = \tan^{-1}\left[\frac{\tan 15}{3}\right] = 5.1°$$

Substituting the preceding values of c'_d and ϕ'_d into Eq. (9.30) gives

$$H = \frac{4c'_d}{\gamma}\left[\frac{\sin \beta \cos \phi'_d}{1 - \cos(\beta - \phi'_d)}\right] = \frac{4 \times 13.33}{17}\left[\frac{\sin 30 \cos 5.1}{1 - \cos(30 - 5.1)}\right] \approx \textbf{16.8 m} \quad \blacksquare$$

9.4 *Analysis of Finite Slope with Circularly Cylindrical Failure Surface—General*

In general, slope failure occurs in one of the following modes (Figure 9.5):

1. When the failure occurs in such a way that the surface of sliding intersects the slope at or above its toe, it is called a *slope failure* (Figure 9.5a). The failure circle is referred to as a *toe circle* if it passes through the toe of the slope, and as a *slope circle* if it passes above the toe of the slope. Under certain circumstances, it is possible to have a shallow slope failure, as shown in Figure 9.5b.
2. When the failure occurs in such a way that the surface of sliding passes at some distance below the toe of the slope, it is called a *base failure* (Figure 9.5c). The failure circle in the case of base failure is called a *midpoint circle*.

Various procedures of stability analysis may, in general, be divided into two major classes:

1. *Mass procedure.* In this case, the mass of the soil above the surface of sliding is taken as a unit. This procedure is useful when the soil that forms the slope is assumed to be homogeneous, although this is hardly the case in most natural slopes.

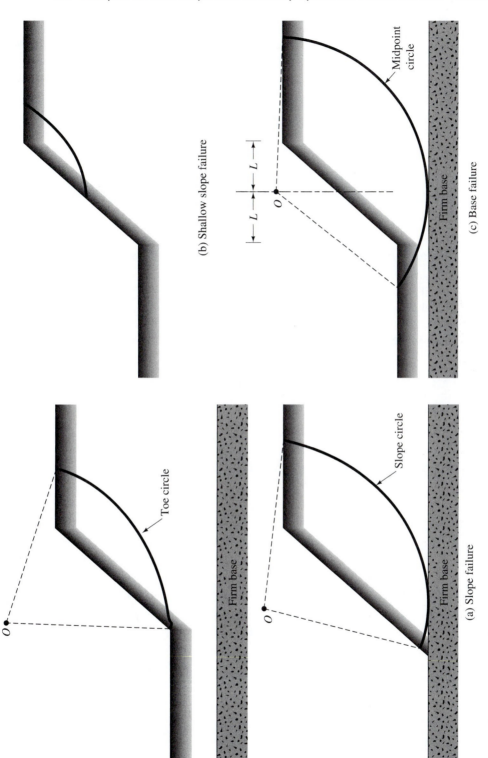

(a) Slope failure

Slope circle

Firm base

(b) Shallow slope failure

Midpoint circle

Toe circle

L L

O

(c) Base failure

Firm base

Figure 9.5 Modes of failure of finite slope

2. *Method of slices.* In this procedure, the soil above the surface of sliding is divided into a number of vertical parallel slices. The stability of each of the slices is calculated separately. This is a versatile technique in which the nonhomogeneity of the soils and pore water pressure can be taken into consideration. It also accounts for the variation of the normal stress along the potential failure surface.

The fundamentals of the analysis of slope stability by mass procedure and method of slices are presented in the following sections.

9.5   *Mass Procedure of Stability Analysis (Circularly Cylindrical Failure Surface)*

Slopes in Homogeneous Clay Soil with $\phi = 0$ (Undrained Condition)

Figure 9.6 shows a slope in a homogeneous soil. The undrained shear strength of the soil is assumed to be constant with depth and may be given by $\tau_f = c_u$. To make the stability analysis, we choose a trial potential curve of sliding AED, which is an arc of a circle that has a radius r. The center of the circle is located at O. Considering the unit length perpendicular to the section of the slope, we can give the total weight of the soil above the curve AED as $W = W_1 + W_2$, where

$$W_1 = (\text{area of } FCDEF)(\gamma)$$

and

$$W_2 = (\text{area of } ABFEA)(\gamma)$$

Note that γ = saturated unit weight of the soil.

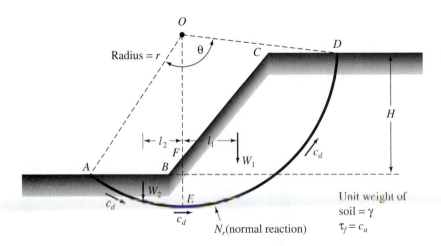

Figure 9.6 Stability analysis of slope in homogeneous clay soil ($\phi = 0$)

Failure of the slope may occur by the sliding of the soil mass. The moment of the driving force about O to cause slope instability is

$$M_d = W_1 l_1 - W_2 l_2 \tag{9.32}$$

where l_1 and l_2 are the moment arms.

The resistance to sliding is derived from the cohesion that acts along the potential surface of sliding. If c_d is the cohesion that needs to be developed, then the moment of the resisting forces about O is

$$M_R = c_d(\widehat{AED})(1)(r) = c_d r^2 \theta \tag{9.33}$$

For equilibrium, $M_R = M_d$; thus,

$$c_d r^2 \theta = W_1 l_1 - W_2 l_2$$

or

$$c_d = \frac{W_1 l_1 - W_2 l_2}{r^2 \theta} \tag{9.34}$$

The factor of safety against sliding may now be found:

$$FS_s = \frac{\tau_f}{c_d} = \frac{c_u}{c_d} \tag{9.35}$$

Note that the potential curve of sliding, AED, was chosen arbitrarily. The critical surface is the one for which the ratio of c_u to c_d is a minimum. In other words, c_d is maximum. To find the critical surface for sliding, a number of trials are made for different trial circles. The minimum value of the factor of safety thus obtained is the factor of safety against sliding for the slope, and the corresponding circle is the critical circle.

Stability problems of this type were solved analytically by Fellenius (1927) and Taylor (1937). For the case of *critical circles*, the developed cohesion can be expressed by the relationship

$$c_d = \gamma H m$$

or

$$\frac{c_d}{\gamma H} = m \tag{9.36}$$

Note that the term m on the right-hand side of the preceding equation is nondimensional and is referred to as the *stability number*. The critical height (that is, $FS_s = 1$) of the slope can be evaluated by substituting $H = H_{cr}$ and $c_d = c_u$ (full mobilization of the undrained shear strength) into Eq. (9.36). Thus,

$$H_{cr} = \frac{c_u}{\gamma m} \tag{9.37}$$

Values of the stability number m for various slope angles β are given in Figure 9.7. Terzaghi and Peck (1967) used the term $\gamma H/c_d$, the reciprocal of m, and called it the *stability factor*. Figure 9.7 should be used carefully. Note that it is valid for slopes of saturated clay and is applicable to only undrained conditions ($\phi = 0$).

In reference to Figure 9.7, consider these issues:

1. For slope angle β greater than 53°, the critical circle is always a toe circle. The location of the center of the critical toe circle may be found with the aid of Figure 9.8.

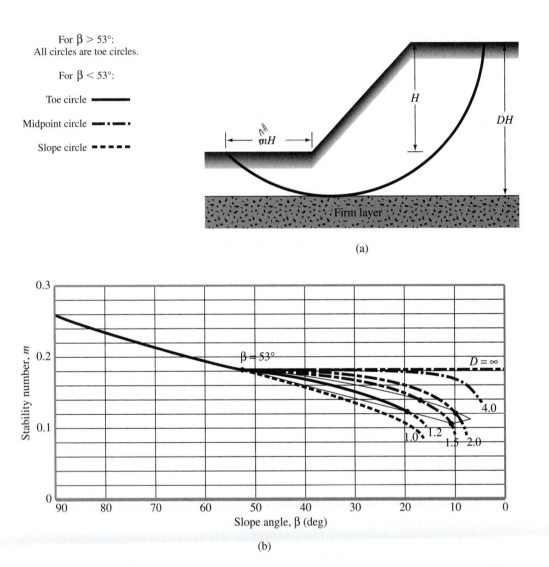

Figure 9.7 (a) Definition of parameters for midpoint circle-type failure; (b) plot of stability number against slope angle (Redrawn from Terzaghi and Peck, 1967)

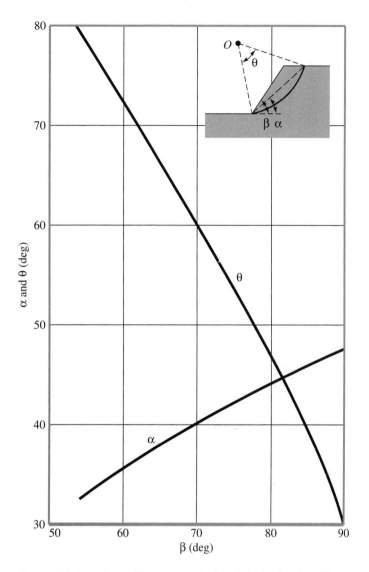

Figure 9.8 Location of the center of critical circles for $\beta > 53°$

2. For $\beta < 53°$, the critical circle may be a toe, slope, or midpoint circle, depending on the location of the firm base under the slope. This is called the *depth function*, which is defined as

$$D = \frac{\text{vertical distance from the top of the slope to the firm base}}{\text{height of the slope}} \quad (9.38)$$

3. When the critical circle is a midpoint circle (that is, the failure surface is tangent to the firm base), its position can be determined with the aid of Figure 9.9.
4. The maximum possible value of the stability number for failure at the midpoint circle is 0.181.

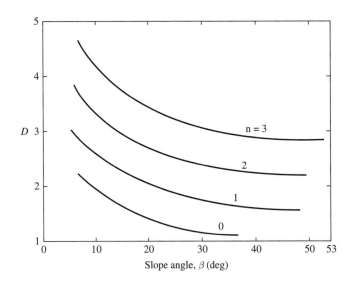

Figure 9.9 Location of midpoint circle (Based on Fellenius, 1927; and Terzaghi and Peck, 1967)

Example 9.2

A cut slope in saturated clay (Figure 9.10) makes an angle of 56° with the horizontal

   a. Determine the maximum depth up to which the cut could be made. Assume that the critical surface for sliding is circularly cylindrical. What will be the nature of the critical circle (that is, toe, slope, or midpoint)?
   b. Referring to part a, determine the distance of the point of intersection of the critical failure circle from the top edge of the slope.

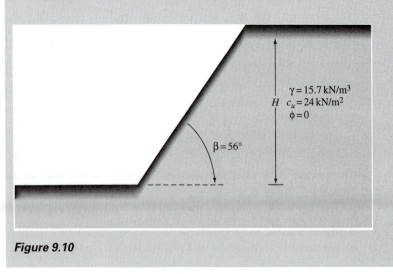

Figure 9.10

c. How deep should the cut be made if a factor of safety of 2 against sliding is required?

Solution

Part a.
Since the slope angle $\beta = 56° > 53°$, the critical circle is a **toe circle**. From Figure 9.7, for $\beta = 56°$, $m = 0.185$. Using Eq. (9.37), we have

$$H_{cr} = \frac{c_u}{\gamma m} = \frac{24}{(15.7)(0.185)} = 8.26 \text{ m} \approx \textbf{8.25 m}$$

Part b.
Refer to Figure 9.11. For the critical circle, we have

$$\overline{BC} = \overline{EF} = \overline{AF} - \overline{AE} = H_{cr}(\cot \alpha - \cot 56°)$$

From Figure 9.8, for $\beta = 56°$, the magnitude of α is 33°, so

$$\overline{BC} = 8.25 \, (\cot 33 - \cot 56) = 7.14 \text{ m} \approx \textbf{7.15 m}$$

Part c.
Developed cohesion is

$$c_d = \frac{c_u}{FS_s} = \frac{24}{2} = 12 \text{ kN/m}^2$$

From Figure 9.7, for $\beta = 56°$, $m = 0.185$. Thus, we have

$$H = \frac{c_d}{\gamma m} = \frac{12}{(15.7)(0.185)} = \textbf{4.13 m} \qquad \blacksquare$$

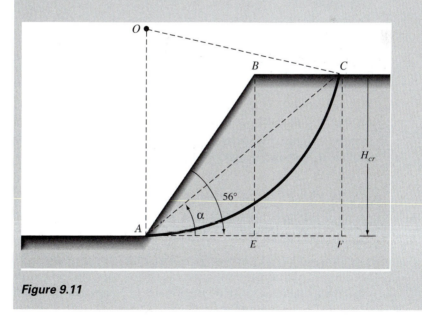

Figure 9.11

Example 9.3

A cut slope was excavated in a saturated clay. The slope made an angle of 40° with the horizontal. Slope failure occurred when the cut reached a depth of 6.1 m. Previous soil explorations showed that a rock layer was located at a depth of 9.15 m below the ground surface. Assume an undrained condition and $\gamma_{sat} = 17.29$ kN/m³.

    a. Determine the undrained cohesion of the clay (use Figure 9.7).
    b. What was the nature of the critical circle?
    c. With reference to the toe of the slope, at what distance did the surface of sliding intersect the bottom of the excavation?

Solution
Part a.
Referring to Figure 9.7, we find

$$D = \frac{9.15}{6.1} = 1.5$$

$$\gamma_{sat} = 17.29 \text{ kN/m}^3$$

$$H_{cr} = \frac{c_u}{\gamma m}$$

From Figure 9.7, for $\beta = 40°$ and $D = 1.5$, $m = 0.175$, so

$$c_u = (H_{cr})(\gamma)(m) = (6.1)(17.29)(0.175) = \mathbf{18.5 \text{ kN/m}^2}$$

Part b.
Midpoint circle

Part c.
From Figure 9.9, for $D = 1.5$ and $\beta = 40°$, $n = 0.9$, so

$$\text{distance} = (n)(H_{cr}) = (0.9)(6.1) = \mathbf{5.49 \text{ m}} \qquad \blacksquare$$

Slopes in Clay Soil with $\phi = 0$; and c_u Increasing with Depth

In many instances the undrained cohesion (c_u) in normally consolidated clay increases with depth as shown in Figure 9.12. Or

$$c_{u(z)} = c_{u(z=0)} + a_0 z \qquad (9.39)$$

where
    $c_{u(z)}$ = undrained shear strength at depth z
    $c_{u(z=0)}$ = undrained shear strength at depth $z = 0$
    a_0 = slope of the line of the plot of $c_{u(z)}$ vs. z

For such a condition, the critical circle will be a toe circle, not a midpoint circle, since the strength increases with depth. Figure 9.13 shows a trial failure circle for this type of case. The moment of the driving force about O can be given as

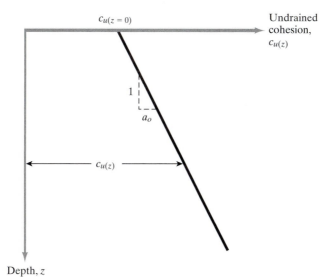
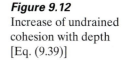

Figure 9.12
Increase of undrained
cohesion with depth
[Eq. (9.39)]

$$M_d = \frac{\gamma H^3}{12}(1 - 2 \cot^2 \beta - 3 \cot \alpha' \cot \beta$$

$$+ \; 3 \cot \beta \cot \lambda + 3 \cot \lambda \cot \alpha') \qquad (9.40)$$

In a similar manner, the moment of the resisting forces about O is

$$M_r = r \int_{-\alpha'}^{+\alpha'} c_{d(z)} r \; d\theta' \qquad (9.41)$$

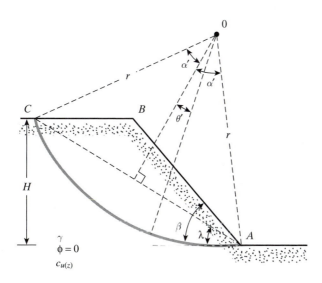

Figure 9.13 Analysis of slope in clay soil ($\phi = 0$ concept) with increasing undrained
shear strength

where

$$c_{d(z)} = c_{d(z=0)} + a_0 z \tag{9.42}$$

The factor of safety against sliding is

$$FS_s = \frac{M_r}{M_d} \tag{9.43}$$

Koppula (1984) has solved this problem in a slightly different form. His solution for obtaining the minimum factor of safety can be expressed as

$$m = \left[\frac{c_{u(z=0)}}{\gamma H}\right]\frac{1}{FS_s} \tag{9.44}$$

where m = stability number, which is also a function of

$$c_R = \frac{a_0 H}{c_{u(z=0)}} \tag{9.45}$$

Table 9.1 gives the values of m for various values of c_R and β, which are slightly different from those expressed by Koppula (1984).

Slopes in Homogeneous Soil with $\phi' > 0$

A slope in a homogeneous soil is shown in Figure 9.14a. The shear strength of the soil is given by

$$\tau_f = c' + \sigma' \tan \phi'$$

The pore water pressure is assumed to be 0. $\widehat{AC}$ is a trial circular arc that passes through the toe of the slope, and O is the center of the circle. Considering unit length perpendicular to the section of the slope, we find

weight of the soil wedge $ABC = W = $ (area of ABC)(γ)

Table 9.1 Variation of m, c_R, and β [Eqs. (9.44) and (9.45).] Based on the Analysis of Koppula

	m					
c_R	1H:1V $\beta = 45°$	1.5H:1V $\beta = 33.69°$	2H:1V $\beta = 26.57°$	3H:1V $\beta = 18.43°$	4H:1V $\beta = 14.04°$	5H:1V $\beta = 11.31°$
0.1	0.158	0.146	0.139	0.130	0.125	0.121
0.2	0.148	0.135	0.127	0.117	0.111	0.105
0.3	0.139	0.126	0.118	0.107	0.0995	0.0937
0.4	0.131	0.118	0.110	0.0983	0.0907	0.0848
0.5	0.124	0.111	0.103	0.0912	0.0834	0.0775
1.0	0.0984	0.086	0.0778	0.0672	0.0600	0.0546
2.0	0.0697	0.0596	0.0529	0.0443	0.0388	0.0347
3.0	0.0541	0.0457	0.0402	0.0331	0.0288	0.0255
4.0	0.0442	0.0371	0.0325	0.0266	0.0229	0.0202
5.0	0.0374	0.0312	0.0272	0.0222	0.0190	0.0167
10.0	0.0211	0.0175	0.0151	0.0121	0.0103	0.0090

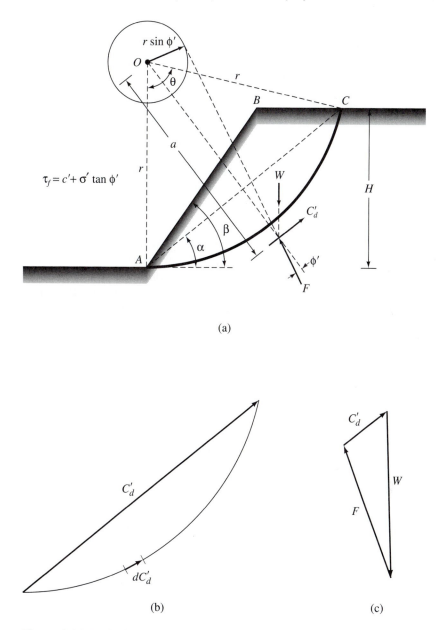

(a)

(b)   (c)

Figure 9.14 Analysis of slopes in homogeneous soils with $\phi' > 0$

For equilibrium, the following other forces are acting on the wedge:

1. C'_d—the resultant of the cohesive force that is equal to the unit cohesion developed times the length of the cord $\overline{AC}$. The magnitude of C'_d is given by (Figure 9.14b).

$$C'_d = c'_d(\overline{AC}) \tag{9.46}$$

C'_d acts in a direction parallel to the cord AC (Figure 9.14b) and at a distance a from the center of the circle O such that

$$C'_d(a) = c'_d(\widehat{AC})r$$

or

$$a = \frac{c'_d(\widehat{AC})r}{C'_d} = \frac{\widehat{AC}}{AC}r \qquad (9.47)$$

2. *F*—the resultant of the normal and frictional forces along the surface of sliding. For equilibrium, the line of action of *F* will pass through the point of intersection of the line of action of W and C'_d.

Now, if we assume the full friction is mobilized ($\phi'_d = \phi'$ or $FS_{\phi'} = 1$), then the line of action of *F* will make an angle ϕ' with a normal to the arc, and thus it will be a tangent to a circle with its center at O and having a radius of $r \sin \phi'$. This circle is called the *friction circle*. Actually, the radius of the friction circle is a little larger than $r \sin \phi'$.

Since the directions of W, C'_d, and F are known and the magnitude of W is known, we can plot a force polygon, as shown in Figure 9.14c. The magnitude of C'_d can be determined from the force polygon. So the unit cohesion developed can be found:

$$c'_d = \frac{C'_d}{AC}$$

Determining the magnitude of c'_d described previously is based on a trial surface of sliding. Several trials must be made to obtain the most critical sliding surface along which the developed cohesion is a maximum. So it is possible to express the maximum cohesion developed along the critical surface as

$$c'_d = \gamma H[f(\alpha, \beta, \theta, \phi')] \qquad (9.48)$$

For critical equilibrium—that is, $FS_{c'} = FS_{\phi'} = FS_s = 1$—we can substitute $H = H_{cr}$ and $c'_d = c'$ into Eq. (9.48):

$$c' = \gamma H_{cr}[f(\alpha, \beta, \theta, \phi')]$$

or

$$\frac{c'}{\gamma H_{cr}} = f(\alpha, \beta, \theta, \phi') = m \qquad (9.49)$$

where $m =$ stability number. The values of m for various values of ϕ' and β are given in Figure 9.15, which is based on the analysis of Taylor (1937). This can be used to determine the factor of safety, F_s, of the homogeneous slope. The procedure to do the analysis is given below:

1. Determine c', ϕ', γ, β and H.
2. Assume several values of ϕ'_d (Note: $\phi'_d \leq \phi'$, such as $\phi'_{d(1)}$, $\phi'_{d(2)}$, ... (Column 1 of Table 9.2).

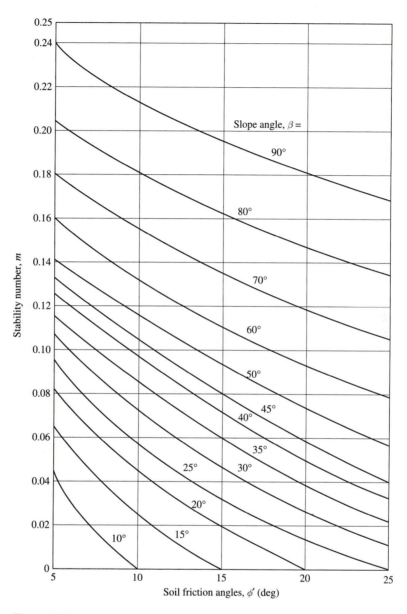

Figure 9.15 Taylor's stability number

3. Determine $FS_{\phi'}$ for each assumed value of ϕ'_d as (Column 2, Table 9.2)

$$FS_{\phi'(1)} = \frac{\tan \phi'}{\tan \phi'_{d(1)}}$$

$$FS_{\phi'(2)} = \frac{\tan \phi'}{\tan \phi'_{d(2)}}$$

Table 9.2 Determination of FS_s by Friction Circle Method

ϕ'_d	$FS_{\phi'} = \dfrac{\tan \phi'}{\tan \phi'_d}$	m	c'_d	$FS_{c'}$
(1)	(2)	(3)	(4)	(5)
$\phi'_{d(1)}$	$\dfrac{\tan \phi'}{\tan \phi'_{d(1)}}$	m_1	$m_1 \gamma H = c'_{d(1)}$	$\dfrac{c'}{c'_{d(1)}} = FS_{c'(1)}$
$\phi'_{d(2)}$	$\dfrac{\tan \phi'}{\tan \phi'_{d(2)}}$	m_2	$m_2 \gamma H = c'_{d(2)}$	$\dfrac{c'}{c'_{d(2)}} = FS_{c'(2)}$

4. For each assumed value of ϕ'_d and β, determine m (that is, $m_1, m_2, m_3, \ldots$) from Figure 9.15 (Column 3, Table 9.2).
5. Determine the developed cohesion for each value of m as (Column 4, Table 9.2)

$$c'_{d(1)} = m_1 \gamma H$$

$$c'_{d(2)} = m_2 \gamma H$$

6. Calculate $FS_{c'}$ for each value of c'_d (Column 5, Table 9.2), or

$$FS_{c'(1)} = \frac{c'}{c'_{d(1)}}$$

$$FS_{c'(2)} = \frac{c'}{c'_{d(2)}}$$

7. Plot a graph of $FS_{\phi'}$ vs. the corresponding $FS_{c'}$ (Figure 9.16) and determine $FS_s = FS_{\phi'} = FS_{c'}$.

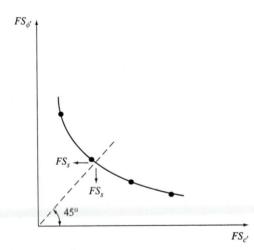

Figure 9.16 Plot of $FS_{\phi'}$ vs. $FS_{c'}$ to determine FS_s

An example of determining FS_s using the procedure just described is given in Example 9.4.

Using Taylor's friction circle method of slope stability (as shown in Example 9.4) Singh (1970) provided graphs of equal factors of safety, FS_s, for various slopes. Using the results of Singh (1970), the variations of $c'/\gamma H$ with factor of safety (FS_s) for various friction angles (ϕ') are plotted in Figure 9.17.

More recently, Michalowski (2002) made a stability analysis of simple slopes using the kinematic approach of limit analysis applied to a rigid rotational collapse mechanism. The failure surface in soil assumed in this study is an arc of a logarithmic spiral (Figure 9.18). The results of this study are summarized in Figure 9.19, from which FS_s can be directly obtained.

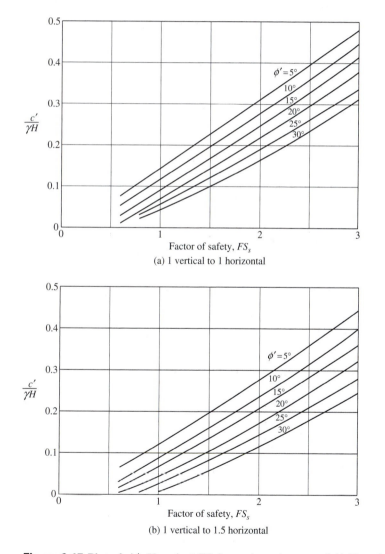

Figure 9.17 Plot of $c'/\gamma H$ against FS_s for various slopes, and ϕ' (Based on Singh, 1970)

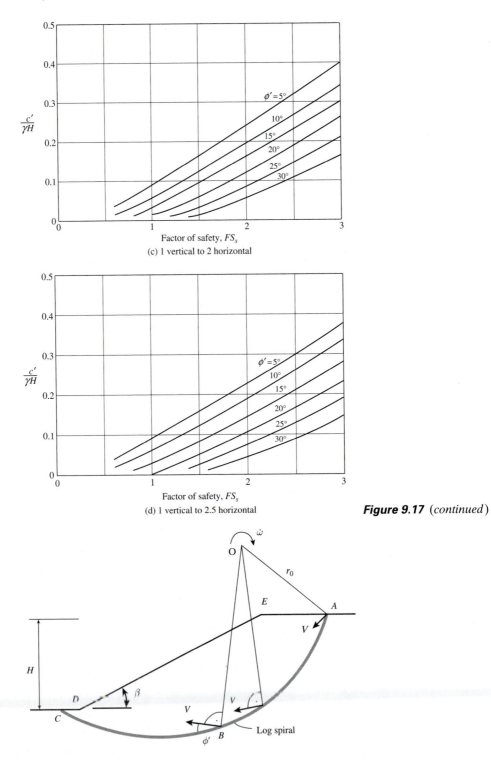

Figure 9.17 (*continued*)

(c) 1 vertical to 2 horizontal

(d) 1 vertical to 2.5 horizontal

Figure 9.18 Stability analysis using rotational collapse mechanism

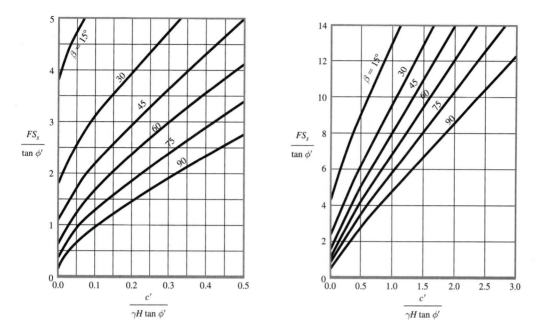

Figure 9.19 Michalowski's analysis for stability of simple slopes

Example 9.4

A slope with $\beta = 45°$ is to be constructed with a soil that has $\phi' = 20°$ and $c' = 24$ kN/m². The unit weight of the compacted soil will be 18.9 kN/m³.

  a. Find the critical height of the slope.
  b. If the height of the slope is 10 m, determine the factor of safety with respect to strength.

Solution
Part a.
We have

$$m = \frac{c'}{\gamma H_{cr}}$$

From Figure 9.15, for $\beta = 45°$ and $\phi' = 20°$, $m = 0.06$. So

$$H_{cr} = \frac{c'}{\gamma m} = \frac{24}{(18.9)(0.06)} = \textbf{21.1 m}$$

Part b.
If we assume that full friction is mobilized, then, referring to Figure 9.15 (for $\beta = 45°$ and $\phi'_d = \phi' = 20°$), we have

$$m = 0.06 = \frac{c'_d}{\gamma H}$$

or

$$c'_d = (0.06)(18.9)(10) = 11.34 \text{ kN/m}^2$$

Thus,

$$FS_{\phi'} = \frac{\tan \phi'}{\tan \phi'_d} = \frac{\tan 20}{\tan 20} = 1$$

and

$$FS_{c'} = \frac{c'}{c'_d} = \frac{24}{11.34} = 2.12$$

Since $FS_{c'} \neq FS_{\phi'}$, this is not the factor of safety with respect to strength.

Now we can make another trial. Let the developed angle of friction, ϕ'_d, be equal to 15°. For $\beta = 45°$ and the friction angle equal to 15°, we find from Figure 9.15

$$m = 0.083 = \frac{c'_d}{\gamma H}$$

or

$$c'_d = (0.083)(18.9)(10) = 15.69 \text{ kN/m}^2$$

For this trial,

$$FS_{\phi'} = \frac{\tan \phi'}{\tan \phi'_d} = \frac{\tan 20}{\tan 15} = 1.36$$

and

$$FS_{c'} = \frac{c'}{c'_d} = \frac{24}{15.69} = 1.53$$

Similar calculations of FS_{ϕ}' and FS_c' for various assumed values of ϕ'_d are given in the following table.

ϕ'_d	tan ϕ'_d	FS_ϕ'	m	c'_d (kN/m²)	FS_c'
20	0.364	1.0	0.06	11.34	2.12
15	0.268	1.36	0.083	15.69	1.53
10	0.176	2.07	0.105	19.85	1.21
5	0.0875	4.16	0.136	25.70	0.93

The values of FS_ϕ' are plotted against their corresponding values of $FS_{c'}$ in Figure 9.20, from which we find

$$FS_{c'} = FS_{\phi'} = FS_s = \mathbf{1.42}$$

Note: We could have found the value of FS_s from Figure 9.17a. Since $\beta = 45°$, it is a slope of 1V:1H. For this slope

$$\frac{c'}{\gamma H} = \frac{24}{(18.9)(10)} = 0.127$$

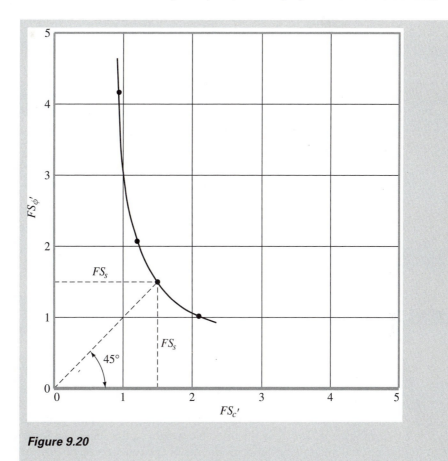

Figure 9.20

From Figure 9.17 a, for' $c'/\gamma H = 0.127$, the value of $FS_s \approx$ **1.4**   ∎

Example 9.5

Solve Example 9.4 using Michalowski's solution.

Solution
Part a.
For critical height (H_{cr}), $FS_s = 1$. Thus

$$\frac{c'}{\gamma H \tan \phi'} = \frac{24}{(18.9)(H_{cr})(\tan 20)} = \frac{3.49}{H_{cr}}$$

$$\frac{FS_s}{\tan \phi'} = \frac{1}{\tan 20} = 2.747$$

$$\beta = 45°$$

From Figure 9.19, for $\beta = 45°$ and $FS_s / \tan \phi' = 2.747$, the value of $c' / \gamma H \tan \phi' \approx 0.17$. So

$$\frac{3.49}{H_{cr}} = 0.17; \quad H_{cr} = \mathbf{20.5 \ m}$$

Part b.

$$\frac{c'}{\gamma H \tan \phi'} = \frac{24}{(18.9)(10)(\tan 20)} = 0.349$$

$$\beta = 45°$$

From Figure 9.19, $FS_s / \tan \phi' = 4$.

$$FS_s = 4 \tan \phi' = (4)(\tan 20) = \mathbf{1.46} \qquad \blacksquare$$

9.6   *Method of Slices*

Stability analysis using the method of slices can be explained by referring to Figure 9.21a, in which AC is an arc of a circle representing the trial failure surface. The soil above the trial failure surface is divided into several vertical slices. The width of each slice need not be the same. Considering unit length perpendicular to the cross-section shown, the forces that act on a typical slice (nth slice) are shown in Figure 9.21b. W_n is the effective weight of the slice. The forces N_r and T_r are the normal and tangential components of the reaction R, respectively. P_n and P_{n+1} are the normal forces that act on the sides of the slice. Similarly, the shearing forces that act on the sides of the slice are T_n and T_{n+1}. For simplicity, the pore water pressure is assumed to be 0. The forces P_n, P_{n+1}, T_n, and T_{n+1} are difficult to determine. However, we can make an approximate assumption that the resultants of P_n and T_n are equal in magnitude to the resultants of P_{n+1} and T_{n+1} and also that their lines of action coincide.

For equilibrium consideration, we have

$$N_r = W_n \cos \alpha_n$$

The resisting shear force can be expressed as

$$T_r = \tau_d(\Delta L_n) = \frac{\tau_f(\Delta L_n)}{FS_s} = \frac{1}{FS_s}[c' + \sigma' \tan \phi']\Delta L_n \qquad (9.50)$$

The effective normal stress, σ', in Eq. (9.50) is equal to

$$\frac{N_r}{\Delta L_n} = \frac{W_n \cos \alpha_n}{\Delta L_n}$$

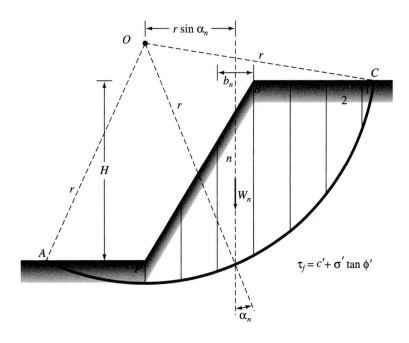

(a)

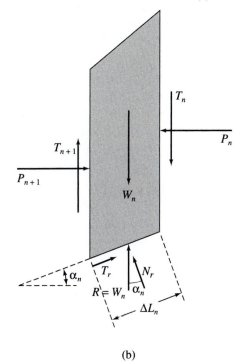

(b)

Figure 9.21 Stability analysis by ordinary method of slices: (a) trial failure surface; (b) forces acting on nth slice

For equilibrium of the trial wedge ABC, the moment of the driving force about O equals the moment of the resisting force about O, or

$$\sum_{n=1}^{n=p} W_n r \sin \alpha_n = \sum_{n=1}^{n=p} \frac{1}{FS_s}\left(c' + \frac{W_n \cos \alpha_n}{\Delta L_n}\tan \phi'\right)(\Delta L_n)(r)$$

or

$$FS_s = \frac{\displaystyle\sum_{n=1}^{n=p}(c'\Delta L_n + W_n \cos \alpha_n \tan \phi')}{\displaystyle\sum_{n=1}^{n=p} W_n \sin \alpha_n} \qquad (9.51)$$

Note: ΔL_n in Eq. (9.51) is approximately equal to $(b_n)/(\cos \alpha_n)$, where b_n = width of the nth slice.

Note that the value of α_n may be either positive or negative. The value of α_n is positive when the slope of the arc is in the same quadrant as the ground slope. To find the minimum factor of safety—that is, the factor of safety for the critical circle—several trials are made by changing the center of the trial circle. This method is generally referred to as the *ordinary method of slices*.

In developing Eq. (9.51), we assumed the pore water pressure to be zero. However, for steady-state seepage through slopes, as is the situation in many practical

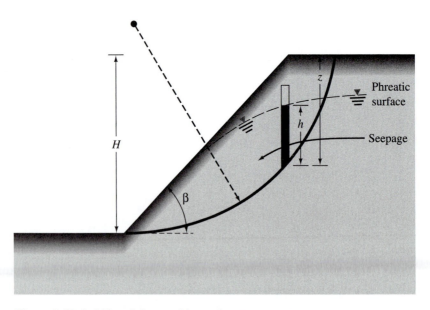

Figure 9.22 Stability of slopes with steady-state seepage

cases, the pore water pressure has to be taken into consideration when effective shear strength parameters are used. So we need to modify Eqs. (9.51) slightly.

Figure 9.22 shows a slope through which there is steady-state seepage. For the nth slice, the average pore water pressure at the bottom of the slice is equal to $u_n = h_n \gamma_w$. The total force caused by the pore water pressure at the bottom of the nth slice is equal to $u_n \Delta L_n$. Thus, Eq. (9.51) for the ordinary method of slices will be modified to read

$$FS_s = \frac{\displaystyle\sum_{n=1}^{n=p} [c' \Delta L_n + (W_n \cos \alpha_n - u_n \Delta L_n)] \tan \phi'}{\displaystyle\sum_{n=1}^{n=p} W_n \sin \alpha_n} \qquad (9.52)$$

Example 9.6

For the slope shown in Figure 9.23, find the factor of safety against sliding for the trial slip surface AC. Use the ordinary method of slices.

Solution
The sliding wedge is divided into seven slices. Other calculations are shown in the table.

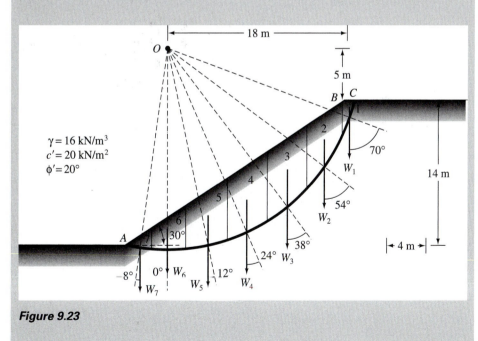

Figure 9.23

Slice no. (1)	W (kN/m) (2)	α_n (deg) (3)	$\sin \alpha_n$ (4)	$\cos \alpha_n$ (5)	ΔL_n (m) (6)	$W_n \sin \alpha_n$ (kN/m) (7)	$W_n \cos \alpha_n$ (kN/m) (8)
1	22.4	70	0.94	0.342	2.924	21.1	6.7
2	294.4	54	0.81	0.588	6.803	238.5	173.1
3	435.2	38	0.616	0.788	5.076	268.1	342.94
4	435.2	24	0.407	0.914	4.376	177.1	397.8
5	390.4	12	0.208	0.978	4.09	81.2	381.8
6	268.8	0	0	1	4	0	268.8
7	66.58	−8	−0.139	0.990	3.232	−9.25	65.9
					Σ Col. 6 = 30.501 m	Σ Col. 7 = 776.75 kN/m	Σ Col. 8 = 1638 kN/m

$$FS_s = \frac{(\Sigma \text{ col. } 6)(c') + (\Sigma \text{ col. } 8)\tan \phi'}{\Sigma \text{ col. } 7}$$

$$= \frac{(30.501)(20) + (1638.04)(\tan 20)}{776.75} = 1.55 \qquad \blacksquare$$

Bishop's Simplified Method of Slices

In 1955, Bishop proposed a more refined solution to the ordinary method of slices. In this method, the effect of forces on the sides of each slice is accounted for to some degree. We can study this method by referring to the slope analysis presented in Figure 9.21. The forces that act on the nth slice shown in Figure 9.21b have been redrawn in Figure 9.24a. Now, let $P_n - P_{n+1} = \Delta P$ and $T_n - T_{n+1} = \Delta T$. Also, we can write

$$T_r = N_r(\tan \phi'_d) + c'_d \Delta L_n = N_r \left(\frac{\tan \phi'}{FS_s} \right) + \frac{c' \Delta L_n}{FS_s} \qquad (9.53)$$

Figure 9.24b shows the force polygon for equilibrium of the nth slice. Summing the forces in the vertical direction gives

$$W_n + \Delta T = N_r \cos \alpha_n + \left[\frac{N_r \tan \phi'}{FS_s} + \frac{c' \Delta L_n}{FS_s} \right] \sin \alpha_n$$

or

$$N_r = \frac{W_n + \Delta T - \dfrac{c' \Delta L_n}{FS_s} \sin \alpha_n}{\cos \alpha_n + \dfrac{\tan \phi' \sin \alpha_n}{FS_s}} \qquad (9.54)$$

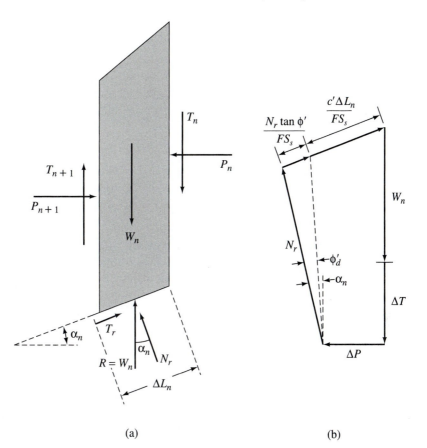

(a)   (b)

Figure 9.24 Bishop's simplified method of slices: (a) forces acting on the *n*th slice; (b) force polygon for equilibrium

For equilibrium of the wedge *ABC* (Figure 9.21a), taking the moment about *O* gives

$$\sum_{n=1}^{n=p} W_n r \sin \alpha_n = \sum_{n=1}^{n=p} T_r r \tag{9.55}$$

where $T_r = \dfrac{1}{FS_s}(c' + \sigma' \tan \phi')\Delta L_n$

$$= \frac{1}{FS_s}(c' \Delta L_n + N_r \tan \phi') \tag{9.56}$$

Substitution of Eqs. (9.54) and (9.56) into Eq. (9.55) gives

$$FS_s = \frac{\displaystyle\sum_{n=1}^{n=p}(c'b_n + W_n \tan \phi' + \Delta T \tan \phi')\dfrac{1}{m_{\alpha(n)}}}{\displaystyle\sum_{n=1}^{n=p} W_n \sin \alpha_n} \tag{9.57}$$

where

$$m_{\alpha(n)} = \cos \alpha_n + \frac{\tan \phi' \sin \alpha_n}{FS_s} \qquad (9.58)$$

For simplicity, if we let $\Delta T = 0$, then Eq. (9.57) becomes

$$FS_s = \frac{\sum\limits_{n=1}^{n=p} (c'b_n + W_n \tan \phi') \dfrac{1}{m_{\alpha(n)}}}{\sum\limits_{n=1}^{n=p} W_n \sin \alpha_n} \qquad (9.59)$$

Note that the term FS_s is present on both sides of Eq. (9.59). Hence, a trial-and-error procedure needs to be adopted to find the value of FS_s. As in the method of ordinary slices, a number of failure surfaces must be investigated to find the critical surface that provides the minimum factor of safety. Figure 9.25 shows the variation of $m_{\alpha(n)}$ [Eq. (9.58)] with α_n and $\tan \phi'/FS_s$.

Bishop's simplified method is probably the most widely used method. When incorporated into computer programs, it yields satisfactory results in most cases. The ordinary method of slices is presented in this chapter as a learning tool. It is rarely used now because it is too conservative.

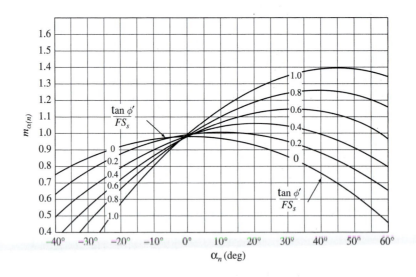

Figure 9.25 Variation of $m_{\alpha(n)}$ with α_n and $\tan \phi'/FS_s$ [Eq. (9.58)]

Similar to Eq. (9.52) for steady-state condition (Figure 9.22), Eq. (9.59) can be modified to the following form:

$$
FS_s = \frac{\sum\limits_{n=1}^{n=p} \left[c'b_n + (W_n - u_n b_n)\tan\phi' \right] \dfrac{1}{m_{(\alpha)n}}}{\sum\limits_{n=1}^{n=p} W_n \sin\alpha_n}
\tag{9.60}
$$

Note that W_n in Eqs. (9.59) and (9.60) is the total weight of the slice. In Eq. (9.60), we have

$$
W_n = \text{total weight of the } n\text{th slice} = \gamma b_n z_n
\tag{9.61}
$$

where
z_n = average height of the nth slice
$u_n = h_n \gamma_w$

So we can let

$$
r_{u(n)} = \frac{u_n}{\gamma z_n} = \frac{h_n \gamma_w}{\gamma z_n}
\tag{9.62}
$$

Note that $r_{u(n)}$ is a nondimensional quantity. Substituting Eqs. (9.61) and (9.62) into Eq. (9.60) and simplifying, we obtain

$$
FS_s = \left[\frac{1}{\sum\limits_{n=1}^{n=p} \dfrac{b_n}{H} \dfrac{z_n}{H} \sin\alpha_n} \right] \times \sum\limits_{n=1}^{n=p} \left\{ \frac{\dfrac{c'}{\gamma H} \dfrac{b_n}{H} + \dfrac{b_n}{H} \dfrac{z_n}{H}[1 - r_{u(n)}]\tan\phi'}{m_{\alpha(n)}} \right\}
\tag{9.63}
$$

For a steady-state seepage condition, a weighted average value of $r_{u(n)}$ can be taken, which is a constant. Let the weighted average value of $r_{u(n)}$ be r_u. For most practical cases, the value of r_u may range up to 0.5. So

$$
FS_s = \left[\frac{1}{\sum\limits_{n=1}^{n=p} \dfrac{b_n}{H} \dfrac{z_n}{H} \sin\alpha_n} \right] \times \sum\limits_{n=1}^{n=p} \left\{ \frac{\dfrac{c'}{\gamma H} \dfrac{b_n}{H} + \dfrac{b_n}{H} \dfrac{z_n}{H}(1 - r_u)\tan\phi'}{m_{\alpha(n)}} \right\}
\tag{9.64}
$$

9.8 Analysis of Simple Slopes with Steady–State Seepage

Several solutions have been developed in the past for stability analysis of simple slopes with steady-state seepage. Following is a partial list of the solutions:

- Bishop and Morgenstern's solution (1960)
- Spencer's solution (1967)
- Cousins' solution (1978)
- Michalowski's solution (2002)

The solutions of Spencer (1967) and Michalowski (2002) will be presented in this section.

Spencer's Solution

Bishop's simplified method of slices described in Section 9.7 satisfies the equations of equilibrium with respect to the moment but not with respect to the forces. Spencer (1967) has provided a method to determine the factor of safety (FS_s) by taking into account the interslice forces (P_n, T_n, P_{n+1}, T_{n+1}, as shown in Figure 9.21), which does satisfy the equations of equilibrium with respect to moment and forces. The details of this method of analysis are beyond the scope of this text; however, the final results of Spencer's work are summarized in this section in Figure 9.26. Note that r_u, as shown in Figure 9.26, is the same as that defined by Eq. (9.64).

In order to use the charts given in Figure 9.26 and to determine the required value of FS_s, the following step-by-step procedure needs to be used.

Step 1: Determine c', γ, H, β, ϕ', and r_u for the given slope.
Step 2: Assume a value of FS_s.
Step 3: Calculate $c'/[FS_{s(\text{assumed})} \gamma H]$.
↑
Step 2
Step 4: With the value of $c'/FS_s \gamma H$ calculated in Step 3 and the slope angle β, enter the proper chart in Figure 9.26 to obtain ϕ'_d. Note that Figures 9.26 a, b, and c, are, respectively, for r_u of 0, 0.25, and 0.5, respectively.
Step 5: Calculate $FS_s = \tan \phi'/\tan \phi'_d$.
↑
Step 4
Step 6: If the values of FS_s as assumed in Step 2 are not the same as those calculated in Step 5, repeat Steps 2, 3, 4, and 5 until they are the same.

Michalowski's Solution

Michalowski (2002) used the kinematic approach of limit analysis similar to that shown in Figures 9.18 and 9.19 to analyze slopes with steady-state seepage. The results of this analysis are summarized in Figure 9.27 for $r_u = 0.25$ and $r_u = 0.5$. Note that Figure 9.19 is applicable for the $r_u = 0$ condition.

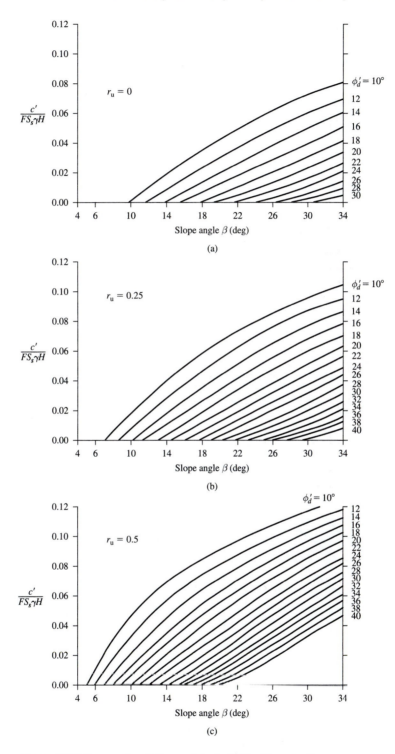

Figure 9.26 Spencer's solution—plot of $c'/FS_s\gamma H$ versus β

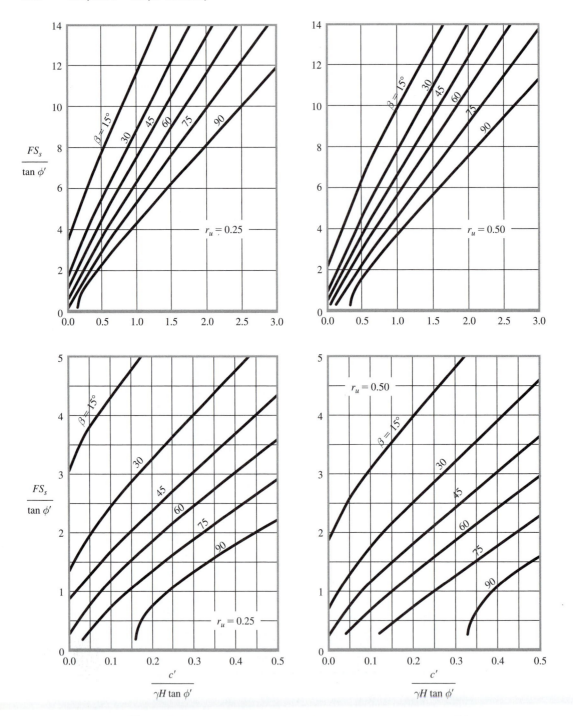

Figure 9.27 Michalowski's solution for steady-state seepage condition

Example 9.7

A given slope under steady-state seepage has the following: $H = 21.62$ m, $\phi' = 25°$, slope: 2H:1V, $c' = 20$ kN/m², $\gamma = 18.5$ kN/m³, $r_u = 0.25$. Determine the factor of safety, FS_s. Use Spencer's method.

Solution
Given: $H = 21.62$ m, $\beta = 26.57°$, $c' = 20$ kN/m², $\gamma = 18.5$ kN/m³, $\phi' = 25°$, and $r_u = 0.25$. Now the following table can be prepared.

β (deg)	$FS_{s(assumed)}$	$\dfrac{c'}{FS_{s\,(assumed)}\gamma H}$	$\phi'_d{}^a$ (deg)	$FS_{s(calculated)} = \dfrac{\tan \phi'}{\tan \phi'_d}$
26.57	1.1	0.0455	18	1.435
26.57	1.2	0.0417	19	1.354
26.57	1.3	0.0385	20	1.281
26.57	1.4	0.0357	21	1.215

ᵃFrom Figure 9.26b

Figure 9.28 shows a plot of $FS_{s(assumed)}$ against $FS_{s(calculated)}$, from which FS_s **1.3.**

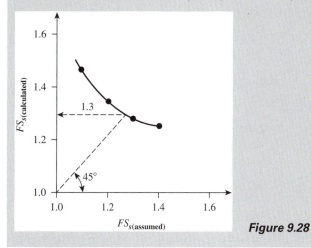

Figure 9.28

■

Example 9.8

Solve Example 9.7 using Michalowski's solution (Figure 9.27).

Solution

$$\frac{c'}{\gamma H \tan \phi'} = \frac{20}{(18.5)(21.62)(\tan 25)} = 0.107$$

For $r_u = 0.25$, from Figure 9.27, $\dfrac{FS_s}{\tan \phi'} \approx 3.1$ So,

$$FS_s = (3.1)(\tan 25) = \textbf{1.45}$$
∎

9.9   Mass Procedure for Stability of Clay Slope with Earthquake Forces

Saturated Clay ($\phi = 0$ Condition)

The stability of saturated clay slopes ($\phi = 0$ condition) with earthquake forces has been analyzed by Koppula (1984). Figure 9.29 shows a clay slope with a potential curve of sliding *AED*, which is an arc of a circle that has radius *r*. The center of the circle is located at *O*. Considering unit length perpendicular to the slope, we consider these forces for stability analysis:

1. Weight of the soil wedge, *W*:

$$W = (\text{area of } ABCDEA)(\gamma)$$

2. Horizontal inertia force, $k_h W$:

$$k_h = \frac{\text{horizontal component of earthquake acceleration}}{g}$$

where g = acceleration from gravity

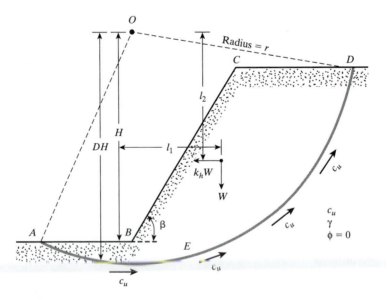

Figure 9.29 Stability analysis of slope in homogeneous clay with earthquake forces ($\phi = 0$ condition)

3. Cohesive force along the surface of sliding, which will have a magnitude of $(\widehat{AED})c_u$

The moment of the driving forces about O can now be given as

$$M_d = Wl_1 + k_h Wl_2 \qquad (9.65)$$

Similarily, the moment of the resisting about O is

$$M_r = (\widehat{AED})c_u r \qquad (9.66)$$

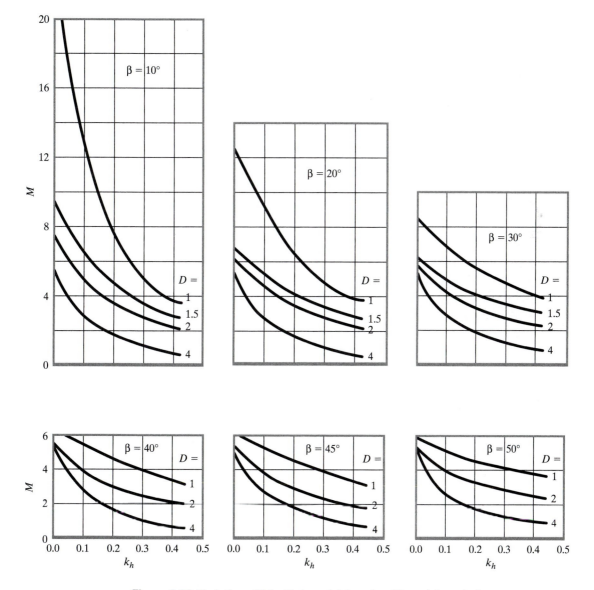

Figure 9.30 Variation of M with k_h and β based on Koppula's analysis

Thus, the factor of safety against sliding is

$$FS_s = \frac{M_r}{M_d} = \frac{(\widehat{AED})(c_u)(r)}{Wl_1 + k_h Wl_2} = \frac{c_u}{\gamma H} M \tag{9.67}$$

where M = stability factor.

The variations of the stability factor M with slope angle β and k_h based on Koppula's (1984) analysis are given in Figures 9.30 and 9.31.

$c'-\phi'$ Soil (Zero Pore Water Pressure)

Similar to those shown in Figures 9.19 and 9.27, Michalowski (2002) solved the stability of slopes for $c'-\phi'$ soils with earthquake forces. This solution used the kinematic approach of limit analysis assuming the failure surface to be an arc of a logarithmic spiral. The results of this solution are shown in Figure 9.32.

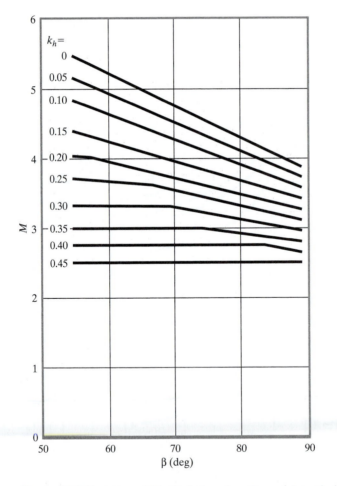

Figure 9.31 Variation of M with k_h based on Koppula's analysis (for $\beta \geq 55°$)

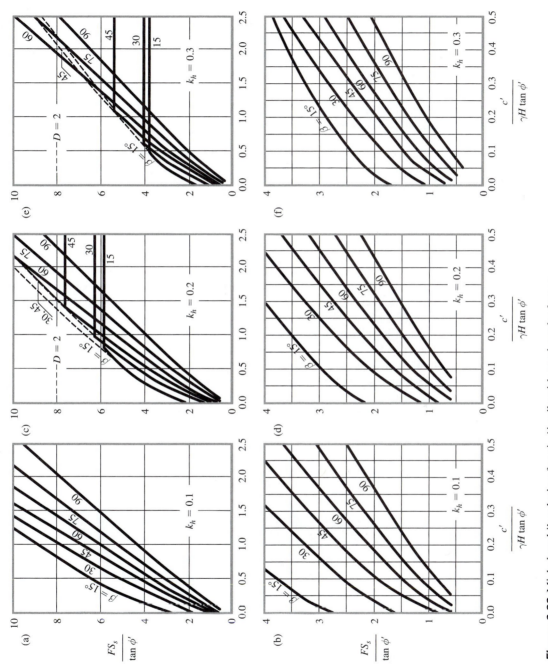

Figure 9.32 Michalowski's solution for $c'-\phi'$ soils with earthquake forces

Problems

9.1   Refer to Figure 9.2. For the infinite slope, given $\gamma = 18$ kN/m³, $c' = 10$ kN/m², $\phi' = 22°$.
   a. If $\beta = 28°$, what will be the height H for critical equilibrium?
   b. If $\beta = 28°$ and $H = 3$ m, what will be the factor of safety of the slope against sliding?
   c. If $\beta = 28°$, find the height H which will have a factor of safety of 2.5 against sliding.

9.2   Refer to the infinite slope described in Problem 9.1. Plot a graph of H_{cr} versus slope angle β (for β varying from 30° to 45°).

9.3   Refer to the infinite slope with seepage shown in Figure 9.3. For the slope, given: $\beta = 20°$, $H = 3$ m. The parameters of the soil are: $G_s = 2.68$, $e = 0.65$, $\phi' = 20°$, $c' = 14.4$ kN/m². Find the factor of safety against sliding along plane AB.

9.4   Repeat Problem 9.3 with the following: $H = 4$ m, $\phi' = 20°$, $c' = 25$ kN/m². $\gamma_{sat} = 18$ kN/m³, $\beta = 45°$.

9.5   A slope is shown in Figure 9.33. AC represents a trial failure plane. For the wedge ABC, find the factor of safety against sliding.

9.6   A finite slope is shown in Figure 9.4. Assuming that the slope failure would occur along a plane (Culmann's assumption), find the height of the slope for critical equilibrium given $\phi' = 10°$, $c' = 12$ kN/m², $\gamma = 17.3$ kN/m³, and $\beta = 50°$.

9.7   Repeat Problem 9.6 with $\phi' = 20°$, $c' = 25$ kN/m², $\gamma = 18$ kN/m³, and $\beta = 45°$.

9.8   Refer to Figure 9.4. Using the soil parameters given in Problem 9.6, find the height of the slope, H, that will have a factor of safety of 2.5 against sliding. Assume that the critical surface for sliding is a plane.

9.9   Refer to Figure 9.4. Given $\phi' = 15°$, $c' = 9.6$ kN/m², $\gamma = 18.0$ kN/m³, $\beta = 60°$, and $H = 2.7$ m, determine the factor of safety with respect to sliding. Assume that the critical surface for sliding is a plane.

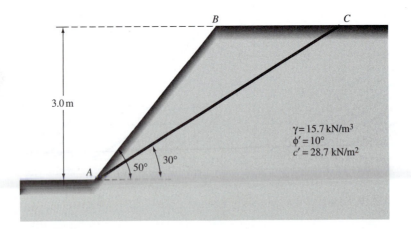

$\gamma = 15.7$ kN/m³
$\phi' = 10°$
$c' = 28.7$ kN/m²

Figure 9.33

9.10 Refer to Problem 9.9. Find the height of the slope, H, that will have $FS_s = 1.5$. Assume that the critical surface for sliding is a plane.

9.11 A cut slope is to be made in a soft clay with its sides rising at an angle of 75° to the horizontal (Figure 9.34). Assume that $c_u = 31.1 \text{ kN/m}^2$ and $\gamma = 17.3 \text{ kN/m}^3$.

    a. Determine the maximum depth up to which the excavation can be carried out.

    b. Find the radius, r, of the critical circle when the factor of safety is equal to 1 (part a).

    c. Find the distance $\overline{BC}$.

9.12 If the cut described in Problem 9.11 is made to a depth of only 3.0 m, what will be the factor of safety of the slope against sliding?

9.13 Using the graph given in Figure 9.7, determine the height of a slope, 1 vertical to $\frac{1}{2}$ horizontal, in saturated clay having an undrained shear strength of 32.6 kN/m^2. The desired factor of safety against sliding is 2. Given $\gamma = 18.9 \text{ kN/m}^3$.

9.14 Refer to Problem 9.13. What should be the critical height of the slope? What will be the nature of the critical circle? Also find the radius of the critical circle.

9.15 For the slope shown in Figure 9.35, find the factor of safety against sliding for the trial surface AC.

9.16 A cut slope was excavated in a saturated clay. The slope angle β is equal to 35° with respect to the horizontal. Slope failure occurred when the cut reached a depth of 8.2 m. Previous soil explorations showed that a rock layer was located at a depth of 11 m below the ground surface. Assume an undrained condition and $\gamma_{sat} = 19.2 \text{ kN/m}^3$.

    a. Determine the undrained cohesion of the clay (use Figure 9.7).

    b. What was the nature of the critical circle?

    c. With reference to the toe of the slope, at what distance did the surface of sliding intersect with the bottom of the excavation?

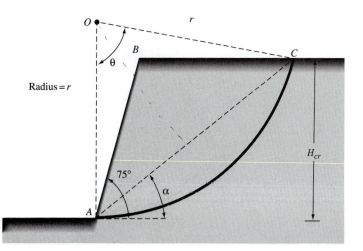

Figure 9.34

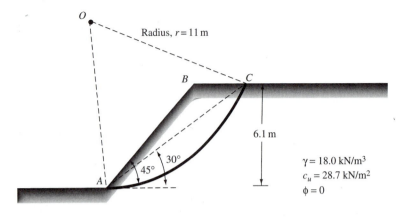

Figure 9.35

9.17 Refer to Figure 9.36. Use Taylor's chart for $\phi' > 0$ (Figure 9.15) to find the critical height of the slope in each case:
   a. $n' = 2$, $\phi' = 15°$, $c' = 31.1 \text{ kN/m}^2$, and $\gamma = 18.0 \text{ kN/m}^3$
   b. $n' = 1$, $\phi' = 25°$, $c' = 24 \text{ kN/m}^2$, and $\gamma = 18.0 \text{ kN/m}^3$
   c. $n' = 2.5$, $\phi' = 12°$, $c' = 25 \text{ kN/m}^2$, and $\gamma = 17 \text{ kN/m}^3$
   d. $n' = 1.5$, $\phi' = 18°$, $c' = 18 \text{ kN/m}^2$, and $\gamma = 16.5 \text{ kN/m}^3$
9.18 Solve Problem 9.17 a, c, and d using Figure 9.26a.
9.19 Referring to Figure 9.36 and using Figure 9.15, find the factor of safety with respect to sliding for the following cases:
   a. $n' = 2.5$, $\phi' = 12°$, $c' = 24 \text{ kN/m}^2$, $\gamma = 17 \text{ kN/m}^3$, and $H = 12 \text{ m}$
   b. $n' = 1.5$, $\phi' = 15°$, $c' = 18 \text{ kN/m}^2$, $\gamma = 18 \text{ kN/m}^3$, and $H = 5 \text{ m}$
9.20 Solve Problem 9.19 using Figure 9.19.
9.21 Referring to Figure 9.37 and using the ordinary method of slices, find the factor of safety against sliding for the trial case $\beta = 45°$, $\phi' = 15°$, $c' = 18 \text{ kN/m}^2$, $\gamma = 17.1 \text{ kN/m}^3$, $H = 5 \text{ m}$, $\alpha = 30°$, and $\theta = 80°$.

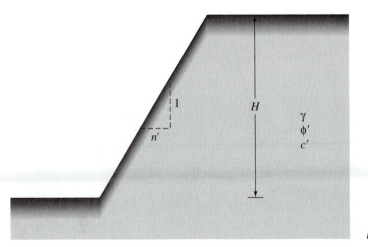

Figure 9.36

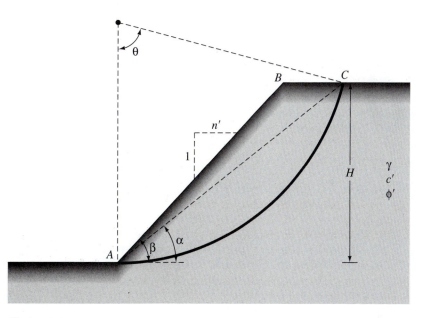

Figure 9.37

9.22 Determine the minimum factor of safety for the steady-state seepage condition of a slope with the following parameters: $H = 6.1$ m, $\beta = 26.57°$, $\phi' = 25°$, $c' = 5.5$ kN/m^2, $\gamma = 18$ kN/m^3, and $r_u = 0.5$. Use Spencer's method.

9.23 Solve Problem 9.22 using Figure 9.27.

References

BISHOP, A. W. (1955). "The Use of Slip Circle in the Stability Analysis of Earth Slopes," *Geotechnique*, Vol. 5, No. 1, 7–17.

BISHOP, A. W., and MORGENSTERN, N. R. (1960). "Stability Coefficients for Earth Slopes," *Geotechnique*, Vol. 10, No. 4, 129–147.

COUSINS, B. F. (1978). "Stability Charts for Simple Earth Slopes," *Journal of the Geotechnical Engineering Division*, ASCE, Vol. 104, No. GT2, 267–279.

CULMANN, C. (1875). *Die Graphische Statik*, Meyer and Zeller, Zurich.

FELLENIUS, W. (1927). *Erdstatische Berechnungen*, revised edition, W. Ernst u. Sons, Berlin.

KOPPULA, S. D. (1984). "Pseudo-Statis Analysis of Clay Slopes Subjected to Earthquakes," *Geotechnique*, Vol. 34, No. 1, 71–79.

MICHALOWSKI, R. L. (2002). "Stability Charts for Uniform Slopes," *Journal of Geotechnical and Geoenvironmental Engineering*, ASCE, Vol. 128, No. 4, 351–355.

SINGH, A. (1970). "Shear Strength and Stability of Man-Made Slopes," *Journal of the Soil Mechanics and Foundations Division*, ASCE, Vol. 96, No. SM6, 1879–1892.

SPENCER, E. (1967). "A Method of Analysis of the Stability of Embankments Assuming Parallel Inter-Slice Forces," *Geotechnique*, Vol. 17, No. 1, 11–26.

TAYLOR, D. W. (1937). "Stability of Earth Slopes," *Journal of the Boston Society of Civil Engineers*, Vol. 24, 197–246.

TERZAGHI, K., and PECK, R. B. (1967). *Soil Mechanics in Engineering Practice*, 2nd ed., Wiley, New York.

10

Subsurface Exploration

The process of identifying the layers of deposits that underlie a proposed structure and their physical characteristics is generally referred to as *subsurface exploration*. The purpose of subsurface exploration is to obtain information that will aid the geotechnical engineer in these tasks:

1. Selecting the type and depth of foundation suitable for a given structure
2. Evaluating the load-bearing capacity of the foundation
3. Estimating the probable settlement of a structure
4. Determining potential foundation problems (for example, expansive soil, collapsible soil, sanitary landfill, and so on)
5. Determining the location of the water table
6. Predicting lateral earth pressure for structures such as retaining walls, sheet pile bulkheads, and braced cuts
7. Establishing construction methods for changing subsoil conditions

Subsurface exploration is also necessary for underground construction and excavation. It may be required when additions or alterations to existing structures are contemplated.

10.1 Subsurface Exploration Program

Subsurface exploration comprises several steps, including collection of preliminary information, reconnaissance, and site investigation.

Collection of Preliminary Information

Information must be obtained regarding the type of structure to be built and its general use. For the construction of buildings, the approximate column loads and their spacing and the local building-code and basement requirements should be known. The construction of bridges requires determining span length and the loading on piers and abutments.

A general idea of the topography and the type of soil to be encountered near and around the proposed site can be obtained from the following sources:

1. U.S. Geological Survey maps
2. State government geological survey maps
3. U.S. Department of Agriculture's Soil Conservation Service county soil reports
4. Agronomy maps published by the agriculture departments of various states
5. Hydrological information published by the U.S. Corps of Engineers, including the records of stream flow, high flood levels, tidal records, and so on
6. Highway department soils manuals published by several states

The information collected from these sources can be extremely helpful to those planning a site investigation. In some cases, substantial savings are realized by anticipating problems that may be encountered later in the exploration program.

Reconnaissance

The engineer should always make a visual inspection of the site to obtain information about these features:

1. The general topography of the site and the possible existence of drainage ditches, abandoned dumps of debris, or other materials. Also, evidence of creep of slopes and deep, wide shrinkage cracks at regularly spaced intervals may be indicative of expansive soils.
2. Soil stratification from deep cuts, such as those made for construction of nearby highways and railroads.
3. Type of vegetation at the site, which may indicate the nature of the soil. For example, a mesquite cover in central Texas may indicate the existence of expansive clays that can cause possible foundation problems.
4. High-water marks on nearby buildings and bridge abutments.
5. Groundwater levels, which can be determined by checking nearby wells.
6. Types of construction nearby and existence of any cracks in walls or other problems.

The nature of stratification and physical properties of the soil nearby can also be obtained from any available soil-exploration reports for nearby existing structures.

Site Investigation

The site investigation phase of the exploration program consists of planning, making test boreholes, and collecting soil samples at desired intervals for subsequent observation and laboratory tests. The approximate required minimum depth of the borings should be predetermined; however, the depth can be changed during the drilling operation, depending on the subsoil encountered. To determine the approximate minimum depth of boring for foundations, engineers may use the rules established by the American Society of Civil Engineers (1972):

1. Determine the net increase of stress, $\Delta\sigma$, under a foundation with depth as shown in Figure 10.1. (The general equations for estimating stress increase are given in Chapter 6.)

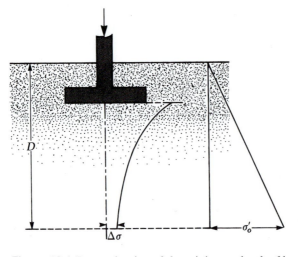

Figure 10.1 Determination of the minimum depth of boring

2. Estimate the variation of the vertical effective stress, σ'_o, with depth.
3. Determine the depth, $D = D_1$, at which the stress increase $\Delta\sigma$ is equal to $\left(\frac{1}{10}\right)q$ (q = estimated net stress on the foundation).
4. Determine the depth, $D = D_2$, at which $\Delta\sigma/\sigma'_o = 0.05$.
5. Unless bedrock is encountered, the smaller of the two depths, D_1 and D_2, just determined is the approximate minimum depth of boring required.

If the preceding rules are used, the depths of boring for a building with a width of 30 m will be approximately as listed in Table 10.1, according to Sowers and Sowers (1970). For hospitals and office buildings, they also use the following rule to determine boring depth:

$$D_b = 3S^{0.7} \quad \text{(for light steel or narrow concrete buildings)} \quad (10.1)$$

$$D_b = 6S^{0.7} \quad \text{(for heavy steel or wide concrete buildings)} \quad (10.2)$$

where
  D_b = depth of boring (m)
  S = number of stories

Table 10.1 Approximate depths of borings for buildings with a width of 30 m

No. of stories	Boring depth (m)
1	3.5
2	6
3	10
4	16
5	24

Table 10.2 Approximate spacing of boreholes

Type of project	Spacing (m)
Multistory building	10–30
One-story industrial plants	20–60
Highways	250–500
Residential subdivision	250–500
Dams and dikes	40–80

When deep excavations are anticipated, the depth of boring should be at least 1.5 times the depth of excavation.

Sometimes subsoil conditions require that the foundation load be transmitted to bedrock. The minimum depth of core boring into the bedrock is about 3 m. If the bedrock is irregular or weathered, the core borings may have to be deeper.

There are no hard and fast rules for borehole spacing. Table 10.2 gives some general guidelines. The spacing can be increased or decreased, depending on the subsoil condition. If various soil strata are more or less uniform and predictable, fewer boreholes are needed than in nonhomogeneous soil strata.

The engineer should also take into account the ultimate cost of the structure when making decisions regarding the extent of field exploration. The exploration cost generally should be 0.1% to 0.5% of the cost of the structure.

10.2 Exploratory Borings in the Field

Soil borings can be made by several methods, including auger boring, wash boring, percussion drilling, and rotary drilling.

Auger boring is the simplest method of making exploratory boreholes. Figure 10.2 shows two types of hand auger: the *post hole auger* and the *helical auger*. Hand augers cannot be used for advancing holes to depths exceeding 3–5 m; however, they can be used for soil exploration work for some highways and small structures. *Portable power-driven helical augers* (30 to 75 mm in diameter) are available for making deeper boreholes. The soil samples obtained from such borings are highly disturbed. In some noncohesive soils or soils that have low cohesion, the walls of the boreholes will not stand unsupported. In such circumstances, a metal pipe is used as a *casing* to prevent the soil from caving in.

When power is available, *continuous-flight augers* are probably the most common method used for advancing a borehole. The power for drilling is delivered by truck- or tractor-mounted drilling rigs. Boreholes up to about 60–70 m can be made easily by this method. Continuous-flight augers are available in sections of about 1–2 m with either a solid or hollow stem. Some of the commonly used solid stem augers have outside diameters of 67 mm, 83 mm, 102 mm, and 114 mm. Hollow stem augers commercially available have dimensions of 64 mm inside diameter (ID) and 158 mm outside

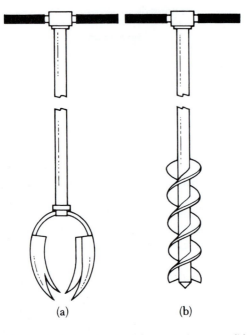

Figure 10.2 Hand tools: (a) post hole auger; (b) helical auger

diameter (OD), 70 mm ID and 178 mm OD, 76 mm ID and 203 mm OD, and 83 mm ID and 229 mm OD.

The tip of the auger is attached to a cutter head. During the drilling operation (Figure 10.3), section after section of auger can be added and the hole extended downward. The flights of the augers bring the loose soil from the bottom of the hole to the surface. The driller can detect changes in soil type by noting changes in the speed and sound of drilling. When solid stem augers are used, the auger must be withdrawn at regular intervals to obtain soil samples and also to conduct other operations such as standard penetration tests. Hollow stem augers have a distinct advantage over solid stem augers in that they do not have to be removed frequently for sampling or other tests. As shown schematically (Figure 10.4), the outside of the hollow stem auger acts like a casing.

The hollow stem auger system includes the following:

Outer component: (a) hollow auger sections, (b) hollow auger cap, and (c) drive cap

Inner component: (a) pilot assembly, (b) center rod column, and (c) rod-to-cap adapter

The auger head contains replaceable carbide teeth. During drilling, if soil samples are to be collected at a certain depth, the pilot assembly and the center rod are removed. The soil sampler is then inserted through the hollow stem of the auger column.

Figure 10.3 Drilling with continuous-flight augers (Courtesy of Danny R. Anderson, Danny R. Anderson Consultants, El Paso, Texas)

Wash boring is another method of advancing boreholes. In this method, a casing about 2–3 m long is driven into the ground. The soil inside the casing is then removed using a chopping bit attached to a drilling rod. Water is forced through the drilling rod and exits at a very high velocity through the holes at the bottom of the chopping bit (Figure 10.5). The water and the chopped soil particles rise in the drill hole and overflow at the top of the casing through a T connection. The washwater is collected in a container. The casing can be extended with additional pieces as the borehole progresses; however, that is not required if the borehole will stay open and not cave in.

Rotary drilling is a procedure by which rapidly rotating drilling bits attached to the bottom of drilling rods cut and grind the soil and advance the borehole. Rotary drilling can be used in sand, clay, and rocks (unless badly fissured). Water, or *drilling mud*, is forced down the drilling rods to the bits, and the return flow forces the cuttings to the surface. Boreholes with diameters of 50–200 mm can be made easily by this technique. The drilling mud is a slurry of water and bentonite. Generally, rotary drilling is used when the soil encountered is likely to cave in. When soil samples are needed, the drilling rod is raised and the drilling bit is replaced by a sampler.

Percussion drilling is an alternative method of advancing a borehole, particularly through hard soil and rock. A heavy drilling bit is raised and lowered to chop the hard soil. The chopped soil particles are brought up by the circulation of water. Percussion drilling may require casing.

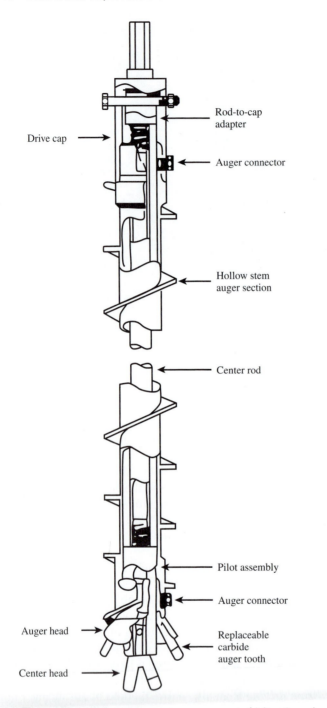

Figure 10.4 Hollow stem auger components (After American Society for Testing and Materials, 2003)

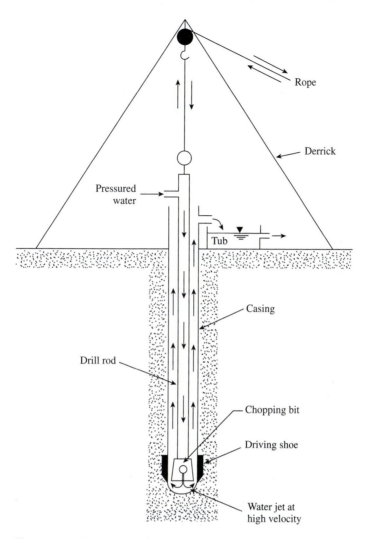

Figure 10.5 Wash boring

10.3   *Procedures for Sampling Soil*

Two types of soil samples can be obtained during subsurface exploration: *disturbed* and *undisturbed*. Disturbed, but representative, samples can generally be used for the following types of laboratory test:

1. Grain-size analysis
2. Determination of liquid and plastic limits
3. Specific gravity of soil solids
4. Organic content determination
5. Classification of soil

Disturbed soil samples, however, cannot be used for consolidation, hydraulic conductivity, or shear strength tests. Undisturbed soil samples must be obtained for these laboratory tests.

Split-Spoon Sampling

Split-spoon samplers can be used in the field to obtain soil samples that are generally disturbed but still representative. A section of a *standard split-spoon sampler* is shown in Figure 10.6a. It consists of a tool-steel driving shoe, a steel tube that is split longitudinally in half, and a coupling at the top. The coupling connects the sampler to the drill rod. The standard split tube has an inside diameter of 34.93 mm and an outside diameter of 50.8 mm; however, samplers that have inside and outside diameters up to 63.5 mm and 76.2 mm, respectively, are also available. When a borehole is extended to a predetermined depth, the drill tools are removed and the sampler is lowered to the bottom of the borehole. The sampler is driven into the soil by hammer blows to the top of the drill rod. The standard weight of the hammer is 623 N and, for each blow, the hammer drops a distance of 762 mm. The number of blows required for spoon penetration of three 152.4-mm intervals is recorded. The numbers of blows required for the last two intervals are added to give the *standard penetration number, N,* at that depth. This number is generally referred to as the *N value* (American Society for Testing and Materials, 2002, Designation D-1586). The sampler is then withdrawn, and the shoe

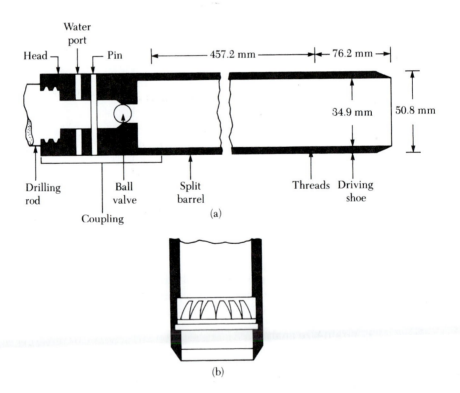

Figure 10.6 (a) Standard split-spoon sampler; (b) spring core catcher

and coupling are removed. The soil sample recovered from the tube is then placed in a glass bottle and transported to the laboratory.

The degree of disturbance for a soil sample is usually expressed as

$$A_R(\%) = \frac{D_o^2 - D_i^2}{D_i^2}(100)$$

(10.3)

where

A_R = area ratio
D_o = outside diameter of the sampling tube
D_i = inside diameter of the sampling tube

When the area ratio is 10% or less, the sample is generally considered to be undisturbed.

Split-spoon samples generally are taken at intervals of about 1.5 m. When the material encountered in the field is sand (particularly fine sand below the water table), sample recovery by a split-spoon sampler may be difficult. In that case, a device such as a *spring core catcher* (Figure 10.6b) may have to be placed inside the split spoon.

At this point, it is important to point out that there are several factors that will contribute to the variation of the standard penetration number N at a given depth for similar soil profiles. These factors include standard penetration test (SPT) hammer efficiency, borehole diameter, sampling method, and rod-length factor (Seed et al., 1985; Skempton, 1986). The two most common types of SPT hammers used in the field are the *safety hammer* and *donut hammer*. They are commonly dropped by a rope with *two wraps around a pulley*.

Based on field observations it appears reasonable to standardize the field standard penetration number based on the input driving energy and its dissipation around the sampler into the surrounding soil, or

$$N_{60} = \frac{N\eta_H\eta_B\eta_S\eta_R}{60}$$

(10.4)

where

N_{60} = standard penetration number corrected for field conditions
N = measured penetration number
η_H = hammer efficiency (%)
η_B = correction for borehole diameter
η_S = sampler correction
η_R = correction for rod length

Based on the recommendations of Seed et al. (1985) and Skempton (1986), the variations of η_H, η_B, η_S, and η_R are summarized in Table 10.3.

Besides obtaining soil samples, standard penetration tests provide several useful correlations. For example, the consistency of clayey soils can often be estimated from the standard penetration number, N_{60}, as shown in Table 10.4. However, correlations for clays require tests to verify that the relationships are valid for the clay deposit being examined.

Table 10.3 Variations of η_H, η_B, η_S, and η_R [Eq. (10.4)]

1. Variation of η_H

Country	Hammer type	Hammer release	η_H (%)
Japan	Donut	Free fall	78
	Donut	Rope and pulley	67
United States	Safety	Rope and pulley	60
	Donut	Rope and pulley	45
Argentina	Donut	Rope and pulley	45
China	Donut	Free fall	60
	Donut	Rope and pulley	50

2. Variation of η_B

Diameter (mm)	η_B
60–120	1
150	1.05
200	1.15

3. Variation of η_S

Variable	η_S
Standard sampler	1.0
With liner for dense sand and clay	0.8
With liner for loose sand	0.9

4. Variation of η_R

Rod length (m)	η_R
>10	1.0
6–10	0.95
4–6	0.85
0–4	0.75

Table 10.4 Consistency of clays and approximate correlation to the standard penetration number, N_{60}

Standard penetration number, N_{60}	Consistency	Unconfined compression strength, q_u (kN/m^2)
0–2	Very soft	0–25
2–5	Soft	25–50
5–10	Medium stiff	50–100
10–20	Stiff	100–200
20–30	Very stiff	200–400
>30	Hard	>400

The literature contains many correlations between the standard penetration number and the undrained shear strength of clay, c_u. Based on the results of undrained triaxial tests conducted on insensitive clays, Stroud (1974) suggested that

$$c_u = KN_{60} \tag{10.5}$$

where
K = constant = 3.5–6.5 kN/m²
N_{60} = standard penetration number obtained from the field

The average value of K is about 4.4 kN/m². Hara et al. (1971) also suggested that

$$c_u(\text{kN/m}^2) = 29N_{60}^{0.72} \tag{10.6}$$

It is important to point out that any correlation between c_u and N_{60} is only approximate.

In granular soils, the N_{60} value is affected by the effective overburden pressure, σ'_o. For that reason, the N_{60} value obtained from field exploration under different effective overburden pressures should be changed to correspond to a standard value of σ'_o; that is,

$$(N_1)_{60} = C_N N_{60} \tag{10.7}$$

where
$(N_1)_{60}$ = corrected N value to a standard value of σ'_o (95.6 kN/m²)
C_N = correction factor
N_{60} = N value obtained from the field

A number of empirical relationships have been proposed for C_N. Some of the relationships are given in Table 10.5. The most commonly cited relationships are those given by Liao and Whitman (1986) and Skempton (1986).

Table 10.5 Empirical relationships for C_N (*Note:* σ'_o is in kN/m²)

Source	C_N
Liao and Whitman (1986)	$9.78\sqrt{\dfrac{1}{\sigma'_o}}$
Skempton (1986)	$\dfrac{2}{1 + 0.01\sigma'_o}$
Seed et al. (1975)	$1 - 1.25 \log\left(\dfrac{\sigma'_o}{95.6}\right)$
Peck et al. (1974)	$0.77 \log\left(\dfrac{1912}{\sigma'_o}\right)$ for $\sigma'_o \geq 25$ kN/m²

An approximate relationship between the corrected standard penetration number and the relative density of sand is given in Table 10.6. These values are approximate, primarily because the effective overburden pressure and the stress history of the soil significantly influence the N_{60} values of sand.

More recently, Hatanaka and Feng (2006) proposed the following relationships between the relative density (D_r) and $(N_1)_{60}$ for fine to medium sand.

$$D_r (\%) = 1.55(N_1)_{60} + 40 \text{ [for } 0 \leqslant (N_1)_{60} \leqslant 25] \qquad (10.8a)$$

$$D_r (\%) = 0.84(N_1)_{60} + 58.8 \text{ [for } 25 \leqslant (N_1)_{60} \leqslant 50] \qquad (10.8b)$$

For fine to medium sands with fines (that is, % passing No. 200 sieve, F_c) between 15% and 20%, the $(N_1)_{60}$ in Eqs. (10.8a and 10.8b) may be modified as

$$(N_1)_{60} = (N_{60} + 12.9)\left(\frac{98}{\sigma'_o}\right)^{0.5} \qquad (10.9)$$

where σ'_o is the vertical effective stress in kN/m^2.

The effective *peak* angle of friction of granular soils, ϕ', was correlated to the corrected standard penetration number by Peck, Hanson, and Thornburn (1974). They gave a correlation between $(N_1)_{60}$ and ϕ' in a graphical form, which can be approximated as (Wolff, 1989)

$$\phi' \text{ (deg)} = 27.1 + 0.3(N_1)_{60} - 0.00054[(N_1)_{60}]^2 \qquad (10.10)$$

Schmertmann (1975) provided a correlation among N_{60}, σ'_o, and ϕ'. The correlation can be approximated as (Kulhawy and Mayne, 1990)

$$\phi' = \tan^{-1}\left[\frac{N_{60}}{12.2 + 20.3\left(\dfrac{\sigma'_o}{p_a}\right)}\right]^{0.34} \qquad (10.11)$$

where
N_{60} = field standard penetration number
σ'_o = effective overburden pressure
p_a = atmospheric pressure in the same unit as σ'_o ($\approx$100 kN/m^2)
ϕ' = soil friction angle (effective)

More recently, Hatanaka and Uchida (1996) provided a simple correlation between ϕ' and $(N_1)_{60}$, which can be expressed as

$$\phi' = \sqrt{20(N_1)_{60}} + 20 \qquad (10.12)$$

Table 10.6 Relation between the corrected N values and the relative density in sands

Standard penetration number, $(N_1)_{60}$	Approximate relative density, D_r (%)
0–5	0–5
5–10	5–30
10–30	30–60
30–50	60–95

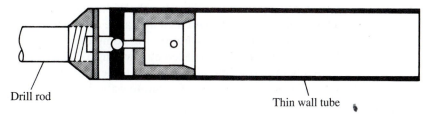

Drill rod

Thin wall tube

Figure 10.7 Thin wall tube

When the standard penetration resistance values are used in the preceding cor-relations to estimate soil parameters, the following qualifications should be noted:

1. The equations are approximate and largely empirical.
2. Because the soil is not homogeneous, the N_{60} values obtained from a given borehole vary widely.
3. In soil deposits that contain large boulders and gravel, standard penetration numbers may be erratic and unreliable.

Although the correlations are approximate, with correct interpretation the standard penetration test provides a good evaluation of soil properties. The primary sources of errors in standard penetration tests are inadequate cleaning of the bore-hole, careless measurement of the blow count, eccentric hammer strikes on the drill rod, and inadequate maintenance of water head in the borehole.

Thin Wall Tube

Thin wall tubes are sometimes called *Shelby tubes*. They are made of seamless steel and are commonly used to obtain undisturbed clayey soils. The commonly used thin wall tube samplers have outside diameters of 50.8 mm and 76.2 mm. The bottom end of the tube is sharpened. The tubes can be attached to drilling rods (Figure 10.7). The drilling rod with the sampler attached is lowered to the bottom of the borehole, and the sampler is pushed into the soil. The soil sample inside the tube is then pulled out. The two ends of the sampler are sealed, and it is sent to the laboratory for testing.

Samples obtained in this manner may be used for consolidation or shear tests. A thin wall tube with a 50.8-mm outside diameter has an inside diameter of about 47.63 mm. The area ratio is

$$A_R(\%) = \frac{D_o^2 - D_i^2}{D_i^2}(100) = \frac{(50.8)^2 - (47.63)^2}{(47.63)^2}(100) = 13.75\%$$

Increasing the diameters of samples increases the cost of obtaining them.

10.4   *Observation of Water Levels*

The presence of a water table near a foundation significantly affects a foundation's load-bearing capacity and settlement. The water level will change seasonally. In many cases, establishing the highest and lowest possible levels of water during the life of a project may become necessary.

If water is encountered in a borehole during a field exploration, that fact should be recorded. In soils with high hydraulic conductivity, the level of water in a borehole will stabilize about 24 hours after completion of the boring. The depth of the water table can then be recorded by lowering a chain or tape into the borehole.

In highly impermeable layers, the water level in a borehole may not stabilize for several weeks. In such cases, if accurate water level measurements are required, a *piezometer* can be used.

The simplest piezometer (Figure 10.8) is a standpipe or Casagrande-type piezometer. It consists of a riser pipe joined to a filter tip that is placed in sand.

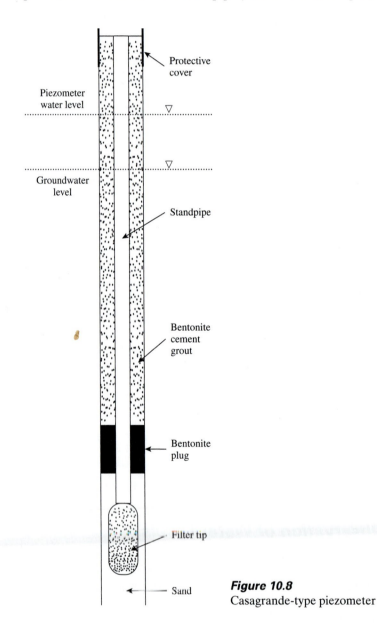

Figure 10.8
Casagrande-type piezometer

Figure 10.9
Components of a Casagrande-type piezometer
(Courtesy of N. Sivakugan, James Cook
University, Australia)

A bentonite seal is placed above the sand to isolate the pore water pressure at the filter tip. The annular space between the riser pipe and the borehole is backfilled with bentonite-cement grout to prevent vertical migration of water. This allows periodic checking until the water level stabilizes. Figure 10.9 shows the components of a Casagrande-type piezometer.

10.5 Vane Shear Test

Fairly reliable results for the *in situ* undrained shear strength, c_u ($\phi = 0$ concept) of soft plastic cohesive soils may be obtained directly from vane shear tests during the drilling operation (ASTM Test Designation 2573). The shear vane usually consists of four thin, equal-sized steel plates welded to a steel torque rod (Figure 10.10a). First, the vane is pushed into the soil. Then torque is applied at the top of the torque rod to rotate the vane at a uniform speed. A cylinder of soil of height h and diameter d will resist the torque until the soil fails. The undrained shear strength of the soil can be calculated as follows.

If T is the maximum torque applied at the head of the torque rod to cause failure, it should be equal to the sum of the resisting moment of the shear force along the side surface of the soil cylinder (M_s) and the resisting moment of the shear force at each end (M_e) (Figure 10.10b):

$$T = M_s + \underbrace{M_e + M_e}_{\text{Two ends}} \tag{10.13}$$

The resisting moment M_s can be given as

$$M_s = \underbrace{(\pi dh)}_{\substack{\text{Surface} \\ \text{area}}} c_u \underbrace{(d/2)}_{\substack{\text{Moment} \\ \text{arm}}} \tag{10.14}$$

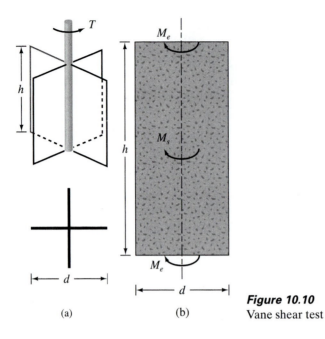

Figure 10.10
Vane shear test

(a)   (b)

The geometry of field vanes—rectangular and tapered—as recommended by ASTM is shown in Figure 10.11. The dimensions of the vanes used in the field are given in Table 10.7. The standard rate of torque application is 0.1°/sec. The maximum torque T applied to cause failure can be given as

$$T = f(c_u, h, \text{ and } d) \tag{10.15}$$

or

$$c_u = \frac{T}{K} \tag{10.16}$$

where T is in N $\cdot$ m, and c_u is in kN/m^2

    K = a constant with a magnitude depending on the dimension and shape of the vane

$$K = \left(\frac{\pi}{10^6}\right)\left(\frac{d^2 h}{2}\right)\left(1 + \frac{d}{3h}\right) \tag{10.17}$$

where
    d = diameter of vane (cm)
    h = measured height of vane (cm)

If $h/d = 2$, Eq. (10.17) yields

$$K = 366 \times 10^{-8} d^3 \tag{10.18}$$
$$\uparrow$$
$$\text{(cm)}$$

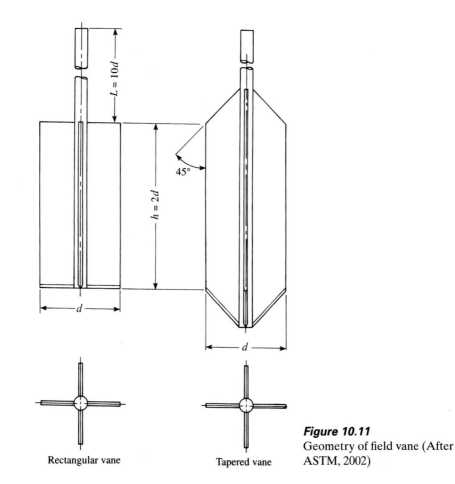

Figure 10.11
Geometry of field vane (After
ASTM, 2002)

Rectangular vane   Tapered vane

Field vane shear tests are moderately rapid and economical and are used
extensively in field soil-exploration programs. The test gives good results in soft and
medium stiff clays, and it is also an excellent test to determine the properties of sensitive clays.

Table 10.7 ASTM recommended dimensions of field vanes*

Casing size	Diameter, d (mm)	Height, h (mm)	Thickness of blade (mm)	Diameter of rod (mm)
AX	38.1	76.2	1.6	12.7
BX	50.8	101.6	1.6	12.7
NX	63.5	127.0	3.2	12.7
101.6 mm[†]	92.1	184.1	3.2	12.7

*Selection of the vane size is directly related to the consistency of the soil being
tested; that is, the softer the soil, the larger the vane diameter should be.
[†]Inside diameter.

Sources of significant error in the field vane shear test are poor calibration of torque measurement and damaged vanes. Other errors may be introduced if the rate of vane rotation is not properly controlled.

Skempton (1957) gave an empirical correction for c_u obtained from field vane shear tests that is of the form

$$\frac{c_{u(\text{VST})}}{\sigma'_o} = 0.11 + 0.0037(PI) \tag{10.19}$$

where

σ'_o = effective overburden pressure
PI = plasticity index, in percent

Figure 10.12 shows a comparison of the variation of c_u with the depth obtained from field vane shear tests, unconfined compression tests, and unconsolidated-undrained triaxial tests for Morgan City recent alluvium (Arman, *et al.*, 1975). It can be seen that the vane shear test values are higher compared to the others.

Bjerrum (1974) also showed that, as the plasticity of soils increases, c_u obtained from vane shear tests may give results that are unsafe for foundation design. For this reason, he suggested the correction

$$c_{u(\text{design})} = \lambda c_{u(\text{vane shear})} \tag{10.20}$$

where

$$\lambda = \text{correction factor} = 1.7 - 0.54 \log(PI) \tag{10.21}$$

PI = plasticity index

More recently, Morris and Williams (1994) gave the correlations of λ as

$$\lambda = 1.18e^{-0.08(PI)} + 0.57 \quad (\text{for } PI > 5) \tag{10.22}$$

and

$$\lambda = 7.01e^{-0.08(LL)} + 0.57 \quad (\text{for } LL > 20) \tag{10.23}$$

where LL = liquid limit (%).

Vane shear tests can be conducted in the laboratory. The laboratory shear vane has dimensions of about 12.7 mm (diameter) and 25.4 mm (height). Figure 10.13 is a photograph of laboratory vane shear equipment. Field shear vanes with the following dimensions are used by the U.S. Bureau of Reclamation:

$$d = 50.8 \text{ mm}; \quad h = 101.6 \text{ mm}$$
$$d = 76.2 \text{ mm}; \quad h = 152.4 \text{ mm}$$
$$d = 101.6 \text{ mm}; \quad h = 203.2 \text{ mm}$$

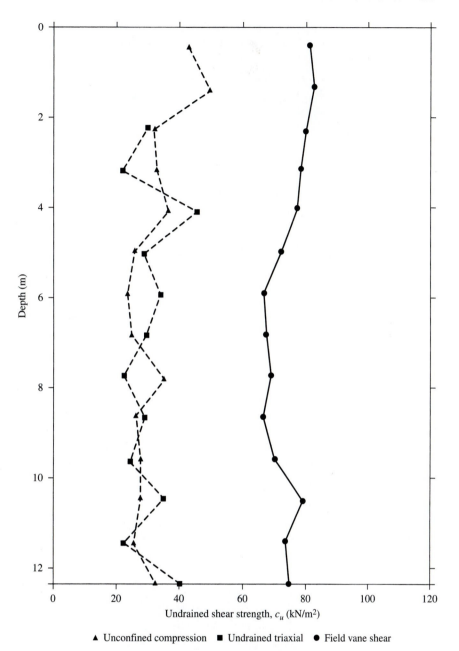

Figure 10.12 Variation of c_u with depth obtained from various tests for Morgan City recent alluvium (Drawn from the test results of Arman, *et al.*, 1975)

Figure 10.13 Laboratory vane shear device (Courtesy of ELE International)

Example 10.1

A soil profile is shown in Figure 10.14. The clay is normally consolidated. Its liquid limit is 60 and its plastic limit is 25. Estimate the unconfined compression strength of the clay at a depth of 10 m measured from the ground surface. Use Skempton's relationship from Eqs. (10.19), (10.20), and (10.21).

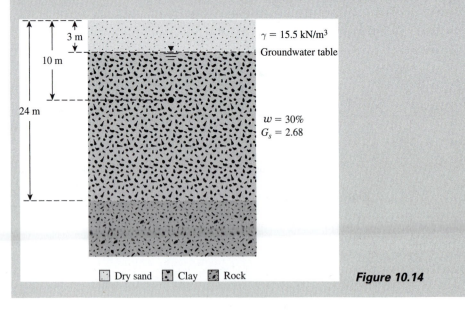

Figure 10.14

Solution

For the saturated clay layer, the void ratio is

$$e = wG_s = (2.68)(0.3) = 0.8$$

The effective unit weight is

$$\gamma'_{clay} = \left(\frac{G_s - 1}{1 + e}\right)\gamma_w = \frac{(2.68 - 1)(9.81)}{1 + 0.8} = 9.16 \text{ kN/m}^3$$

The effective stress at a depth of 10 m from the ground surface is

$$\sigma'_o = 3\gamma_{sand} + 7\gamma'_{clay} = (3)(15.5) + (7)(9.16)$$
$$= 110.62 \text{ kN/m}^2$$

From Eq. (10.19),

$$\frac{c_{u(VST)}}{\sigma'_O} = 0.11 + 0.0037(PI)$$

$$\frac{c_{u(VST)}}{110.62} = 0.11 + 0.0037(60 - 25)$$

Hence

$$c_{u(VST)} = 26.49 \text{ kN/m}^2$$

From Eqs. (10.20) and (10.21), we get

$$c_u = \lambda c_{u(VST)}$$
$$= [1.7 - 0.54 \log(PI)]c_{u(VST)}$$
$$= [1.7 - 0.54 \log(60 - 25)]26.49 = 22.95 \text{ kN/m}^2$$

So the unconfined compression strength is

$$q_u = 2c_u = (2)(22.95) = \textbf{45.9 kN/m}^2 \qquad \blacksquare$$

10.6   *Cone Penetration Test*

The cone penetration test (CPT), originally known as the Dutch cone penetration test, is a versatile sounding method that can be used to determine the materials in a soil profile and estimate their engineering properties. This test is also called the *static penetration test,* and no boreholes are necessary to perform it. In the original version, a 60° cone with a base area of 10 cm^2 was pushed into the ground at a steady rate of about 20 mm/sec, and the resistance to penetration (called the point resistance) was measured.

The cone penetrometers in use at present measure (a) the *cone resistance, q_c,* to penetration developed by the cone, which is equal to the vertical force applied to the cone divided by its horizontally projected area; and (b) the *frictional resistance, f_c,* which is the resistance measured by a sleeve located above the cone with the local soil surrounding it. The frictional resistance is equal to the vertical force applied to the sleeve divided by its surface area—actually, the sum of friction and adhesion.

Generally, two types of penetrometers are used to measure q_c and f_c:

1. *Mechanical friction-cone penetrometer* (Figure 10.15). In this case, the penetrometer tip is connected to an inner set of rods. The tip is first advanced about 40 mm, thus giving the cone resistance. With further thrusting, the tip engages the friction sleeve. As the inner rod advances, the rod force is equal to the sum of the vertical forces on the cone and the sleeve. Subtracting the force on the cone gives the side resistance.

2. *Electric friction-cone penetrometer* (Figure 10.16). In this case, the tip is attached to a string of steel rods. The tip is pushed into the ground at the rate of 20 mm/sec. Wires from the transducers are threaded through the center of the rods and continuously give the cone and side resistances.

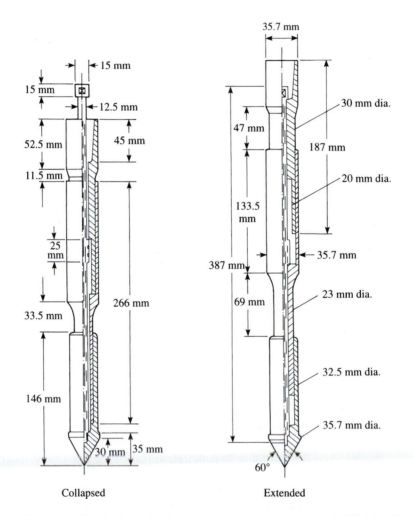

Figure 10.15 Mechanical friction-cone penetrometer (After ASTM, 2002)

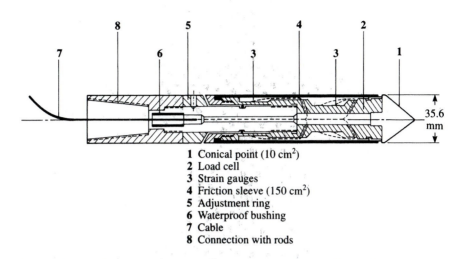

1 Conical point (10 cm²)
2 Load cell
3 Strain gauges
4 Friction sleeve (150 cm²)
5 Adjustment ring
6 Waterproof bushing
7 Cable
8 Connection with rods

Figure 10.16 Electric friction-cone penetrometer (After ASTM, 1997)

Figure 10.17 shows the results of penetrometer tests in a soil profile with friction measurement by an electric friction-cone penetrometer.

Several correlations that are useful in estimating the properties of soils encountered during an exploration program have been developed for the cone resistance, q_c, and the friction ratio, F_r, obtained from the cone penetration tests. The friction ratio, F_r, is defined as

$$F_r = \frac{\text{frictional resistance}}{\text{cone resistance}} = \frac{f_c}{q_c} \qquad (10.24)$$

In a more recent study on several soils in Greece, Anagnostopoulos et al. (2003) expressed F_r as

$$F_r(\%) = 1.45 - 1.36 \log D_{50} \text{ (electric cone)} \qquad (10.25)$$

and

$$F_r(\%) = 0.7811 - 1.611 \log D_{50} \text{ (mechanical cone)} \qquad (10.26)$$

where D_{50} = size through which 50% of soil will pass through (mm).

The D_{50} for soils based on which Eqs. (10.25) and (10.26) have been developed ranged from 0.001 mm to about 10 mm.

Correlation between Relative Density (D_r) and q_c for Sand

Lancellotta (1983) and Jamiolkowski et al. (1985) showed that the relative density of *normally consolidated sand, D_r*, and q_c can be correlated according to the formula

$$D_r(\%) = A + B \log_{10}\left(\frac{q_c}{\sqrt{\sigma'_o}}\right) \qquad (10.27)$$

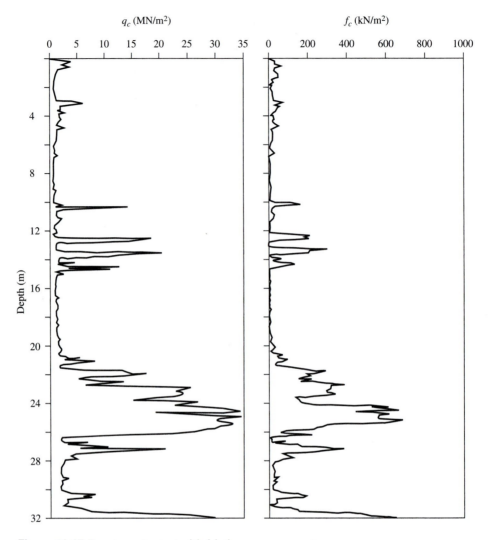

Figure 10.17 Penetrometer test with friction measurement

The preceding relationship can be rewritten as (Kulhawy and Mayne, 1990)

$$D_r(\%) = 68\left[\log\left(\frac{q_c}{\sqrt{p_o \cdot \sigma_0'}}\right) - 1\right]$$ (10.28)

where

p_a = atmospheric pressure

σ_o' = vertical effective stress

Baldi et al. (1982), and Robertson and Campanella (1983) recommended an empirical relationship shown in Figure 10.18 between vertical effective stress (σ'_o), relative density (D_r), and for *normally consolidated sand* (q_c).

Kulhawy and Mayne (1990) proposed the following relationship to correlate D_r, q_c, and the vertical effective stress σ'_o:

$$D_r = \sqrt{\left[\frac{1}{305Q_c\text{OCR}^{1.8}}\right]\left[\frac{\dfrac{q_c}{p_a}}{\left(\dfrac{\sigma'_o}{p_a}\right)^{0.5}}\right]} \qquad (10.29)$$

In this equation,

OCR = overconsolidation ratio
p_a = atmospheric pressure
Q_c = compressibility factor

The recommended values of Q_c are as follows:

Highly compressible sand = 0.91
Moderately compressible sand = 1.0
Low compressible sand = 1.09

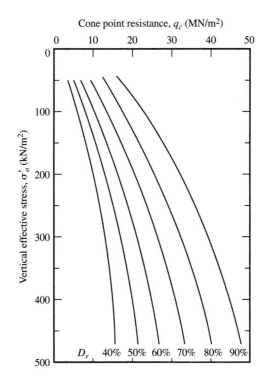

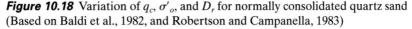

Figure 10.18 Variation of q_c, σ'_o, and D_r for normally consolidated quartz sand (Based on Baldi et al., 1982, and Robertson and Campanella, 1983)

Correlation between q_c and Drained Friction Angle (ϕ') for Sand

On the basis of experimental results, Robertson and Campanella (1983) suggested the variation of D_r, σ'_o, and ϕ' for normally consolidated quartz sand. This relationship can be expressed as (Kulhawy and Mayne, 1990)

$$\phi' = \tan^{-1}\left[0.1 + 0.38 \log\left(\frac{q_c}{\sigma'_o}\right)\right] \tag{10.30}$$

Based on the cone penetration tests on the soils in the Venice Lagoon (Italy), Ricceri et al. (2002) proposed a similar relationship for soil with classifications of ML and SP-SM as

$$\phi' = \tan^{-1}\left[0.38 + 0.27 \log\left(\frac{q_c}{\sigma'_o}\right)\right] \tag{10.31}$$

In a more recent study, Lee et al. (2004) developed a correlation between ϕ', q_c, and the horizontal effective stress (σ'_h) in the form

$$\phi' = 15.575\left(\frac{q_c}{\sigma'_h}\right)^{0.1714} \tag{10.32}$$

Correlation between q_c and N_{60}

Figure 10.19 shows a plot of q_c (kN/m^2)/N_{60} (N_{60} = standard penetration number) against the mean grain size (D_{50} in mm) for various types of soil. This was developed from field test results by Robertson and Campanella (1983).

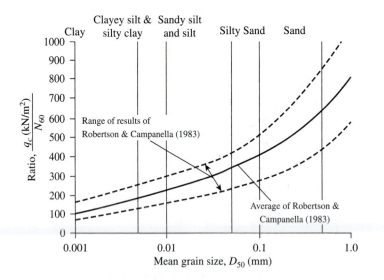

Figure 10.19 General range of variation of q_c/N_{60} for various types of soil

Anagnostopoulos et al. (2003) provided a similar relationship correlating q_c, N_{60}, and D_{50}. Or

$$\frac{\left(\dfrac{q_c}{p_a}\right)}{N_{60}} = 7.6429\, D_{50}^{0.26}$$

(10.33)

where p_a = atmospheric pressure (same unit as q_c).

Correlations of Soil Types

Robertson and Campanella (1983) provided the correlations shown in Figure 10.20 between q_c and the friction ratio [Eq. (10.24)] to identify various types of soil encountered in the field.

Correlations for Undrained Shear Strength (c_u), Preconsolidation Pressure (σ'_c), and Overconsolidation Ratio (OCR) for Clays

The undrained shear strength, c_u, can be expressed as

$$c_u = \frac{q_c - \sigma_o^*}{N_K}$$

(10.34)

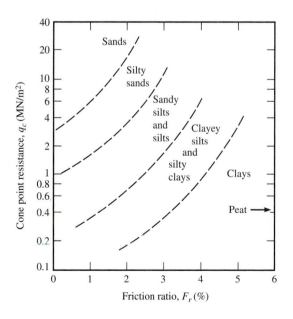

Figure 10.20 Robertson and Campanella correlation (1983) of q_c, F_r, and the soil type

where

σ_o = total vertical stress

N_K = bearing capacity factor

The bearing capacity factor, N_K, may vary from 11 to 19 for normally consolidated clays and may approach 25 for overconsolidated clay. According to Mayne and Kemper (1988),

$$N_K = 15 \text{ (for electric cone)}$$

and

$$N_K = 20 \text{ (for mechanical cone)}$$

Based on tests in Greece, Anagnostopoulos et al. (2003) determined

$$N_K = 17.2 \text{ (for electric cone)}$$

and

$$N_K = 18.9 \text{ (for mechanical cone)}$$

These field tests also showed that

$$c_u = \frac{f_c}{1.26} \text{ (for mechanical cones)} \tag{10.35}$$

and

$$c_u = f_c \text{ (for electrical cones)} \tag{10.36}$$

Mayne and Kemper (1988) provided correlations for preconsolidation pressure (σ'_c) and overconsolidation ratio (OCR) as

$$\sigma'_c = 0.243(q_c)^{0.96}$$
$$\uparrow \qquad \uparrow$$
$$\text{MN/m}^2 \quad \text{MN/m}^2 \tag{10.37}$$

and

$$\text{OCR} = 0.37\left(\frac{q_c - \sigma_o}{\sigma'_o}\right)^{1.01} \tag{10.38}$$

where σ_o and σ'_o = total and effective stress, respectively.

10.7   *Pressuremeter Test (PMT)*

The pressuremeter test is an *in situ* test conducted in a borehole. It was originally developed by Menard (1956) to measure the strength and deformability of soil. It has also been adopted by ASTM as Test Designation 4719. The Menard-type PMT essentially consists of a probe with three cells. The top and bottom ones are *guard cells* and the middle one is the *measuring cell*, as shown schematically in Figure 10.21a. The test is conducted in a pre-bored hole. The pre-bored hole should have a diameter that is between 1.03 and 1.2 times the nominal diameter of the probe. The probe that is most commonly used has a diameter of 58 mm and a length of 420 mm. The probe cells can be expanded by either liquid or gas. The guard cells are expanded to reduce the end-condition effect on the measuring cell. The measuring cell has a volume, V_o, of 535 cm^3. Table 10.8 lists the probe diameters and the diameters of the boreholes as recommended by ASTM.

To conduct a test, the measuring cell volume, V_o, is measured and the probe is inserted into the borehole. Pressure is applied in increments, and the volumatic expansion of the cell is measured. This process is continued until the soil fails or until the pressure limit of the device is reached. The soil is considered to have failed when the total volume of the expanded cavity, V, is about twice the volume of the original cavity. After the completion of the test, the probe is deflated and advanced for testing at another depth.

The results of the pressuremeter test are expressed in a graphical form of pressure versus volume in Figure 10.21b. In this figure, Zone I represents the reloading portion during which the soil around the borehole is pushed back into the initial state (that is, the state it was in before drilling). The pressure, p_o, represents the *in situ* total horizontal stress. Zone II represents a pseudo-elastic zone in which the relation of cell volume to cell pressure is practically linear. The pressure,

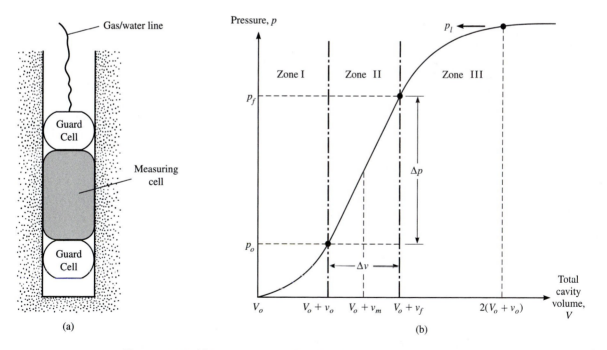

Figure 10.21 (a) Pressuremeter; (b) plot of pressure versus total cavity volume

Table 10.8 Probe and borehole diameter for pressuremeter test

Probe diameter (mm)	Borehole diameter	
	Nominal (mm)	Maximum (mm)
44	45	53
58	60	70
74	76	89

p_f, represents the creep, or yield, pressure. The zone marked III is the plastic zone. The pressure, p_l, represents the limit pressure.

The pressuremeter modulus, E_p, of the soil is determined using the theory of expansion of an infinitely thick cylinder. Thus,

$$E_p = 2(1 + \mu_s)(V_o + v_m)\left(\frac{\Delta p}{\Delta v}\right) \tag{10.39}$$

where

$$v_m = \frac{v_o + v_f}{2}$$

$\Delta p = p_f - p_o$
$\Delta v = v_f - v_o$
μ_s = Poisson's ratio (which may be assumed to be 0.33)

The limit pressure, p_l, is usually obtained by extrapolation and not by direct measurement.

To overcome the difficulty of preparing the borehole to the proper size, self-boring pressuremeters (SBPMT) have also been developed. The details concerning SBPMTs can be found in the work of Baguelin et al. (1978).

Ohya et al. (1982) (see also Kulhawy and Mayne, 1990) correlated E_p with field standard penetration numbers, N_{60}, for sand and clay as follows:

$$\text{clay:} \quad E_p(\text{kN/m}^2) = 1930(N_{60})^{0.63} \tag{10.40}$$

$$\text{sand:} \quad E_p(\text{kN/m}^2) = 908(N_{60})^{0.66} \tag{10.41}$$

10.8 Dilatometer Test

The use of the flat-plate dilatometer test (DMT) is relatively recent (Marchetti, 1980; Schmertmann, 1986). The equipment essentially consists of a flat plate measuring 220 mm (length) × 95 mm (width) × 14 mm (thickness). A thin, flat, circular expandable steel membrane with a diameter of 60 mm is located flush at the center on one side of the plate (Figure 10.22a). The dilatometer probe is inserted into the ground using a cone penetrometer testing rig (Figure 10.22b). Gas and electric lines extend from the surface control box through the penetrometer rod into the blade.

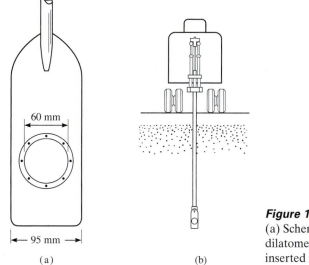

60 mm

95 mm

(a)

(b)

Figure 10.22
(a) Schematic diagram of a flat-plate
dilatometer; (b) dilatometer probe
inserted into ground

At the required depth, high-pressure nitrogen gas is used to inflate the membrane.
Two pressure readings are taken:

1. The pressure A to "lift off" the membrane
2. The pressure B at which the membrane expands 1.1 mm into the surrounding
   soil

The A and B readings are corrected as follows (Schmertmann, 1986):

$$\text{contact stress, } p_o = 1.05(A + \Delta A - Z_m) - 0.05(B - \Delta B - Z_m) \quad (10.42)$$

$$\text{expansion stress, } p_1 = B - Z_m - \Delta B \quad (10.43)$$

where
ΔA = vacuum pressure required to keep the membrane in contact with its seating
ΔB = air pressure required inside the membrane to deflect it outward to a
center expansion of 1.1 mm
Z_m = gauge pressure deviation from 0 when vented to atmospheric pressure

The test is normally conducted at depths 200 to 300 mm apart. The result of a given
test is used to determine three parameters:

1. Material index, $I_D = \dfrac{p_1 - p_o}{p_o - u_o}$

2. Horizontal stress index, $K_D = \dfrac{p_o - u_o}{\sigma'_o}$

3. Dilatometer modulus, $E_D \ (\text{kN/m}^2) = 34.7[p_1 \ (\text{kN/m}^2) - p_o \ (\text{kN/m}^2)]$

where
u_o = pore water pressure
σ'_o = *in situ* vertical effective stress

Figure 10.23 shows an assembly of equipment necessary for the dilatometer test.

Marchetti (1980) conducted several dilatometer tests in Porto Tolle, Italy. The subsoil consisted of recent, normally consolidated delta deposits of the Po River. A thick layer of silty clay was found below a depth of about 3 m ($c' = 0$; $\phi' \approx 28°$). The results obtained from the dilatometer tests were correlated with several soil properties (Marchetti, 1980). Some of these correlations are given here:

$$K_o = \left(\frac{K_D}{1.5}\right)^{0.47} - 0.6 \qquad (10.44)$$

$$OCR = (0.5K_D)^{1.6} \qquad (10.45)$$

$$\frac{c_u}{\sigma'_o} = 0.22 \qquad \text{(for normally consolidated clay)} \qquad (10.46)$$

$$\left(\frac{c_u}{\sigma'_o}\right)_{OC} = \left(\frac{c_u}{\sigma'_o}\right)_{NC}(0.5K_D)^{1.25} \qquad (10.47)$$

$$E_s = (1 - \mu_s^2)E_D \qquad (10.48)$$

where
K_o = coefficient of at-rest earth pressure
OCR = overconsolidation ratio
OC = overconsolidated soil
NC = normally consolidated soil
E_s = modulus of elasticity

Schmertmann (1986) also provided a correlation between the material index, I_D, and the dilatometer modulus, E_D, for determination of soil description and unit weight, γ. This relationship is shown in Figure 10.24.

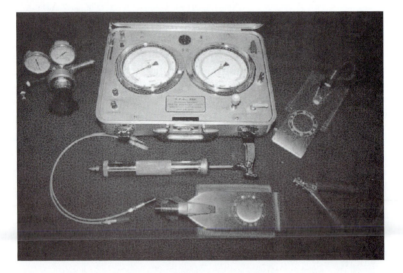

Figure 10.23 Assembly of equipment for dilatometer test (Courtesy of N. Sivakugan, James Cook University, Australia)

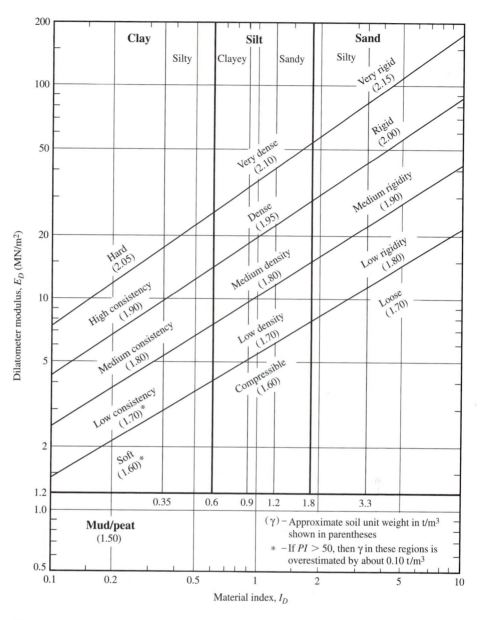

Figure 10.24 Chart for determination of soil description and unit weight (After Schmertmann, 1986). *Note:* 1 t/m³ = 9.81 kN/m³.

10.9   *Coring of Rocks*

When a rock layer is encountered during a drilling operation, rock coring may be necessary. For coring of rocks, a *core barrel* is attached to a drilling rod. A *coring bit* is attached to the bottom of the core barrel (Fig. 10.25). The cutting elements may

Table 10.9 Standard size and designation of casing, core barrel, and compatible drill rod

Casing and core barrel designation	Outside diameter of core barrel bit (mm)	Drill rod designation	Outside diameter of drill rod (mm)	Diameter of borehole (mm)	Diameter of core sample (mm)
EX	36.51	E	33.34	38.1	22.23
AX	47.63	A	41.28	50.8	28.58
BX	58.74	B	47.63	63.5	41.28
NX	74.61	N	60.33	76.2	53.98

be diamond, tungsten, carbide, or others. Table 10.9 summarizes the various types of core barrel and their sizes, as well as the compatible drill rods commonly used for foundation exploration. The coring is advanced by rotary drilling. Water is circulated through the drilling rod during coring, and the cutting is washed out.

Two types of core barrel are available: the *single-tube core barrel* (Figure 10.25a) and the *double-tube core barrel* (Figure 10.25b). Rock cores obtained by single-tube

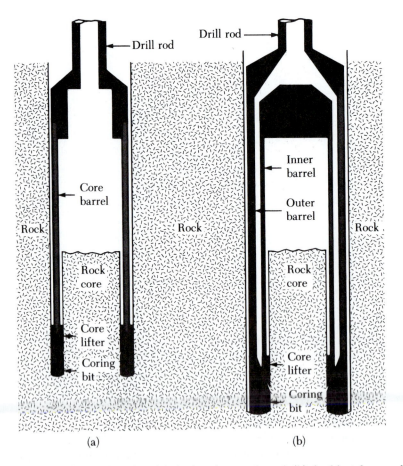

Figure 10.25 Rock coring: (a) single-tube core barrel; (b) double-tube core barrel

Table 10.10 Relationship between
in situ rock quality and *RQD*

RQD	Rock quality
0–0.25	Very poor
0.25–0.5	Poor
0.5–0.75	Fair
0.75–0.9	Good
0.9–1	Excellent

core barrels can be highly disturbed and fractured because of torsion. Rock cores smaller than the BX size tend to fracture during the coring process.

When the core samples are recovered, the depth of recovery should be properly recorded for further evaluation in the laboratory. Based on the length of the rock core recovered from each run, the following quantities may be calculated for a general evaluation of the rock quality encountered:

$$\text{recovery ratio} = \frac{\text{length of core recovered}}{\text{theoretical length of rock cored}} \qquad (10.49)$$

rock quality designation $(RQD) =$

$$\frac{\Sigma \text{ length of recovered pieces equal to or larger than 101.6 mm}}{\text{theoretical length of rock cored}} \qquad (10.50)$$

A recovery ratio of 1 indicates the presence of intact rock; for highly fractured rocks, the recovery ratio may be 0.5 or less. Table 10.10 presents the general relationship (Deere, 1963) between the *RQD* and the *in situ* rock quality.

10.10 *Preparation of Boring Logs*

The detailed information gathered from each borehole is presented in a graphical form called the *boring log*. As a borehole is advanced downward, the driller generally should record the following information in a standard log:

1. Name and address of the drilling company
2. Driller's name
3. Job description and number
4. Number and type of boring and boring location
5. Date of boring
6. Subsurface stratification, which can be obtained by visual observation of the soil brought out by auger, split-spoon sampler, and thin wall Shelby tube sampler
7. Elevation of water table and date observed, use of casing and mud losses, and so on
8. Standard penetration resistance and the depth
9. Number, type, and depth of soil sample collected
10. In case of rock coring, type of core barrel used and, for each run, the actual length of coring, length of core recovery, and the *RQD*

This information should never be left to memory because not recording the data often results in erroneous boring logs.

After completing all the necessary laboratory tests, the geotechnical engineer prepares a finished log that includes notes from the driller's field log and the results of tests conducted in the laboratory. Figure 10.26 shows a typical boring log. These logs should be attached to the final soil exploration report submitted to the client. Note that Figure 10.26 also lists the classifications of the soils in the left-hand column, along with the description of each soil (based on the Unified Soil Classification System).

Boring Log

Name of the Project   Two-story apartment building

Location   Johnson & Olive St.   Date of Boring   March 2, 2006

Boring No.   3   Type of   Hollow stem auger   Ground Elevation   60.8 m
                       Boring

Soil description	Depth (m)	Soil sample type and number	N_{60}	w_n (%)	Comments
Light brown clay (fill)					
Silty sand (SM)	1 — 2	SS-1	9	8.2	
°G.W.T. ▽ 3.5 m	3 — 4	SS-2	12	17.6	$LL = 38$ $PI = 11$
Light gray clayey silt (ML)	5	ST-1		20.4	$LL = 36$ $q_u = 112 \text{ kN/m}^2$
	6	SS-3	11	20.6	
Sand with some gravel (SP)	7				
End of boring @ 8 m	8	SS-4	27	9	

N_{60} = standard penetration number (blows/305 mm)
w_n = natural moisture content
LL = liquid limit; PI = plasticity index
q_u = unconfined compression strength
SS = split-spoon sample; ST = Shelby tube sample

°Groundwater table observed after 1 week of drilling

Figure 10.26 A typical boring log

10.11   *Soil Exploration Report*

At the end of all soil exploration programs, the soil and/or rock specimens collected in the field are subject to visual observation and appropriate laboratory testing. After all the required information has been compiled, a soil exploration report is prepared for use by the design office and for reference during future construction work. Although the details and sequence of information in the report may vary to some degree, depending on the structure under consideration and the person compiling the report, each report should include the following items:

1. The scope of the investigation
2. A description of the proposed structure for which the subsoil exploration has been conducted
3. A description of the location of the site, including structure(s) nearby, drainage conditions of the site, nature of vegetation on the site and surrounding it, and any other feature(s) unique to the site
4. Geological setting of the site
5. Details of the field exploration—that is, number of borings, depths of borings, type of boring, and so on
6. General description of the subsoil conditions as determined from soil specimens and from related laboratory tests, standard penetration resistance and cone penetration resistance, and so on
7. Water table conditions
8. Foundation recommendations, including the type of foundation recommended, allowable bearing pressure, and any special construction procedure that may be needed; alternative foundation design procedures should also be discussed in this portion of the report
9. Conclusions and limitations of the investigations

The following graphical presentations should be attached to the report:

1. Site location map
2. A plan view of the location of the borings with respect to the proposed structures and those existing nearby
3. Boring logs
4. Laboratory test results
5. Other special graphical presentations

The exploration reports should be well planned and documented. They will help in answering questions and solving foundation problems that may arise later during design and construction.

Problems

10.1   A Shelby tube has an outside diameter of 50.8 mm and an inside diameter of 47.6 mm.
    a. What is the area ratio of the tube?
    b. If the outside diameter remains the same, what should be the inside diameter of the tube to give an area ratio of 10%?

10.2 A soil profile is shown in Figure 10.27 along with the standard penetration numbers in the clay layer. Use Eqs. 10.6 to determine and plot the variation of c_u with depth.

10.3 The average value of the field standard penetration number in a saturated clay layer is 6. Estimate the unconfined compression strength of the clay. Use Eq. (10.5) ($K \approx 4.2 \text{ kN/m}^2$).

10.4 The table gives the variation of the field standard penetration number, N_{60}, in a sand deposit:

Depth (m)	N_{60}
1.5	5
3	7
4.5	9
6	8
7.5	12
9	11

The groundwater table is located at a depth of 5.5 m. The dry unit weight of sand from 0 to a depth of 5.5 m is 18.08 kN/m³, and the saturated unit weight of sand for depths of 5.5 to 10.5 m is 19.34 kN/m³. Use the relationship of Liao and Whitman given in Table 10.5 to calculate the corrected penetration numbers.

10.5 The field standard penetration numbers for a deposit of dry sand are given below. For the sand, given $\gamma = 18.7 \text{ kN/m}^3$. Determine the variation of $(N_1)_{60}$ with depth. Use Skempton's correction factor given in Table 10.5.

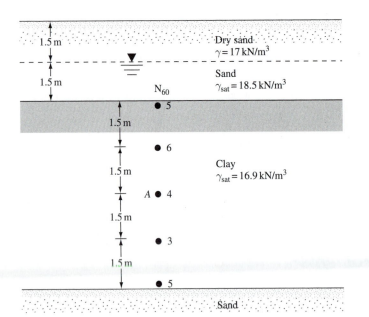

Figure 10.27

Depth (m)	N_{60}
1.5	9
3.0	9
4.5	12
6.0	12
7.5	16

10.6 For the soil profile described in Problem 10.5, estimate an average peak soil friction angle. Use Eq. (10.10).

10.7 The table gives the standard penetration numbers determined from a sandy soil deposit in the field:

Depth (m)	Unit weight of soil (kN/m³)	N_{60}
3.0	16.7	7
4.5	16.7	9
6.0	16.7	11
7.5	18.6	16
9.0	18.6	18
10.5	18.6	20
12	18.6	22

Using Eq. (10.11), determine the variation of the peak soil friction angle, ϕ'. Estimate an average value of ϕ' for the design of a shallow foundation. *Note:* For depth greater than 6 m, the unit weight of soil is 18.6 kN/m³.

10.8 Redo Problem 10.7 using Skempton's relationship given in Table 10.5 and Eq. (10.12).

10.9 The details for a soil deposit in sand are given in the table:

Depth (m)	Effective overburden pressure (kN/m²)	Field standard penetration number, N_{60}
3.0	55.1	9
4.5	82.7	11
6.0	97.3	12

The sand deposit has an average of 18% fines. Use Eqs. (10.8) and (10.9) and estimate the average relative density of sand between the depths of 3 m and 6 m.

10.10 Refer to Figure 10.27. Vane shear tests were conducted in the clay layer. The vane dimensions were 63.5 mm (d) $\times$ 127 mm (h). For the test at A, the torque required to cause failure was 0.051 N $\cdot$ m. For the clay, liquid limit was 46 and plastic limit was 21. Estimate the undrained cohesion of the clay for use in the design by using each equation:

a. Bjerrum's λ relationship (Eq. 10.21)

b. Morris and Williams' λ and PI relationship (Eq. 10.22)

c. Morris and Williams' λ and LL relationship (Eq. 10.23)

10.11 a. A vane shear test was conducted in a saturated clay. The height and diameter of the vane were 101.6 mm and 50.8 mm, respectively. During the test, the maximum torque applied was 0.0168 N · m. Determine the undrained shear strength of the clay.

b. The clay soil described in part (a) has a liquid limit of 64 and a plastic limit of 29. What would be the corrected undrained shear strength of the clay for design purposes? Use Bjerrum's relationship for λ (Eq. 10.21).

10.12 In a deposit of normally consolidated dry sand, a cone penetration test was conducted. The table gives the results:

Depth (m)	Point resistance of cone, q_c (MN/m²)
1.5	2.05
3.0	4.23
4.5	6.01
6.0	8.18
7.5	9.97
9.0	12.42

Assume the dry unit weight of sand is 15.5 kN/m³.

a. Estimate the average peak friction angle, ϕ', of the sand. Use Eq. (10.30).
b. Estimate the average relative density of the sand. Use Eq. (10.29) and $Q_c = 1$.

10.13 Refer to the soil profile shown in Figure 10.28. Assume the cone penetration resistance, q_c, at A as determined by an electric friction-cone penetrometer is 0.6 MN/m².

a. Determine the undrained cohesion, c_u.
b. Find the overconsolidation ratio, *OCR*.

10.14 Consider a pressuremeter test in a soft saturated clay.
Measuring cell volume, $V_o = 535$ cm³
$p_o = 42.4$ kN/m² $v_o = 46$ cm³
$p_f = 326.5$ kN/m² $v_f = 180$ cm³

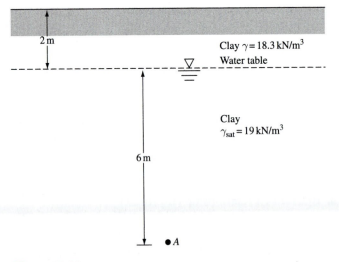

Clay γ = 18.3 kN/m³
Water table

2 m

Clay
$\gamma_{sat} = 19$ kN/m³

6 m

● A

Figure 10.28

Assuming Poisson's ratio, μ_s, to be 0.5 and referring to Figure 10.21, calculate the pressuremeter modulus, E_p.

10.15 A dilatometer test was conducted in a clay deposit. The groundwater table was located at a depth of 3 m below the ground surface. At a depth of 8 m below the ground surface, the contact pressure, p_o, was 280 kN/m^2 and the expansion stress, p_1, was 350 kN/m^2.

    a. Determine the coefficient of at-rest earth pressure, K_o.

    b. Find the overconsolidation ratio, *OCR*.

    c. What is the modulus of elasticity, E_s?

    Assume σ'_o at a depth of 8 m to be 95 kN/m^2 and $\mu_s = 0.35$.

10.16 During a field exploration, coring of rock was required. The core barrel was advanced 1.5 m during the coring. The length of the core recovered was 0.98 m. What was the recovery ratio?

References

AMERICAN SOCIETY FOR TESTING AND MATERIALS (2003). *Annual Book of ASTM Standards*, Vol. 04.09, West Conshohocken, PA.

AMERICAN SOCIETY FOR TESTING AND MATERIALS (2002). *Annual Book of ASTM Standards*, Vol. 04.08, West Conshohocken, PA.

AMERICAN SOCIETY FOR TESTING AND MATERIALS (1997). *Annual Book of ASTM Standards*, Vol. 04.08, West Conshohocken, PA.

AMERICAN SOCIETY OF CIVIL ENGINEERS (1972). "Subsurface Investigation for Design and Construction of Foundations of Buildings," *Journal of the Soil Mechanics and Foundations Division*, American Society of Civil Engineers, Vol. 98. No. SM5, 481–490.

ANAGNOSTOPOULOS, A., KOUKIS, G., SABATAKAKIS, N., and TSIAMBAOS, G. (2003). "Empirical Correlations of Soil Parameters Based on Cone Penetration Tests (CPT) for Greek Soils," *Geotechnical and Geological Engineering*, Vol. 21, No. 4, 377–387.

ARMAN, A., POPLIN, J. K., and AHMAD, N. (1975). "Study of Vane Shear," *Proceedings*, Conference on *In Situ* Measurement and Soil Properties, ASCE, Vol. 1, 93–120.

BAGUELIN, F., JÉZÉQUEL, J. F., and SHIELDS, D. H. (1978). *The Pressuremeter and Foundation Engineering*, Trans Tech Publications, Clausthal.

BALDI, G., BELLOTTI, R., GHIONNA, V., and JAMIOLKOWSKI, M. (1982). "Design Parameters for Sands from CPT," *Proceedings*, Second European Symposium on Penetration Testing, Amsterdam, Vol. 2, 425–438.

BJERRUM, L. (1974). "Problems of Soil Mechanics and Construction on Soft Clays," Norwegian Geotechnical Institute, *Publications No. 110*, Oslo.

DEERE, D. U. (1963). "Technical Description of Rock Cores for Engineering Purposes," *Felsmechanik und Ingenieurgeologie*, Vol. 1, No. 1, 16–22.

HARA, A., OHATA, T., and NIWA, M. (1971). "Shear Modulus and Shear Strength of Cohesive Soils," *Soils and Foundations*, Vol. 14, No. 3, 1–12.

HATANAKA, M., and FENG, L. (2006). "Estimating Relative Density of Sandy Soils," *Soils and Foundation*, Vol. 46, No. 3, 299 313.

HATANAKA, M., and UCHIDA, A. (1996). "Empirical Correlation Between Penetration Resistance and Internal Friction Angle of Sandy Soils," *Soils and Foundations*, Vol. 36, No. 4, 1–10.

JAMIOLKOWSKI, M., LADD, C. C., GERMAINE, J. T., and LANCELLOTTA, R. (1985). "New Developments in Field and Laboratory Testing of Soils," *Proceedings, 11th International Conference on Soil Mechanics and Foundation Engineering*, Vol. 1, 57–153.

KULHAWY, F. H., and MAYNE, P. W. (1990). *Manual on Estimating Soil Properties for Foundation Design*, Electric Power Research Institute, Palo Alto, CA.

LANCELLOTTA, R. (1983). *Analisi di Affidabilità in Ingegneria Geotecnica*, Atti Istituto Scienza Construzioni, No. 625, Politecnico di Torino.

LEE, J., SALGADO, R., and CARRARO, A. H. (2004). "Stiffness Degradation and Shear Strength of Silty Sand," *Canadian Geotechnical Journal*, Vol. 41, No. 5, 831–843.

LIAO, S. S. C., and WHITMAN, R. V. (1986). "Overburden Correction Factors for SPT in Sand," *Journal of Geotechnical Engineering*, American Society of Civil Engineers, Vol. 112, No. 3, 373–377.

MARCHETTI, S. (1980). "*In Situ* Test by Flat Dilatometer," *Journal of Geotechnical Engineering Division*, ASCE, Vol. 106, GT3, 299–321.

MAYNE, P. W., and KEMPER, J. B. (1988). "Profiling OCR in Stiff Clays by CPT and SPT," *Geotechnical Testing Journal*, ASTM, Vol. 11, No. 2, 139–147.

MENARD, L. (1956). *An Apparatus for Measuring the Strength of Soils in Place*, M.S. Thesis, University of Illinois, Urbana, IL.

MORRIS, P. M., and WILLIAMS, D. J. (1994). "Effective Stress Vane Shear Strength Correction Factor Correlations," *Canadian Geotechnical Journal*, Vol. 31, No. 3, 335–342.

OHYA, S., IMAI, T., and MATSUBARA, M. (1982). "Relationships Between N Value by SPT and LLT Pressuremeter Results," *Proceedings*, 2nd European Symposium on Penetration Testing, Amsterdam, Vol. 1, 125–130.

PECK, R. B., HANSON, W. E., and THORNBURN, T. H. (1974). *Foundation Engineering*, 2nd ed., Wiley, New York.

RICCERI, G., SIMONINI, P., and COLA, S. (2002). "Applicability of Piezocone and Dilatometer to Characterize the Soils of the Venice Lagoon" *Geotechnical and Geological Engineering*, Vol. 20, No. 2, 89–121.

ROBERTSON, P. K., and CAMPANELLA, R. G. (1983). "Interpretation of Cone Penetration Tests. Part I: Sand," *Canadian Geotechnical Journal*, Vol. 20, No. 4, 718–733.

SCHMERTMANN, J. H. (1975). "Measurement of *In Situ* Shear Strength," *Proceedings*, Specialty Conference on *In Situ* Measurement of Soil Properties, ASCE, Vol. 2, 57–138.

SCHMERTMANN, J. H. (1986). "Suggested Method for Performing the Flat Dilatometer Test," *Geotechnical Testing Journal*, ASTM, Vol. 9, No. 2, 93–101.

SEED, H. B., ARANGO, I., and CHAN, C. K. (1975). "Evaluation of Soil Liquefaction Potential During Earthquakes," *Report No. EERC 75–28*, Earthquake Engineering Research Center, University of California, Berkeley.

SEED, H. B., TOKIMATSU, K., HARDER, L. F., and CHUNG, R. M. (1985). "Influence of SPT Procedures in Soil Liquefaction Resistance Evaluations," *Journal of Geotechnical Engineering*, ASCE, Vol. 111, No. 12, 1425–1445.

SKEMPTON, A. W. (1957). "Discussion: The Planning and Design of New Hong Kong Airport," *Proceedings*, Institute of Civil Engineers, London, Vol. 7, 305–307.

SKEMPTON, A. W. (1986). "Standard Penetration Test Procedures and the Effect in Sands of Overburden Pressure, Relative Density, Particle Size, Aging and Overconsolidation," *Geotechnique*, Vol. 36, No. 3, 425–447.

SOWERS, G. B., and SOWERS, G. F. (1970). *Introductory Soil Mechanics and Foundations*, 3rd ed., Macmillan, New York.

STROUD, M. (1974). "SPT in Insensitive Clays," *Proceedings*, European Symposium on Penetration Testing, Vol. 2.2, 367–375.

WOLFF, T. F. (1989). "Pile Capacity Prediction Using Parameter Functions," in *Predicted and Observed Axial Behavior of Piles, Results of a Pile Prediction Symposium*, sponsored by Geotechnical Engineering Division, ASCE, Evanston, IL, June 1989, ASCE Geotechnical Special Publication No. 23, 96–106.

11

Lateral Earth Pressure

Retaining structures, such as retaining walls, basement walls, and bulkheads, are commonly encountered in foundation engineering, and they may support slopes of earth masses. Proper design and construction of these structures require a thorough knowledge of the lateral forces that act between the retaining structures and the soil masses being retained. These lateral forces are caused by lateral earth pressure. This chapter is devoted to the study of various earth pressure theories.

11.1   Earth Pressure at Rest

Let us consider the mass of soil shown in Figure 11.1. The mass is bounded by a frictionless wall AB that extends to an infinite depth. A soil element located at a depth z is subjected to *effective* vertical and horizontal pressures of σ_o' and σ_h', respectively. For this case, since the soil is dry, we have

$$\sigma_o' = \sigma_o$$

and

$$\sigma_h' = \sigma_h$$

where σ_o and σ_h = *total* vertical and horizontal pressures, respectively. Also, note that there are no shear stresses on the vertical and horizontal planes.

If the wall AB is static—that is, if it does not move either to the right or to the left of its initial position—the soil mass will be in a state of *elastic equilibrium*; that is, the horizontal strain is 0. The ratio of the effective horizontal stress to the vertical stress is called the *coefficient of earth pressure at rest, K_o,* or

$$K_o = \frac{\sigma_h'}{\sigma_o'} \tag{11.1}$$

Since $\sigma_o' = \gamma z$, we have

$$\sigma_h' = K_o(\gamma z) \tag{11.2}$$

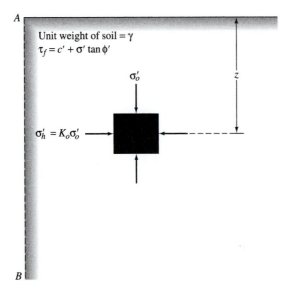

A

Unit weight of soil $= \gamma$
$\tau_f = c' + \sigma' \tan \phi'$

σ'_o

z

$\sigma'_h = K_o \sigma'_o$

B

Figure 11.1
Earth pressure at rest

For coarse-grained soils, the coefficient of earth pressure at rest can be estimated by the empirical relationship (Jaky, 1944)

$$K_o = 1 - \sin \phi' \qquad (11.3)$$

where ϕ' = drained friction angle.
For overconsolidated coarse-grained soil, Eq. (11.3) can be modified as (Mayne and Kulhawy, 1982)

$$K_o = (1 - \sin \phi')(OCR)^{\sin \phi'} \qquad (11.4)$$

where OCR = overconsolidation ratio. The overconsolidation ratio was defined in Chapter 7 as

$$OCR = \frac{\text{preconsolidation pressure}}{\text{present effective overburden pressure}} \qquad (11.5)$$

For fine-grained, normally consolidated soils, Massarsch (1979) suggested the following equation for K_o:

$$K_o = 0.44 + 0.42 \left[\frac{PI\,(\%)}{100} \right] \qquad (11.6)$$

For overconsolidated clays, the coefficient of earth pressure at rest can be approximated as

$$K_{o(\text{overconsolidated})} = K_{o(\text{normally consolidated})} \sqrt{OCR} \qquad (11.7)$$

The magnitude of K_o in most soils ranges between 0.5 and 1.0, with perhaps higher values for heavily overconsolidated clays.

Figure 11.2 shows the distribution of earth pressure at rest on a wall of height H. The total force per unit length of the wall, P_o, is equal to the area of the pressure diagram, so

$$P_o = \frac{1}{2} K_o \gamma H^2 \tag{11.8}$$

Earth Pressure at Rest for Partially Submerged Soil

Figure 11.3a shows a wall of height H. The groundwater table is located at a depth H_1 below the ground surface, and there is no compensating water on the other side of the wall. For $z \leq H_1$, the total lateral earth pressure at rest can be given as $\sigma'_h = K_o \gamma z$. The variation of σ'_h with depth is shown by triangle ACE in Figure 11.3a. However, for $z \geq H_1$ (that is, below the groundwater table), the pressure

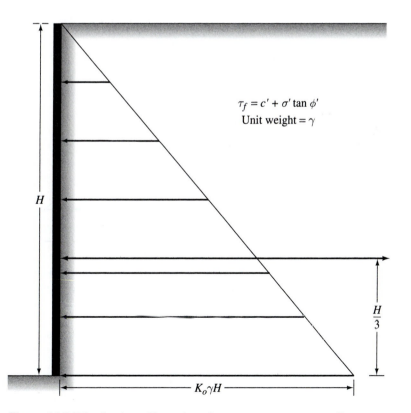

$$\tau_f = c' + \sigma' \tan \phi'$$
$$\text{Unit weight} = \gamma$$

$K_o \gamma H$

Figure 11.2 Distribution of lateral earth pressure at rest on a wall

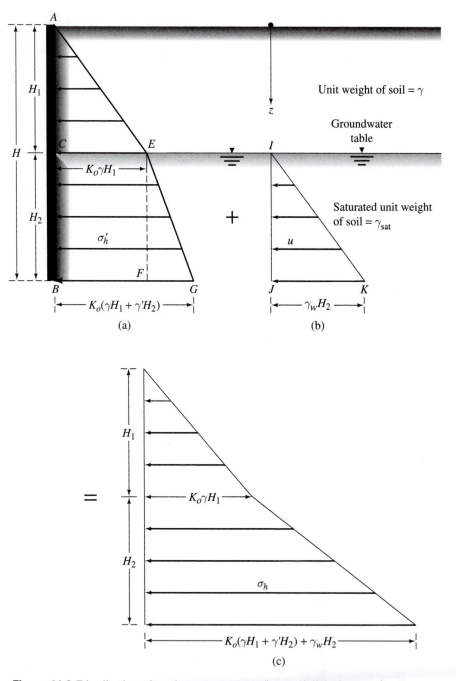

Figure 11.3 Distribution of earth pressure at rest for partially submerged soil

on the wall is found from the effective stress and pore water pressure components in the following manner:

$$\text{effective vertical pressure} = \sigma'_o = \gamma H_1 + \gamma'(z - H_1) \qquad (11.9)$$

where $\gamma' = \gamma_{sat} - \gamma_w$ = effective unit weight of soil. So, the effective lateral pressure at rest is

$$\sigma'_h = K_o\sigma'_o = K_o[\gamma H_1 + \gamma'(z - H_1)] \qquad (11.10)$$

The variation of σ'_h with depth is shown by $CEGB$ in Figure 11.3a. Again, the lateral pressure from pore water is

$$u = \gamma_w(z - H_1) \qquad (11.11)$$

The variation of u with depth is shown in Figure 11.3b.

Hence, the total lateral pressure from earth and water at any depth $z \geq H_1$ is equal to

$$\sigma_h = \sigma'_h + u$$
$$= K_o[\gamma H_1 + \gamma'(z - H_1)] + \gamma_w(z - H_1) \qquad (11.12)$$

The force per unit length of the wall can be found from the sum of the areas of the pressure diagrams in Figures 11.3a and b and is equal to

$$P_o = \underbrace{\frac{1}{2}K_o\gamma H_1^2}_{\substack{\text{Area} \\ ACE}} + \underbrace{K_o\gamma H_1 H_2}_{\substack{\text{Area} \\ CEFB}} + \underbrace{\frac{1}{2}(K_o\gamma' + \gamma_w)H_2^2}_{\substack{\text{Areas} \\ EFG \text{ and } IJK}} \qquad (11.13)$$

or

$$P_o = \frac{1}{2}K_o[\gamma H_1^2 + 2\gamma H_1 H_2 + \gamma'H_2^2] + \frac{1}{2}\gamma_w H_2^2 \qquad (11.14)$$

11.2 Rankine's Theory of Active and Passive Earth Pressures

The term *plastic equilibrium* in soil refers to the condition in which every point in a soil mass is on the verge of failure. Rankine (1857) investigated the stress conditions in soil at a state of plastic equilibrium. This section deals with Rankine's theory of earth pressure.

Rankine's Active State

Figure 11.4a shows the same soil mass that was illustrated in Figure 11.1. It is bounded by a frictionless wall AB that extends to an infinite depth. The vertical and horizontal effective principal stresses on a soil element at a depth z are σ'_o, and σ'_h, respectively. As we saw in Section 11.1, if the wall AB is not allowed to move at all, then $\sigma'_h = K_o\sigma'_o$. The stress condition in the soil element can be represented by the

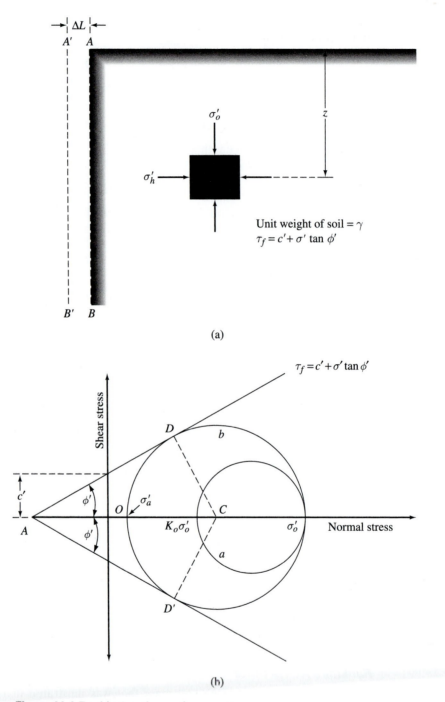

Figure 11.4 Rankine's active earth pressure

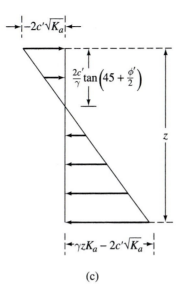

$$-2c'\sqrt{K_a}$$

$$\frac{2c'}{\gamma}\tan\left(45+\frac{\phi'}{2}\right)$$

z

$$\gamma z K_a - 2c'\sqrt{K_a}$$

(c)

$$45+\frac{\phi'}{2} \qquad 45+\frac{\phi'}{2}$$

(d)

Figure 11.4 (*continued*)

Mohr's circle *a* in Figure 11.4b. However, if the wall *AB* is allowed to move away from the soil mass gradually, then the horizontal effective principal stress will decrease. Ultimately a state will be reached at which the stress condition in the soil element can be represented by the Mohr's circle *b*, the state of plastic equilibrium, and failure of the soil will occur. This state is *Rankine's active state*, and the pressure σ'_o on the vertical plane (which is a principal plane) is *Rankine's active earth pressure*. Following is the derivation for expressing σ'_a in terms of γ, z, c', and ϕ'. From Figure 11.4b, we have

$$\sin \phi' = \frac{CD}{AC} = \frac{CD}{AO + OC}$$

but

$$CD = \text{radius of the failure circle} = \frac{\sigma'_o - \sigma'_a}{2}$$

$$AO = c' \cot \phi'$$

and

$$OC = \frac{\sigma'_o + \sigma'_a}{2}$$

so

$$\sin \phi' = \frac{\dfrac{\sigma'_o - \sigma'_a}{2}}{c' \cot \phi' + \dfrac{\sigma'_o + \sigma'_a}{2}}$$

or

$$c' \cos \phi' + \frac{\sigma'_o + \sigma'_a}{2} \sin \phi' = \frac{\sigma'_o - \sigma'_a}{2}$$

or

$$\sigma'_a = \sigma'_o \frac{1 - \sin \phi'}{1 + \sin \phi'} - 2c' \frac{\cos \phi'}{1 + \sin \phi'} \tag{11.15}$$

But

$$\sigma'_o = \text{vertical effective overburden pressure} = \gamma z$$

$$\frac{1 - \sin \phi'}{1 + \sin \phi'} = \tan^2\left(45 - \frac{\phi'}{2}\right)$$

and

$$\frac{\cos \phi'}{1 + \sin \phi'} = \tan\left(45 - \frac{\phi'}{2}\right)$$

Substituting the above into Eq. (11.15), we get

$$\sigma_a' = \gamma z \tan^2\left(45 - \frac{\phi'}{2}\right) - 2c' \tan\left(45 - \frac{\phi'}{2}\right) \tag{11.16}$$

The variation of σ_a' with depth is shown in Figure 11.4c. For cohesionless soils, $c' = 0$ and

$$\sigma_a' = \sigma_o' \tan^2\left(45 - \frac{\phi'}{2}\right) \tag{11.17}$$

The ratio of σ_a' to σ_o' is called the *coefficient of Rankine's active earth pressure, K_a*, or

$$K_a = \frac{\sigma_a'}{\sigma_o'} = \tan^2\left(45 - \frac{\phi'}{2}\right) \tag{11.18}$$

Again, from Figure 11.4b, we can see that the failure planes in the soil make $\pm(45 + \phi'/2)$-degree angles with the direction of the major principal plane—that is, the horizontal. These failure planes are called *slip planes*. The slip planes are shown in Figure 11.4d.

Rankine's Passive State

Rankine's passive state is illustrated in Figure 11.5. *AB* is a frictionless wall (Figure 11.5a) that extends to an infinite depth. The initial stress condition on a soil element is represented by the Mohr's circle *a* in Figure 11.5b. If the wall is gradually pushed into the soil mass, the effective principal stress σ_h' will increase. Ultimately the wall will reach a state at which the stress condition in the soil element can be represented by the Mohr's circle *b*. At this time, failure of the soil will occur. This is referred to as *Rankine's passive state*. The effective lateral earth pressure σ_p', which is the major principal stress, is called *Rankine's passive earth pressure*. From Figure 11.5b, it can be shown that

$$\sigma_p' = \sigma_o' \tan^2\left(45 + \frac{\phi'}{2}\right) + 2c' \tan\left(45 + \frac{\phi'}{2}\right)$$

$$= \gamma z \tan^2\left(45 + \frac{\phi'}{2}\right) + 2c' \tan\left(45 + \frac{\phi'}{2}\right) \tag{11.19}$$

The derivation is similar to that for Rankine's active state.

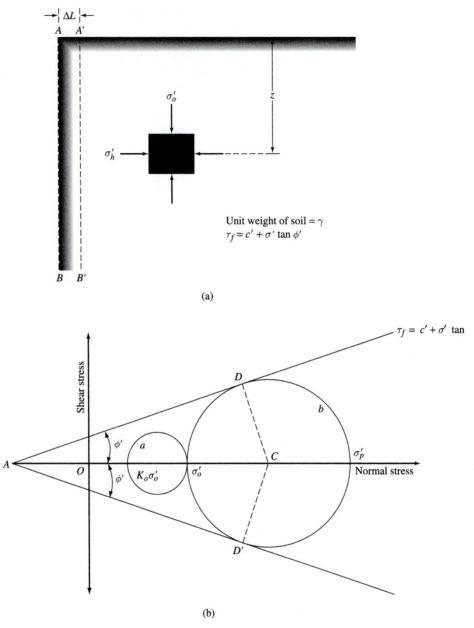

(a)

(b)

Figure 11.5 Rankine's passive earth pressure

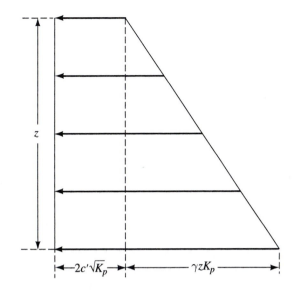

(c)

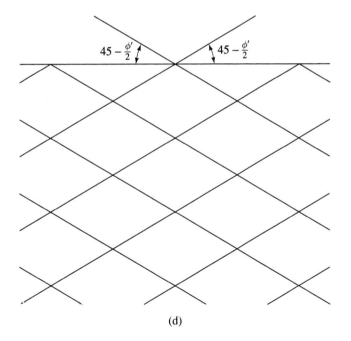

(d)

Figure 11.5 (*continued*)

Figure 11.5c shows the variation of passive pressure with depth. For cohesionless soils ($c' = 0$), we have

$$\sigma'_p = \sigma'_o \tan^2\left(45 + \frac{\phi'}{2}\right)$$

or

$$\frac{\sigma'_p}{\sigma'_o} = K_p = \tan^2\left(45 + \frac{\phi'}{2}\right) \qquad (11.20)$$

K_p in the preceding equation is referred to as the *coefficient of Rankine's passive earth pressure*.

The points D and D' on the failure circle (Figure 11.5b) correspond to the slip planes in the soil. For Rankine's passive state, the slip planes make $\pm(45 - \phi'/2)$- degree angles with the direction of the minor principal plane—that is, in the horizontal direction. Figure 11.5d shows the distribution of slip planes in the soil mass.

Effect of Wall Yielding

From the preceding discussion we know that sufficient movement of the wall is necessary to achieve a state of plastic equilibrium. However, the distribution of lateral earth pressure against a wall is very much influenced by the manner in which the wall actually yields. In most simple retaining walls, movement may occur by simple translation or, more frequently, by rotation about the bottom.

For preliminary theoretical analysis, let us consider a frictionless retaining wall represented by a plane AB, as shown in Figure 11.6a. If the wall AB rotates sufficiently about its bottom to a position $A'B$, then a triangular soil mass ABC' adjacent to the wall will reach Rankine's active state. Since the slip planes in Rankine's active state make angles of $\pm(45 + \phi'/2)$ degrees with the major principal plane, the soil mass in the state of plastic equilibrium is bounded by the plane BC', which makes an angle of $(45 + \phi'/2)$ degrees with the horizontal. The soil inside the zone ABC' undergoes the same unit deformation in the horizontal direction everywhere, which is equal to $\Delta L_a/L_a$. The lateral earth pressure on the wall at any depth z from the ground surface can be calculated by Eq. (11.16).

In a similar manner, if the frictionless wall AB (Figure 11.6b) rotates sufficiently into the soil mass to a position $A''B$, then the triangular mass of soil ABC'' will reach Rankine's passive state. The slip plane BC'' bounding the soil wedge that is at a state of plastic equilibrium makes an angle of $(45 - \phi'/2)$ degrees with the horizontal. Every point of the soil in the triangular zone ABC'' undergoes the same unit deformation in the horizontal direction, which is equal to $\Delta L_p/L_p$. The passive pressure on the wall at any depth z can be evaluated by using Eq. (11.19).

Typical values of the minimum wall tilt (ΔL_a and ΔL_p) required for achieving Rankine's state are given in Table 11.1.

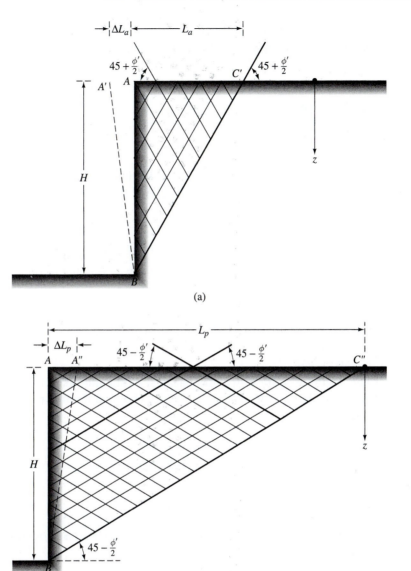

Figure 11.6 Rotation of frictionless wall about the bottom

Table 11.1 Typical values of $\Delta L_a/H$ and $\Delta L_p/H$ for Rankine's state

Soil type	$\Delta L_a/H$	$\Delta L_p/H$
Loose sand	0.001–0.002	0.01
Dense sand	0.0005–0.001	0.005
Soft clay	0.02	0.04
Stiff clay	0.01	0.02

11.3 Diagrams for Lateral Earth Pressure Distribution against Retaining Walls

Backfill—Cohesionless Soil with Horizontal Ground Surface

Active Case

Figure 11.7a shows a retaining wall with cohesionless soil backfill that has a horizontal ground surface. The unit weight and the angle of friction of the soil are γ and ϕ', respectively. For Rankine's active state, the earth pressure at any depth against the retaining wall can be given by Eq. (11.16):

$$\sigma_a = \sigma'_a = K_a \gamma z \quad (\textit{Note: } c' = 0)$$

σ_a increases linearly with depth, and at the bottom of the wall, it will be

$$\sigma_a = K_a \gamma H \tag{11.21}$$

The total force, P_a, per unit length of the wall is equal to the area of the pressure diagram, so

$$P_a = \frac{1}{2} K_a \gamma H^2 \tag{11.22}$$

Passive Case

The lateral pressure distribution against a retaining wall of height H for Rankine's passive state is shown in Figure 11.7b. The lateral earth pressure at any depth z [Eq. (11.20), $c' = 0$] is

$$\sigma_p = \sigma'_p = K_p \gamma H \tag{11.23}$$

The total force, P_p, per unit length of the wall is

$$P_p = \frac{1}{2} K_p \gamma H^2 \tag{11.24}$$

Backfill—Partially Submerged Cohesionless Soil Supporting Surcharge

Active Case

Figure 11.8a shows a frictionless retaining wall of height H and a backfill of cohesionless soil. The groundwater table is located at a depth of H_1 below the ground surface, and the backfill is supporting a surcharge pressure of q per unit area. From Eq. (11.18), we know that the effective active earth pressure at any depth can be given by

$$\sigma'_a = K_a \sigma'_o \tag{11.25}$$

where σ'_o and σ'_a are the effective vertical pressure and lateral pressure, respectively. At $z = 0$,

$$\sigma_o = \sigma'_o = q \tag{11.26}$$

and

$$\sigma_a = \sigma'_a = K_a q \tag{11.27}$$

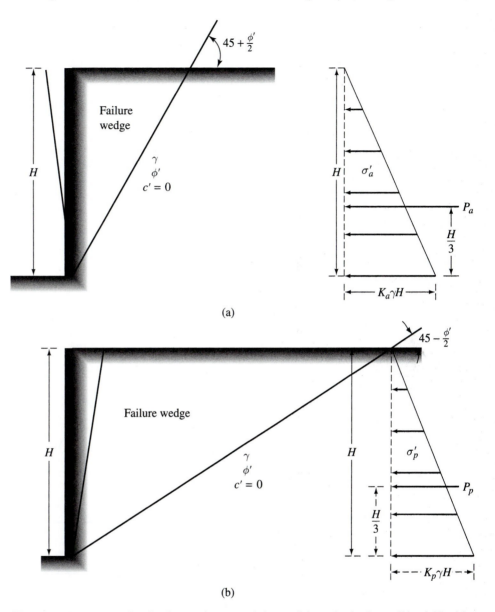

Figure 11.7 Pressure distribution against a retaining wall for cohesionless soil backfill with horizontal ground surface: (a) Rankine's active state; (b) Rankine's passive state

At depth $z = H_1$,

$$\sigma_o = \sigma'_o = (q + \gamma H_1) \qquad (11.28)$$

and

$$\sigma_a = \sigma'_a = K_a(q + \gamma H_1) \qquad (11.29)$$

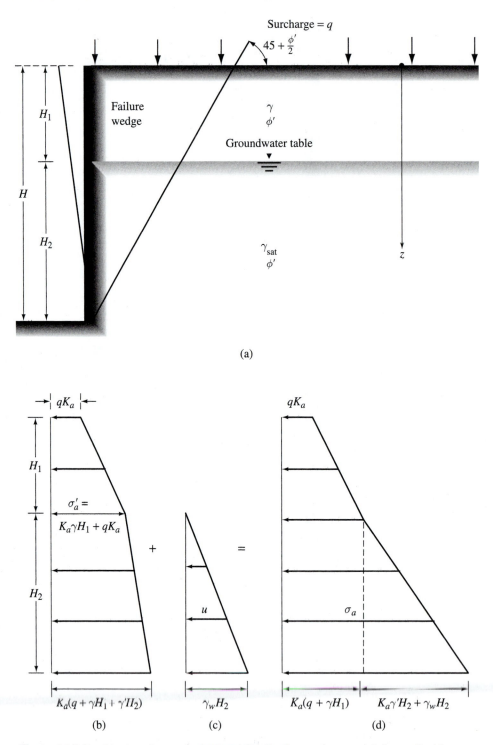

Figure 11.8 Rankine's active earth-pressure distribution against a retaining wall with partially submerged cohesionless soil backfill supporting a surcharge

At depth $z = H$,

$$\sigma'_o = (q + \gamma H_1 + \gamma' H_2) \tag{11.30}$$

and

$$\sigma'_a = K_a(q + \gamma H_1 + \gamma' H_2) \tag{11.31}$$

where $\gamma' = \gamma_{sat} - \gamma_w$. The variation of σ'_a with depth is shown in Figure 11.8b.

The lateral pressure on the wall from the pore water between $z = 0$ and H_1 is 0, and for $z > H_1$, it increases linearly with depth (Figure 11.8c). At $z = H$,

$$u = \gamma_w H_2$$

The total lateral pressure, σ_a, diagram (Figure 11.8d) is the sum of the pressure diagrams shown in Figures 11.8b and c. The total active force per unit length of the wall is the area of the total pressure diagram. Thus,

$$P_a = K_a q H + \frac{1}{2} K_a \gamma H_1^2 + K_a \gamma H_1 H_2 + \frac{1}{2}(K_a \gamma' + \gamma_w) H_2^2 \tag{11.32}$$

Passive Case

Figure 11.9a shows the same retaining wall as in Figure 11.8a. Rankine's passive pressure (effective) at any depth against the wall can be given by Eq. (11.20):

$$\sigma'_p = K_p \sigma'_o$$

Using the preceding equation, we can determine the variation of σ'_p with depth, as shown in Figure 11.9b. The variation of the pressure on the wall from water with depth is shown in Figure 11.9c. Figure 11.9d shows the distribution of the total pressure, σ_p, with depth. The total lateral passive force per unit length of the wall is the area of the diagram given in Figure 11.9d, or

$$P_p = K_p q H + \frac{1}{2} K_p \gamma H_1^2 + K_p \gamma H_1 H_2 + \frac{1}{2}(K_p \gamma' + \gamma_w) H_2^2 \tag{11.33}$$

Backfill—Cohesive Soil with Horizontal Backfill

Active Case

Figure 11.10a shows a frictionless retaining wall with a cohesive soil backfill. The active pressure against the wall at any depth below the ground surface can be expressed as [Eq. (11.15)]

$$\sigma'_a = K_a \gamma z - 2c' \sqrt{K_a}$$

The variation of $K_a \gamma z$ with depth is shown in Figure 11.10b, and the variation of $2c' \sqrt{K_a}$ with depth is shown in Figure 11.10c. Note that $2c' \sqrt{K_a}$ is not a function of z, and hence Figure 11.10c is a rectangle. The variation of the net value of σ'_a with depth is plotted in Figure 11.10d. Also note that, because of the effect of cohesion, σ'_a is negative in the upper part of the retaining wall. The depth z_o at which the active pressure becomes equal to 0 can be found from Eq. (11.16) as

$$K_a \gamma z_o - 2c' \sqrt{K_a} = 0$$

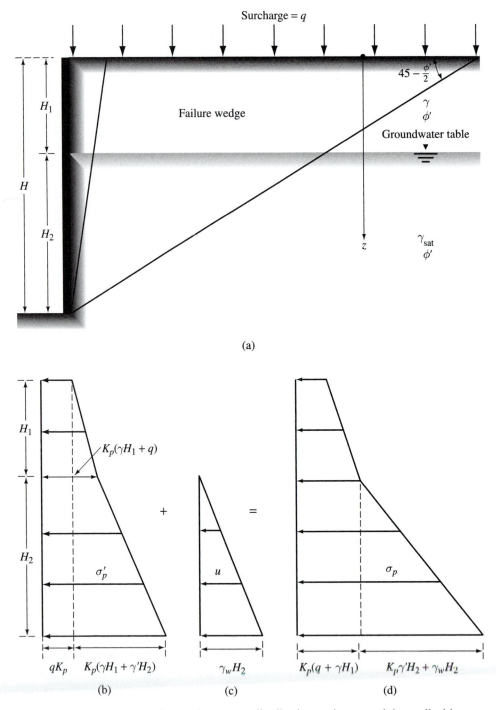

Figure 11.9 Rankine's passive earth-pressure distribution against a retaining wall with partially submerged cohesionless soil backfill supporting a surcharge

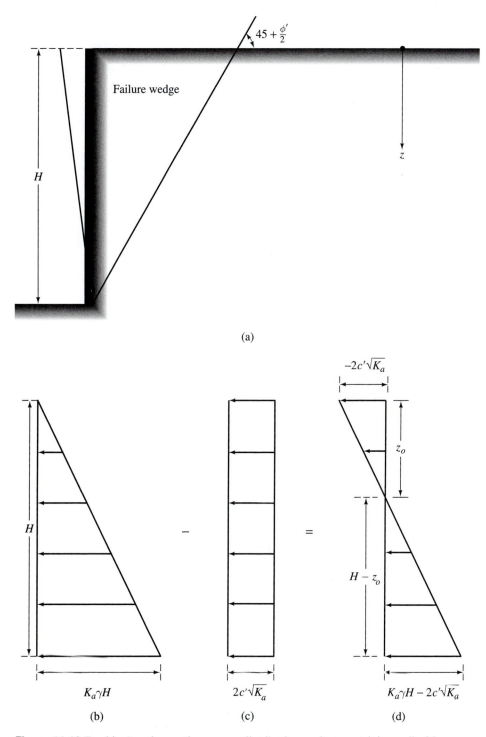

Figure 11.10 Rankine's active earth-pressure distribution against a retaining wall with cohesive soil backfill

or

$$z_o = \frac{2c'}{\gamma\sqrt{K_a}} \tag{11.34}$$

For the undrained condition—that is, $\phi = 0$, $K_a = \tan^2 45 = 1$, and $c = c_u$ (undrained cohesion)—we have

$$z_o = \frac{2c_u}{\gamma} \tag{11.35}$$

So, with time, tensile cracks at the soil–wall interface will develop up to a depth of z_o.

The total active force per unit length of the wall can be found from the area of the total pressure diagram (Figure 11.10d), or

$$P_a = \frac{1}{2}K_a\gamma H^2 - 2\sqrt{K_a}cH \tag{11.36}$$

For $\phi = 0$ condition,

$$P_a = \frac{1}{2}\gamma H^2 - 2c_uH \tag{11.37}$$

For calculation of the total active force, it is common practice to take the tensile cracks into account. Since there is no contact between the soil and the wall up to a depth of z_o after the development of tensile cracks, the active pressure distribution against the wall between $z = 2c'/(\gamma\sqrt{K_a})$ and H (Figure 11.10d) only is considered. In that case,

$$P_a = \frac{1}{2}(K_a\gamma H - 2\sqrt{K_a}c')\left(H - \frac{2c'}{\gamma\sqrt{K_a}}\right)$$
$$= \frac{1}{2}K_a\gamma H^2 - 2\sqrt{K_a}c'H + 2\frac{c'^2}{\gamma} \tag{11.38}$$

For the $\phi = 0$ condition,

$$P_a = \frac{1}{2}\gamma H^2 - 2c_uH + 2\frac{c_u^2}{\gamma} \tag{11.39}$$

Note that, in Eq. (11.39), γ is the saturated unit weight of the soil.

Passive Case

Figure 11.11a shows the same retaining wall with backfill similar to that considered in Figure 11.10a. Rankine's passive pressure against the wall at depth z can be given by [Eq. (11.19)]

$$\sigma'_p = K_p\gamma z + 2\sqrt{K_p}c'$$

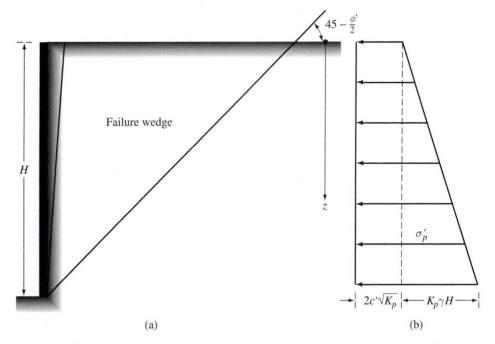

Figure 11.11 Rankine's passive earth-pressure distribution against a retaining wall with cohesive soil backfill

At $z = 0$,

$$\sigma_p = \sigma'_p = 2\sqrt{K_p}c' \tag{11.40}$$

and at $z = H$,

$$\sigma_p = \sigma'_p = K_p\gamma H + 2\sqrt{K_p}c' \tag{11.41}$$

The variation of $\sigma_p = \sigma'_p$ with depth is shown in Figure 11.11b. The passive force per unit length of the wall can be found from the area of the pressure diagrams as

$$P_p = \frac{1}{2}K_p\gamma H^2 + 2\sqrt{K_p}c'H \tag{11.42}$$

For the $\phi = 0$ condition, $K_p = 1$ and

$$P_p = \frac{1}{2}\gamma H^2 + 2c_uH \tag{11.43}$$

In Eq. (11.43), γ is the saturated unit weight of the soil.

Example 11.1

If the retaining wall shown in Figure 11.12 is restrained from moving, what will be the lateral force per unit length of the wall? Use $\phi' = 20°$.

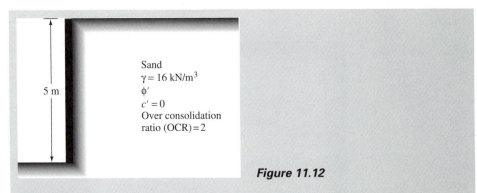

Figure 11.12

Solution

If the wall is restrained from moving, the backfill will exert at-rest earth pressure. Thus,

$$\sigma'_h = \sigma_h = K_o\sigma'_o = K_o(\gamma z) \qquad \text{[Eq. (11.2)]}$$

From Eq. (11.4),

$$K_o = (1 - \sin \phi')(OCR)^{\sin \phi'} = (1 - \sin 20)(2)^{\sin 20} = 0.834$$

and at $z = 0$, $\sigma'_h = 0$; at 5 m, $\sigma'_h = (0.834)(5)(16) = 66.72 \text{ kN/m}^2$.

The total pressure distribution diagram will be similar to that shown in Figure 11.2.

$$P_o = \frac{1}{2}(5)(66.72) = \textbf{166.8 kN/m}$$ ∎

Example 11.2

Calculate the Rankine active and passive forces per unit length of the wall shown in Figure 11.12, and also determine the location of the resultant. Use $\phi' = 32°$

Solution

To determine the active force, since $c' = 0$, we have

$$\sigma'_a = K_a\sigma'_o = K_a\gamma z$$

$$K_a = \frac{1 - \sin \phi'}{1 + \sin \phi'} = \frac{1 - \sin 32°}{1 + \sin 32°} = 0.307$$

At $z = 0$, $\sigma'_a = 0$; at $z = 5$ m, $\sigma'_a = (0.307)(16)(5) = 24.56 \text{ kN/m}^2$.

The active pressure distribution diagram will be similar to that shown in Figure 11.7a.

$$\text{Active force, } P_a = \frac{1}{2}(5)(24.56)$$

$$= \textbf{61.4 kN/m}$$

The total pressure distribution is triangular, and so P_a will act at a distance of $5/3 = 1.67$ m above the bottom of the wall.

To determine the passive force, we are given $c' = 0$, so

$$\sigma'_p = \sigma_p = K_p \sigma'_o = K_p \gamma z$$

$$K_p = \frac{1 + \sin \phi'}{1 - \sin \phi'} = \frac{1 + 0.53}{1 - 0.53} = 3.26$$

At $z = 0$, $\sigma'_p = 0$; at $z = 5$ m, $\sigma'_p = 3.26(16)(5) = 260.8$ kN/m².

The total passive pressure distribution against the wall will be as shown in Figure 11.7b.

$$P_p = \frac{1}{2}(5)(260.8) = \mathbf{652 \text{ kN/m}}$$

The resultant will act at a distance of $5/3 = 1.67$ m above the bottom of the wall. ∎

Example 11.3

A retaining wall that has a soft, saturated clay backfill is shown in Figure 11.13. For the undrained condition ($\phi = 0$) of the backfill, determine the following values:

a. The maximum depth of the tensile crack
b. P_a before the tensile crack occurs
c. P_a after the tensile crack occurs

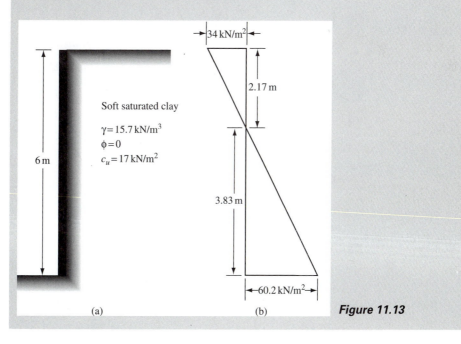

34 kN/m²

2.17 m

Soft saturated clay

$\gamma = 15.7$ kN/m³
$\phi = 0$
$c_u = 17$ kN/m²

6 m

3.83 m

60.2 kN/m²

(a)   (b)   *Figure 11.13*

Solution

For $\phi = 0$, $K_a = \tan^2 45 = 1$ and $c = c_u$. From Eq. (11.16), for the undrained condition, we have

$$\sigma_a = \gamma z - 2c_u$$

At $z = 0$,

$$\sigma_a = -2c_u = -(2)(17) = -34 \text{ kN/m}^2$$

At $z = 6$ m,

$$\sigma_a = (15.7)(6) - (2)(17) = 60.2 \text{ kN/m}^2$$

The variation of σ_a with depth is shown in Figure 11.13b.

Part a.

From Eq. (11.35), the depth of the tensile crack equals

$$z_o = \frac{2c_u}{\gamma} = \frac{(2)(17)}{15.7} = \textbf{2.17 m}$$

Part b.

Before the tensile crack occurs [Eq. (11.37)],

$$P_a = \frac{1}{2}\gamma H^2 - 2c_u H$$

or

$$P_a = \frac{1}{2}(15.7)(6)^2 - 2(17)(6) = \textbf{78.6 kN/m}$$

Part c.

After the tensile crack occurs,

$$P_a = \frac{1}{2}(6 - 2.17)(60.2) = \textbf{115.3 kN/m}$$

Note: The preceding P_a can also be obtained by substituting the proper values into Eq. (11.39).   ∎

Example 11.4

A retaining wall is shown in Figure 11.14. Determine Rankine's active force, P_a, per unit length of the wall. Also determine the location of the resultant.

Solution

Given $c' = 0$, we know that $\sigma'_a = K_a \sigma'_o$. For the upper layer of the soil, the Rankine active earth pressure coefficient is

$$K_a = K_{a(1)} = \frac{1 - \sin 30°}{1 + \sin 30°} = \frac{1}{3}$$

For the lower layer,

$$K_a = K_{a(2)} = \frac{1 - \sin 35°}{1 + \sin 35°} = \frac{0.4264}{1.5736} = 0.271$$

At $z = 0$, $\sigma'_o = 0$. At $z = 1.2$ m (just inside the bottom of the upper layer), $\sigma'_o = (1.2)(16.5) = 19.8$ kN/m². So

$$\sigma'_a = K_{a(1)}\sigma'_o = \frac{1}{3}(19.8) = 6.6 \text{ kN/m}^2$$

Again, at $z = 1.2$ m (in the lower layer), $\sigma'_o = (1.2)(16.5) = 19.8$ kN/m², and

$$\sigma'_a = K_{a(2)}\sigma'_o = (0.271)(19.8) = 5.37 \text{ kN/m}^2$$

At $z = 6$ m,

$$\sigma'_o = (1.2)(16.5) + (4.8)(19.2 - 9.81) = 64.87 \text{ kN/m}^2$$
$$\uparrow$$
$$\gamma_w$$

and

$$\sigma'_a = K_{a(2)}\sigma'_o = (0.271)(64.87) = 17.58 \text{ kN/m}^2$$

The variation of σ'_a with depth is shown in Figure 11.14b.
    The lateral pressures from the pore water are as follows:

- At $z = 0$, $u = 0$
- At $z = 1.2$ m, $u = 0$
- At $z = 6$ m, $u = (4.8)(\gamma_w) = (4.8)(9.81) = 47.1$ kN/m²

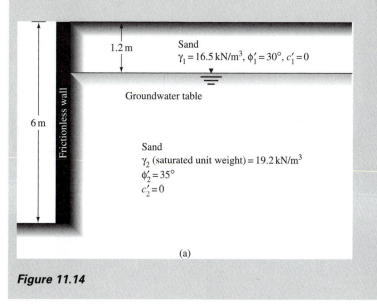

(a)

Figure 11.14

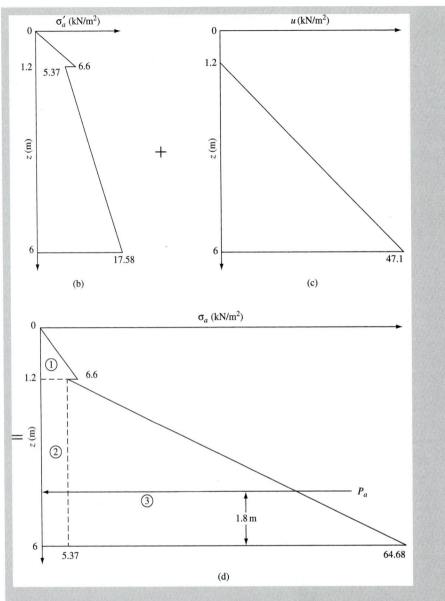

Figure 11.14 *(continued)*

The variation of u with depth is shown in Figure 11.14c, and the variation for σ_a (total active pressure) is shown in Figure 11.14d. Thus,

$$P_a = \left(\frac{1}{2}\right)(6.6)(1.2) + (4.8)(5.37) + \left(\frac{1}{2}\right)(4.8)(64.68 - 5.37)$$

$$= 3.96 + 25.78 + 142.34 = \textbf{172.08 kN/m}$$

The location of the resultant can be found by taking the moment about the bottom of the wall. Thus,

$$\bar{z} = \frac{3.96\left(4.8 + \dfrac{1.2}{3}\right) + (25.78)(2.4) + (142.34)\left(\dfrac{4.8}{3}\right)}{172.08} = \mathbf{1.8\ m} \quad \blacksquare$$

Example 11.5

A frictionless retaining wall is shown in Figure 11.15a. Find the passive resistance (P_p) due to the backfill, and the location of the resultant passive force.

Solution
Passive Resistance
  Given: $\phi' = 26°$,

$$K_p = \frac{1 + \sin \phi'}{1 - \sin \phi'} = \frac{1 + \sin 26°}{1 - \sin 26°} = \frac{1.4384}{0.5616} = 2.56$$

From Eq. (11.19),

$$\sigma'_p = K_p \sigma'_o + 2\sqrt{K_p}\,c'$$

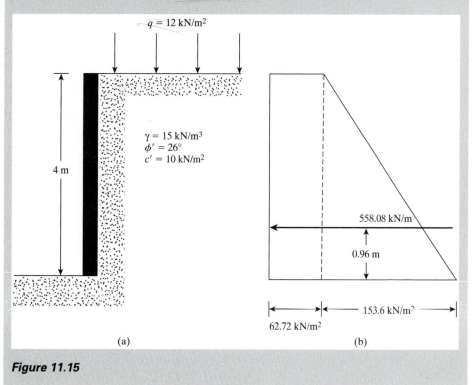

$q = 12$ kN/m^2

$\gamma = 15$ kN/m^3
$\phi' = 26°$
$c' = 10$ kN/m^2

4 m

558.08 kN/m

0.96 m

153.6 kN/m^2

62.72 kN/m^2

(a)   (b)

Figure 11.15

At $z = 0$, $\sigma'_o = 12$ kN/m^2

$$\sigma'_p = \sigma_p = (2.56)(12) + 2\sqrt{2.56}\,(10)$$
$$= 30.72 + 32 = 62.72 \text{ kN/m}^2$$

Again, at $z = 4$ m, $\sigma'_o = (12 + 4 \times 15) = 72$ kN/m^2. So

$$\sigma'_p = \sigma_p = (2.56)(72) + 2\sqrt{2.56}\,(10)$$
$$= 216.32 \text{ kN/m}^2$$

The pressure distribution is shown in Figure 11.15b. The passive resistance per unit length of wall:

$$P_p = (62.72)(4) + \frac{1}{2}(4)(153.6)$$
$$= 250.88 + 307.2 = \textbf{558.08 kN/m}$$

Location of Resultant
Taking the moment of the pressure diagram about the bottom of the wall,

$$\bar{z} = \frac{(30.72 + 32)\left(\dfrac{4}{2}\right) + \dfrac{1}{2}(153.6)(4)\left(\dfrac{4}{3}\right)}{558.08}$$

$$= \frac{125.44 + 409.6}{558.08} = \textbf{0.96 m}$$

∎

11.4   Rankine Active and Passive Pressure with Sloping Backfill

In Section 11.2, we considered retaining walls with vertical backs and horizontal backfills. In some cases, however, the backfill may be continuously sloping at an angle α with the horizontal as shown in Figure 11.16 for active pressure case. In such cases, the directions of Rankine's active or passive pressures are no longer horizontal. Rather, the directions of pressure are inclined at an angle α with the horizontal. If the backfill is a granular soil with a drained friction angle ϕ', and $c' = 0$, then

$$\sigma'_a = \gamma z K_a$$

where

$$K_a = \text{Rankine's active pressure coefficient}$$
$$= \cos\alpha \frac{\cos\alpha - \sqrt{\cos^2\alpha - \cos^2\phi'}}{\cos\alpha + \sqrt{\cos^2\alpha - \cos^2\phi'}} \qquad (11.44)$$

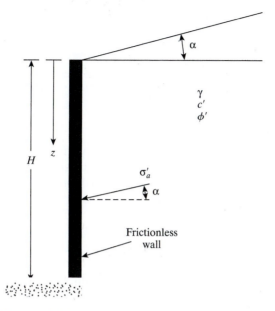

Figure 11.16 Frictionless vertical retaining wall with sloping backfill

The active force per unit length of the wall can be given as

$$P_a = \frac{1}{2}K_a\gamma H^2 \tag{11.45}$$

The line of action of the resultant acts at a distance of $H/3$ measured from the bottom of the wall. Table 11.2 gives the values of K_a for various combinations of α and ϕ'.

In a similar manner, the *Rankine passive earth pressure* for a wall of height H with a granular sloping backfill can be represented by the equation

$$P_p = \frac{1}{2}\gamma H^2 K_p \tag{11.46}$$

Table 11.2 Values of K_a [Eq. (11.44)]

	ϕ' (deg) →						
↓ α (deg)	28	30	32	34	36	38	40
0	0.361	0.333	0.307	0.283	0.260	0.238	0.217
5	0.366	0.337	0.311	0.286	0.262	0.240	0.219
10	0.380	0.350	0.321	0.294	0.270	0.246	0.225
15	0.409	0.373	0.341	0.311	0.283	0.258	0.235
20	0.461	0.414	0.374	0.338	0.306	0.277	0.250
25	0.573	0.494	0.434	0.385	0.343	0.307	0.275

where

$$K_p = \cos \alpha \frac{\cos \alpha + \sqrt{\cos^2 \alpha - \cos^2 \phi'}}{\cos \alpha - \sqrt{\cos^2 \alpha - \cos^2 \phi'}} \qquad (11.47)$$

is the passive earth pressure coefficient.

As in the case of the active force, the resultant force P_p is inclined at an angle α with the horizontal and intersects the wall at a distance of $H/3$ measured from the bottom of the wall. The values of K_p (passive earth pressure coefficient) for various values of α and ϕ' are given in Table 11.3.

c'-ɸ' Soil

The preceding analysis can be extended to the determination of the active and passive Rankine earth pressure for an inclined backfill with a c'-ϕ' soil. The details of the mathematical derivation are given by Mazindrani and Ganjali (1997). For a c'-ϕ' backfill, the active pressure is given by

$$\sigma'_a = \gamma z K_a = \gamma z K''_a \cos \alpha \qquad (11.48)$$

where K_a = Rankine active earth pressure coefficient and

$$K''_a = \frac{K_a}{\cos \alpha} \qquad (11.49)$$

The passive pressure is given by

$$\sigma'_p = \gamma z K_p = \gamma z K''_p \cos \alpha \qquad (11.50)$$

where K_p = Rankine passive earth pressure coefficient and

$$K''_p = \frac{K_p}{\cos \alpha} \qquad (11.51)$$

Table 11.3 Passive Earth Pressure Coefficient, K_p [Eq. (11.47)]

↓ α (deg)	ϕ' (deg) →						
	28	30	32	34	36	38	40
0	2.770	3.000	3.255	3.537	3.852	4.204	4.599
5	2.715	2.943	3.196	3.476	3.788	4.136	4.527
10	2.551	2.775	3.022	3.295	3.598	3.937	4.316
15	2.284	2.502	2.740	3.003	3.293	3.615	3.977
20	1.918	2.132	2.362	2.612	2.886	3.189	3.526
25	1.434	1.664	1.894	2.135	2.394	2.676	2.987

Also,

$$K''_a, K''_p = \frac{1}{\cos^2 \phi'}$$

$$\times \left\{ 2 \cos^2 \alpha + 2\left(\frac{c'}{\gamma z}\right)\cos \phi' \sin \phi' \right.$$

$$\left. \pm \sqrt{\left[4 \cos^2 \alpha(\cos^2 \alpha - \cos^2 \phi') + 4\left(\frac{c'}{\gamma z}\right)^2 \cos^2 \phi' + 8\left(\frac{c'}{\gamma z}\right)\cos^2 \alpha \sin \phi' \cos \phi'\right]} \right\} - 1$$

(11.52)

Tables 11.4 and 11.5 give, respectively, the variations of K''_a and K''_p with α, $c'/\gamma z$, and ϕ'.

For the *active* case, the depth of the tensile crack can be given as

$$z_o = \frac{2c'}{\gamma}\sqrt{\frac{1 + \sin \phi'}{1 - \sin \phi'}}$$

(11.53)

Table 11.4 Variation of K''_a with α, $c'/\gamma z$, and ϕ'

α (deg)	ϕ' (deg)	$c'/\gamma z$			
		0	0.025	0.050	0.100
0	15	0.589	0.550	0.512	0.435
	20	0.490	0.455	0.420	0.351
	25	0.406	0.374	0.342	0.278
	30	0.333	0.305	0.276	0.218
	35	0.271	0.245	0.219	0.167
	40	0.217	0.194	0.171	0.124
5	15	0.607	0.566	0.525	0.445
	20	0.502	0.465	0.429	0.357
	25	0.413	0.381	0.348	0.283
	30	0.339	0.309	0.280	0.221
	35	0.275	0.248	0.222	0.169
	40	0.220	0.196	0.173	0.126
10	15	0.674	0.621	0.571	0.477
	20	0.539	0.497	0.456	0.377
	25	0.438	0.402	0.366	0.296
	30	0.355	0.323	0.292	0.230
	35	0.286	0.258	0.230	0.175
	40	0.228	0.203	0.179	0.130
15	15	1.000	0.776	0.683	0.546
	20	0.624	0.567	0.514	0.417
	25	0.486	0.443	0.401	0.321
	30	0.386	0.350	0.315	0.246
	35	0.307	0.276	0.246	0.186
	40	0.243	0.216	0.190	0.337

Table 11.5 Variation of K_p'' with α, $c'/\gamma z$, and ϕ'

α (deg)	ϕ' (deg)	$c'/\gamma z$			
		0	0.025	0.050	0.100
0	15	1.698	1.764	1.829	1.959
	20	2.040	2.111	2.182	2.325
	25	2.464	2.542	2.621	2.778
	30	3.000	3.087	3.173	3.346
	35	3.690	3.786	3.882	4.074
	40	4.599	4.706	4.813	5.028
5	15	1.674	1.716	1.783	1.916
	20	1.994	2.067	2.140	2.285
	25	2.420	2.499	2.578	2.737
	30	2.954	3.042	3.129	3.303
	35	3.641	3.738	3.834	4.027
	40	5.545	4.652	4.760	4.975
10	15	1.484	1.564	1.641	1.788
	20	1.854	1.932	2.010	2.162
	25	2.285	2.368	2.450	2.614
	30	2.818	2.907	2.996	3.174
	35	3.495	3.593	3.691	3.887
	40	4.383	4.491	4.600	4.817
15	15	1.000	1.251	1.370	1.561
	20	1.602	1.696	1.786	1.956
	25	2.058	2.147	2.236	2.409
	30	2.500	2.684	2.777	2.961
	35	3.255	3.356	3.456	3.656
	40	4.117	4.228	4.338	4.558

Example 11.6

Refer to Figure 11.16. Given that $H = 6.1$ m, $\alpha = 5°$, $\gamma = 16.5$ kN/m³, $\phi' = 20°$, $c' = 10$ kN/m², determine the Rankine active force P_a on the retaining wall after the tensile crack occurs.

Solution
From Eq. (11.53), the depth of tensile crack is

$$z_o = \frac{2c'}{\gamma}\sqrt{\frac{1 + \sin\phi'}{1 - \sin\phi'}} = \frac{(2)(10)}{16.5}\sqrt{\frac{1 + \sin 20}{1 - \sin 20}} = 1.73 \text{ m}$$

So

- At $z = 0$, $\sigma_a' = 0$

- At $z = 6.1$ m, $\sigma_a' = \gamma z K_a'' \cos\alpha$

$$\frac{c'}{\gamma z} = \frac{10}{(16.5)(6.1)} \approx 0.1$$

From Table 11.4, for $\alpha = 5°$ and $c'/\gamma z = 0.1$, the magnitude of $K''_a = 0.357$. So

$$\sigma'_a = (16.5)(6.1)(0.357)(\cos 5°) = 35.8 \text{ kN/m}^2$$

Hence,

$$P_a = \frac{1}{2}(H - z_o)(35.8) = \frac{1}{2}(6.1 - 1.73)(35.8) = \textbf{78.2 kN/m} \qquad \blacksquare$$

11.5 *Retaining Walls with Friction*

So far in our study of active and passive earth pressures, we have considered the case of frictionless walls. In reality, retaining walls are rough, and shear forces develop between the face of the wall and the backfill. To understand the effect of wall friction on the failure surface, let us consider a rough retaining wall AB with a horizontal granular backfill, as shown in Figure 11.17.

In the active case (Figure 11.17a), when the wall AB moves to a position $A'B$, the soil mass in the active zone will be stretched outward. This will cause a downward motion of the soil relative to the wall. This motion causes a downward shear on the wall (Figure 11.17b), and it is called *positive wall friction in the active case*. If δ' is the angle of friction between the wall and the backfill, then the resultant active force, P_a, will be inclined at an angle δ' to the normal drawn to the back face of the retaining wall. Advanced studies show that the failure surface in the backfill can be represented by BCD, as shown in Figure 11.17a. The portion BC is curved, and the portion CD of the failure surface is a straight line. Rankine's active state exists in the zone ACD.

Under certain conditions, if the wall shown in Figure 11.17a is forced downward relative to the backfill, then the direction of the active force, P_a, will change as shown in Figure 11.17c. This is a situation of *negative wall friction in the active case* $(-\delta')$. Figure 11.17c also shows the nature of the failure surface in the backfill.

The effect of wall friction for the passive state is shown in Figures 11.17d and e. When the wall AB is pushed to a position $A'B$ (Figure 11.17d), the soil in the passive zone will be compressed. The result is an upward motion relative to the wall. The upward motion of the soil will cause an upward shear on the retaining wall (Figure 11.17e). This is referred to as *positive wall friction in the passive case*. The resultant passive force, P_p, will be inclined at an angle δ' to the normal drawn to the back face of the wall. The failure surface in the soil has a curved lower portion BC and a straight upper portion CD. Rankine's passive state exists in the zone ACD.

If the wall shown in Figure 11.17d is forced upward relative to the backfill, then the direction of the passive force, P_p, will change as shown in Figure 11.17f. This is *negative wall friction in the passive case* $(-\delta')$. Figure 11.17f also shows the nature of the failure surface in the backfill under such a condition.

For practical considerations, in the case of loose granular backfill, the angle of wall friction δ' is taken to be equal to the angle of friction of the soil, ϕ'. For dense granular backfills, δ' is smaller than ϕ' and is in the range $\phi'/2 \leq \delta' \leq (2/3)\phi'$.

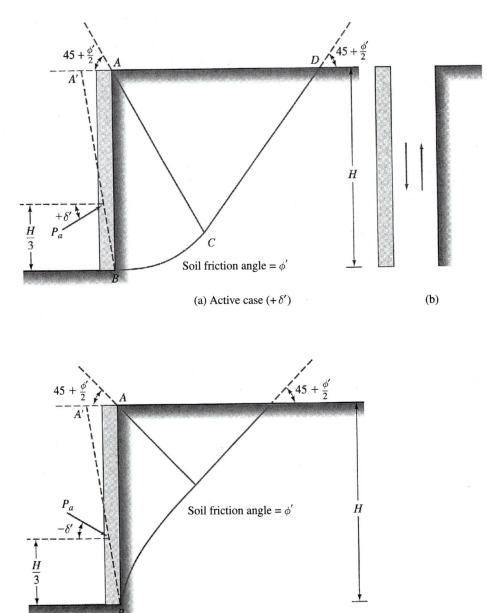

(a) Active case $(+\delta')$

(b)

(c) Active case $(-\delta')$

Figure 11.17 Effect of wall friction on failure surface

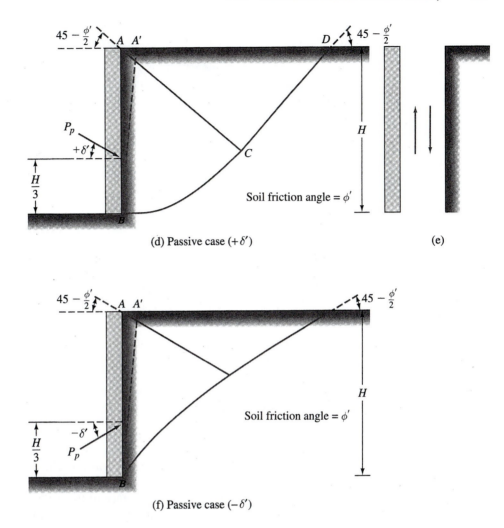

(d) Passive case $(+\delta')$

(e)

(f) Passive case $(-\delta')$

Figure 11.17 (*continued*)

Coulomb's Earth Pressure Theory

More than 200 years ago, Coulomb (1776) presented a theory for active and passive earth pressures against retaining walls. In this theory, Coulomb assumed that the failure surface is a plane. The wall friction was taken into consideration. The general principles of the derivation of Coulomb's earth pressure theory for a cohesionless backfill (shear strength defined by the equation $\tau_f = \sigma' \tan \phi'$) are given in this section.

Active Case

Let AB (Figure 11.18a) be the back face of a retaining wall supporting a granular soil, the surface of which is constantly sloping at an angle α with the horizontal. BC

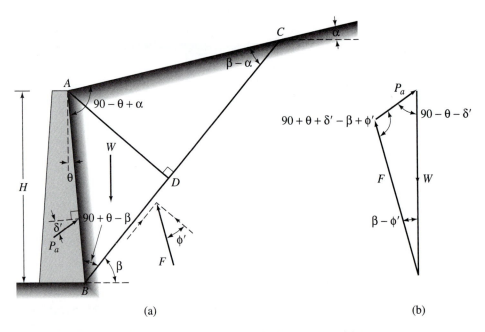

Figure 11.18 Coulomb's active pressure: (a) trial failure wedge; (b) force polygon

is a trial failure surface. In the stability consideration of the probable failure wedge *ABC*, the following forces are involved (per unit length of the wall):

1. W, the effective weight of the soil wedge.
2. F, the resultant of the shear and normal forces on the surface of failure, *BC*. This is inclined at an angle ϕ' to the normal drawn to the plane *BC*.
3. P_a, the active force per unit length of the wall. The direction of P_a is inclined at an angle δ' to the normal drawn to the face of the wall that supports the soil. δ' is the angle of friction between the soil and the wall.

The force triangle for the wedge is shown in Figure 11.18b. From the law of sines, we have

$$\frac{W}{\sin(90 + \theta + \delta' - \beta + \phi')} = \frac{P_a}{\sin(\beta - \phi')} \tag{11.53}$$

or

$$P_a = \frac{\sin(\beta - \phi')}{\sin(90 + \theta + \delta' - \beta + \phi')} W \tag{11.54}$$

The preceding equation can be written in the form

$$P_a = \frac{1}{2}\gamma H^2 \left[\frac{\cos(\theta - \beta)\cos(\theta - \alpha)\sin(\beta - \phi')}{\cos^2\theta \sin(\beta - \alpha)\sin(90 + \theta + \delta' - \beta + \phi')} \right] \tag{11.55}$$

where γ = unit weight of the backfill. The values of γ, H, θ, α, ϕ', and δ' are constants, and β is the only variable. To determine the critical value of β for maximum P_a, we have

$$\frac{dP_a}{d\beta} = 0 \tag{11.56}$$

After solving Eq. (11.56), when the relationship of β is substituted into Eq. (11.55), we obtain Coulomb's active earth pressure as

$$P_a = \frac{1}{2} K_a \gamma H^2 \tag{11.57}$$

where K_a is Coulomb's active earth pressure coefficient, given by

$$K_a = \frac{\cos^2(\phi' - \theta)}{\cos^2\theta \cos(\delta' + \theta)\left[1 + \sqrt{\dfrac{\sin(\delta' + \phi')\sin(\phi' - \alpha)}{\cos(\delta' + \theta)\cos(\theta - \alpha)}}\right]^2} \tag{11.58}$$

Note that when $\alpha = 0°$, $\theta = 0°$, and $\delta' = 0°$, Coulomb's active earth pressure coefficient becomes equal to $(1 - \sin\phi')/(1 + \sin\phi')$, which is the same as Rankine's earth pressure coefficient given earlier in this chapter.

The variation of the values of K_a for retaining walls with a vertical back ($\theta = 0$) and horizontal backfill ($\alpha = 0$) is given in Table 11.6. From this table note that for a given value of ϕ', the effect of wall friction is to reduce somewhat the active earth pressure coefficient.

Tables 11.7 and 11.8 give the values of K_a [Eq. (11.58)] for $\delta' = \frac{2}{3}\phi'$ and $\delta' = \phi'/2$. These tables may be useful in retaining wall design (see Chapter 13).

Table 11.6 Values of K_a [Eq. (11.58)] for $\theta = 0°$, $\alpha = 0°$ - coulombs

	δ' (deg) $\rightarrow$					
↓ ϕ' (deg)	0	5	10	15	20	25
28	0.3610	0.3448	0.3330	0.3251	0.3203	0.3186
30	0.3333	0.3189	0.3085	0.3014	0.2973	0.2956
32	0.3073	0.2945	0.2853	0.2791	0.2755	0.2745
34	0.2827	0.2714	0.2633	0.2579	0.2549	0.2542
36	0.2596	0.2497	0.2426	0.2379	0.2354	0.2350
38	0.2379	0.2292	0.2230	0.2190	0.2169	0.2167
40	0.2174	0.2089	0.2045	0.2011	0.1994	0.1995
42	0.1982	0.1916	0.1870	0.1841	0.1828	0.1831

Table 11.7 Values of K_a [Eq. (11.58)]. *Note:* $\delta' = \frac{2}{3}\phi'$

α (deg)	ϕ' (deg)	θ (deg)					
		0	5	10	15	20	25
0	28	0.3213	0.3588	0.4007	0.4481	0.5026	0.5662
	29	0.3091	0.3467	0.3886	0.4362	0.4908	0.5547
	30	0.2973	0.3349	0.3769	0.4245	0.4794	0.5435
	31	0.2860	0.3235	0.3655	0.4133	0.4682	0.5326
	32	0.2750	0.3125	0.3545	0.4023	0.4574	0.5220
	33	0.2645	0.3019	0.3439	0.3917	0.4469	0.5117
	34	0.2543	0.2916	0.3335	0.3813	0.4367	0.5017
	35	0.2444	0.2816	0.3235	0.3713	0.4267	0.4919
	36	0.2349	0.2719	0.3137	0.3615	0.4170	0.4824
	37	0.2257	0.2626	0.3042	0.3520	0.4075	0.4732
	38	0.2168	0.2535	0.2950	0.3427	0.3983	0.4641
	39	0.2082	0.2447	0.2861	0.3337	0.3894	0.4553
	40	0.1998	0.2361	0.2774	0.3249	0.3806	0.4468
	41	0.1918	0.2278	0.2689	0.3164	0.3721	0.4384
	42	0.1840	0.2197	0.2606	0.3080	0.3637	0.4302
5	28	0.3431	0.3845	0.4311	0.4843	0.5461	0.6190
	29	0.3295	0.3709	0.4175	0.4707	0.5325	0.6056
	30	0.3165	0.3578	0.4043	0.4575	0.5194	0.5926
	31	0.3039	0.3451	0.3916	0.4447	0.5067	0.5800
	32	0.2919	0.3329	0.3792	0.4324	0.4943	0.5677
	33	0.2803	0.3211	0.3673	0.4204	0.4823	0.5558
	34	0.2691	0.3097	0.3558	0.4088	0.4707	0.5443
	35	0.2583	0.2987	0.3446	0.3975	0.4594	0.5330
	36	0.2479	0.2881	0.3338	0.3866	0.4484	0.5221
	37	0.2379	0.2778	0.3233	0.3759	0.4377	0.5115
	38	0.2282	0.2679	0.3131	0.3656	0.4273	0.5012
	39	0.2188	0.2582	0.3033	0.3556	0.4172	0.4911
	40	0.2098	0.2489	0.2937	0.3458	0.4074	0.4813
	41	0.2011	0.2398	0.2844	0.3363	0.3978	0.4718
	42	0.1927	0.2311	0.2753	0.3271	0.3884	0.4625
10	28	0.3702	0.4164	0.4686	0.5287	0.5992	0.6834
	29	0.3548	0.4007	0.4528	0.5128	0.5831	0.6672
	30	0.3400	0.3857	0.4376	0.4974	0.5676	0.6516
	31	0.3259	0.3713	0.4230	0.4826	0.5526	0.6365
	32	0.3123	0.3575	0.4089	0.4683	0.5382	0.6219
	33	0.2993	0.3442	0.3953	0.4545	0.5242	0.6078
	34	0.2868	0.3314	0.3822	0.4412	0.5107	0.5942
	35	0.2748	0.3190	0.3696	0.4283	0.4976	0.5810
	36	0.2633	0.3072	0.3574	0.4158	0.4849	0.5682
	37	0.2522	0.2957	0.3456	0.4037	0.4726	0.5558
	38	0.2415	0.2846	0.3342	0.3920	0.4607	0.5437
	39	0.2313	0.2740	0.3231	0.3807	0.4491	0.5321
	40	0.2214	0.2636	0.3125	0.3697	0.4379	0.5207
	41	0.2119	0.2537	0.3021	0.3590	0.4270	0.5097
	42	0.2027	0.2441	0.2921	0.3487	0.4164	0.4990

Table 11.7 (*continued*)

α (deg)	ϕ' (deg)	θ (deg)					
		0	**5**	**10**	**15**	**20**	**25**
15	28	0.4065	0.4585	0.5179	0.5868	0.6685	0.7670
	29	0.3881	0.4397	0.4987	0.5672	0.6483	0.7463
	30	0.3707	0.4219	0.4804	0.5484	0.6291	0.7265
	31	0.3541	0.4049	0.4629	0.5305	0.6106	0.7076
	32	0.3384	0.3887	0.4462	0.5133	0.5930	0.6895
	33	0.3234	0.3732	0.4303	0.4969	0.5761	0.6721
	34	0.3091	0.3583	0.4150	0.4811	0.5598	0.6554
	35	0.2954	0.3442	0.4003	0.4659	0.5442	0.6393
	36	0.2823	0.3306	0.3862	0.4513	0.5291	0.6238
	37	0.2698	0.3175	0.3726	0.4373	0.5146	0.6089
	38	0.2578	0.3050	0.3595	0.4237	0.5006	0.5945
	39	0.2463	0.2929	0.3470	0.4106	0.4871	0.5805
	40	0.2353	0.2813	0.3348	0.3980	0.4740	0.5671
	41	0.2247	0.2702	0.3231	0.3858	0.4613	0.5541
	42	0.2146	0.2594	0.3118	0.3740	0.4491	0.5415
20	28	0.4602	0.5205	0.5900	0.6714	0.7689	0.8880
	29	0.4364	0.4958	0.5642	0.6445	0.7406	0.8581
	30	0.4142	0.4728	0.5403	0.6195	0.7144	0.8303
	31	0.3935	0.4513	0.5179	0.5961	0.6898	0.8043
	32	0.3742	0.4311	0.4968	0.5741	0.6666	0.7799
	33	0.3559	0.4121	0.4769	0.5532	0.6448	0.7569
	34	0.3388	0.3941	0.4581	0.5335	0.6241	0.7351
	35	0.3225	0.3771	0.4402	0.5148	0.6044	0.7144
	36	0.3071	0.3609	0.4233	0.4969	0.5856	0.6947
	37	0.2925	0.3455	0.4071	0.4799	0.5677	0.6759
	38	0.2787	0.3308	0.3916	0.4636	0.5506	0.6579
	39	0.2654	0.3168	0.3768	0.4480	0.5342	0.6407
	40	0.2529	0.3034	0.3626	0.4331	0.5185	0.6242
	41	0.2408	0.2906	0.3490	0.4187	0.5033	0.6083
	42	0.2294	0.2784	0.3360	0.4049	0.4888	0.5930

Table 11.8 Values of K_a [Eq. (11.58)]. *Note:* $\delta' = \phi'/2$

α (deg)	ϕ' (deg)	θ (deg)					
		0	**5**	**10**	**15**	**20**	**25**
0	28	0.3264	0.3629	0.4034	0.4490	0.5011	0.5616
	29	0.3137	0.3502	0.3907	0.4363	0.4886	0.5492
	30	0.3014	0.3379	0.3784	0.4241	0.4764	0.5371
	31	0.2896	0.3260	0.3665	0.4121	0.4645	0.5253
	32	0.2782	0.3145	0.3549	0.4005	0.4529	0.5137
	33	0.2671	0.3033	0.3436	0.3892	0.4415	0.5025
	34	0.2564	0.2925	0.3327	0.3782	0.4305	0.4915

(*continued*)

Table 11.8 (*continued*)

α (deg)	φ' (deg)	θ (deg) 0	5	10	15	20	25
	35	0.2461	0.2820	0.3221	0.3675	0.4197	0.4807
	36	0.2362	0.2718	0.3118	0.3571	0.4092	0.4702
	37	0.2265	0.2620	0.3017	0.3469	0.3990	0.4599
	38	0.2172	0.2524	0.2920	0.3370	0.3890	0.4498
	39	0.2081	0.2431	0.2825	0.3273	0.3792	0.4400
	40	0.1994	0.2341	0.2732	0.3179	0.3696	0.4304
	41	0.1909	0.2253	0.2642	0.3087	0.3602	0.4209
	42	0.1828	0.2168	0.2554	0.2997	0.3511	0.4117
5	28	0.3477	0.3879	0.4327	0.4837	0.5425	0.6115
	29	0.3337	0.3737	0.4185	0.4694	0.5282	0.5972
	30	0.3202	0.3601	0.4048	0.4556	0.5144	0.5833
	31	0.3072	0.3470	0.3915	0.4422	0.5009	0.5698
	32	0.2946	0.3342	0.3787	0.4292	0.4878	0.5566
	33	0.2825	0.3219	0.3662	0.4166	0.4750	0.5437
	34	0.2709	0.3101	0.3541	0.4043	0.4626	0.5312
	35	0.2596	0.2986	0.3424	0.3924	0.4505	0.5190
	36	0.2488	0.2874	0.3310	0.3808	0.4387	0.5070
	37	0.2383	0.2767	0.3199	0.3695	0.4272	0.4954
	38	0.2282	0.2662	0.3092	0.3585	0.4160	0.4840
	39	0.2185	0.2561	0.2988	0.3478	0.4050	0.4729
	40	0.2090	0.2463	0.2887	0.3374	0.3944	0.4620
	41	0.1999	0.2368	0.2788	0.3273	0.3840	0.4514
	42	0.1911	0.2276	0.2693	0.3174	0.3738	0.4410
10	28	0.3743	0.4187	0.4688	0.5261	0.5928	0.6719
	29	0.3584	0.4026	0.4525	0.5096	0.5761	0.6549
	30	0.3432	0.3872	0.4368	0.4936	0.5599	0.6385
	31	0.3286	0.3723	0.4217	0.4782	0.5442	0.6225
	32	0.3145	0.3580	0.4071	0.4633	0.5290	0.6071
	33	0.3011	0.3442	0.3930	0.4489	0.5143	0.5920
	34	0.2881	0.3309	0.3793	0.4350	0.5000	0.5775
	35	0.2757	0.3181	0.3662	0.4215	0.4862	0.5633
	36	0.2637	0.3058	0.3534	0.4084	0.4727	0.5495
	37	0.2522	0.2938	0.3411	0.3957	0.4597	0.5361
	38	0.2412	0.2823	0.3292	0.3833	0.4470	0.5230
	39	0.2305	0.2712	0.3176	0.3714	0.4346	0.5103
	40	0.2202	0.2604	0.3064	0.3597	0.4226	0.4979
	41	0.2103	0.2500	0.2956	0.3484	0.4109	0.4858
	42	0.2007	0.2400	0.2850	0.3375	0.3995	0.4740
15	28	0.4095	0.4594	0.5159	0.5812	0.6579	0.7498
	29	0.3908	0.4402	0.4964	0.5611	0.6373	0.7284
	30	0.3730	0.4220	0.4777	0.5419	0.6175	0.7080
	31	0.3560	0.4046	0.4598	0.5235	0.5985	0.6884
	32	0.3398	0.3880	0.4427	0.5059	0.5803	0.6695
	33	0.3244	0.3721	0.4262	0.4889	0.5627	0.6513
	34	0.3097	0.3568	0.4105	0.4726	0.5458	0.6338

Table 11.8 (*continued*)

α (deg)	ϕ' (deg)	θ (deg)					
		0	5	10	15	20	25
	35	0.2956	0.3422	0.3953	0.4569	0.5295	0.6168
	36	0.2821	0.3282	0.3807	0.4417	0.5138	0.6004
	37	0.2692	0.3147	0.3667	0.4271	0.4985	0.5846
	38	0.2569	0.3017	0.3531	0.4130	0.4838	0.5692
	39	0.2450	0.2893	0.3401	0.3993	0.4695	0.5543
	40	0.2336	0.2773	0.3275	0.3861	0.4557	0.5399
	41	0.2227	0.2657	0.3153	0.3733	0.4423	0.5258
	42	0.2122	0.2546	0.3035	0.3609	0.4293	0.5122
20	28	0.4614	0.5188	0.5844	0.6608	0.7514	0.8613
	29	0.4374	0.4940	0.5586	0.6339	0.7232	0.8313
	30	0.4150	0.4708	0.5345	0.6087	0.6968	0.8034
	31	0.3941	0.4491	0.5119	0.5851	0.6720	0.7772
	32	0.3744	0.4286	0.4906	0.5628	0.6486	0.7524
	33	0.3559	0.4093	0.4704	0.5417	0.6264	0.7289
	34	0.3384	0.3910	0.4513	0.5216	0.6052	0.7066
	35	0.3218	0.3736	0.4331	0.5025	0.5851	0.6853
	36	0.3061	0.3571	0.4157	0.4842	0.5658	0.6649
	37	0.2911	0.3413	0.3991	0.4668	0.5474	0.6453
	38	0.2769	0.3263	0.3833	0.4500	0.5297	0.6266
	39	0.2633	0.3120	0.3681	0.4340	0.5127	0.6085
	40	0.2504	0.2982	0.3535	0.4185	0.4963	0.5912
	41	0.2381	0.2851	0.3395	0.4037	0.4805	0.5744
	42	0.2263	0.2725	0.3261	0.3894	0.4653	0.5582

Passive Case

Figure 11.19a shows a retaining wall with a sloping cohesionless backfill similar to that considered in Figure 11.18a. The force polygon for equilibrium of the wedge *ABC* for the passive state is shown in Figure 11.19b. P_p is the notation for the passive force. Other notations used are the same as those for the active case considered in this section. In a procedure similar to the one we followed in the active case, we get

$$P_p = \frac{1}{2} K_p \gamma H^2 \tag{11.59}$$

where K_p = coefficient of passive earth pressure for Coulomb's case, or

$$K_p = \frac{\cos^2(\phi' + \theta)}{\cos^2 \theta \cos(\delta' - \theta) \left[1 - \sqrt{\frac{\sin(\phi' - \delta') \sin(\phi' + \alpha)}{\cos(\delta' - \theta) \cos(\alpha - \theta)}} \right]^2} \tag{11.60}$$

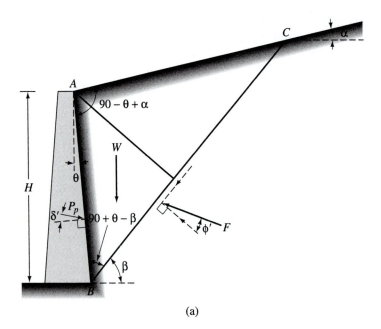

(a)

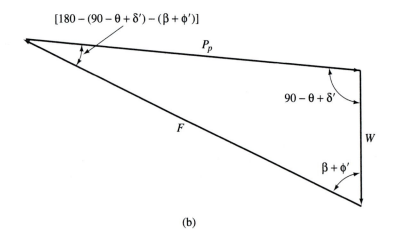

(b)

Figure 11.19 Coulomb's passive pressure: (a) trial failure wedge; (b) force polygon

For a frictionless wall with the vertical back face supporting granular soil backfill with a horizontal surface (that is, $\theta = 0°$, $\alpha = 0°$, and $\delta' = 0°$), Eq. (11.60) yields

$$K_p = \frac{1 + \sin \phi'}{1 - \sin \phi'} = \tan^2\left(45 + \frac{\phi'}{2}\right)$$

This is the same relationship that was obtained for the passive earth pressure coefficient in Rankine's case given by Eq. (11.20).

Table 11.9 Values of K_p [Eq. (11.60)] for $\theta = 0°$ and $\alpha = 0°$

↓ ϕ' (deg)	δ' (deg) →				
	0	5	10	15	20
15	1.698	1.900	2.130	2.405	2.735
20	2.040	2.313	2.636	3.030	3.525
25	2.464	2.830	3.286	3.855	4.597
30	3.000	3.506	4.143	4.977	6.105
35	3.690	4.390	5.310	6.854	8.324
40	4.600	5.590	6.946	8.870	11.772

The variation of K_p with ϕ' and δ' (for $\theta = 0$ and $\alpha = 0$) is given in Table 11.9. It can be observed from this table that, for given values of α and ϕ', the value of K_p increases with the wall friction. *Note that making the assumption that the failure surface is a plane in Coulomb's theory grossly overestimates the passive resistance of walls, particularly for $\delta' > \phi'/2$. This error is somewhat unsafe for all design purposes.*

11.7 Passive Pressure Assuming Curved Failure Surface in Soil

As mentioned in Section 11.6, Coulomb's theory overestimates the passive resistance for $\delta' > \phi'/2$. Several studies have been conducted in the past to obtain K_p assuming curved failure surface in soil. In this section, the solution given by Caquot and Kerisel (1948) will be presented.

Figure 11.20 shows a retaining wall with an inclined back and a horizontal backfill. For this case, the passive pressure per unit length of the wall can be calculated as,

$$P_p = \frac{1}{2} \gamma H_1^2 K_p \qquad (11.61)$$

where K_p = the passive pressure coefficient

For definition of H_1, refer to Figure 11.20. The variation of K_p determined by Caquot and Kerisel (1948) also is shown in Figure 11.20. It is important to note that the K_p values shown are for $\delta'/\phi' = 1$. If $\delta'/\phi' \neq 1$, the following procedure must be used to determine K_p.

1. Assume δ' and ϕ'.
2. Calculate δ'/ϕ'.
3. Using the ratio of δ'/ϕ' (Step 2), determine the reduction factor, R, from Table 11.10.
4. Determine K_p from Figure 11.20 for $\delta'/\phi' = 1$.
5. Calculate K_p for the required δ'/ϕ' as

$$K_p = (R)[K_{p(\delta'/\phi' = 1)}] \qquad (11.62)$$

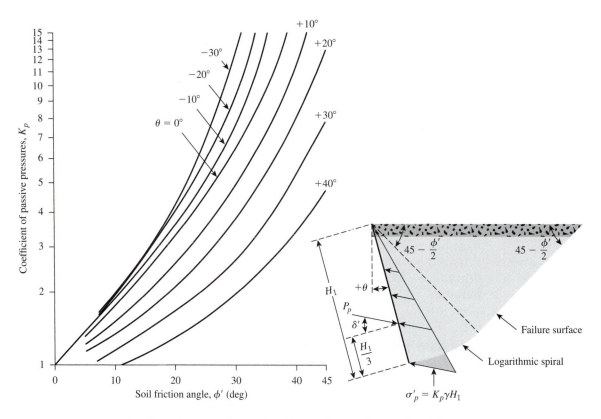

Figure 11.20 Caquot and Kerisel's solution for K_p [Eq. (11.61)]

Figure 11.21 shows a vertical retaining wall with an inclined granular backfill. For this case,

$$P_p = \frac{1}{2}\gamma H^2 K_p \qquad (11.63)$$

Table 11.10 Caquot and Kerisel's Reduction Factor, R, for Passive Pressure Calculation

ϕ'	δ'/ϕ'							
	0.7	0.6	0.5	0.4	0.3	0.2	0.1	0.0
10	0.978	0.962	0.946	0.929	0.912	0.898	0.881	0.864
15	0.961	0.934	0.907	0.881	0.854	0.830	0.803	0.775
20	0.939	0.901	0.862	0.824	0.787	0.752	0.716	0.678
25	0.912	0.860	0.808	0.759	0.711	0.666	0.620	0.574
30	0.878	0.811	0.746	0.686	0.627	0.574	0.520	0.467
35	0.836	0.752	0.674	0.603	0.536	0.475	0.417	0.362
40	0.783	0.682	0.592	0.512	0.439	0.375	0.316	0.262
45	0.718	0.600	0.500	0.414	0.339	0.276	0.221	0.174

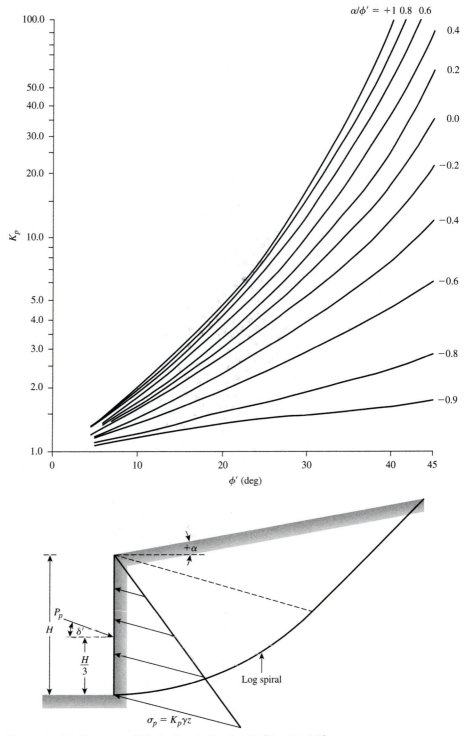

Figure 11.21 Caquot and Kerisel's solution for K_p [Eq. (11.63)]

Caquot and Kerisel's solution (1948) for K_p to use in Eq. (11.63) is given in Figure 11.21 for $\delta'/\phi' = 1$. In order to use Figure 11.21, the following steps need to be taken:

Step 1: Determine α/ϕ' (note the sign of α)

Step 2: Knowing ϕ' and α/ϕ', use Figure 11.21 to determine K_p for $\delta'/\phi' = 1$

Step 3: Calculate δ'/ϕ'

Step 4: Go to Table 11.10 to determine the reduction factor, R

Step 5: $K_p = (R) [K_{p(\delta'/\phi' = 1)}]$ (11.64)

Problems

11.1 Assuming that the wall shown in Figure 11.22 is restrained from yielding, find the magnitude and location of the resultant lateral force per unit length of the wall for the following cases:

a. $H = 5$ m, $\gamma = 14.4$ kN/m^3, $\phi' = 31°$, $OCR = 2.5$

b. $H = 4$ m, $\gamma = 13.4$ kN/m^3, $\phi' = 28°$, $OCR = 1.5$

11.2 Figure 11.22 shows a retaining wall with cohesionless soil backfill. For the following cases, determine the total active force per unit length of the wall for Rankine's state and the location of the resultant.

a. $H = 4.5$ m, $\gamma = 17.6$ kN/m^3, $\phi' = 36°$

b. $H = 5$ m, $\gamma = 17.0$ kN/m^3, $\phi' = 38°$

c. $H = 4$ m, $\gamma = 19.95$ kN/m^3, $\phi' = 42°$

11.3 From Figure 11.22, determine the passive force, P_p, per unit length of the wall for Rankine's case. Also state Rankine's passive pressure at the bottom of the wall. Consider the following cases:

a. $H = 2.45$ m, $\gamma = 16.67$ kN/m^3, $\phi' = 33°$

b. $H = 4$ m, $\rho = 1800$ kg/m^3, $\phi' = 38°$

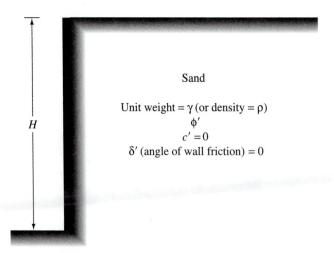

Sand

Unit weight $= \gamma$ (or density $= \rho$)

ϕ'

$c' = 0$

δ' (angle of wall friction) $= 0$

H

Figure 11.22

11.4 A retaining wall is shown in Figure 11.23. Determine Rankine's active force, P_a, per unit length of the wall and the location of the resultant in each of the following cases:

    a. $H = 6$ m, $H_1 = 2$ m, $\gamma_1 = 16$ kN/m³, $\gamma_2 = 19$ kN/m³, $\phi'_1 = 32°$, $\phi'_2 = 36°$, $q = 15$ kN/m²

    b. $H = 5$ m, $H_1 = 1.5$ m, $\gamma_1 = 17.2$ kN/m³, $\gamma_2 = 20.4$ kN/m³, $\phi'_1 = 30°$, $\phi'_2 = 34°$, $q = 19.15$ kN/m²

11.5 A retaining wall 6 m high with a vertical back face retains a homogeneous saturated soft clay. The saturated unit weight of the clay is 19 kN/m³. Laboratory tests showed that the undrained shear strength, c_u, of the clay is 16.8 kN/m².

    a. Do the necessary calculations and draw the variation of Rankine's active pressure on the wall with depth.

    b. Find the depth up to which a tensile crack can occur.

    c. Determine the total active force per unit length of the wall before the tensile crack occurs.

    d. Determine the total active force per unit length of the wall after the tensile crack occurs. Also find the location of the resultant.

11.6 Redo Problem 11.5 assuming that the backfill is supporting a surcharge of 9.6 kN/m².

11.7 Repeat Problem 11.5 with the following values:

$$\text{height of wall} = 6 \text{ m}$$

$$\gamma_{\text{sat}} = 19.8 \text{ kN/m}^3$$

$$c_u = 14.7 \text{ kN/m}^2$$

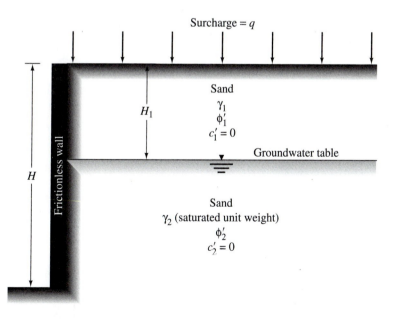

Figure 11.23

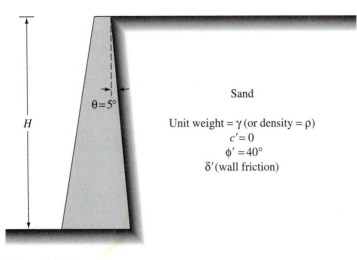

Figure 11.24

11.8 A retaining wall 6 m high with a vertical back face has c'-ϕ' soil for backfill. For the backfill, $\gamma = 18.1$ kN/m³, $c' = 29$ kN/m², and $\phi' = 18°$. Taking the existence of the tensile crack into consideration, determine the active force, P_a, per unit length of the wall for Rankine's active state.

11.9 For the wall described in Problem 11.8, determine the passive force, P_p, per unit length for Rankine's passive state.

11.10 A retaining wall is shown in Figure 11.24. The height of the wall is 6 m, and the unit weight of the backfill is 18.9 kN/m³. Calculate the active force, P_a, on the wall using Coulomb's equation for the following values of the angle of wall friction:
  a. $\delta' = 0°$
  b. $\delta' = 20°$
  c. $\delta' = 26.7°$
Comment on the direction and location of the resultant.

11.11 For the wall described in Problem 11.10, determine the passive force, P_p, per unit length of the wall using the Caquot and Kerisel solution.

References

CAQUOT, A., and KERISEL, J. (1948). *Tables for the Calculation of Passive Pressure, Active Pressure, and Bearing Capacity of Foundations*, Gauthier-Villars, Paris.

COULOMB, C. A. (1776). "Essai sur une Application des Règles de Maximis et Minimis à quelques Problèmes de Statique, relatifs a l'Architecture," *Mem. Roy. des Sciences*, Paris, Vol. 3, 38.

JAKY, J. (1944). "The Coefficient of Earth Pressure at Rest," *Journal of the Society of Hungarian Architects and Engineers*, Vol. 7, 355–358.

MASSARSCH, K. R. (1979). "Lateral Earth Pressure in Normally Consolidated Clay," *Proceedings of the Seventh European Conference on Soil Mechanics and Foundation Engineering*, Brighton, England, Vol. 2, 245–250.

MAYNE, P. W., and KULHAWY, F. H. (1982). "K_o—OCR Relationships in Soil," *Journal of the Geotechnical Division*, ASCE, Vol. 108, No. 6, 851–872.

MAZINDRANI, Z. H., and GANJALI, M. H. (1997). "Lateral Earth Problem of Cohesive Backfill with Inclined Surface," *Journal of Geotechnical and Geoenvironmental Engineering*, ASCE, Vol. 123, No. 2, 110–112.

RANKINE, W. M. J. (1857). "On Stability on Loose Earth," *Philosophic Transactions of Royal Society*, London, Part I, 9–27.

12

Shallow Foundations—
Bearing Capacity and Settlement

The lowest part of a structure is generally referred to as the *foundation*. Its function is to transfer the load of the structure to the soil on which it is resting. A properly designed foundation is one that transfers the load throughout the soil without overstressing the soil. Overstressing the soil can result in either excessive settlement or shear failure of the soil, both of which cause damage to the structure. Thus, geotechnical and structural engineers who design foundations must evaluate the bearing capacity of soils.

Depending on the structure and soil encountered, various types of foundations are used. A *spread footing* is simply an enlargement of a load-bearing wall or column that makes it possible to spread the load of the structure over a larger area of the soil. In soil with low load-bearing capacity, the size of the spread footings required is impracticably large. In that case, it is more economical to construct the entire structure over a concrete pad. This is called a *mat foundation*.

Pile and *drilled shaft foundations* are used for heavier structures when great depth is required for supporting the load. Piles are structural members made of timber, concrete, or steel that transmit the load of the superstructure to the lower layers of the soil. According to how they transmit their load into the subsoil, piles can be divided into two categories: friction piles and end-bearing piles. In the case of friction piles, the superstructure load is resisted by the shear stresses generated along the surface of the pile. In the end-bearing pile, the load carried by the pile is transmitted at its tip to a firm stratum.

In the case of drilled shafts, a shaft is drilled into the subsoil and is then filled with concrete. A metal casing may be used while the shaft is being drilled. The casing may be left in place or withdrawn during the placing of concrete. Generally, the diameter of a drilled shaft is much larger than that of a pile. The distinction between piles and drilled shafts becomes hazy at an approximate diameter of 1 m, and then the definitions and nomenclature are inaccurate.

Spread footings and mat foundations are generally referred to as shallow foundations, and pile and drilled shaft foundations are classified as deep foundations. In a more general sense, shallow foundations are those foundations that have a depth-of-embedment-to-width ratio of approximately less than four. When the depth-of-embedment-to-width ratio of a foundation is greater than four, it may be classified as a deep foundation.

In this chapter, we discuss the soil-bearing capacity for shallow foundations. As mentioned before, for a foundation to function properly, (1) the settlement of soil caused by the load must be within the tolerable limit, and (2) shear failure of the soil supporting the foundation must not occur. Compressibility of soil due to consolidation was introduced in Chapter 7. This chapter introduces the load-carrying capacity of shallow foundations based on the criterion of shear failure in soil; and also the elastic settlement.

$$P = q_{all} B^2 \qquad\qquad \Delta T^1 = .1 \frac{1}{P}$$

ULTIMATE BEARING CAPACITY OF SHALLOW FOUNDATIONS

12.1 *General Concepts*

Consider a strip (i.e., theoretically length is infinity) foundation resting on the surface of a dense sand or stiff cohesive soil, as shown in Figure 12.1a, with a width of B. Now, if load is gradually applied to the foundation, settlement will increase. The variation of the load per unit area on the foundation, q, with the foundation settlement is also shown in Figure 12.1a. At a certain point—when the load per unit area equals q_u—a sudden failure in the soil supporting the foundation will take place, and the failure surface in the soil will extend to the ground surface. This load per unit area, q_u, is usually referred to as the *ultimate bearing capacity of the foundation*. When this type of sudden failure in soil takes place, it is called *general shear failure*.

If the foundation under consideration rests on sand or clayey soil of medium compaction (Figure 12.1b), an increase of load on the foundation will also be accompanied by an increase of settlement. However, in this case the failure surface in the soil will gradually extend outward from the foundation, as shown by the solid lines in Figure 12.1b. When the load per unit area on the foundation equals $q_{u(1)}$, the foundation movement will be accompanied by sudden jerks. A considerable movement of the foundation is then required for the failure surface in soil to extend to the ground surface (as shown by the broken lines in Figure 12.1b). The load per unit area at which this happens is the *ultimate bearing capacity, q_u*. Beyond this point, an increase of load will be accompanied by a large increase of foundation settlement. The load per unit area of the foundation, $q_{u(1)}$, is referred to as the *first failure load* (Vesic, 1963). Note that a peak value of q is not realized in this type of failure, which is called *local shear failure* in soil.

If the foundation is supported by a fairly loose soil, the load–settlement plot will be like the one in Figure 12.1c. In this case, the failure surface in soil will not extend to the ground surface. Beyond the ultimate failure load, q_u, the load-settlement plot will be steep and practically linear. This type of failure in soil is called *punching shear failure*.

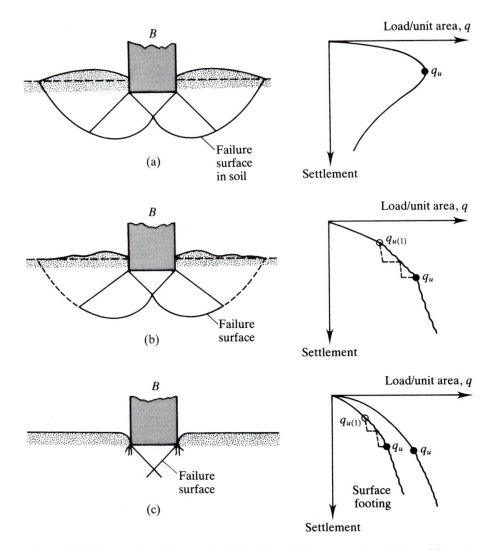

Figure 12.1 Nature of bearing capacity failure in soil: (a) general shear failure; (b) local shear failure; (c) punching shear failure

Based on experimental results, Vesic (1963) proposed a relationship for the mode of bearing capacity failure of foundations resting on sands. Figure 12.2 shows this relationship, which involves the following notation:

D_r = relative density of sand
D_f = depth of foundation measured from the ground surface
B = width of foundation
L = length of foundation

From Figure 12.2 it can be seen that

$$\text{Nature of failure in soil} = f\left(D_r, \frac{D_f}{B}, \frac{B}{L}\right) \tag{12.1}$$

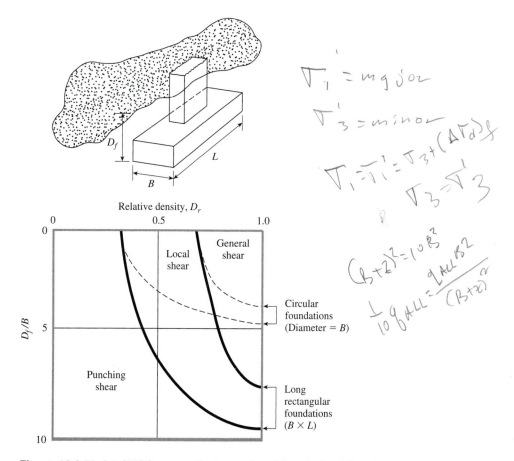

Figure 12.2 Vesic's (1963) test results for modes of foundation failure in sand

For foundations at a shallow depth (that is, small D_f/B^*), the ultimate load may occur at a foundation settlment of 4% to 10% of B. This condition occurs with general shear failure in soil; however, with local or punching shear failure, the ultimate load may occur in settlements of 15% to 25% of the width of foundation (B). Note that

$$B^* = \frac{2BL}{B + L} \tag{12.2}$$

12.2 *Ultimate Bearing Capacity Theory*

Terzaghi (1943) was the first to present a comprehensive theory for evaluating the ultimate bearing capacity of rough shallow foundations. According to this theory, a foundation is *shallow* if the depth, D_f (Figure 12.3), of the foundation is less than or equal to the width of the foundation. Later investigators, however, have suggested

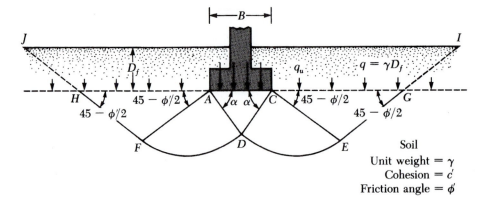

Figure 12.3 Bearing capacity failure in soil under a rough rigid continuous foundation

that foundations with D_f equal to 3 to 4 times the width of the foundation may be defined as *shallow foundations*.

Terzaghi suggested that for a *continuous*, or *strip*, *foundation* (that is, the width-to-length ratio of the foundation approaches 0), the failure surface in soil at ultimate load may be assumed to be similar to that shown in Figure 12.3. (Note that this is the case of general shear failure as defined in Figure 12.1a.) The effect of soil above the bottom of the foundation may also be assumed to be replaced by an equivalent surcharge, $q = \gamma D_f$ (where γ = unit weight of soil). The failure zone under the foundation can be separated into three parts (see Figure 12.3):

1. The *triangular zone ACD* immediately under the foundation
2. The *radial shear zones ADF* and *CDE*, with the curves *DE* and *DF* being arcs of a logarithmic spiral
3. Two *triangular Rankine passive zones AFH* and *CEG*

The angles *CAD* and *ACD* are assumed to be equal to the soil friction angle (that is, $\alpha = \phi'$). Note that, with the replacement of the soil above the bottom of the foundation by an equivalent surcharge q, the shear resistance of the soil along the failure surfaces *GI* and *HJ* was neglected.

Using the equilibrium analysis, Terzaghi expressed the ultimate bearing capacity in the form

$$q_u = c'N_c + qN_q + \frac{1}{2}\gamma BN_\gamma \quad \text{(strip foundation)} \qquad (12.3)$$

where

c' = cohesion of soil
γ = unit weight of soil
$q = \gamma D_f$
N_c, N_q, N_γ = bearing capacity factors that are nondimensional and are only functions of the soil friction angle, ϕ'

For square and circular footings, Terzaghi suggested the following equations for ultimate soil-bearing capacity:

The square footing is

$$q_u = 1.3c'N_c + qN_q + 0.4\gamma BN_\gamma \qquad (12.4)$$

The circular footing is

$$q_u = 1.3c'N_c + qN_q + 0.3\gamma BN_\gamma \qquad (12.5)$$

where B = diameter of the footing.

The variation of N_c, N_q, and N_γ with ϕ' is given in Table 12.1.

Table 12.1 Terzaghi's Bearing Capacity Factors—N_c, N_q and N_γ—Eqs. (12.3), (12.4), and (12.5).

ϕ' (deg)	N_c	N_q	N^a_γ	ϕ' (deg)	N_c	N_q	N^a_γ
0	5.70	1.00	0.00	26	27.09	14.21	9.84
1	6.00	1.10	0.01	27	29.24	15.90	11.60
2	6.30	1.22	0.04	28	31.61	17.81	13.70
3	6.62	1.35	0.06	29	34.24	19.98	16.18
4	6.97	1.49	0.10	30	37.16	22.46	19.13
5	7.34	1.64	0.14	31	40.41	25.28	22.65
6	7.73	1.81	0.20	32	44.04	28.52	26.87
7	8.15	2.00	0.27	33	48.09	32.23	31.94
8	8.60	2.21	0.35	34	52.64	36.50	38.04
9	9.09	2.44	0.44	35	57.75	41.44	45.41
10	9.61	2.69	0.56	36	63.53	47.16	54.36
11	10.16	2.98	0.69	37	70.01	53.80	65.27
12	10.76	3.29	0.85	38	77.50	61.55	78.61
13	11.41	3.63	1.04	39	85.97	70.61	95.03
14	12.11	4.02	1.26	40	95.66	81.27	115.31
15	12.86	4.45	1.52	41	106.81	93.85	140.51
16	13.68	4.92	1.82	42	119.67	108.75	171.99
17	14.60	5.45	2.18	43	134.58	126.50	211.56
18	15.12	6.04	2.59	44	151.95	147.74	261.60
19	16.56	6.70	3.07	45	172.28	173.28	325.34
20	17.69	7.44	3.64	46	196.22	204.19	407.11
21	18.92	8.26	4.31	47	224.55	241.80	512.84
22	20.27	9.19	5.09	48	258.28	287.85	650.67
23	21.75	10.23	6.00	49	298.71	344.63	831.99
24	23.36	11.40	7.08	50	347.50	415.14	1072.80
25	25.13	12.72	8.34				

[a] From Kumbhojkar (1993)

Modification to Terzaghi's Bearing Capacity Equation

Based on laboratory and field studies of bearing capacity, the basic nature of the failure surface in soil suggested by Terzaghi now appears to be correct (Vesic, 1973). However, the angle α shown in Figure 12.3 is closer to $45 + \phi'/2$ than to ϕ', as was originally assumed by Terzaghi. With $\alpha = 45 + \phi'/2$, the relations for N_c and N_q can be derived as

$$N_q = \tan^2\left(45 + \frac{\phi'}{2}\right)e^{\pi \tan \phi'} \tag{12.6}$$

$$N_c = (N_q - 1)\cot \phi' \tag{12.7}$$

The equation for N_c given by Eq. (12.7) was originally derived by Prandtl (1921), and the relation for N_q [Eq. (12.6)] was presented by Reissner (1924). Caquot and Kerisel (1953) and Vesic (1973) gave the relation for N_γ as

$$N_\gamma = 2(N_q + 1)\tan \phi' \tag{12.8}$$

Table 12.2 shows the variation of the preceding bearing capacity factors with soil friction angles.

The form of Eq. (12.3), which is for a strip foundation subjected to vertical loading, can be generalized by taking into consideration the following:

a. The shearing resistance along the failure surface in soil above the bottom of the foundation (portion of the failure surface marked as *GI* and *HJ* in Figure 12.3);
b. The width-to-length ratio of rectangular foundations; and
c. Load inclination.

The ultimate bearing capacity equation will thus take the form (Meyerhof, 1963)

$$q_u = c'N_cF_{cs}F_{cd}F_{ci} + qN_qF_{qs}F_{qd}F_{qi} + \frac{1}{2}\gamma BN_\gamma F_{\gamma s}F_{\gamma d}F_{\gamma i} \tag{12.9}$$

where

$$c' = \text{cohesion}$$
$$q = \text{effective stress at the level of the bottom of the foundation}$$
$$\gamma = \text{unit weight of soil}$$
$$B = \text{width of foundation} \,(= \text{diameter for a circular foundation})$$
$$F_{cs}, F_{qs}, F_{\gamma s} = \text{shape factors}$$
$$F_{cd}, F_{qd}, F_{\gamma d} = \text{depth factors}$$
$$F_{ci}, F_{qi}, F_{\gamma i} = \text{load inclination factors}$$
$$N_c, N_q, N_\gamma = \text{bearing capacity factors [Eqs. (12.6), (12.7) and (12.8)]}$$

Table 12.2 Bearing capacity factors [Eqs. (12.6), (12.7), and (12.8)]

ϕ'	N_c	N_q	N_γ	ϕ'	N_c	N_q	N_γ
0	5.14	1.00	0.00	23	18.05	8.66	8.20
1	5.38	1.09	0.07	24	19.32	9.60	9.44
2	5.63	1.20	0.15	25	20.72	10.66	10.88
3	5.90	1.31	0.24	26	22.25	11.85	12.54
4	6.19	1.43	0.34	27	23.94	13.20	14.47
5	6.49	1.57	0.45	28	25.80	14.72	16.72
6	6.81	1.72	0.57	29	27.86	16.44	19.34
7	7.16	1.88	0.71	30	30.14	18.40	22.40
8	7.53	2.06	0.86	31	32.67	20.63	25.99
9	7.92	2.25	1.03	32	35.49	23.18	30.22
10	8.35	2.47	1.22	33	38.64	26.09	35.19
11	8.80	2.71	1.44	34	42.16	29.44	41.06
12	9.28	2.97	1.69	35	46.12	33.30	48.03
13	9.81	3.26	1.97	36	50.59	37.75	56.31
14	10.37	3.59	2.29	37	55.63	42.92	66.19
15	10.98	3.94	2.65	38	61.35	48.93	78.03
16	11.63	4.34	3.06	39	67.87	55.96	92.25
17	12.34	4.77	3.53	40	75.31	64.20	109.41
18	13.10	5.26	4.07	41	83.86	73.90	130.22
19	13.93	5.80	4.68	42	93.71	85.38	155.55
20	14.83	6.40	5.39	43	105.11	99.02	186.54
21	15.82	7.07	6.20	44	118.37	115.31	224.64
22	16.88	7.82	7.13	45	133.88	134.88	271.76

The relationships for the shape factors, depth factors, and inclination factors *recommended for use* are given in Table 12.3.

Net Ultimate Bearing Capacity

The net ultimate bearing capacity is defined as the ultimate pressure per unit area of the foundation that can be supported by the soil in excess of the pressure caused by

Table 12.3 Shape, depth, and inclination factors recommended for use

Factor	Relationship	Source
Shape*	$F_{cs} = 1 + \dfrac{B}{L}\dfrac{N_q}{N_c}$	De Beer (1970)
	$F_{qs} = 1 + \dfrac{B}{L}\tan\phi'$	
	$F_{\gamma s} = 1 - 0.4\dfrac{B}{L}$	
	where L = length of the foundation ($L > B$)	

(*continued*)

Table 12.3 (continued)

Factor	Relationship	Source
Depth[†]	Condition (a): $D_f/B \leq 1$	Hansen (1970)
	$F_{cd} = 1 + 0.4 \dfrac{D_f}{B}$	
	$F_{qd} = 1 + 2 \tan \phi'(1 - \sin \phi')^2 \dfrac{D_f}{B}$	
	$F_{\gamma d} = 1$	
	Condition (b): $D_f/B > 1$	
	$F_{cd} = 1 + (0.4) \tan^{-1}\left(\dfrac{D_f}{B}\right)$	
	$F_{qd} = 1 + 2 \tan \phi'(1 - \sin \phi')^2 \tan^{-1}\left(\dfrac{D_f}{B}\right)$	
	$F_{\gamma d} = 1$	
Inclination	$F_{ci} = F_{qi} = \left(1 - \dfrac{\beta°}{90°}\right)^2$	Meyerhof (1963); Hanna and Meyerhof (1981)
	$F_{\gamma i} = \left(1 - \dfrac{\beta}{\phi'}\right)^2$	
	where β = inclination of the load on the foundation with respect to the vertical	

*These shape factors are empirical relations based on extensive laboratory tests.
[†]The factor $\tan^{-1}(D_f/B)$ is in radians.

the surrounding soil at the foundation level. If the difference between the unit weight of concrete used in the foundation and the unit weight of soil surrounding the foundation is assumed to be negligible, then

$$q_{\text{net}(u)} = q_u - q \qquad (12.10)$$

where $q_{\text{net}(u)}$ = net ultimate bearing capacity.

12.3 Modification of Bearing Capacity Equations for Water Table

Equations (12.3), (12.4), (12.5) and (12.9) were developed for determining the ultimate bearing capacity based on the assumption that the water table is located well below the foundation. However, if the water table is close to the foundation, some

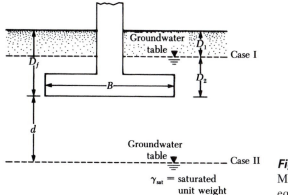

Figure 12.4
Modification of bearing capacity equations for water table

modifications of the bearing capacity equation are necessary, depending on the location of the water table (see Figure 12.4).

Case I: If the water table is located so that $0 \leq D_1 \leq D_f$, the factor q in the bearing capacity equations takes the form

$$q = \text{effective surcharge} = D_1\gamma + D_2(\gamma_{sat} - \gamma_w) \qquad (12.11)$$

where
$$\gamma_{sat} = \text{saturated unit weight of soil}$$
$$\gamma_w = \text{unit weight of water}$$

Also, the value of γ in the last term of the equations has to be replaced by $\gamma' = \gamma_{sat} - \gamma_w$.

Case II: For a water table located so that $0 \leq d \leq B$,

$$q = \gamma D_f \qquad (12.12)$$

The factor γ in the last term of the bearing capacity equations must be replaced by the factor

$$\bar{\gamma} = \gamma' + \frac{d}{B}(\gamma - \gamma') \qquad (12.13)$$

The preceding modifications are based on the assumption that there is no seepage force in the soil.

Case III: When the water table is located so that $d \geq B$, the water will have no effect on the ultimate bearing capacity.

12.4   *The Factor of Safety*

Calculating the gross allowable load-bearing capacity of shallow foundations requires the application of a factor of safety (*FS*) to the gross ultimate bearing capacity, or

$$q_{all} = \frac{q_u}{FS} \qquad (12.14)$$

However, some practicing engineers prefer to use a factor of safety of

$$\text{net stress increase on soil} = \frac{\text{net ultimate bearing capacity}}{FS} \qquad (12.15)$$

The net ultimate bearing capacity was defined in Eq. (12.10) as

$$q_{\text{net}(u)} = q_u - q$$

Substituting this equation into Eq. (12.15) yields
net stress increase on soil

= load from the superstructure per unit area of the foundation

$$= q_{\text{all(net)}} = \frac{q_u - q}{FS} \qquad (12.16)$$

The factor of safety defined by Eq. (12.16) may be at least 3 in all cases.

Example 12.1

A square foundation is 1.5 m × 1.5 m in plan. The soil supporting the foundation has a friction angle $\phi' = 20°$, and $c' = 15.2 \text{ kN/m}^2$. The unit weight of soil, γ, is 17.8 kN/m^3. Determine the allowable gross load on the foundation with a factor of safety (FS) of 4. Assume that the depth of the foundation (D_f) is 1 meter and use Eq. (12.4) and Table 12.1.

Solution
From Eq. (12.4),

$$q_u = 1.3c' N_c + qN_q + 0.4\gamma BN_\gamma$$

From Table 12.1, for $\phi' = 20°$,

$$N_c = 17.69$$
$$N_q = 7.44$$
$$N_\gamma = 3.64$$

Thus,

$$q_u = (1.3)(15.2)(17.69) + (1 \times 17.8)(7.44) + (0.4)(17.8)(1.5)(3.64)$$
$$= 349.55 + 132.43 + 38.87 = 520.85 \approx 521 \text{ kN/m}^2$$

So the allowable load per unit area of the foundation is

$$q_{\text{all}} = \frac{q_u}{FS} = \frac{521}{4} = 130.25 \text{ kN/m}^2 \approx \textbf{130 kN/m}^2$$

Example 12.2

A square footing is shown in Figure 12.5. Determine the safe gross load (factor of safety of 3) that the footing can carry. Use Eq. (12.9).

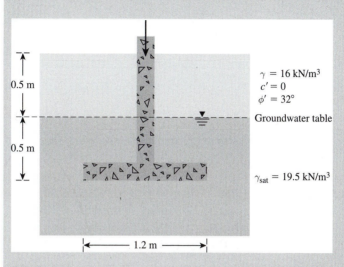

Figure 12.5

Solution

From Eq. (12.9) with $c' = 0$, $F_{ci} = F_{qi} = F_{\gamma i} = 1$ (vertical loading),

$$q_u = qN_qF_{qs}F_{qd} + \tfrac{1}{2}\gamma BN_\gamma F_{\gamma s}F_{\gamma d}$$

For $\phi' = 32°$, Table 12.2 given $N_q = 23.18$ and $N_\gamma = 30.22$.

$$F_{qs} = 1 + \left(\frac{B}{L}\right)\tan\phi' = 1 + \left(\frac{1.2}{1.2}\right)\tan 32 = 1.625$$

$$F_{qd} = 1 + 2\tan\phi'(1 - \sin\phi')^2\frac{D_f}{B} = 1 + 2\tan 32(1 - \sin 32)^2\left(\frac{1}{1.2}\right) = 1.23$$

$$F_{\gamma s} = 1 - 0.4\left(\frac{B}{L}\right) = 1 - 0.4\left(\frac{1.2}{1.2}\right) = 0.6$$

$$F_{\gamma d} = 1$$

$$q = (0.5)(16) + (0.5)(19.5 - 9.81) = 12.845 \text{ kN/m}^2$$

Thus,

$$q_u = (12.845)(23.18)(1.625)(1.23) + \tfrac{1}{2}(19.5 - 9.81)(1.2)(30.22)(0.6)(1)$$

$$= 700.54 \text{ kN/m}^2$$

$$q_{all} = \frac{q_u}{3} = \frac{700.54}{3} = 233.51 \text{ kN/m}^2$$

$$Q = q_{all} B^2 = (233.51)(1.2 \times 1.2) \approx \mathbf{336 \text{ kN}} \qquad \blacksquare$$

Example 12.3

A square column foundation to be constructed on a sandy soil has to carry a gross allowable total load of 150 kN. The depth of the foundation will be 0.7 m. The load will be inclined at an angle of 20° to the vertical (Figure 12.6). The standard penetration resistances, N_{60}, obtained from field exploration are listed in the table.

Depth (m)	N_{60}
1.5	3
3.0	6
4.5	9
6	10
7.5	10
9	8

Assume that the unit weight of the soil is 18 kN/m³. Determine the width of the foundation, B. Use a factor of safety of 3, and Eq. (12.9).

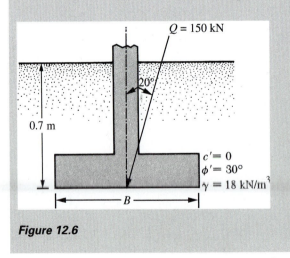

Figure 12.6

Solution

The standard penetration resistances can be corrected by using Eq. (10.7) and the Liao and Whitman equation given in Table 10.5. See the following table.

Depth (m)	Effective overburden pressure, σ'_o (kN/m²)	C_N	N_{60}	$(N_1)_{60} = N_{60}C_N$
1.5	27	1.88	3	≈6
3.0	54	1.33	6	≈8
4.5	81	1.09	9	≈10
6	108	0.94	10	≈9
7.5	135	0.84	10	≈8
9	162	0.77	8	≈6

The average corrected $(N_1)_{60}$ value obtained is about 8. Now, referring to Eq. (10.10), we can conservatively assume the soil friction angle ϕ' to be about 30°. With $c' = 0$, the ultimate bearing capacity [Eq. (12.9)] becomes

$$q_u = qN_qF_{qs}F_{qd}F_{qi} + \frac{1}{2}\gamma BN_\gamma F_{\gamma s}F_{\gamma d}F_{\gamma i}$$

$$q = (0.7)(18) = 12.6 \text{ kN/m}^2$$

$$\gamma = 18 \text{ kN/m}^3$$

From Table 12.2, for $\phi' = 30°$, we find

$$N_q = 18.4$$

$$N_\gamma = 22.4$$

From Table 12.3,

$$F_{qs} = 1 + \left(\frac{B}{L}\right)\tan\phi' = 1 + 0.577 = 1.577$$

$$F_{\gamma s} = 1 - 0.4\left(\frac{B}{L}\right) = 0.6$$

$$F_{qd} = 1 + 2\tan\phi'(1 - \sin\phi')^2\frac{D_f}{B} = 1 + \frac{(0.289)(0.7)}{B} = 1 + \frac{0.202}{B}$$

$$F_{\gamma d} = 1$$

$$F_{qi} = \left(1 - \frac{\beta°}{90°}\right)^2 = \left(1 - \frac{20}{90}\right)^2 = 0.605$$

$$F_{\gamma i} = \left(1 - \frac{\beta°}{\phi'}\right)^2 = \left(1 - \frac{20}{30}\right)^2 = 0.11$$

Hence,

$$q_u = (12.6)(18.4)(1.577)\left(1 + \frac{0.202}{B}\right)(0.605) + (0.5)(18)(B)(22.4)(0.6)(1)(0.11)$$

$$= 221.2 + \frac{44.68}{B} + 13.3B \tag{a}$$

Thus,

$$q_{all} = \frac{q_u}{3} = 73.73 + \frac{14.89}{B} + 4.43B \tag{b}$$

For Q = total allowable load = $q_{all} \times B^2$ or

$$q_{all} = \frac{150}{B^2} \tag{c}$$

Equating the right-hand sides of Eqs. (b) and (c) gives

$$\frac{150}{B^2} = 73.73 + \frac{14.89}{B} + 4.43B$$

By trial and error, we find $B \approx$ **1.3 m**. ∎

12.5 Eccentrically Loaded Foundations

As with the base of a retaining wall, there are several instances in which foundations are subjected to moments in addition to the vertical load, as shown in Figure 12.7a. In such cases, the distribution of pressure by the foundation on the soil is not uniform. The distribution of nominal pressure is

$$q_{max} = \frac{Q}{BL} + \frac{6M}{B^2L} \tag{12.17}$$

and

$$q_{min} = \frac{Q}{BL} - \frac{6M}{B^2L} \tag{12.18}$$

where
Q = total vertical load
M = moment on the foundation

The exact distribution of pressure is difficult to estimate.

The factor of safety for such types of loading against bearing capacity failure can be evaluated using the procedure suggested by Meyerhof (1953), which is generally referred to as the *effective area* method. The following is Meyerhof's

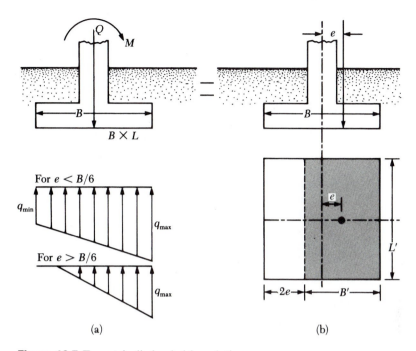

Figure 12.7 Eccentrically loaded foundations

step-by-step procedure for determining the ultimate load that the soil can support and the factor of safety against bearing capacity failure.

1. Figure 12.7b shows a force system equivalent to that shown in Figure 12.6a. The distance e is the eccentricity, or

$$e = \frac{M}{Q} \tag{12.19}$$

Substituting Eq. (12.19) in Eqs. (12.17) and (12.18) gives

$$q_{max} = \frac{Q}{BL}\left(1 + \frac{6e}{B}\right) \tag{12.20}$$

and

$$q_{min} = \frac{Q}{BL}\left(1 - \frac{6e}{B}\right) \tag{12.21}$$

Note that, in these equations, when the eccentricity e becomes $B/6$, q_{min} is 0. For $e > B/6$, q_{min} will be negative, which means that tension will develop. Because soil cannot take any tension, there will be a separation between the foundation and the soil underlying it. The nature of the pressure distribution on the soil will be as shown in Figure 12.7a. The value of q_{max} then is

$$q_{max} = \frac{4Q}{3L(B - 2e)} \tag{12.22}$$

2. Determine the effective dimensions of the foundation as

$$B' = \text{effective width} = B - 2e$$

$$L' = \text{effective length} = L$$

Note that, if the eccentricity were in the direction of the length of the foundation, then the value of L' would be equal to $L - 2e$. The value of B' would equal B. The smaller of the two dimensions (that is, L' and B') is the effective width of the foundation.

3. Use Eq. (12.9) for the ultimate bearing capacity as

$$q'_u = c'N_cF_{cs}F_{cd}F_{ci} + qN_qF_{qs}F_{qd}F_{qi} + \tfrac{1}{2}\gamma B'N_\gamma F_{\gamma s}F_{\gamma d}F_{\gamma i} \tag{12.23}$$

To evaluate F_{cs}, F_{qs}, and $F_{\gamma s}$, use Table 12.2 with *effective length* and *effective width* dimensions instead of L and B, respectively. To determine F_{cd}, F_{qd}, and $F_{\gamma d}$, use Table 12.2 (*do not* replace B with B').

4. The total ultimate load that the foundation can sustain is

$$Q_{ult} = q'_u \overbrace{(B')(L')}^{A'} \tag{12.24}$$

where $A' = $ effective area.

5. The factor of safety against bearing capacity failure is

$$FS = \frac{Q_{ult}}{Q} \tag{12.25}$$

Foundations with Two-Way Eccentricity

Consider a situation in which a foundation is subjected to a vertical ultimate load Q_{ult} and a moment M, as shown in Figures 12.8a and b. For this case, the components of the moment, M, about the x and y axes can be determined as M_x and M_y, respectively (Figure 12.8c). This condition is equivalent to a load Q_{ult} placed eccentrically on the foundation with $x = e_B$ and $y = e_L$ (Figure 12.8d). Note that

$$e_B = \frac{M_y}{Q_{ult}} \tag{12.26}$$

and

$$e_L = \frac{M_x}{Q_{ult}} \tag{12.27}$$

If Q_{ult} is needed, it can be obtained as follows [Eq. (12.24)]:

$$Q_{ult} = q'_u A'$$

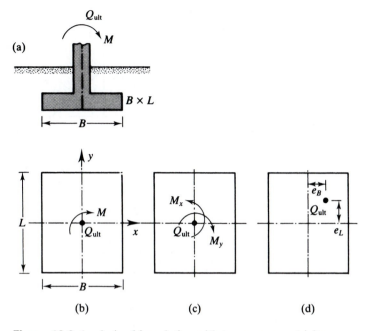

Figure 12.8 Analysis of foundation with two-way eccentricity

where, from Eq. (12.23),

$$q'_u = c'N_cF_{cs}F_{cd}F_{ci} + qN_qF_{qs}F_{qd}F_{qi} + \frac{1}{2}\gamma B'N_\gamma F_{\gamma s}F_{\gamma d}F_{\gamma i}$$

and

$$A' = \text{effective area} = B'L'$$

As before, to evaluate F_{cs}, F_{qs}, and $F_{\gamma s}$ (Table 12.3), we use the effective length (L') and effective width (B') dimensions instead of L and B, respectively. To calculate F_{cd}, F_{qd}, and $F_{\gamma d}$, we use Table 12.3; however, we do not replace B with B'. When we determine the effective area (A'), effective width (B'), and effective length (L'), four possible cases may arise (Higher and Anders, 1985). The effective area is such that its centroid coincides with the load.

Case I: $e_L/L \geq \frac{1}{6}$ and $e_B/B \geq \frac{1}{6}$. The effective area for this condition is shown in Figure 12.9a, or

$$A' = \frac{1}{2}B_1L_1 \tag{12.28}$$

where

$$B_1 = B\left(1.5 - \frac{3e_B}{B}\right) \tag{12.29}$$

$$L_1 = L\left(1.5 - \frac{3e_L}{L}\right) \tag{12.30}$$

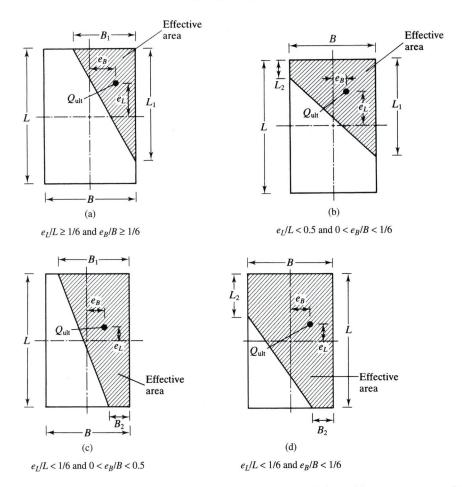

Figure 12.9 Definition of effective area for load on foundation with two-way eccentricity

The effective length, L', is the larger of the two dimensions—that is, B_1 or L_1. So, the effective width is

$$B' = \frac{A'}{L'} \tag{12.31}$$

Case II: $e_L/L < 0.5$ and $0 < e_B/B < \frac{1}{6}$. The effective area for this case is shown in Figure 12.9b.

$$A' = \frac{1}{2}(L_1 + L_2)B \tag{12.32}$$

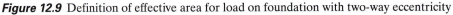

The magnitudes of L_1 and L_2 can be determined from Figure 12.10. The effective width is

$$B' = \frac{A'}{L_1 \text{ or } L_2 \text{ (whichever is larger)}} \tag{12.33}$$

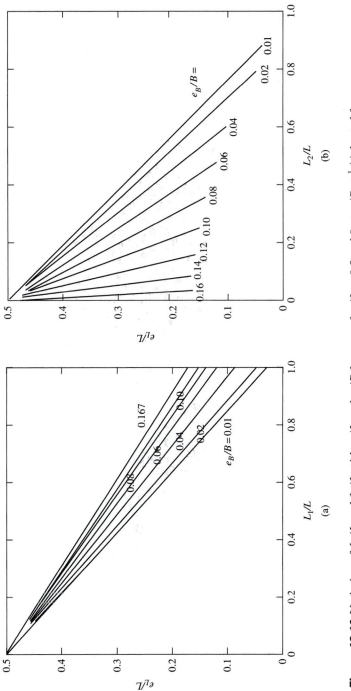

Figure 12.10 Variation of L_1/L and L_2/L with e_L/L and e_B/B for case of $e_L/L < 0.5$ and $0 < e_B/B < \frac{1}{6}$ (Adapted from Highter and Anders, 1985)

The effective length is

$$L' = L_1 \text{ or } L_2 \text{ (whichever is larger)} \tag{12.34}$$

Case III: $e_L/L < \frac{1}{6}$ and $0 < e_B/B < 0.5$. The effective area is shown in Figure 12.9c.

$$A' = \frac{1}{2}(B_1 + B_2)L \tag{12.35}$$

The effective width is

$$B' = \frac{A'}{L} \tag{12.36}$$

The effective length is

$$L' = L \tag{12.37}$$

The magnitudes of B_1 and B_2 can be determined from Figure 12.11.

Case IV: $e_L/L < \frac{1}{6}$ and $e_B/B < \frac{1}{6}$. Figure 12.9d shows the effective area for this case. The ratios of B_2/B and L_2/L (and hence B_2 and L_2) can be obtained from Figure 12.12. The effective area is then

$$A' = L_2B + \frac{1}{2}(B + B_2)(L - L_2) \tag{12.38}$$

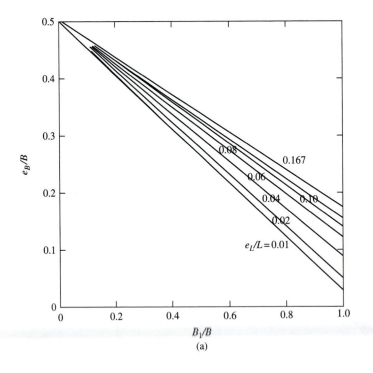

(a)

Figure 12.11 Variation of B_1/B and B_2/B with e_L/L and e_B/B for case of $e_L/L < \frac{1}{6}$ and $0 < e_B/B < 0.5$ (Adapted from Highter and Anders, 1985)

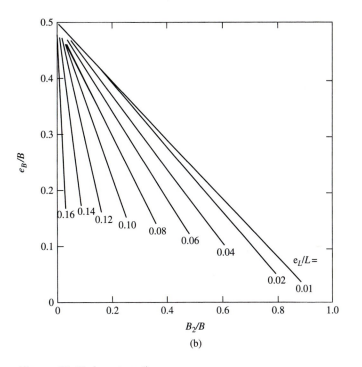

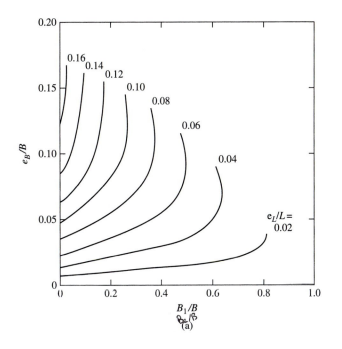

Figure 12.11 (*continued*)

Figure 12.12 Variation of B_2/B and L_2/L with e_B/B and e_L/L for the case of $e_L/L < \frac{1}{6}$ and $e_B/B < \frac{1}{6}$ (Adapted from Highter and Anders, 1985)

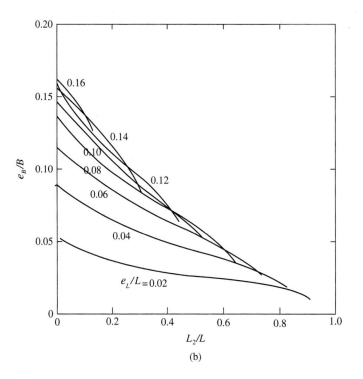

Figure 12.12 (*continued*)

The effective width is

$$B' = \frac{A'}{L} \tag{12.39}$$

The effective length is

$$L' = L \tag{12.40}$$

Example 12.4

A continuous foundation is shown in Figure 12.13. Assume that the load eccentricity $e = 0.15$ m. Determine the ultimate load, Q_{ult}.

Solution
With $c' = 0$, Eq. 12.23 becomes

$$q'_u = qN_qF_{qs}F_{qd}F_{qi} + \frac{1}{2}\gamma B' N_\gamma F_{\gamma s}F_{\gamma d}F_{\gamma i}$$

$$q = (1.2)(17.3) = 20.76 \text{ kN/m}^2$$

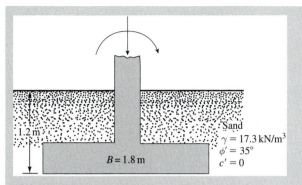

Figure 12.13

For $\phi' = 35°$, from Table 12.2, we find $N_q = 33.3$ and $N_\gamma = 48.03$. We have

$$B' = 1.8 - (2)(0.15) = 1.5 \text{ m}$$

Because it is a continuous foundation, B'/L' is 0. Hence, $F_{qs} = 1$ and $F_{\gamma s} = 1$, and

$$F_{qi} = F_{\gamma i} = 1$$

From Table 12.3, we have

$$F_{qd} = 1 + 2 \tan \phi'(1 - \sin \phi')^2 \frac{D_f}{B} = 1 + 0.255\left(\frac{1.2}{1.8}\right) = 1.17$$

$$F_{\gamma d} = 1$$

$$q'_u = (20.76)(33.3)(1)(1.17)(1) + \left(\frac{1}{2}\right)(17.3)(1.5)(48.03)(1)(1)(1) = 1432 \text{ kN/m}^2$$

Hence,

$$Q_{\text{ult}} = (B')(1)(q'_u) = (1.5)(1)(1432) = \textbf{2148 kN/m} \qquad \blacksquare$$

Example 12.5

A square foundation is shown in Figure 12.14, with $e_L = 0.3$ m and $e_B = 0.15$ m. Assume two-way eccentricity and determine the ultimate load, Q_{ult}.

Solution

$$\frac{e_L}{L} = \frac{0.3}{1.5} - 0.2$$

$$\frac{e_B}{B} = \frac{0.15}{1.5} = 0.1$$

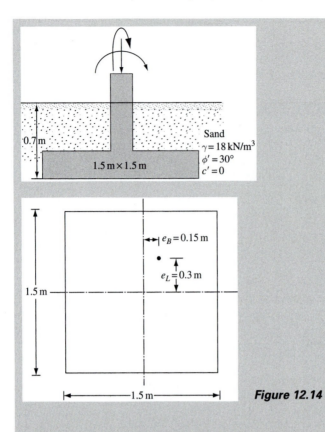

Sand
$\gamma = 18 \text{ kN/m}^3$
$\phi' = 30°$
$c' = 0$

0.7 m

1.5 m × 1.5 m

$e_B = 0.15$ m

$e_L = 0.3$ m

1.5 m

1.5 m

Figure 12.14

This case is similar to that shown in Figure 12.9b. From Figure 12.10, for $e_L/L = 0.2$ and $e_B/B = 0.1$, we have

$$\frac{L_1}{L} \approx 0.85; \quad L_1 = (0.85)(1.5) = 1.275 \text{ m}$$

and

$$\frac{L_2}{L} \approx 0.21; \quad L_2 = (0.21)(1.5) = 0.315 \text{ m}$$

From Eq. (12.32),

$$A' = \frac{1}{2}(L_1 + L_2)B = \frac{1}{2}(1.275 + 0.315)(1.5) = 1.193 \text{ m}^2$$

From Eq. (12.34),

$$L' = L_1 = 1.275 \text{ m}$$

From Eq. (12.33),

$$B' = \frac{A'}{L_1} = \frac{1.193}{1.275} = 0.936 \text{ m}$$

Note, from Eq. (12.23), for $c' = 0$, we have

$$q'_u = qN_qF_{qs}F_{qd}F_{qi} + \frac{1}{2}\gamma B' N_\gamma F_{\gamma s}F_{\gamma d}F_{\gamma i}$$

$$q = (0.7)(18) = 12.6 \text{ kN/m}^2$$

For $\phi' = 30°$, from Table 12.2, $N_q = 18.4$ and $N_\gamma = 22.4$. Thus,

$$F_{qs} = 1 + \left(\frac{B'}{L'}\right)\tan \phi' = 1 + \left(\frac{0.936}{1.275}\right)\tan 30° = 1.424$$

$$F_{\gamma s} = 1 - 0.4\left(\frac{B'}{L'}\right) = 1 - 0.4\left(\frac{0.936}{1.275}\right) = 0.706$$

$$F_{qd} = 1 + 2\tan \phi'(1 - \sin \phi')^2\frac{D_f}{B} = 1 + \frac{(0.289)(0.7)}{1.5} = 1.135$$

$$F_{\gamma d} = 1$$

So

$$Q_{\text{ult}} = A'q'_u = A'\left(qN_qF_{qs}F_{qd} + \frac{1}{2}\gamma B' N_\gamma F_{\gamma s}F_{\gamma d}\right)$$

$$= (1.193)[(12.6)(18.4)(1.424)(1.135) + (0.5)(18)(0.936)(22.4)(0.706)(1)]$$

$$= 605.95 \text{ kN} \qquad\blacksquare$$

SETTLEMENT OF SHALLOW FOUNDATIONS

12.6 *Types of Foundation Settlement*

As was discussed in Chapter 7, foundation settlement is made up of *elastic* (or immediate) settlement, S_e, and consolidation settlement, S_c. The procedure for calculating the consolidation settlement of foundations was also explained in Chapter 7. The methods for estimating elastic settlement will be elaborated upon in the following sections.

It is important to point out that, theoretically at least, a foundation could be considered fully flexible or fully rigid. A uniformly loaded, perfectly flexible foundation resting on an elastic material such as saturated clay will have a sagging profile, as shown in Figure 12.15a, because of elastic settlement. However, if the foundation is rigid and is resting on an elastic material such as clay, it will undergo uniform settlement and the contact pressure will be redistributed (Figure 12.15b).

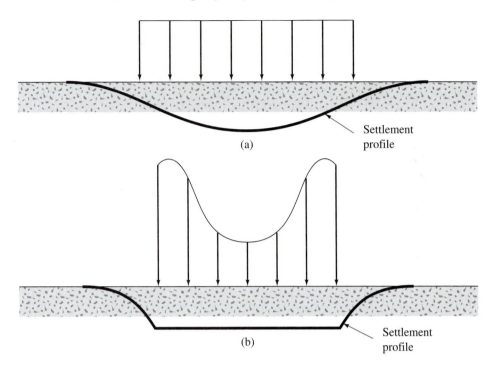

Figure 12.15 Elastic settlement profile and contact pressure in clay: (a) flexible foundation; (b) rigid foundation

12.7   *Elastic Settlement*

Figure 12.16 shows a shallow foundation subjected to a net force per unit area equal to q_o. Let the Poisson's ratio and the modulus of elasticity of the soil supporting it be μ_s and E_s, respectively. Theoretically, if the foundation is perfectly flexible, the settlement may be expressed as

$$S_e = q_o(\alpha B') \frac{1 - \mu_s^2}{E_s} I_s I_f \tag{12.41}$$

where

q_o = net applied pressure on the foundation
μ_s = Poisson's ratio of soil
E_s = average modulus of elasticity of the soil under the foundation measured
     from $z = 0$ to about $z = 4B$
$B' = B/2$ for center of foundation
   = B for corner of foundation
I_s = shape factor (Steinbrenner, 1934)

$$= F_1 + \frac{1 - 2\mu_s}{1 - \mu_s} F_2 \tag{12.42}$$

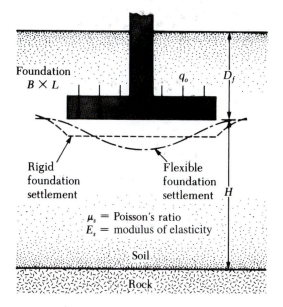

Figure 12.16
Elastic settlement of flexible and rigid foundations

$$F_1 = \frac{1}{\pi}(A_0 + A_1) \tag{12.43}$$

$$F_2 = \frac{n'}{2\pi} \tan^{-1} A_2 \tag{12.44}$$

$$A_0 = m' \ln \frac{(1 + \sqrt{m'^2 + 1})\sqrt{m'^2 + n'^2}}{m'(1 + \sqrt{m'^2 + n'^2 + 1})} \tag{12.45}$$

$$A_1 = \ln \frac{(m' + \sqrt{m'^2 + 1})\sqrt{1 + n'^2}}{m' + \sqrt{m'^2 + n'^2 + 1}} \tag{12.46}$$

$$A_2 = \frac{m'}{n'\sqrt{m'^2 + n'^2 + 1}} \tag{12.47}$$

$$I_f = \text{depth factor (Fox, 1948)} = f\left(\frac{D_f}{B}, \mu_s, \text{ and } \frac{L}{B}\right) \tag{12.48}$$

α = factor that depends on the location on the foundation where settlement is being calculated

- For calculation of settlement at the *center* of the foundation:

$$\alpha = 4$$

$$m' = \frac{L}{B}$$

$$n' = \frac{H}{\left(\dfrac{B}{2}\right)}$$

- For calculation of settlement at a *corner* of the foundation:

$$\alpha = 1$$

$$m' = \frac{L}{B}$$

$$n' = \frac{H}{B}$$

The variations of F_1 and F_2 [Eqs. (12.43) and (12.44)] with m' and n' are given in Tables 12.4 and 12.5. Also the variation of I_f with D_f/B and μ_s is given in Figure 12.17. Note that *when $D_f = 0$, the value of $I_f = 1$ in all cases.*

Table 12.4 Variation of F_1 with m' and n'

	m'									
n'	1.0	1.2	1.4	1.6	1.8	2.0	2.5	3.0	3.5	4.0
0.25	0.014	0.013	0.012	0.011	0.011	0.011	0.010	0.010	0.010	0.010
0.50	0.049	0.046	0.044	0.042	0.041	0.040	0.038	0.038	0.037	0.037
0.75	0.095	0.090	0.087	0.084	0.082	0.080	0.077	0.076	0.074	0.074
1.00	0.142	0.138	0.134	0.130	0.127	0.125	0.121	0.118	0.116	0.115
1.25	0.186	0.183	0.179	0.176	0.173	0.170	0.165	0.161	0.158	0.157
1.50	0.224	0.224	0.222	0.219	0.216	0.213	0.207	0.203	0.199	0.197
1.75	0.257	0.259	0.259	0.258	0.255	0.253	0.247	0.242	0.238	0.235
2.00	0.285	0.290	0.292	0.292	0.291	0.289	0.284	0.279	0.275	0.271
2.25	0.309	0.317	0.321	0.323	0.323	0.322	0.317	0.313	0.308	0.305
2.50	0.330	0.341	0.347	0.350	0.351	0.351	0.348	0.344	0.340	0.336
2.75	0.348	0.361	0.369	0.374	0.377	0.378	0.377	0.373	0.369	0.365
3.00	0.363	0.379	0.389	0.396	0.400	0.402	0.402	0.400	0.396	0.392
3.25	0.376	0.394	0.406	0.415	0.420	0.423	0.426	0.424	0.421	0.418
3.50	0.388	0.408	0.422	0.431	0.438	0.442	0.447	0.447	0.444	0.441
3.75	0.399	0.420	0.436	0.447	0.454	0.460	0.467	0.458	0.466	0.464
4.00	0.408	0.431	0.448	0.460	0.469	0.476	0.484	0.487	0.486	0.484
4.25	0.417	0.440	0.458	0.472	0.481	0.484	0.495	0.514	0.515	0.515
4.50	0.424	0.450	0.469	0.484	0.495	0.503	0.516	0.521	0.522	0.522
4.75	0.431	0.458	0.478	0.494	0.506	0.515	0.530	0.536	0.539	0.539
5.00	0.437	0.465	0.487	0.503	0.516	0.526	0.543	0.551	0.554	0.554
5.25	0.443	0.472	0.494	0.512	0.526	0.537	0.555	0.564	0.568	0.569
5.50	0.448	0.478	0.501	0.520	0.534	0.546	0.566	0.576	0.581	0.584
5.75	0.453	0.483	0.508	0.527	0.542	0.555	0.576	0.588	0.594	0.597
6.00	0.457	0.489	0.514	0.534	0.550	0.563	0.585	0.598	0.606	0.609
6.25	0.461	0.493	0.519	0.540	0.557	0.570	0.594	0.609	0.617	0.621
6.50	0.465	0.498	0.524	0.546	0.563	0.577	0.603	0.618	0.627	0.632
6.75	0.468	0.502	0.529	0.551	0.569	0.584	0.610	0.627	0.637	0.643
7.00	0.471	0.506	0.533	0.556	0.575	0.590	0.618	0.635	0.646	0.653
7.25	0.474	0.509	0.538	0.561	0.580	0.596	0.625	0.643	0.655	0.662
7.50	0.477	0.513	0.541	0.565	0.585	0.601	0.631	0.650	0.663	0.671
7.75	0.480	0.516	0.545	0.569	0.589	0.606	0.637	0.658	0.671	0.680
8.00	0.482	0.519	0.549	0.573	0.594	0.611	0.643	0.664	0.678	0.688
8.25	0.485	0.522	0.552	0.577	0.598	0.615	0.648	0.670	0.685	0.695

Table 12.4 (*continued*)

n′	1.0	1.2	1.4	1.6	1.8	2.0	2.5	3.0	3.5	4.0
8.50	0.487	0.524	0.555	0.580	0.601	0.619	0.653	0.676	0.692	0.703
8.75	0.489	0.527	0.558	0.583	0.605	0.623	0.658	0.682	0.698	0.710
9.00	0.491	0.529	0.560	0.587	0.609	0.627	0.663	0.687	0.705	0.716
9.25	0.493	0.531	0.563	0.589	0.612	0.631	0.667	0.693	0.710	0.723
9.50	0.495	0.533	0.565	0.592	0.615	0.634	0.671	0.697	0.716	0.719
9.75	0.496	0.536	0.568	0.595	0.618	0.638	0.675	0.702	0.721	0.735
10.00	0.498	0.537	0.570	0.597	0.621	0.641	0.679	0.707	0.726	0.740
20.00	0.529	0.575	0.614	0.647	0.677	0.702	0.756	0.797	0.830	0.858
50.00	0.548	0.598	0.640	0.678	0.711	0.740	0.803	0.853	0.895	0.931
100.00	0.555	0.605	0.649	0.688	0.722	0.753	0.819	0.872	0.918	0.956

n′	4.5	5.0	6.0	7.0	8.0	9.0	10.0	25.0	50.0	100.0
0.25	0.010	0.010	0.010	0.010	0.010	0.010	0.010	0.010	0.010	0.010
0.50	0.036	0.036	0.036	0.036	0.036	0.036	0.036	0.036	0.036	0.036
0.75	0.073	0.073	0.072	0.072	0.072	0.072	0.071	0.071	0.071	0.071
1.00	0.114	0.113	0.112	0.112	0.112	0.111	0.111	0.110	0.110	0.110
1.25	0.155	0.154	0.153	0.152	0.152	0.151	0.151	0.150	0.150	0.150
1.50	0.195	0.194	0.192	0.191	0.190	0.190	0.189	0.188	0.188	0.188
1.75	0.233	0.232	0.229	0.228	0.227	0.226	0.225	0.223	0.223	0.223
2.00	0.269	0.267	0.264	0.262	0.261	0.260	0.259	0.257	0.256	0.256
2.25	0.302	0.300	0.296	0.294	0.293	0.291	0.291	0.287	0.287	0.287
2.50	0.333	0.331	0.327	0.324	0.322	0.321	0.320	0.316	0.315	0.315
2.75	0.362	0.359	0.355	0.352	0.350	0.348	0.347	0.343	0.342	0.342
3.00	0.389	0.386	0.382	0.378	0.376	0.374	0.373	0.368	0.367	0.367
3.25	0.415	0.412	0.407	0.403	0.401	0.399	0.397	0.391	0.390	0.390
3.50	0.438	0.435	0.430	0.427	0.424	0.421	0.420	0.413	0.412	0.411
3.75	0.461	0.458	0.453	0.449	0.446	0.443	0.441	0.433	0.432	0.432
4.00	0.482	0.479	0.474	0.470	0.466	0.464	0.462	0.453	0.451	0.451
4.25	0.516	0.496	0.484	0.473	0.471	0.471	0.470	0.468	0.462	0.460
4.50	0.520	0.517	0.513	0.508	0.505	0.502	0.499	0.489	0.487	0.487
4.75	0.537	0.535	0.530	0.526	0.523	0.519	0.517	0.506	0.504	0.503
5.00	0.554	0.552	0.548	0.543	0.540	0.536	0.534	0.522	0.519	0.519
5.25	0.569	0.568	0.564	0.560	0.556	0.553	0.550	0.537	0.534	0.534
5.50	0.584	0.583	0.579	0.575	0.571	0.568	0.585	0.551	0.549	0.548
5.75	0.597	0.597	0.594	0.590	0.586	0.583	0.580	0.565	0.583	0.562
6.00	0.611	0.610	0.608	0.604	0.601	0.598	0.595	0.579	0.576	0.575
6.25	0.623	0.623	0.621	0.618	0.615	0.611	0.608	0.592	0.589	0.588
6.50	0.635	0.635	0.634	0.631	0.628	0.625	0.622	0.605	0.601	0.600
6.75	0.646	0.647	0.646	0.644	0.641	0.637	0.634	0.617	0.613	0.612
7.00	0.656	0.658	0.658	0.656	0.653	0.650	0.647	0.628	0.624	0.623
7.25	0.666	0.669	0.669	0.668	0.665	0.662	0.659	0.640	0.635	0.634
7.50	0.676	0.679	0.680	0.679	0.676	0.673	0.670	0.651	0.646	0.645
7.75	0.685	0.688	0.690	0.689	0.687	0.684	0.681	0.661	0.656	0.655
8.00	0.694	0.697	0.700	0.700	0.698	0.695	0.692	0.672	0.666	0.665

(*continued*)

Table 12.4 (*continued*)

					m′					
n′	4.5	5.0	6.0	7.0	8.0	9.0	10.0	25.0	50.0	100.0
8.25	0.702	0.706	0.710	0.710	0.708	0.705	0.703	0.682	0.676	0.675
8.50	0.710	0.714	0.719	0.719	0.718	0.715	0.713	0.692	0.686	0.684
8.75	0.717	0.722	0.727	0.728	0.727	0.725	0.723	0.701	0.695	0.693
9.00	0.725	0.730	0.736	0.737	0.736	0.735	0.732	0.710	0.704	0.702
9.25	0.731	0.737	0.744	0.746	0.745	0.744	0.742	0.719	0.713	0.711
9.50	0.738	0.744	0.752	0.754	0.754	0.753	0.751	0.728	0.721	0.719
9.75	0.744	0.751	0.759	0.762	0.762	0.761	0.759	0.737	0.729	0.727
10.00	0.750	0.758	0.766	0.770	0.770	0.770	0.768	0.745	0.738	0.735
20.00	0.878	0.896	0.925	0.945	0.959	0.969	0.977	0.982	0.965	0.957
50.00	0.962	0.989	1.034	1.070	1.100	1.125	1.146	1.265	1.279	1.261
100.00	0.990	1.020	1.072	1.114	1.150	1.182	1.209	1.408	1.489	1.499

Table 12.5 Variation of F_2 with $m′$ and $n′$

					m′					
n′	1.0	1.2	1.4	1.6	1.8	2.0	2.5	3.0	3.5	4.0
0.25	0.049	0.050	0.051	0.051	0.051	0.052	0.052	0.052	0.052	0.052
0.50	0.074	0.077	0.080	0.081	0.083	0.084	0.086	0.086	0.0878	0.087
0.75	0.083	0.089	0.093	0.097	0.099	0.101	0.104	0.106	0.107	0.108
1.00	0.083	0.091	0.098	0.102	0.106	0.109	0.114	0.117	0.119	0.120
1.25	0.080	0.089	0.096	0.102	0.107	0.111	0.118	0.122	0.125	0.127
1.50	0.075	0.084	0.093	0.099	0.105	0.110	0.118	0.124	0.128	0.130
1.75	0.069	0.079	0.088	0.095	0.101	0.107	0.117	0.123	0.128	0.131
2.00	0.064	0.074	0.083	0.090	0.097	0.102	0.114	0.121	0.127	0.131
2.25	0.059	0.069	0.077	0.085	0.092	0.098	0.110	0.119	0.125	0.130
2.50	0.055	0.064	0.073	0.080	0.087	0.093	0.106	0.115	0.122	0.127
2.75	0.051	0.060	0.068	0.076	0.082	0.089	0.102	0.111	0.119	0.125
3.00	0.048	0.056	0.064	0.071	0.078	0.084	0.097	0.108	0.116	0.122
3.25	0.045	0.053	0.060	0.067	0.074	0.080	0.093	0.104	0.112	0.119
3.50	0.042	0.050	0.057	0.064	0.070	0.076	0.089	0.100	0.109	0.116
3.75	0.040	0.047	0.054	0.060	0.067	0.073	0.086	0.096	0.105	0.113
4.00	0.037	0.044	0.051	0.057	0.063	0.069	0.082	0.093	0.102	0.110
4.25	0.036	0.042	0.049	0.055	0.061	0.066	0.079	0.090	0.099	0.107
4.50	0.034	0.040	0.046	0.052	0.058	0.063	0.076	0.086	0.096	0.104
4.75	0.032	0.038	0.044	0.050	0.055	0.061	0.073	0.083	0.093	0.101
5.00	0.031	0.036	0.042	0.048	0.053	0.058	0.070	0.080	0.090	0.098
5.25	0.029	0.035	0.040	0.046	0.051	0.056	0.067	0.078	0.087	0.095
5.50	0.028	0.033	0.039	0.044	0.049	0.054	0.065	0.075	0.084	0.092
5.75	0.027	0.032	0.037	0.042	0.047	0.052	0.063	0.073	0.082	0.090
6.00	0.026	0.031	0.036	0.040	0.045	0.050	0.060	0.070	0.079	0.087
6.25	0.025	0.030	0.034	0.039	0.044	0.048	0.058	0.068	0.077	0.085
6.50	0.024	0.029	0.033	0.038	0.042	0.046	0.056	0.066	0.075	0.083
6.75	0.023	0.028	0.032	0.036	0.041	0.045	0.055	0.064	0.073	0.080
7.00	0.022	0.027	0.031	0.035	0.039	0.043	0.053	0.062	0.071	0.078
7.25	0.022	0.026	0.030	0.034	0.038	0.042	0.051	0.060	0.069	0.076

Table 12.5 (*continued*)

n'	m' 1.0	1.2	1.4	1.6	1.8	2.0	2.5	3.0	3.5	4.0
7.50	0.021	0.025	0.029	0.033	0.037	0.041	0.050	0.059	0.067	0.074
7.75	0.020	0.024	0.028	0.032	0.036	0.039	0.048	0.057	0.065	0.072
8.00	0.020	0.023	0.027	0.031	0.035	0.038	0.047	0.055	0.063	0.071
8.25	0.019	0.023	0.026	0.030	0.034	0.037	0.046	0.054	0.062	0.069
8.50	0.018	0.022	0.026	0.029	0.033	0.036	0.045	0.053	0.060	0.067
8.75	0.018	0.021	0.025	0.028	0.032	0.035	0.043	0.051	0.059	0.066
9.00	0.017	0.021	0.024	0.028	0.031	0.034	0.042	0.050	0.057	0.064
9.25	0.017	0.020	0.024	0.027	0.030	0.033	0.041	0.049	0.056	0.063
9.50	0.017	0.020	0.023	0.026	0.029	0.033	0.040	0.048	0.055	0.061
9.75	0.016	0.019	0.023	0.026	0.029	0.032	0.039	0.047	0.054	0.060
10.00	0.016	0.019	0.022	0.025	0.028	0.031	0.038	0.046	0.052	0.059
20.00	0.008	0.010	0.011	0.013	0.014	0.016	0.020	0.024	0.027	0.031
50.00	0.003	0.004	0.004	0.005	0.006	0.006	0.008	0.010	0.011	0.013
100.00	0.002	0.002	0.002	0.003	0.003	0.003	0.004	0.005	0.006	0.006

n'	m' 4.5	5.0	6.0	7.0	8.0	9.0	10.0	25.0	50.0	100.0
0.25	0.053	0.053	0.053	0.053	0.053	0.053	0.053	0.053	0.053	0.053
0.50	0.087	0.087	0.088	0.088	0.088	0.088	0.088	0.088	0.088	0.088
0.75	0.109	0.109	0.109	0.110	0.110	0.110	0.110	0.111	0.111	0.111
1.00	0.121	0.122	0.123	0.123	0.124	0.124	0.124	0.125	0.125	0.125
1.25	0.128	0.130	0.131	0.132	0.132	0.133	0.133	0.134	0.134	0.134
1.50	0.132	0.134	0.136	0.137	0.138	0.138	0.139	0.140	0.140	0.140
1.75	0.134	0.136	0.138	0.140	0.141	0.142	0.142	0.144	0.144	0.145
2.00	0.134	0.136	0.139	0.141	0.143	0.144	0.145	0.147	0.147	0.148
2.25	0.133	0.136	0.140	0.142	0.144	0.145	0.146	0.149	0.150	0.150
2.50	0.132	0.135	0.139	0.142	0.144	0.146	0.147	0.151	0.151	0.151
2.75	0.130	0.133	0.138	0.142	0.144	0.146	0.147	0.152	0.152	0.153
3.00	0.127	0.131	0.137	0.141	0.144	0.145	0.147	0.152	0.153	0.154
3.25	0.125	0.129	0.135	0.140	0.143	0.145	0.147	0.153	0.154	0.154
3.50	0.122	0.126	0.133	0.138	0.142	0.144	0.146	0.153	0.155	0.155
3.75	0.119	0.124	0.131	0.137	0.141	0.143	0.145	0.154	0.155	0.155
4.00	0.116	0.121	0.129	0.135	0.139	0.142	0.145	0.154	0.155	0.156
4.25	0.113	0.119	0.127	0.133	0.138	0.141	0.144	0.154	0.156	0.156
4.50	0.110	0.116	0.125	0.131	0.136	0.140	0.143	0.154	0.156	0.156
4.75	0.107	0.113	0.123	0.130	0.135	0.139	0.142	0.154	0.156	0.157
5.00	0.105	0.111	0.120	0.128	0.133	0.137	0.140	0.154	0.156	0.157
5.25	0.102	0.108	0.118	0.126	0.131	0.136	0.139	0.154	0.156	0.157
5.50	0.099	0.106	0.116	0.124	0.130	0.134	0.138	0.154	0.156	0.157
5.75	0.097	0.103	0.113	0.122	0.128	0.133	0.136	0.154	0.157	0.157
6.00	0.094	0.101	0.111	0.120	0.126	0.131	0.135	0.153	0.157	0.157
6.25	0.092	0.098	0.109	0.118	0.124	0.129	0.134	0.153	0.157	0.158
6.50	0.090	0.096	0.107	0.116	0.122	0.128	0.132	0.153	0.157	0.158
6.75	0.087	0.094	0.105	0.114	0.121	0.126	0.131	0.153	0.157	0.158
7.00	0.085	0.092	0.103	0.112	0.119	0.125	0.129	0.152	0.157	0.158

(*continued*)

Table 12.5 (*continued*)

	m'									
n'	4.5	5.0	6.0	7.0	8.0	9.0	10.0	25.0	50.0	100.0
7.25	0.083	0.090	0.101	0.110	0.117	0.123	0.128	0.152	0.157	0.158
7.50	0.081	0.088	0.099	0.108	0.115	0.121	0.126	0.152	0.156	0.158
7.75	0.079	0.086	0.097	0.106	0.114	0.120	0.125	0.151	0.156	0.158
8.00	0.077	0.084	0.095	0.104	0.112	0.118	0.124	0.151	0.156	0.158
8.25	0.076	0.082	0.093	0.102	0.110	0.117	0.122	0.150	0.156	0.158
8.50	0.074	0.080	0.091	0.101	0.108	0.115	0.121	0.150	0.156	0.158
8.75	0.072	0.078	0.089	0.099	0.107	0.114	0.119	0.150	0.156	0.158
9.00	0.071	0.077	0.088	0.097	0.105	0.112	0.118	0.149	0.156	0.158
9.25	0.069	0.075	0.086	0.096	0.104	0.110	0.116	0.149	0.156	0.158
9.50	0.068	0.074	0.085	0.094	0.102	0.109	0.115	0.148	0.156	0.158
9.75	0.066	0.072	0.083	0.092	0.100	0.107	0.113	0.148	0.156	0.158
10.00	0.065	0.071	0.082	0.091	0.099	0.106	0.112	0.147	0.156	0.158
20.00	0.035	0.039	0.046	0.053	0.059	0.065	0.071	0.124	0.148	0.156
50.00	0.014	0.016	0.019	0.022	0.025	0.028	0.031	0.071	0.113	0.142
100.00	0.007	0.008	0.010	0.011	0.013	0.014	0.016	0.039	0.071	0.113

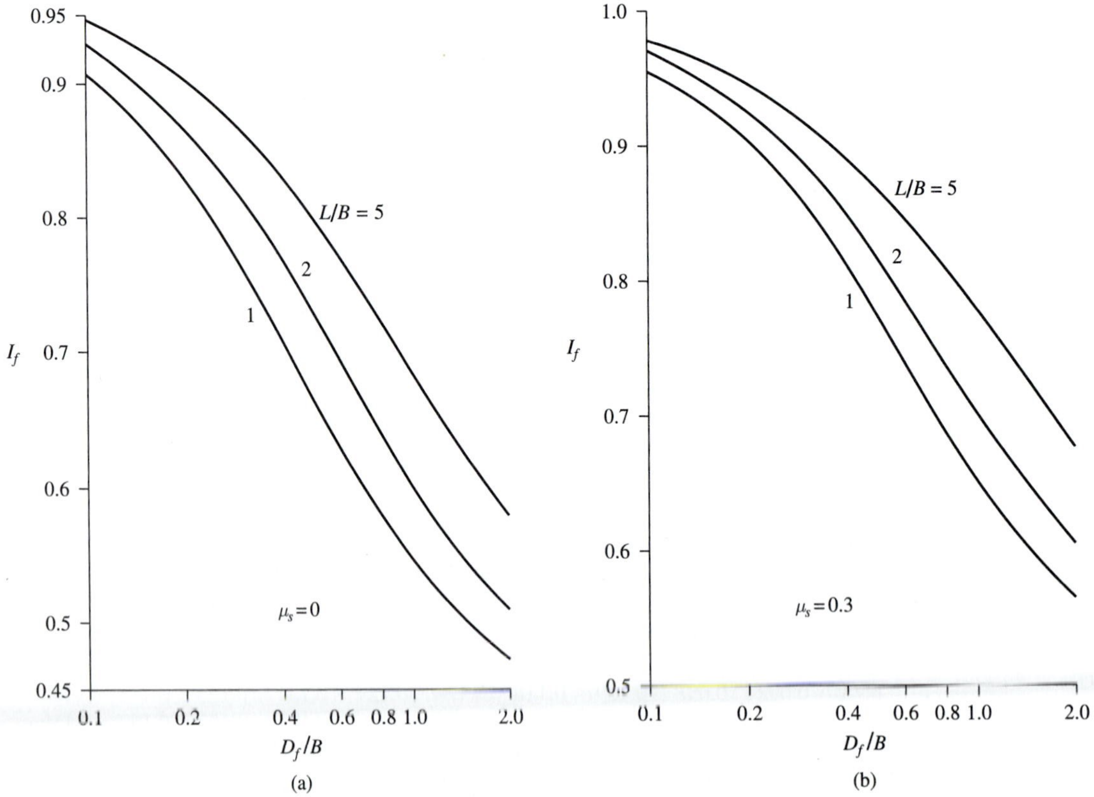

Figure 12.17 Variation of I_f with D_f/B, L/B, and μ_s

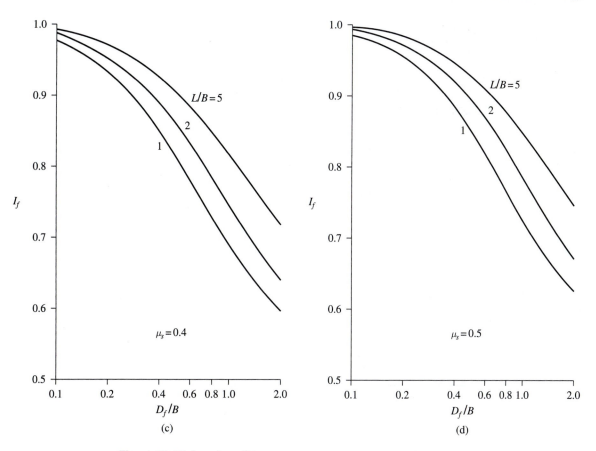

Figure 12.17 (*continued*)

The elastic settlement of a *rigid foundation* can be estimated as

$$S_{e(\text{rigid})} \approx 0.93 S_{e(\text{flexible, center})} \qquad (12.49)$$

Due to the nonhomogeneous nature of soil deposits, the magnitude of E_s may vary with depth. For that reason, Bowles (1987) recommended using a weighted average value of E_s in Eq. (12.41), or

$$E_s = \frac{\sum E_{s(i)} \Delta z}{\bar{z}} \qquad (12.50)$$

where

$\quad E_{s(i)} = $ soil modulus of elasticity within a depth Δz
$\quad \bar{z} = H$ or $5B$, whichever is smaller

Example 12.6

A rigid shallow foundation 1 m × 2 m is shown in Figure 12.18. Calculate the elastic settlement at the center of the foundation.

Solution

Given $B = 1$ m and $L = 2$ m. Note that $\bar{z} = 5$ m $= 5B$. From Eq. (12.50),

$$E_s = \frac{\Sigma E_{s(i)} \Delta z}{\bar{z}}$$

$$= \frac{(10{,}000)(2) + (8{,}000)(1) + (12{,}000)(2)}{5} = 10{,}400 \text{ kN/m}^2$$

For the *center of the foundation,*

$$\alpha = 4$$

$$m' = \frac{L}{B} = \frac{2}{1} = 2$$

$$n' = \frac{H}{\left(\dfrac{B}{2}\right)} = \frac{5}{\left(\dfrac{1}{2}\right)} = 10$$

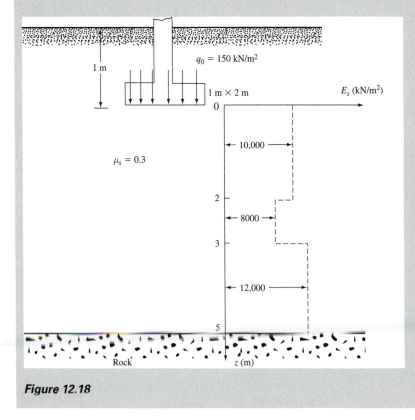

Figure 12.18

From Tables 12.4 and 12.5, $F_1 = 0.641$ and $F_2 = 0.031$. From Eq. (12.42),

$$I_s = F_1 + \frac{2 - \mu_s}{1 - \mu_s} F_2$$

$$= 0.641 + \frac{2 - 0.3}{1 - 0.3}(0.031) = 0.716$$

Again, $\dfrac{D_f}{B} = \dfrac{1}{1} = 1, \dfrac{L}{B} = 2, \mu_s = 0.3$. From Figure 12.17b, $I_f = 0.709$. Hence,

$$S_{e(\text{flexible})} = q_o(\alpha B') \frac{1 - \mu_s^2}{E_s} I_s I_f$$

$$= (150)\left(4 \times \frac{1}{2}\right)\left(\frac{1 - 0.3^2}{10{,}400}\right)(0.716)(0.709) = 0.0133 \text{ m} = 13.3 \text{ mm}$$

Since the foundation is rigid, from Eq. (12.49),

$$S_{e(\text{rigid})} = (0.93)(13.3) = \textbf{12.4 mm}$$ ∎

12.8 Range of Material Parameters for Computing Elastic Settlement

Section 12.7 presented the equation for calculating the elastic settlement of foundations. The equation contains the elastic parameters, such as E_s and μ_s. If the laboratory test results for these parameters are not available, certain realistic assumptions have to be made. Table 12.6 gives the approximate range of the elastic parameters for various soils.

Table 12.6 Elastic parameters of various soils

Type of soil	Modulus of elasticity, E_s (MN/m^2)	Poisson's ratio, μ_s
Loose sand	10–25	0.20–0.40
Medium dense sand	15–30	0.25–0.40
Dense sand	35–55	0.30–0.45
Silty sand	10–20	0.20–0.40
Sand and gravel	70–170	0.15–0.35
Soft clay	4–20	
Medium clay	20–40	0.20–0.50
Stiff clay	40–100	

12.9 Settlement of Sandy Soil: Use of Strain Influence Factor

The settlement of granular soils can also be evaluated by the use of a semiempirical *strain influence factor* proposed by Schmertmann et al. (1978). According to this method, the settlement is

$$S_e = C_1 C_2 (\bar{q} - q) \sum_0^{z_2} \frac{I_z}{E_s} \Delta z \qquad (12.51)$$

where

I_z = strain influence factor

C_1 = a correction factor for the depth of foundation embedment = $1 - 0.5$
$\quad [q/(\bar{q} - q)]$

C_2 = a correction factor to account for creep in soil
$\quad = 1 + 0.2 \log$ (time in years/0.1)

$\bar{q}$ = stress at the level of the foundation

$q = \gamma D_f$

The recommended variation of the strain influence factor I_z for square ($L/B = 1$) or circular foundations and for foundations with $L/B \geq 10$ is shown in Figure 12.19. The

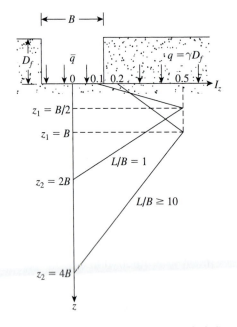

Figure 12.19 Variation of the strain influence factor, I_z

I_z diagrams for $1 < L/B < 10$ can be interpolated. The procedure to calculate elastic settlement using Eq. (12.51) is given here (Figure 12.20).

Step 1. Plot the foundation and the variation of I_z with depth to scale (Figure 12.20a).

Step 2. Using the correlation from standard penetration resistance (N_{60}) or cone penetration resitance (q_c), plot the actual variation of E_s with depth (Figure 12.20b). Schmertmann et al. (1978) suggested $E_s \approx 3.5q_c$.

Step 3. Approximate the actual variation of E_s into a number of layers of soil having a constant E_s, such as $E_{s(1)}$, $E_{s(2)}$, ... , $E_{s(i)}$, ... $E_{s(n)}$ (Figure 12.20b).

Step 4. Divide the soil layer from $z = 0$ to $z = z_2$ into a number of layers by drawing horizontal lines. The number of layers will depend on the break in continuity in the I_z and E_s diagrams.

Step 5. Prepare a table (such as Table 12.7) to obtain $\sum \dfrac{I_z}{E_s} \Delta z$.

Step 6. Calculate C_1 and C_2.

Step 7. Calculate S_e from Eq. (12.51).

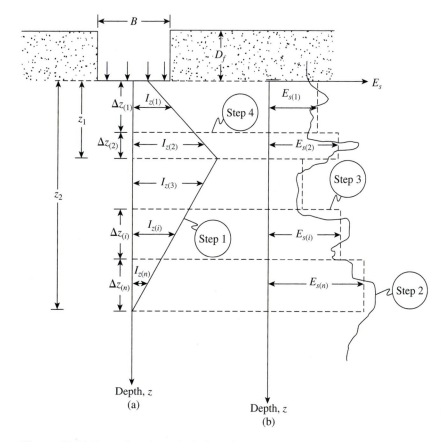

Figure 12.20 Procedure for calculation of S_e using the strain influence factor

Table 12.7 Calculation of $\Sigma \dfrac{I_z}{E_s} \Delta z$

Layer No.	Δz	E_s	I_z at the middle of the layer	$\dfrac{I_z}{E_s} \Delta z$
1	$\Delta z_{(1)}$	$E_{s(1)}$	$I_{z(1)}$	$\dfrac{I_{z(1)}}{E_{s(1)}} \Delta z_1$
2	$\Delta z_{(2)}$	$E_{s(2)}$	$I_{z(2)}$	
⋮	⋮	⋮	⋮	
i	$\Delta z_{(i)}$	$E_{s(i)}$	$I_{z(i)}$	$\dfrac{I_{z(i)}}{E_{s(i)}} \Delta z_i$
⋮	⋮	⋮	⋮	⋮
n	$\Delta z_{(n)}$	$E_{s(n)}$	$I_{z(n)}$	$\dfrac{I_{z(n)}}{E_{s(n)}} \Delta z_n$
				$\Sigma \dfrac{I_z}{E_s} \Delta z$

A Case History of the Calculation of S_e Using the Strain Influence Factor

Schmertmann (1970) provided a case history of a rectangular foundation (a Belgian bridge pier) having $L = 23$ m and $B = 2.6$ m and being supported by a granular soil deposit. For this foundation, we may assume that $L/B \approx 10$ for plotting the strain influence factor diagram. Figure 12.21 shows the details of the foundation, along with the approximate variation of the cone penetration resistance, q_c, with depth. For this foundation [see Eq. (12.51)], note that

$$\bar{q} = 178.54 \text{ kN/m}^2$$

$$q = 31.39 \text{ kN/m}^2$$

$$C_1 = 1 - 0.5 \frac{1}{\bar{q} - q} = 1 - (0.5)\left(\frac{31.39}{178.54 - 31.39} \right) = 0.893$$

and

$$C_2 = 1 + 0.2 \log\left(\frac{t \text{ yr}}{0.1} \right)$$

For $t = 5$ yr,

$$C_2 = 1 + 0.2 \log\left(\frac{5}{0.1} \right) = 1.34$$

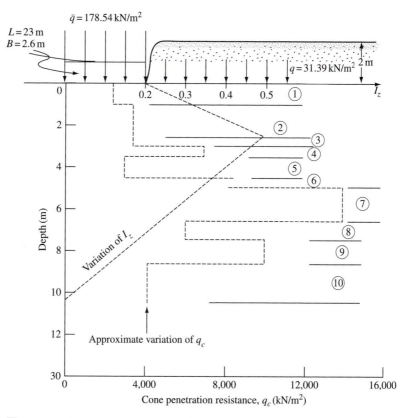

Figure 12.21 Variation of I_z and q_c below the foundation

The following table shows the calculation of $\sum_0^{z_2} (I_z/E_s) \, \Delta z$ in conjunction with Figure 12.21:

Layer	Δz (m)	q_c (kN/m²)	E_s^a (kN/m²)	z to the center of the layer (m)	I_z at the center of the layer	$(I_z/E_s) \, \Delta z$ (m²/kN)
1	1	2,450	8,575	0.5	0.258	3.00×10^{-5}
2	1.6	3,430	12,005	1.8	0.408	5.43×10^{-5}
3	0.4	3,430	12,005	2.8	0.487	1.62×10^{-5}
4	0.5	6,870	24,045	3.25	0.458	0.95×10^{-5}
5	1.0	2,950	10,325	4.0	0.410	3.97×10^{-5}
6	0.5	8,340	29,190	4.75	0.362	0.62×10^{-5}
7	1.5	14,000	49,000	5.75	0.298	0.91×10^{-5}
8	1	6,000	21,000	7.0	0.247	1.17×10^{-5}
9	1	10,000	35,000	8.0	0.154	0.44×10^{-5}
10	1.9	4,000	14,000	9.45	0.062	0.84×10^{-5}
	Σ 10.4 m = 4B					Σ 18.95 $\times 10^{-5}$

[a] $E_s \approx 3.5 q_c$

Hence, the elastic settlement is calculated as

$$S_e = C_1 C_2 (\bar{q} - q) \, \Sigma \frac{I_z}{E_s} \Delta z$$

$$= (0.893)(1.34)(178.54 - 31.39)(18.95 \times 10^{-5})$$

$$= 0.03336 \text{ m} \approx 33 \text{ mm}$$

After five years, the actual *maximum* settlement observed for the foundation was about 39 mm.

12.10 Allowable Bearing Pressure in Sand Based on Settlement Consideration

Meyerhof (1956) proposed a correlation for the *net allowable bearing pressure* for foundations with the standard penetration resistance, N_{60}. The net allowable pressure can be defined as

$$q_{all(net)} = q_{all} - \gamma D_f \tag{12.52}$$

Since Meyerhof proposed his original correlation, researchers have observed that its results are rather conservative. Later, Meyerhof (1965) suggested that the net allowable bearing pressure should be increased by about 50%. Bowles (1977) proposed that the modified form of the bearing pressure equations be expressed as

$$q_{net}(\text{kN/m}^2) = \frac{N_{60}}{0.05} F_d \left(\frac{S_e}{25} \right) \quad \text{(for } B \leq 1.22 \text{ m)} \tag{12.53}$$

and

$$q_{net}(\text{kN/m}^2) = \frac{N_{60}}{0.08} \left(\frac{B + 0.3}{B} \right)^2 F_d \left(\frac{S_e}{25} \right) \quad \text{(for } B > 1.22 \text{ m)} \tag{12.54}$$

where
F_d = depth factor = $1 + 0.33(D_f/B) \leq 1.33$ (12.55)
S_e = tolerable settlement (mm)
B = width (m)

The empirical relations just presented may raise some questions. For example, which value of the standard penetration number should be used? What is the effect of the water table on the net allowable bearing capacity? The design value of N_{60} should be determined by taking into account the N_{60} values for a depth of $2B$ to $3B$, measured from the bottom of the foundation. Many engineers are also of the opinion that the N_{60} value should be reduced somewhat if the water table is close to the foundation. However, the author believes that this reduction is not required because the penetration resistance reflects the location of the water table.

12.11 *Common Types of Mat Foundations*

Mat foundations are shallow foundations. This type of foundation, which is sometimes referred to as a *raft foundation*, is a combined footing that may cover the entire area under a structure supporting several columns and walls. Mat foundations are sometimes preferred for soils that have low load-bearing capacities but that will have to support high column and/or wall loads. Under some conditions, spread footings would have to cover more than half the building area, and mat foundations might be more economical. Several types of mat foundations are currently used. Some of the common types are shown schematically in Figure 12.22 and include the following:

1. Flat plate (Figure 12.22a). The mat is of uniform thickness.
2. Flat plate thickened under columns (Figure 12.22b).

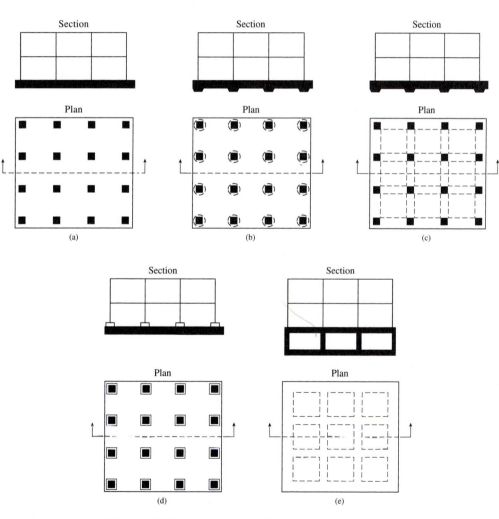

Figure 12.22 Common types of mat foundations

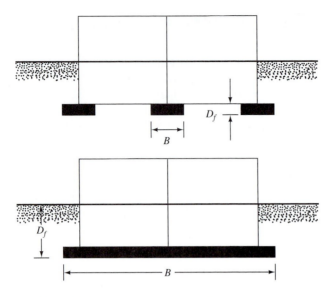

Figure 12.23 Isolated foundation and mat foundation comparison (B = width, D_f = depth)

3. Beams and slab (Figure 12.22c). The beams run both ways, and the columns are located at the intersection of the beams.
4. Flat plates with pedestals (Figure 12.22d).
5. Slab with basement walls as a part of the mat (Figure 12.22e). The walls act as stiffeners for the mat.

Mats may be supported by piles. The piles help in reducing the settlement of a structure built over highly compressible soil. Where the water table is high, mats are often placed over piles to control buoyancy.

Figure 12.23 shows the difference between the depth D_f and the width B of isolated foundations and mat foundation.

12.12   *Bearing Capacity of Mat Foundations*

The *gross ultimate bearing capacity* of a mat foundation can be determined by the same equation used for shallow foundations, or

$$q_u = c'N_cF_{cs}F_{cd}F_{ci} + qN_qF_{qs}F_{qd}F_{qi} + \frac{1}{2}\gamma BN_\gamma F_{\gamma s}F_{\gamma d}F_{\gamma i} \qquad (12.9)$$

Tables 12.2 and 12.3 give the proper values of the bearing capacity factors and the shape, depth, and load inclination factors. The term B in Eq. (12.9) is the smallest dimension of the mat.

The *net ultimate bearing capacity* is

$$q_{net(u)} = q_u - q \qquad (12.10)$$

A suitable factor of safety should be used to calculate the net *allowable* bearing capacity. For mats on clay, the factor of safety should not be less than 3 under dead load and maximum live load. However, under the most extreme conditions, the factor of safety should be at least 1.75 to 2. For mats constructed over sand, a factor of safety of 3 should normally be used. Under most working conditions, the factor of safety against bearing capacity failure of mats on sand is very large.

For saturated clays with $\phi = 0$ and vertical loading condition, Eq. (12.9) gives

$$q_u = c_u N_c F_{cs} F_{cd} + q \tag{12.56}$$

where c_u = undrained cohesion. (*Note:* $N_c = 5.14$, $N_q = 1$, and $N_\gamma = 0$.) From Table 12.3, for $\phi = 0$,

$$F_{cs} = 1 + \left(\frac{B}{L}\right)\left(\frac{N_q}{N_c}\right) = 1 + \left(\frac{B}{L}\right)\left(\frac{1}{5.14}\right) = 1 + \frac{0.195B}{L}$$

and

$$F_{cd} = 1 + 0.4\left(\frac{D_f}{B}\right)$$

Substitution of the preceding shape and depth factors into Eq. (12.56) yields

$$q_u = 5.14c_u\left(1 + \frac{0.195B}{L}\right)\left(1 + 0.4\frac{D_f}{B}\right) + q \tag{12.57}$$

Hence, the net ultimate bearing capacity is

$$q_{net(u)} = q_u - q = 5.14c_u\left(1 + \frac{0.195B}{L}\right)\left(1 + 0.4\frac{D_f}{B}\right) \tag{12.58}$$

For $FS = 3$, the net allowable soil bearing capacity becomes

$$q_{all(net)} = \frac{q_{net(u)}}{FS} = 1.713c_u\left(1 + \frac{0.195B}{L}\right)\left(1 + 0.4\frac{D_f}{B}\right) \tag{12.59}$$

The net allowable bearing capacity for mats constructed over granular soil deposits can be adequately determined from the standard penetration resistance numbers. From Eq. (12.54), for shallow foundations, we have

$$q_{net}(\text{kN/m}^2) = \frac{N_{60}}{0.08}\left(\frac{B + 0.3}{B}\right)^2 F_d\left(\frac{S_e}{25}\right)$$

where
  N_{60} = standard penetration resistance
  B = width (m)
  $F_d = 1 + 0.33(D_f/B) \leq 1.33$
  S_e = settlement, (mm)

When the width, B, is large, the preceding equation can be approximated as

$$q_{net}(kN/m^2) = \frac{N_{60}}{0.08}F_d\left(\frac{S_e}{25}\right)$$

$$= \frac{N_{60}}{0.08}\left[1 + 0.33\left(\frac{D_f}{B}\right)\right]\left[\frac{S_e(mm)}{25}\right] \qquad (12.60)$$

$$\le 16.63\,N_{60}\left[\frac{S_e(mm)}{25}\right]$$

Note that the original Eq. (12.54) was for a settlement of 25 mm, with a differential settlement of about 19 mm. However, the widths of the mat foundations are larger than the isolated spread footings. The depth of significant stress increase in the soil below a foundation depends on the foundation width. Hence, for a mat foundation, the depth of the zone of influence is likely to be much larger than that of a spread footing. Thus, the loose soil pockets under a mat may be more evenly distributed, resulting in a smaller differential settlement. Hence, the customary assumption is that, for a maximum mat settlement of 50 mm, the differential settlement would be 19 mm. Using this logic and conservatively assuming that F_d equals 1, we can approximate Eq. (12.60) as

$$q_{all(net)} = q_{net}(kN/m^2) \approx 25N_{60} \qquad (12.61)$$

The net pressure applied on a foundation (Figure 12.24) may be expressed as

$$q = \frac{Q}{A} - \gamma D_f \qquad (12.62)$$

where
Q = dead weight of the structure and the live load
A = area of the raft

Hence, in all cases, q should be less than or equal to $q_{all(net)}$.

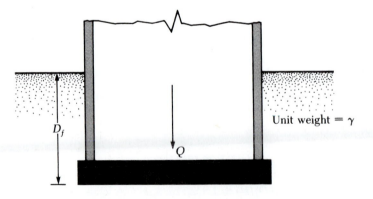

Figure 12.24 Definition of net pressure on soil caused by a mat foundation

Example 12.7

Determine the net ultimate bearing capacity of a mat foundation measuring 12 m × 8 m on a saturated clay with $c_u = 80$ kN/m², $\phi = 0$, and $D_f = 2$ m.

Solution
From Eq. (12.58), we have

$$q_{net(u)} = 5.14c_u\left[1 + \left(\frac{0.195B}{L}\right)\right]\left[1 + 0.4\left(\frac{D_f}{B}\right)\right]$$

$$= (5.14)(80)\left[1 + \left(\frac{0.195 \times 8}{12}\right)\right]\left[1 + 0.4\left(\frac{2}{8}\right)\right]$$

$$= \textbf{512 kN/m}^2 \qquad\blacksquare$$

Example 12.8

What will be the net allowable bearing capacity of a mat foundation with dimensions of 13 m × 9 m constructed over a sand deposit? Here, $D_f = 2$ m, allowable settlement = 25 mm, and average penetration number $N_{60} = 10$.

Solution
From Eq. (12.60), we have

$$q_{all(net)} = \frac{N_{60}}{0.08}\left[1 + 0.33\left(\frac{D_f}{B}\right)\right]\left[\frac{S_e}{25}\right] \le 16.63\,N_{60}\left[\frac{S_e}{25}\right]$$

$$= \frac{10}{0.08}\left[1 + \frac{(0.33)(2)}{9}\right]\left(\frac{25}{25}\right) \approx \textbf{134 kN/m}^2 \qquad\blacksquare$$

12.13   *Compensated Foundations*

The settlement of a mat foundation can be reduced by decreasing the net pressure increase on soil and by increasing the depth of embedment, D_f. This increase is particularly important for mats on soft clays, where large consolidation settlements are expected. From Eq. (12.62), the net average applied pressure on soil is

$$q - \frac{Q}{A} - \gamma D_f$$

For no increase of the net soil pressure on soil below a mat foundation, q should be 0. Thus,

$$D_f = \frac{Q}{A\gamma} \qquad (12.63)$$

This relation for D_f is usually referred to as the depth of embedment of a *fully compensated foundation*.

The factor of safety against bearing capacity failure for partially compensated foundations (that is, $D_f < Q/A\gamma$) may be given as

$$FS = \frac{q_{net(u)}}{q} = \frac{q_{net(u)}}{\dfrac{Q}{A} - \gamma D_f} \tag{12.64}$$

For saturated clays, the factor of safety against bearing capacity failure can thus be obtained by substituting Eq. (12.58) into Eq. (12.64):

$$FS = \frac{5.14c_u\left(1 + \dfrac{0.195B}{L}\right)\left(1 + 0.4\dfrac{D_f}{B}\right)}{\dfrac{Q}{A} - \gamma D_f} \tag{12.65}$$

Example 12.9

Refer to Figure 12.24. The mat has dimensions of 40 m × 20 m, and the live load and dead load on the mat are 200 MN. The mat is placed over a layer of soft clay that has a unit weight of 17.5 kN/m³. Find D_f for a fully compensated foundation.

Solution

From Eq. (12.63), we have

$$D_f = \frac{Q}{A\gamma} = \frac{200 \times 10^3\,\text{kN}}{(40 \times 20)(17.5)} = \textbf{14.29 m} \qquad \blacksquare$$

Example 12.10

Refer to Example 12.9. For the clay, $c_u = 60$ kN/m². If the required factor of safety against bearing capacity failure is 3, determine the depth of the foundation.

Solution

From Eq. (12.65), we have

$$FS = \frac{5.14c_u\left(1 + \dfrac{0.195B}{L}\right)\left(1 + 0.4\dfrac{D_f}{B}\right)}{\dfrac{Q}{A} - \gamma D_f}$$

Here, $FS = 3$, $c_u = 60$ kN/m², $B/L = 20/40 = 0.5$, and $Q/A = (200 \times 10^3)/$
$(40 \times 20) = 250$ kN/m². Substituting these values into Eq. (12.65) yields

$$3 = \frac{(5.14)(60)[1 + (0.195)(0.5)]\left[1 + 0.4\left(\dfrac{D_f}{20}\right)\right]}{250 - (17.5)D_f}$$

$$750 - 52.5D_f = 338.47 + 6.77D_f$$

$$411.53 = 59.27D_f$$

or

$$D_f \approx 6.9 \text{ m} \qquad \blacksquare$$

Problems

12.1 For a continuous foundation, given the following: $\gamma = 18.2$ kN/m³, $\phi' = 20°$, $c' = 14.2$ kN/m², $D_f = 0.5$ m, $B = 1.2$ m. Determine the gross allowable bearing capacity. Use $FS = 4$. Use Eq. (12.3) and Table 12.1.

12.2 A square column foundation is shown in Figure 12.25. With the following, determine the safe gross allowable load, Q_{all}, the foundation can carry: $\gamma = 17.66$ kN/m³, $\gamma_{\text{sat}} = 19.42$ kN/m³, $c' = 23.94$ kN/m², $\phi' = 25°$, $B = 1.8$ m, $D_f = 1.2$ m, $D_1 = 2$ m. Use $FS = 3$. Use Eq. (12.4) and Table 12.1.

12.3 A square column foundation is 2 m × 2 m in plan. The design conditions are $D_f = 1.5$ m, $\gamma = 15.9$ kN/m³, $\phi' = 34°$, and $c' = 0$. Determine the allowable gross vertical load that the column could carry ($FS = 3$). Use Eq. (12.9).

12.4 For the foundation given in Problem 12.4, what will be the gross allowable load-bearing capacity if the load is inclined at an angle of 10° to the vertical?

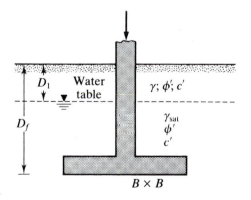

Figure 12.25

12.5 A square foundation ($B \times B$) has to be constructed as shown in Figure 12.25. Assume that $\gamma = 16.5$ kN/m³, $\gamma_{sat} = 18.6$ kN/m³, $c' = 0$, $D_f = 1.2$ m, and $D_1 = 0.6$ m. The gross allowable load, Q_{all}, with $FS = 3$ is 670 kN. The field standard penetration resistance, N_{60}, values are given in the table.

Depth (m)	N_{60}
1.5	4
3.0	6
4.5	6
6.0	10
7.5	5

Determine the size of the foundation. Use Eq. (12.4) and Table 12.1.

12.6 A column foundation is 3 m × 2 m in plan. For $D_f = 1.2$ m, $c' = 20$ kN/m², $\phi' = 24°$, and $\gamma = 17.5$ kN/m³, what is the net ultimate load per unit area that the column could carry? Use Eq. (12.9).

12.7 A square foundation is shown in Figure 12.26. Use an FS of 6 and determine the size of the foundation.

12.8 An eccentrically loaded foundation is shown in Figure 12.27. Determine the ultimate load, Q_u, that the foundation can carry.

12.9 Refer to Figure 12.8 for a foundation with a two-way eccentricity. The soil conditions are $\gamma = 18$ kN/m³, $\phi' = 35°$, and $c' = 0$. The design criteria are $D_f = 1$ m, $B = 1.5$ m, $L = 2$ m, $e_B = 0.3$ m, and $e_L = 0.364$ m. Determine the gross ultimate load that the foundation could carry.

12.10 Repeat Problem 12.9 for $e_L = 0.4$ m and $e_B = 0.19$ m.

12.11 Refer to Figure 12.16. A foundation that is 3 m × 2 m in plan is resting on a sand deposit. The net load per unit area at the level of the foundation, q_o, is 200 kN/m². For the sand, $\mu_s = 0.3$, $E_s = 22$ MN/m², $D_f = 0.9$ m, and $H = 12$ m.

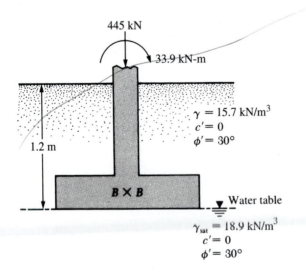

445 kN

33.9 kN-m

$\gamma = 15.7$ kN/m³
$c' = 0$
$\phi' = 30°$

1.2 m

B × B

Water table

$\gamma_{sat} = 18.9$ kN/m³
$c' = 0$
$\phi' = 30°$

Figure 12.26

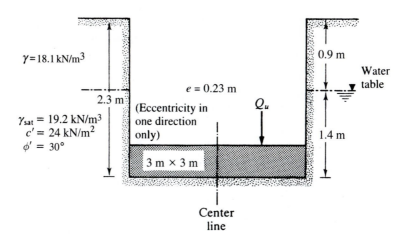

Figure 12.27

Assume that the foundation is rigid and determine the elastic settlement that the foundation would undergo. Use Eqs. (12.41) and (12.49).

12.12 Repeat Problem 12.11 for foundation criteria of size = 1.8 m × 1.8 m, q_o = 190 kN/m², D_f = 1 m, and H = 9 m; and soil conditions of μ_s = 0.4, E_s = 20,000 kN/m², and γ = 17.2 kN/m³.

12.13 A continuous foundation on a deposit of sand layer is shown in Figure 12.28 along with the variation of the modulus of elasticity of the soil (E_s). Assuming

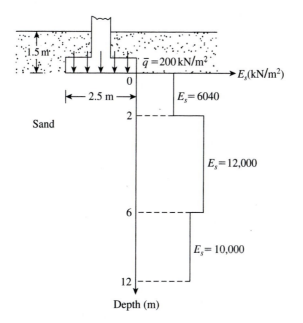

Figure 12.28

$\gamma = 18$ kN/m^2 and time(t) for $C_2 = 10$ years, calculate the elastic settlement of the foundation using the strain influence factor.

12.14 Following are the results of standard penetration tests in a granular soil deposit.

Depth (m)	Standard penetration number, N_{60}
1.5	10
3.0	12
4.5	9
6.0	14
7.5	16

What will be the net allowable bearing capacity of a foundation planned to be 1.5 m ✳ 1.5 m? Let $D_f = 1$ m and the allowable settlement = 25 mm, and use the relationships presented in Section 12.10.

12.15 A shallow square foundation for a column is to be constructed. It must carry a net vertical load of 1000 kN. The soil supporting the foundation is sand. The standard penetration numbers (N_{60}) obtained from field exploration are as follows:

Depth (m)	N_{60}
2	4
4	7
6	12
8	12
10	16
12	13
14	12
16	14
18	18

The groundwater table is located at a depth of 12 m. The unit weight of soil above the water table is 15.7 kN/m^3, and the saturated unit weight of soil below the water table is 18.8 kN/m^3. Assume that the depth of the foundation will be 1.5 m and the tolerable settlement is 25 mm. Determine the size of the foundation.

12.16 A square column foundation is shown in Figure 12.29. Determine the average increase in pressure in the clay layer below the center of the foundation. Use Eqs. (6.45) and (7.46).

12.17 Estimate the consolidation settlement of the clay layer shown in Figure 12.29 from the results of Problem 12.16.

12.18 A mat foundation measuring 14 m × 9 m has to be constructed on a saturated clay. For the clay, $c_u = 93$ kN/m^2 and $\phi = 0$. The depth, D_f, for the mat foundation is 2 m. Determine the net ultimate bearing capacity.

12.19 Repeat Problem 12.18 with the following:
- Mat foundation: $B = 8$ m, $L = 20$ m, and $D_f = 2$ m
- Clay: $\phi = 0$ and $c_u = 130$ kN/m^2

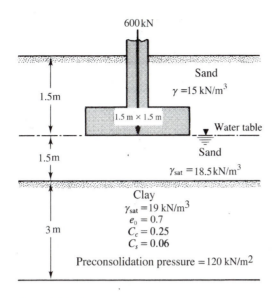

600 kN

Sand
$\gamma = 15$ kN/m³

1.5m

1.5 m × 1.5 m

▼ Water table

1.5m

Sand
$\gamma_{sat} = 18.5$ kN/m³

Clay
$\gamma_{sat} = 19$ kN/m³
$e_0 = 0.7$
$C_c = 0.25$
$C_s = 0.06$

3 m

Preconsolidation pressure = 120 kN/m²

Figure 12.29

12.20 The table gives the results of a standard penetration test in the field (sandy soil):

Depth (m)	Field value of N_{60}
2	8
4	10
6	12
8	9
10	14

Estimate the net allowable bearing capacity of a mat foundation 6 m × 5 m in plan. Here, $D_f = 1.5$ m and allowable settlement = 50 mm. Assume that the unit weight of soil, $\gamma = 18.5$ kN/m³.

12.21 Consider a mat foundation with dimensions of 18 m × 12 m. The combined dead and live load on the mat is 44.5 MN. The mat is to be placed on a clay with $c_u = 40.7$ kN/m² and $\gamma = 17.6$ kN/m³. Find the depth, D_f, of the mat for a fully compensated foundation.

12.22 For the mat in Problem 12.21, what will be the depth, D_f, of the mat for $FS = 3$ against bearing capacity failure?

References

BOWLES, J. E. (1977). *Foundation Analysis and Design*, 2nd ed., McGraw-Hill, New York.

BOWLES, J. E. (1987). "Elastic Foundation Settlement on Sand Deposits," *Journal of Geotechnical Engineering*, ASCE, Vol. 113, No. 8, 846–860.

CAQUOT, A., and KERISEL, J. (1953). "Sur le terme de surface dans le calcul des fondations en milieu pulverulent," *Proceedings*, Third International Conference on Soil Mechanics and Foundation Engineering, Zürich, Vol. I, 336–337.

DE BEER, E. E. (1970). "Experimental Determination of the Shape Factors and Bearing Capacity Factors of Sand," *Geotechnique*, Vol. 20, No. 4, 387–411.

FOX, E. N. (1948). "The Mean Elastic Settlement of a Uniformaly Loaded Area at a Depth Below the Ground Surface," *Proceedings*, 2nd International Conference on Soil Mechanics and Foundation Engineering, Rotterdam, Vol. 1, pp. 129–132.

HANNA, A. M., and MEYERHOF, G. G. (1981). "Experimental Evaluation of Bearing Capacity of Footings Subjected to Inclined Loads," *Canadian Geotechnical Journal*, Vol. 18, No. 4, 599–603.

HANSEN, J. B. (1970). "A Revised and Extended Formula for Bearing Capacity," Danish Geotechnical Institute, *Bulletin 28*, Copenhagen.

HIGHTER, W. H., and ANDERS, J. C. (1985). "Dimensioning Footings Subjected to Eccentric Loads," *Journal of Geotechnical Engineering*, American Society of Civil Engineers, Vol. 111, No. GT5, 659–665.

KUMBHOJKAR, A. S. (1993). "Numerical Evaluation of Terzaghi's N_γ," *Journal of Geotechnical Engineering*, American Society of Civil Engineers, Vol. 119, No. 3, 598–607.

MEYERHOF, G. G. (1953). "The Bearing Capacity of Foundations Under Eccentric and Inclined Loads," *Proceedings*, Third International Conference on Soil Mechanics and Foundation Engineering, Zürich, Vol. 1, 440–445.

MEYERHOF, G. G. (1956). "Penetration Tests and Bearing Capacity of Cohesionless Soils," *Journal of the Soil Mechanics and Foundations Division*, American Society of Civil Engineers, Vol. 82, No. SM1, 1–19.

MEYERHOF, G. G. (1963). "Some Recent Research on the Bearing Capacity of Foundations," *Canadian Geotechnical Journal*, Vol. 1, No. 1, 16–26.

MEYERHOF, G. G. (1965). "Shallow Foundations," *Journal of the Soil Mechanics and Foundations Division*, ASCE, Vol. 91, No. SM2, 21–31.

PRANDTL, L. (1921). "über die Eindringungsfestigkeit (Härte) plastischer Baustoffe und die Festigkeit von Schneiden," *Zeitschrift für angewandte Mathematik und Mechanik*, Vol. 1, No. 1, 15–20.

REISSNER, H. (1924). "Zum Erddruckproblem," *Proceedings*, First International Congress of Applied Mechanics, Delft, 295–311.

SCHMERTMANN, J. H. (1970). "Static Cone to Compute Settlement Over Sand," *Journal of the Soil Mechanics and Foundations Division*, American Society of Civil Engineers, Vol. 96, No. SM3, 1011–1043.

SCHMERTMANN, J. H., and HARTMAN, J. P. (1978). "Improved Strain Influence Factor Diagrams," *Journal of the Geotechnical Engineering Division*, American Society of Civil Engineers, Vol. 104, No. GT8, 113–1135.

STEINBRENNER, W. (1934). "Tafeln zur Setzungsberechnung," *Die Strasse*, Vol. 1, pp. 121–124.

TERZAGHI, K. (1943). *Theoretical Soil Mechanics*, Wiley, New York.

TERZAGHI, K., and PECK, R. B. (1967). *Soil Mechanics in Engineering Practice*, 2nd ed., Wiley, New York.

VESIC, A. S. (1963). "Bearing Capacity of Deep Foundations in Sand," *Highway Research Record No. 39*, National Academy of Sciences, 112–153.

VESIC, A. S. (1973). "Analysis of Ultimate Loads of Shallow Foundations," *Journal of the Soil Mechanics and Foundations Division*, American Society of Civil Engineers, Vol. 99, No. SM1, 45–73.

13

Retaining Walls and Braced Cuts

The general principles of lateral earth pressure were presented in Chapter 11. Those principles can be extended to the analysis and design of earth-retaining structures such as retaining walls and braced cuts. *Retaining walls* provide permanent lateral support to *vertical* or *near-vertical* slopes of soil. Also, at times, construction work requires ground excavations with vertical or near-vertical faces—for example, basements of buildings in developed areas or underground transportation facilities at shallow depths below the ground surface (cut-and-cover type of construction). The vertical faces of the cuts should be protected by *temporary bracing systems* to avoid failure that may be accompanied by considerable settlement or by bearing capacity failure of nearby foundations. These cuts are called *braced cuts*. This chapter is divided into two parts: The first part discusses the analysis of retaining walls, and the second part presents the analysis of braced cuts.

RETAINING WALLS

13.1 Retaining Walls—General

Retaining walls are commonly used in construction projects and may be grouped into four classifications:

1. Gravity retaining walls
2. Semigravity retaining walls
3. Cantilever retaining walls
4. Counterfort retaining walls

Gravity retaining walls (Figure 13.1a) are constructed with plain concrete or stone masonry. They depend on their own weight and any soil resting on the masonry for stability. This type of construction is not economical for high walls.

In many cases, a small amount of steel may be used for the construction of gravity walls, thereby minimizing the size of wall sections. Such walls are generally referred to as *semigravity retaining walls* (Figure 13.1b).

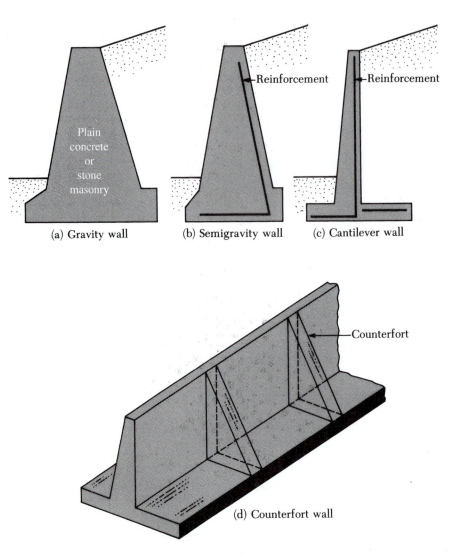

Figure 13.1 Types of retaining wall

Cantilever retaining walls (Figure 13.1c) are made of reinforced concrete that consists of a thin stem and a base slab. This type of wall is economical to a height of about 8 m.

Counterfort retaining walls (Figure 13.1d) are similar to cantilever walls. At regular intervals, however, they have thin vertical concrete slabs known as *counterforts* that tie the wall and the base slab together. The purpose of the counterforts is to reduce the shear and the bending moments.

To design retaining walls properly, an engineer must know the basic soil parameters—that is, the *unit weight, angle of friction,* and *cohesion* – of the soil retained behind the wall and the soil below the base slab. Knowing the properties of the soil behind the wall enables the engineer to determine the lateral pressure distribution that has to be considered in the design.

The design of a retaining wall proceeds in two phases. First, with the lateral earth pressure known, the structure as a whole is checked for *stability*, including checking for possible *overturning, sliding,* and *bearing capacity* failures. Second, each component of the structure is checked for *adequate strength*, and the *steel reinforcement* of each component is determined.

13.2 Proportioning Retaining Walls

When designing retaining walls, an engineer must assume some of the dimensions, called *proportioning*, to check trial sections for stability. If the stability checks yield undesirable results, the sections can be changed and rechecked. Figure 13.2 shows the general proportions of various retaining wall components that can be used for initial checks.

Note that the top of the stem of any retaining wall should be no less than about 0.3 m wide for proper placement of concrete. The depth, D, to the bottom of the base slab should be a minimum of 0.6 m. However, the bottom of the base slab should be positioned below the seasonal frost line.

For counterfort retaining walls, the general proportion of the stem and the base slab is the same as for cantilever walls. However, the counterfort slabs may be about 0.3 m thick and spaced at center-to-center distances of $0.3H$ to $0.7H$.

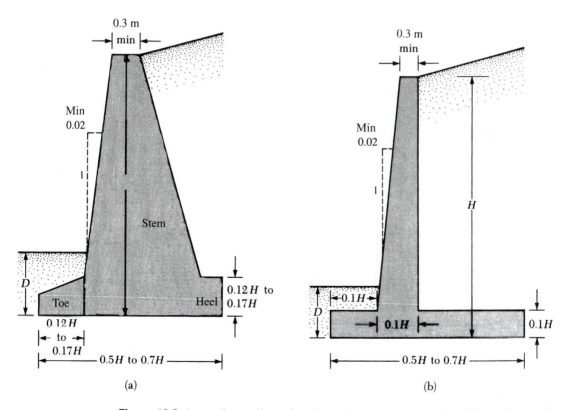

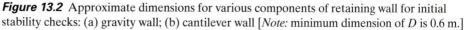

Figure 13.2 Approximate dimensions for various components of retaining wall for initial stability checks: (a) gravity wall; (b) cantilever wall [*Note:* minimum dimension of D is 0.6 m.]

13.3 Application of Lateral Earth Pressure Theories to Design

Chapter 11 presented the fundamental theories for calculating lateral earth pressure. To use these theories in design, an engineer must make several simple assumptions. In the case of cantilever walls, using the Rankine earth pressure theory for stability checks involves drawing a vertical line AB through point A, as shown in Figure 13.3a (located at the edge of the heel of the base slab). The Rankine active condition is assumed to exist along the vertical plane AB. Rankine's active earth pressure equations may then be used to calculate the lateral pressure on the face AB. In the analysis of stability for the wall, the force $P_{a(Rankine)}$, the weight of soil above the heel, W_s, and the weight of the concrete, W_c, all should be taken into consideration. The assumption for the development of Rankine's active pressure along the soil face AB is theoretically correct if the shear zone bounded by the line AC is not obstructed by the stem of the wall. The angle, η, that the line AC makes with the vertical is

$$\eta = 45 + \frac{\alpha}{2} - \frac{\phi_1'}{2} - \sin^{-1}\left(\frac{\sin \alpha}{\sin \phi_1'}\right) \qquad (13.1)$$

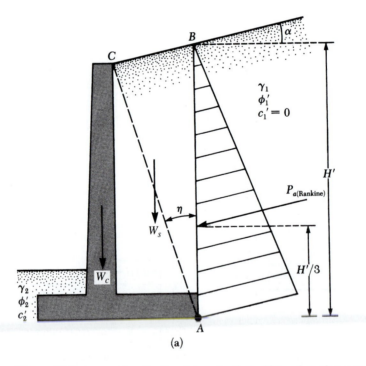

(a)

Figure 13.3 Assumption for the determination of lateral earth pressure: (a) cantilever wall; (b) and (c) gravity wall

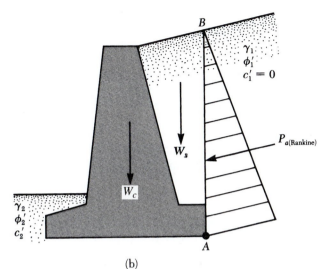

(b)

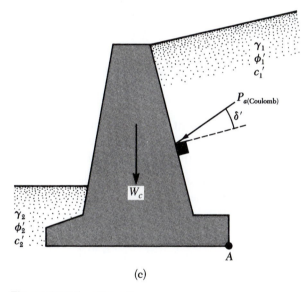

(c)

Figure 13.3 (*continued*)

For gravity walls, a similar type of analysis may be used, as shown in Figure 13.3b. However, Coulomb's theory also may be used, as shown in Figure 13.3c. If *Coulomb's active pressure theory* is used, the only forces to be considered are $P_{a(\text{Coulomb})}$ and the weight of the wall, W_c.

In the case of ordinary retaining walls, water table problems and hence hydrostatic pressure are not encountered. Facilities for drainage from the soils retained are always provided.

To check the stability of a retaining wall, the following steps are taken:

1. Check for *overturning* about its toe.
2. Check for *sliding failure* along its base.
3. Check for *bearing capacity failure* of the base.
4. Check for *settlement*.
5. Check for *overall stability*.

The following sections describe the procedure for checking for overturning, sliding, and bearing capacity failure. The principles of investigation for settlement were covered in Chapters 7 and 12 and will not be repeated here.

13.4   *Check for Overturning*

Figure 13.4 shows the forces that act on a cantilever and a gravity retaining wall, based on the assumption that the Rankine active pressure is acting along a vertical plane AB drawn through the heel. P_p is the Rankine passive pressure; recall that its magnitude is [from Eq. (11.42) with $\gamma = \gamma_2$, $c' = c_2'$, and $H = D$]

$$P_p = \frac{1}{2} K_p \gamma_2 D^2 + 2c_2' \sqrt{K_p} D \tag{13.2}$$

where

γ_2 = unit weight of soil in front of the heel and under the base slab
K_p = Rankine's passive earth pressure coefficient = $\tan^2 (45 + \phi'_2/2)$
c_2', ϕ_2' = cohesion and soil friction angle, respectively

The factor of safety against overturning about the toe—that is, about point C in Figure 13.4—may be expressed as

$$FS_{(\text{overturning})} = \frac{\Sigma M_R}{\Sigma M_O} \tag{13.3}$$

where

ΣM_O = sum of the moments of forces tending to overturn about point C
ΣM_R = sum of the moments of forces tending to resist overturning about point C

The overturning moment is

$$\Sigma M_O = P_h \left(\frac{H'}{3} \right) \tag{13.4}$$

where $P_h = P_a \cos \alpha$.

When calculating the resisting moment, ΣM_R (neglecting P_p), we can prepare a table such as Table 13.1. The weight of the soil above the heel and the weight of the concrete (or masonry) are both forces that contribute to the resisting moment. Note

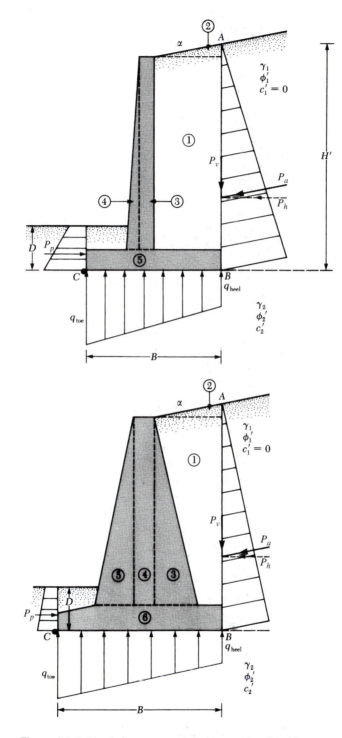

Figure 13.4 Check for overturning; assume that Rankine pressure is valid

Table 13.1 Procedure for calculation of ΣM_R

Section (1)	Area (2)	Weight/unit length of wall (3)	Moment arm measured from C (4)	Moment about C (5)
1	A_1	$W_1 = \gamma_1 \times A_1$	X_1	M_1
2	A_2	$W_2 = \gamma_1 \times A_2$	X_2	M_2
3	A_3	$W_3 = \gamma_c \times A_3$	X_3	M_3
4	A_4	$W_4 = \gamma_c \times A_4$	X_4	M_4
5	A_5	$W_5 = \gamma_c \times A_5$	X_5	M_5
6	A_6	$W_6 = \gamma_c \times A_6$	X_6	M_6
		P_v	B	M_v
		ΣV		ΣM_R

Note: γ_1 = unit weight of backfill
γ_c = unit weight of concrete

that the force P_v also contributes to the resisting moment. P_v is the vertical component of the active force P_a, or

$$P_v = P_a \sin \alpha \tag{13.5}$$

The moment of the force P_v about C is

$$M_v = P_v B = P_a \sin \alpha B \tag{13.6}$$

where B = width of the base slab.

Once ΣM_R is known, the factor of safety can be calculated as

$$FS_{\text{(overturning)}} = \frac{M_1 + M_2 + M_3 + M_4 + M_5 + M_6 + M_v}{P_a \cos \alpha (H'/3)} \tag{13.7}$$

The usual minimum desirable value of the factor of safety with respect to overturning is 1.5 to 2.

Some designers prefer to determine the factor of safety against overturning with

$$FS_{\text{(overturning)}} = \frac{M_1 + M_2 + M_3 + M_4 + M_5 + M_6}{P_a \cos \alpha (H'/3) - M_v} \tag{13.8}$$

13.5   Check for Sliding along the Base

The factor of safety against sliding may be expressed by the equation

$$FS_{\text{(sliding)}} = \frac{\Sigma F_{R'}}{\Sigma F_d} \tag{13.9}$$

where
$\Sigma F_{R'}$ = sum of the horizontal resisting forces
ΣF_d = sum of the horizontal driving forces

Figure 13.5 indicates that the shear strength of the soil below the base slab may be represented as

$$\tau_f = \sigma' \tan \phi_2' + c_2'$$

Thus, the maximum resisting force that can be derived from the soil per unit length of the wall along the bottom of the base slab is

$$R' = \tau_f(\text{area of cross section}) = \tau_f(B \times 1) = B\sigma' \tan \phi_2' + Bc_2'$$

However,

$$B\sigma' = \text{sum of the vertical force} = \sum V \text{ (see Table 13.1)}$$

so

$$R' = \left(\sum V\right)\tan \phi_2' + Bc_2'$$

Figure 13.5 shows that the passive force, P_p, is also a horizontal resisting force. The expression for P_p is given in Eq. (13.2). Hence,

$$\sum F_{R'} = \left(\sum V\right)\tan \phi_2' + Bc_2' + P_p \tag{13.10}$$

The only horizontal force that will tend to cause the wall to slide (*driving force*) is the horizontal component of the active force P_a, so

$$\sum F_d = P_a \cos \alpha \tag{13.11}$$

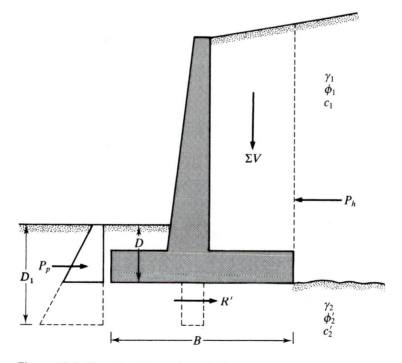

Figure 13.5 Check for sliding along the base

Combining Eqs. (13.9), (13.10), and (13.11) yields

$$FS_{(sliding)} = \frac{(\Sigma V)\tan \phi_2' + Bc_2' + P_p}{P_a \cos \alpha} \qquad (13.12)$$

A minimum factor of safety of 1.5 against sliding is generally required.

In many cases, the passive force, P_p, is ignored when calculating the factor of safety with respect to sliding. The friction angle, ϕ_2', is also reduced in several instances for safety. The reduced soil friction angle may be on the order of one-half to two-thirds of the angle ϕ_2'. In a similar manner, the cohesion c_2' may be reduced to the value of $0.5c_2'$ to $0.67c_2'$. Thus,

$$FS_{(sliding)} = \frac{(\Sigma V)\tan(k_1\phi_2') + Bk_2c_2' + P_p}{P_a \cos \alpha} \qquad (13.13)$$

where k_1 and k_2 are in the range of $\frac{1}{2}$ to $\frac{2}{3}$.

In some instances, certain walls may not yield a desired factor of safety of 1.5. To increase their resistance to sliding, a base key may be used. Base keys are illustrated by broken lines in Figure 13.5. The passive force at the toe *without the key* is

$$P_p = \frac{1}{2}\gamma_2 D^2 K_p + 2c_2'D\sqrt{K_p}$$

However, if a key is included, the passive force per unit length of the wall becomes (*note:* $D = D_1$)

$$P_p = \frac{1}{2}\gamma_2 D_1^2 K_p + 2c_2'D_1\sqrt{K_p}$$

where $K_p = \tan^2(45 + \phi_2'/2)$. Because $D_1 > D$, a key obviously will help increase the passive resistance at the toe and hence, the factor of safety against sliding. Usually the base key is constructed below the stem, and some main steel is run into the key.

13.6   *Check for Bearing Capacity Failure*

The vertical pressure transmitted to the soil by the base slab of the retaining wall should be checked against the ultimate bearing capacity of the soil. The nature of variation of the vertical pressure transmitted by the base slab into the soil is shown in Figure 13.6. Note that q_{toe} and q_{heel} are the *maximum* and the *minimum* pressures occurring at the ends of the toe and heel sections, respectively. The magnitudes of q_{toe} and q_{heel} can be determined in the following manner.

The sum of the vertical forces acting on the base slab is ΣV (see col. 3, Table 13.1), and the horizontal force is $P_a \cos \alpha$. Let R be the resultant force, or

$$\overrightarrow{R} = \overrightarrow{\Sigma V} + \overrightarrow{(P_a \cos \alpha)} \qquad (13.14)$$

The net moment of these forces about point C (Figure 13.6) is

$$M_{net} = \Sigma M_R - \Sigma M_O \qquad (13.15)$$

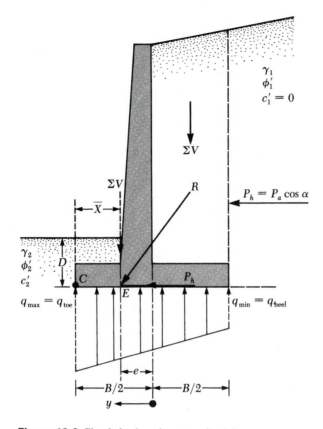

Figure 13.6 Check for bearing capacity failure

The values of ΣM_R and ΣM_O were previously determined [see col. 5, Table 13.1, and Eq. (13.4).] Let the line of action of the resultant, R, intersect the base slab at E, as shown in Figure 13.6. The distance CE then is

$$\overline{CE} = \overline{X} = \frac{M_{\text{net}}}{\Sigma V} \tag{13.16}$$

Hence, the eccentricity of the resultant, R, may be expressed as

$$e = \frac{B}{2} - \overline{CE} \tag{13.17}$$

The pressure distribution under the base slab may be determined by using the simple principles of mechanics of materials:

$$q = \frac{\Sigma V}{A} \pm \frac{M_{\text{net}} y}{I} \tag{13.18}$$

where

M_{net} = moment = $(\Sigma V)e$

I = moment of inertia per unit length of the base section = $\frac{1}{12}(1)(B^3)$

For maximum and minimum pressures, the value of y in Eq. (13.18) equals $B/2$. Substituting the preceding values into Eq. (13.18) gives

$$q_{max} = q_{toe} = \frac{\Sigma V}{(B)(1)} + \frac{e(\Sigma V)\frac{B}{2}}{\left(\frac{1}{12}\right)(B^3)} = \frac{\Sigma V}{B}\left(1 + \frac{6e}{B}\right) \qquad (13.19)$$

Similarly,

$$q_{min} = q_{heel} = \frac{\Sigma V}{B}\left(1 - \frac{6e}{B}\right) \qquad (13.20)$$

Note that ΣV includes the soil weight, as shown in Table 13.1, and that, when the value of the eccentricity, e, becomes greater than $B/6$, q_{min} becomes negative [Eq. (13.20)]. Thus, there will be some tensile stress at the end of the heel section. This stress is not desirable because the tensile strength of soil is very small. If the analysis of a design shows that $e > B/6$, the design should be reproportioned and calculations redone.

The relationships for the ultimate bearing capacity of a shallow foundation were discussed in Chapter 12. Recall that

$$q_u = c_2' N_c F_{cd} F_{ci} + q N_q F_{qd} F_{qi} + \frac{1}{2}\gamma_2 B' N_\gamma F_{\gamma d} F_{\gamma i} \qquad (13.21)$$

where

$q = \gamma_2 D$

$B' = B - 2e$

$F_{cd} = 1 + 0.4\dfrac{D}{B'}$

$F_{qd} = 1 + 2 \tan \phi_2'(1 - \sin \phi_2')^2 \dfrac{D}{B'}$

$F_{\gamma d} = 1$

$F_{ci} = F_{qi} = \left(1 - \dfrac{\psi^\circ}{90^\circ}\right)^2$

$F_{\gamma i} = \left(1 - \dfrac{\psi^\circ}{\phi_2'^\circ}\right)^2$

$\psi^\circ = \tan^{-1}\left(\dfrac{P_a \cos \alpha}{\Sigma V}\right)$

Note that the shape factors F_{cs}, F_{qs}, and $F_{\gamma s}$ given in Chapter 12 are all equal to 1 because they can be treated as a continuous foundation. For this reason, the shape factors are not shown in Eq. (13.21).

Once the ultimate bearing capacity of the soil has been calculated using Eq. (13.21), the factor of safety against bearing capacity failure can be determined:

$$FS_{\text{(bearing capacity)}} = \frac{q_u}{q_{max}} \tag{13.22}$$

Generally, a factor of safety of 3 is required. In Chapter 12, we noted that the ultimate bearing capacity of shallow foundations occurs at a settlement of about 10% of the foundation width. In the case of retaining walls, the width B is large. Hence, the ultimate load q_u will occur at a fairly large foundation settlement. A factor of safety of 3 against bearing capacity failure may not ensure, in all cases, that settlement of the structure will be within the tolerable limit. Thus, this situation needs further investigation.

Example 13.1

The cross section of a cantilever retaining wall is shown in Figure 13.7. Calculate the factors of safety with respect to overturning, sliding, and bearing capacity.

Solution
Referring to Figure 13.7, we find

$$H' = H_1 + H_2 + H_3 = 2.6 \tan 10° + 6 + 0.7$$
$$= 0.458 + 6 + 0.7 = 7.158 \text{ m}$$

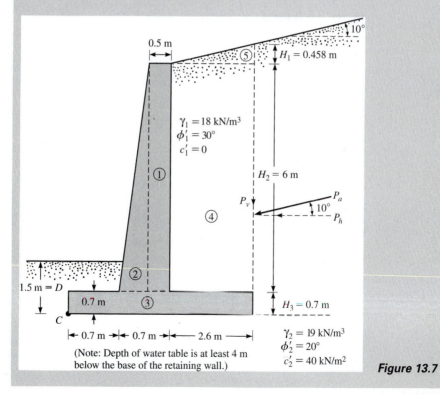

(Note: Depth of water table is at least 4 m below the base of the retaining wall.)

Figure 13.7

The Rankine active force per unit length of wall is

$$P_a = \frac{1}{2}\gamma_1 H'^2 K_a$$

For $\phi_1' = 30°$ and $\alpha = 10°$, K_a is equal to 0.350 (Table 11.2). Thus,

$$P_a = \frac{1}{2}(18)(7.158)^2(0.35) = 161.4 \text{ kN/m}$$

$$P_v = P_a \sin 10° = 161.4(\sin 10°) = 28.03 \text{ kN/m}$$

$$P_h = P_a \cos 10° = 161.4(\cos 10°) = 158.95 \text{ kN/m}$$

Factor of Safety against Overturning

The following table can now be prepared to determine the resisting moment.

Section no.*	Area (m²)	Weight/unit length (kN/m)†	Moment arm from point C (m)	Moment (kN-m/m)
1	$6 \times 0.5 = 3$	70.74	1.15	81.35
2	$\frac{1}{2}(0.2)6 = 0.6$	14.15	0.833	11.79
3	$4 \times 0.7 = 2.8$	66.02	2.0	132.04
4	$6 \times 2.6 = 15.6$	280.80	2.7	758.16
5	$\frac{1}{2}(2.6)(0.458) = 0.595$	10.71	3.13	33.52
		$P_v = 28.03$	4.0	112.12
		$\Sigma V = 470.45$		$\Sigma 1128.98 = \Sigma M_R$

*For section numbers, refer to Figure 13.7.
†$\gamma_{\text{concrete}} = 23.58 \text{ kN/m}^3$

For the overturning moment, we get

$$M_O = P_h\left(\frac{H'}{3}\right) = 158.95\left(\frac{7.158}{3}\right) = 379.25 \text{ kN-m/m}$$

Hence,

$$FS_{\text{(overturning)}} = \frac{\Sigma M_R}{M_O} = \frac{1128.98}{379.25} = 2.98 > 2\text{—OK}$$

Factor of Safety against Sliding

From Eq. (13.13), we have

$$FS_{\text{(sliding)}} = \frac{(\Sigma V)\tan(k_1\phi_1') + Bk_2c_2' + P_p}{P_a \cos \alpha}$$

Let $k_1 = k_2 = \frac{2}{3}$. Also,

$$P_p = \frac{1}{2}K_p\gamma_2 D^2 + 2c_2'\sqrt{K_p}\,D$$

$$K_p = \tan^2\left(45 + \frac{\phi_2'}{2}\right) = \tan^2(45 + 10) = 2.04$$

$$D = 1.5 \text{ m}$$

So

$$P_p = \frac{1}{2}(2.04)(19)(1.5)^2 + 2(40)(\sqrt{2.04})(1.5)$$

$$= 43.61 + 171.39 = 215 \text{ kN/m}$$

Hence,

$$FS_{(\text{sliding})} = \frac{(470.45)\tan\left(\dfrac{2 \times 20}{3}\right) + (4)\left(\dfrac{2}{3}\right)(40) + 215}{158.95}$$

$$= \frac{111.5 + 106.67 + 215}{158.95} = \mathbf{2.73 > 1.5\text{—OK}}$$

Note: For some designs, the depth, D, for passive pressure calculation may be taken to be *equal to the thickness of the base slab.*

Factor of Safety against Bearing Capacity Failure

Combining Eqs. (13.15), (13.16), and (13.17), we have

$$e = \frac{B}{2} - \frac{\Sigma M_R - M_O}{\Sigma V} = \frac{4}{2} - \frac{1128.98 - 379.25}{470.45}$$

$$= 0.406 \text{ m} < \frac{B}{6} = \frac{4}{6} = 0.666 \text{ m}$$

Again, from Eqs. (13.19) and (13.20),

$$q_{\substack{\text{toe} \\ \text{heel}}} = \frac{\Sigma V}{B}\left(1 \pm \frac{6e}{B}\right) = \frac{470.45}{4}\left(1 \pm \frac{6 \times 0.406}{4}\right) = 189.2 \text{ kN/m}^2 \text{ (toe)}$$

$$= 45.99 \text{ kN/m}^2 \text{ (heel)}$$

The ultimate bearing capacity of the soil can be determined from Eq. (13.21):

$$q_u = c_2' N_c F_{cd} F_{ci} + q N_q F_{qd} F_{qi} + \frac{1}{2}\gamma_2 B' N_\gamma F_{\gamma d} F_{\gamma i}$$

For $\phi_2' = 20°$, we find $N_c = 14.83$, $N_q = 6.4$, and $N_\gamma = 5.39$ (Table 12.2). Also,

$$q = \gamma_2 D = (19)(1.5) = 28.5 \text{ kN/m}^2$$

$$B' = B - 2e = 4 - 2(0.406) = 3.188 \text{ m}$$

$$F_{cd} = 1 + 0.4\left(\frac{D}{B'}\right) = 1 + 0.4\left(\frac{1.5}{3.188}\right) = 1.188$$

$$F_{qd} = 1 + 2\tan\phi_2'(1 - \sin\phi_2')^2\left(\frac{D}{B'}\right) = 1 + 0.315\left(\frac{1.5}{3.188}\right) = 1.148$$

$$F_{\gamma d} = 1$$

$$F_{ci} = F_{qi} = \left(1 - \frac{\psi^\circ}{90^\circ}\right)^2$$

$$\psi = \tan^{-1}\left(\frac{P_a\cos\alpha}{\Sigma V}\right) = \tan^{-1}\left(\frac{158.95}{470.45}\right) = 18.67^\circ$$

So

$$F_{ci} = F_{qi} = \left(1 - \frac{18.67}{90}\right)^2 = 0.628$$

$$F_{\gamma i} = \left(1 - \frac{\psi}{\phi_2'}\right)^2 = \left(1 - \frac{18.67}{20}\right)^2 \approx 0$$

Hence,

$$q_u = (40)(14.83)(1.188)(0.628) + (28.5)(6.4)(1.148)(0.628)$$

$$+ \frac{1}{2}(19)(5.93)(3.188)(1)(0)$$

$$= 442.57 + 131.50 + 0 = 574.07 \text{ kN/m}^2$$

$$FS_{(bearing\ capacity)} = \frac{q_u}{q_{toe}} = \frac{574.07}{189.2} = 3.03 > 3\text{—OK}$$ ∎

Example 13.2

A gravity retaining wall is shown in Figure 13.8. Use $\delta' = \frac{2}{3}\phi_1'$ and Coulomb's active earth pressure theory. Determine these values:

a. The factor of safety against overturning
b. The factor of safety against sliding
c. The pressure on the soil at the toe and heel

Solution

$$H' = 5 + 1.5 = 6.5 \text{ m}$$

Coulomb's active force

$$P_a = \frac{1}{2}\gamma_1 H'^2 K_a$$

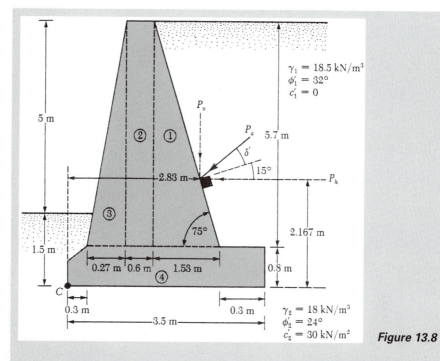

Figure 13.8

With $\alpha = 0°$, $\theta = 15°$, $\delta' = \frac{2}{3}\phi_1'$, and $\phi_1 = 32°$, we find $K_a = 0.4023$ (Table 11.7). So

$$P_a = \frac{1}{2}(18.5)(6.5)^2(0.4023) = 157.22 \text{ kN/m}$$

$$P_h = P_a \cos\left(15 + \frac{2}{3}\phi_1'\right) = 157.22 \cos 36.33 = 126.65 \text{ kN/m}$$

$$P_v = P_a \sin\left(15 + \frac{2}{3}\phi_1'\right) = 157.22 \sin 36.33 = 93.14 \text{ kN/m}$$

Part a: Factor of Safety against Overturning
Referring to Figure 13.8, we can prepare the following table:

Area no.	Area (m²)	Weight/unit length (kN/m)*	Moment arm from point C (m)	Moment (kN-m/m)
1	$\frac{1}{2}(5.7)(1.53) = 4.36$	102.81	2.18	224.13
2	$(0.6)(5.7) = 3.42$	80.64	1.37	110.48
3	$\frac{1}{2}(0.27)(5.7) = 0.77$	18.16	0.98	17.80
4	$\approx(3.5)(0.8) = 2.8$	66.02	1.75	115.52
		$P_v = 93.14$	2.83	263.59
		$\Sigma V = 360.77 \text{ kN/m}$		$\Sigma M_R = 731.54 \text{ kN-m/m}$

*$\gamma_{\text{concrete}} = 23.58 \text{ kN/m}^3$

We have for the overturning moment

$$M_O = P_h\left(\frac{H'}{3}\right) = 126.65(2.167) = 274.45 \text{ kN-m/m}$$

Hence,

$$FS_{(overturning)} = \frac{\Sigma M_R}{\Sigma M_O} = \frac{731.54}{274.45} = \mathbf{2.665} > 2\text{—OK}$$

Part b: Factor of Safety against Sliding

$$FS_{(sliding)} = \frac{(\Sigma V)\tan\left(\frac{2}{3}\phi_2'\right) + \frac{2}{3}c_2'B + P_p}{P_h}$$

$$P_p = \frac{1}{2}K_p\gamma_2 D^2 + 2c_2'\sqrt{K_p}D$$

$$K_p = \tan^2\left(45 + \frac{24}{2}\right) = 2.37$$

Hence,

$$P_p = \frac{1}{2}(2.37)(18)(1.5)^2 + 2(30)(1.54)(1.5) = 186.59 \text{ kN/m}$$

So

$$FS_{(sliding)} = \frac{360.77\tan\left(\frac{2}{3}\times 24\right) + \frac{2}{3}(30)(3.5) + 186.59}{126.65}$$

$$= \frac{103.45 + 70 + 186.59}{126.65} = \mathbf{2.84}$$

If P_p is ignored, the factor of safety would be **1.37**.

Part c: Pressure on Soil at Toe and Heel
From Eqs. (13.15), (13.16), and (13.17), we have

$$e = \frac{B}{2} - \frac{\Sigma M_R - \Sigma M_O}{\Sigma V} = \frac{3.5}{2} - \frac{731.54 - 274.45}{360.77} = 0.483 < \frac{B}{6} = 0.583$$

$$q_{toe} = \frac{\Sigma V}{B}\left[1 + \frac{6e}{B}\right] = \frac{360.77}{3.5}\left[1 + \frac{(6)(0.483)}{3.5}\right] = \mathbf{188.43 \text{ kN/m}^2}$$

$$q_{heel} = \frac{V}{B}\left[1 - \frac{6e}{B}\right] = \frac{360.77}{3.5}\left[1 - \frac{(6)(0.483)}{3.5}\right] = \mathbf{17.73 \text{ kN/m}^2} \qquad \blacksquare$$

Mechanically Stabilized Retaining Walls

More recently, soil reinforcement has been used in the construction and design of foundations, retaining walls, embankment slopes, and other structures. Depending on the type of construction, the reinforcements may be galvanized metal strips, geotextiles, geogrids, or geocomposites. Sections 13.7 and 13.8 provide a general overview of soil reinforcement and various reinforcement materials.

Reinforcement materials such as metallic strips, geotextiles, and geogrids are now being used to reinforce the backfill of retaining walls, which are generally referred to as *mechanically stabilized retaining walls*. The general principles for designing these walls are given in Sections 13.9 through 13.14.

13.7 Soil Reinforcement

The use of reinforced earth is a recent development in the design and construction of foundations and earth-retaining structures. *Reinforced earth* is a construction material made from soil that has been strengthened by tensile elements such as metal rods or strips, nonbiodegradable fabrics (geotextiles), geogrids, and the like. The fundamental idea of reinforcing soil is not new; in fact, it goes back several centuries. However, the present concept of systematic analysis and design was developed by a French engineer, H. Vidal (1966). The French Road Research Laboratory has done extensive research on the applicability and the beneficial effects of the use of reinforced earth as a construction material. This research has been documented in detail by Darbin (1970), Schlosser and Long (1974), and Schlosser and Vidal (1969). The tests that were conducted involved the use of metallic strips as reinforcing material.

Retaining walls with reinforced earth have been constructed around the world since Vidal began his work. The first reinforced-earth retaining wall with metal strips as reinforcement in the United States was constructed in 1972 in southern California.

The beneficial effects of soil reinforcement derive from (a) the soil's increased tensile strength and (b) the shear resistance developed from the friction at the soil-reinforcement interfaces. Such reinforcement is comparable to that of concrete structures. Currently, most reinforced-earth design is done with *free-draining granular soil only*. Thus, the effect of pore water development in cohesive soils, which, in turn, reduces the shear strength of the soil, is avoided.

13.8 Considerations in Soil Reinforcement

Metal Strips

In most instances, galvanized steel strips are used as reinforcement in soil. However, galvanized steel is subject to corrosion. The rate of corrosion depends on several environmental factors. Binquet and Lee (1975) suggested that the average rate of corrosion of galvanized steel strips varies between 0.025 and 0.050 mm/yr. So, in the actual design of reinforcement, allowance must be made for the rate of corrosion. Thus,

$$t_c = t_{\text{design}} + r \text{ (life span of structure)}$$

where

t_c = actual thickness of reinforcing strips to be used in construction
t_{design} = thickness of strips determined from design calculations
r = rate of corrosion

Further research needs to be done on corrosion-resistant materials such as fiber-glass before they can be used as reinforcing strips.

Nonbiodegradable Fabrics

Nonbiodegradable fabrics are generally referred to as *geotextiles*. Since 1970, the use of geotextiles in construction has increased greatly around the world. The fabrics are usually made from petroleum products—polyester, polyethylene, and polypropy-lene. They may also be made from fiberglass. Geotextiles are not prepared from nat-ural fabrics, because they decay too quickly. Geotextiles may be woven, knitted, or nonwoven.

Woven geotextiles are made of two sets of parallel filaments or strands of yarn systematically interlaced to form a planar structure. *Knitted geotextiles* are formed by interlocking a series of loops of one or more filaments or strands of yarn to form a planar structure. *Nonwoven geotextiles* are formed from filaments or short fibers arranged in an oriented or random pattern in a planar structure. These filaments or short fibers are arranged into a loose web in the beginning and then are bonded by one or a combination of the following processes:

1. *Chemical bonding*—by glue, rubber, latex, a cellulose derivative, or the like
2. *Thermal bonding*—by heat for partial melting of filaments
3. *Mechanical bonding*—by needle punching

Needle-punched nonwoven geotextiles are thick and have high in-plane permeability. Geotextiles have four primary uses in foundation engineering:

1. *Drainage:* The fabrics can rapidly channel water from soil to various outlets, thereby providing a higher soil shear strength and hence stability.
2. *Filtration:* When placed between two soil layers, one coarse grained and the other fine grained, the fabric allows free seepage of water from one layer to the other. However, it protects the fine-grained soil from being washed into the coarse-grained soil.
3. *Separation:* Geotextiles help keep various soil layers separate after construc-tion and during the projected service period of the structure. For example, in the construction of highways, a clayey subgrade can be kept separate from a granular base course.
4. *Reinforcement:* The tensile strength of geofabrics increases the load-bearing capacity of the soil.

Geogrids

Geogrids are high-modulus polymer materials, such as polypropylene and polyethylene, and are prepared by tensile drawing. Netlon, Ltd., of the United Kingdom was the first producer of geogrids. In 1982, the Tensar Corporation, presently Tensar International, introduced geogrids into the United States.

The major function of geogrids is *reinforcement*. Geogrids are relatively stiff netlike materials with openings called *apertures* that are large enough to allow interlocking with the surrounding soil or rock to perform the function of reinforcement or segregation (or both).

Geogrids generally are of two types: (a) uniaxial and (b) biaxial. Figures 13.9a and 13.9b show these two types of geogrids, which are produced by Tensar International. Uniaxial TENSAR grids are manufactured by stretching a punched sheet of extruded high-density polyethylene in one direction under carefully controlled conditions. The process aligns the polymer's long-chain molecules in the direction of draw and results in a product with high one-directional tensile strength and a high modulus. Biaxial TENSAR grids are manufactured by stretching the punched sheet of polypropylene in two orthogonal directions. This process results in a product with high tensile strength and a high modulus in two perpendicular directions. The resulting grid apertures are either square or rectangular.

The commercial geogrids currently available for soil reinforcement have nominal rib thicknesses of about 0.5 to 1.5 mm and junctions of about 2.5 to 5 mm. The grids used for soil reinforcement usually have apertures that are rectangular or elliptical. The dimensions of the apertures vary from about 25 to 150 mm. Geogrids are manufactured so that the open areas of the grids are greater than 50% of the total area. They develop reinforcing strength at low strain levels, such as 2%.

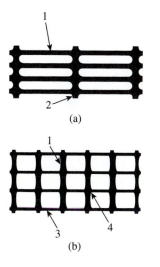

Figure 13.9 Geogrids: (a) uniaxial; (b) biaxial (*Note:* 1—longitudinal rib; 2—transverse bar; 3—transverse rib; 4—junction)

13.9   *General Design Considerations*

The general design procedure of any mechanically stabilized retaining wall can be divided into two parts:

1. Satisfying *internal stability* requirements
2. Checking the *external stability* of the wall

The internal stability checks involve determining tension and pullout resistance in the reinforcing elements and ascertaining the integrity of facing elements. The external stability checks include checks for overturning, sliding, and bearing capacity failure. The sections that follow will discuss the retaining-wall design procedures for use with metallic strips, geotextiles, and geogrids.

13.10   *Retaining Walls with Metallic Strip Reinforcement*

Reinforced-earth walls are flexible walls. Their main components are

1. *Backfill*, which is granular soil
2. *Reinforcing strips*, which are thin, wide strips placed at regular intervals, and
3. *A cover* or *skin*, on the front face of the wall

Figure 13.10 is a diagram of a reinforced-earth retaining wall. Note that, at any depth, the reinforcing strips or ties are placed with a horizontal spacing of S_H center to center; the vertical spacing of the strips or ties is S_V center to center. The skin can be constructed with sections of relatively flexible thin material. Lee et al. (1973) showed that, with a conservative design, a 5 mm-thick galvanized steel skin would be enough to hold a wall about 14 to 15 m high. In most cases, precast concrete slabs can

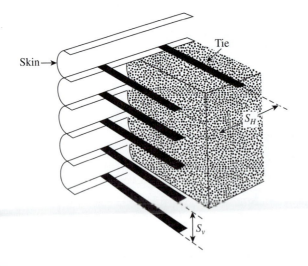

Figure 13.10 Reinforced-earth retaining wall

also be used as skin. The slabs are grooved to fit into each other so that soil cannot flow out between the joints. When metal skins are used, they are bolted together, and reinforcing strips are placed between the skins.

The simplest and most common method for the design of ties is the *Rankine method*. We discuss this procedure next.

Calculation of Active Horizontal Pressure

Figure 13.11 shows a retaining wall with a granular backfill having a unit weight of γ_1 and a friction angle of ϕ'_1. Below the base of the retaining wall, the *in situ* soil has been excavated and recompacted, with granular soil used as backfill. Below the backfill, the *in situ* soil has a unit weight of γ_2, friction angle of ϕ'_2, and cohesion of c'_2. The retaining wall has reinforcement ties at depths $z = 0, S_V, 2S_V, \ldots NS_V$. The height of the wall is $NS_V = H$.

According to the Rankine active pressure theory,

$$\sigma'_a = \sigma'_o K_a - 2c'\sqrt{K_a}$$

where σ'_a = Rankine active pressure effective at any depth z.

For dry granular soils with no surcharge at the top, $c' = 0$, $\sigma'_o = \gamma_1 z$, and $K_a = \tan^2(45 - \phi'_1/2)$. Thus,

$$\sigma'_a = \gamma_1 z K_a \tag{13.23}$$

At the bottom of the wall (that is, at $z = H$),

$$\sigma'_a = \gamma H K_a$$

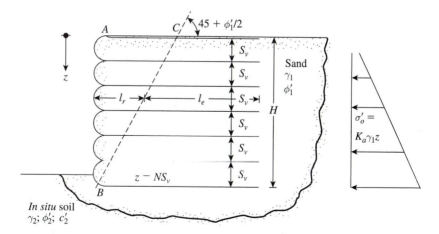

Figure 13.11 Analysis of a reinforced-earth retaining wall

Tie Force

The tie force *per unit length of the wall* developed at any depth z (see Figure 13.11) is

$$T = \text{active earth pressure at depth } z$$

$$\times \text{ area of the wall to be supported by the tie}$$

$$= (\sigma_a')(S_V S_H) \tag{13.24}$$

Factor of Safety against Tie Failure

The reinforcement ties at each level, and thus the walls, could fail by either (a) tie breaking or (b) tie pullout.

The factor of safety against *tie breaking* may be determined as

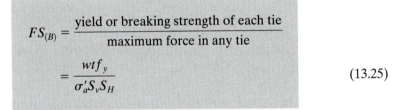

$$FS_{(B)} = \frac{\text{yield or breaking strength of each tie}}{\text{maximum force in any tie}}$$

$$= \frac{wtf_y}{\sigma_a' S_V S_H} \tag{13.25}$$

where

w = width of each tie
t = thickness of each tie
f_y = yield or breaking strength of the tie material

A factor of safety of about 2.5 to 3 is generally recommended for ties at all levels.

Reinforcing ties at any depth z will fail by pullout if the fractional resistance developed along the surfaces of the ties is less than the force to which the ties are being subjected. The *effective length* of the ties along which the frictional resistance is developed may be conservatively taken as the length that extends *beyond the limits of the Rankine active failure zone*, which is the zone *ABC* in Figure 13.11. Line *BC* makes an angle of $45 + \phi_1'/2$ with the horizontal. Now, the maximum friction force that can be realized for a tie at depth z is

$$F_R = 2l_e w \sigma_o' \tan \phi_\mu' \tag{13.26}$$

where

l_e = effective length
σ_o' = effective vertical pressure at a depth z
ϕ_μ' = soil–tie friction angle

Thus, the factor of safety against *tie pullout* at any depth z is

$$FS_{(P)} = \frac{F_R}{T} \tag{13.27}$$

Substituting Eqs. (13.24) and (13.26) into Eq. (13.27) yields

$$FS_{(P)} = \frac{2l_e w \sigma'_o \tan \phi'_\mu}{\sigma'_a S_V S_H}$$

(13.28)

Total Length of Tie

The total length of ties at any depth is

$$L = l_r + l_e$$

(13.29)

where

l_r = length within the Rankine failure zone
l_e = effective length

For a given $FS_{(P)}$ from Eq. (13.28),

$$l_e = \frac{FS_{(P)} \sigma'_a S_V S_H}{2w \sigma'_o \tan \phi'_\mu}$$

(13.30)

Again, at any depth z,

$$l_r = \frac{(H - z)}{\tan\left(45 + \dfrac{\phi'_1}{2}\right)}$$

(13.31)

So, combining Eqs. (13.29), (13.30), and (13.31) gives

$$L = \frac{(H - z)}{\tan\left(45 + \dfrac{\phi'_1}{2}\right)} + \frac{FS_{(P)} \sigma'_a S_V S_H}{2w \sigma'_o \tan \phi'_\mu}$$

(13.32)

13.11   Step-by-Step-Design Procedure Using Metallic Strip Reinforcement

Following is a step-by-step procedure for the design of reinforced-earth retaining walls.

General

Step 1. Determine the height of the wall, H, and also the properties of the granular backfill material, such as unit weight (γ_1) and angle of friction (ϕ'_1).

Step 2. Obtain the soil-tie friction angle, ϕ'_μ, and also the required values of $FS_{(B)}$ and $FS_{(P)}$.

Internal Stability

Step 3.   Assume values for horizontal and vertical tie spacing. Also, assume the width of reinforcing strip, w, to be used.

Step 4.   Calculate σ'_a from Eq. (13.23)

Step 5.   Calculate the tie forces at various levels from Eq. (13.24).

Step 6.   For the known values of $FS_{(B)}$, calculate the thickness of ties, t, required to resist the tie breakout:

$$T = \sigma'_a S_V S_H = \frac{wtf_y}{FS_{(B)}}$$

or

$$t = \frac{(\sigma'_a S_V S_H)[FS_{(B)}]}{wf_y} \tag{13.33}$$

The convention is to keep the magnitude of t the same at all levels, so σ'_a in Eq. (13.33) should equal $\sigma'_{a(\max)}$.

Step 7.   For the known values of ϕ'_μ and $FS_{(P)}$, determine the length L of the ties at various levels from Eq. (13.32).

Step 8.   The magnitudes of S_V, S_H, t, w, and L may be changed to obtain the most economical design.

External Stability

Step 9.   Check for *overturning*, using Figure 13.12 as a guide. Taking the moment about B yields the overturning moment for the unit length of the wall:

$$M_o = P_a z' \tag{13.34}$$

Here,

$$P_a = \text{active force} = \int_0^H \sigma'_a dz$$

The resisting moment per unit length of the wall is

$$M_R = W_1 x_1 + W_2 x_2 + \cdots \tag{13.35}$$

where
$$W_1 = (\text{area } AFEGI)\,(1)\,(\gamma_1)$$
$$W_2 = (\text{area } FBDE)\,(1)\,(\gamma_1)$$
$$\vdots$$

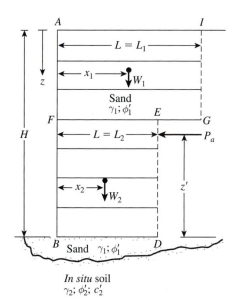

Figure 13.12 Stability check for the retaining wall

So,

$$FS_{(overturning)} = \frac{M_R}{M_o}$$

$$= \frac{W_1 x_1 + W_2 x_2 + \cdots}{\left(\int_0^H \sigma_a' dz \right) z'} \tag{13.36}$$

Step 10. The check for *sliding* can be done by using Eq. (13.13), or

$$FS_{(sliding)} = \frac{(W_1 + W_2 + \cdots)[\tan(k\phi_1')]}{P_a} \tag{13.37}$$

where $k \approx \frac{2}{3}$.

Step 11. Check for ultimate bearing capacity failure, which can be given as

$$q_u = c_2' N_c + \frac{1}{2}\gamma_2 L_2' N_\gamma \tag{13.38}$$

The bearing capacity factors N_c and N_γ correspond to the soil friction angle ϕ'_2. (See Table 12.2.) In Eq. (13.38), L'_2 is the effective length; that is,

$$L'_2 = L_2 - 2e \qquad (13.39)$$

where e = eccentricity given by

$$e = \frac{L_2}{2} - \frac{M_R - M_O}{\Sigma V} \qquad (13.40)$$

in which $\Sigma V = W_1 + W_2 \ldots$
The vertical stress at $z = H$ is

$$\sigma'_{o(H)} = \gamma_1 H \qquad (13.41)$$

So the factor of safety against bearing capacity failure is

$$FS_{(\text{bearing capacity})} = \frac{q_{\text{ult}}}{\sigma'_{o(H)}} \qquad (13.42)$$

Generally, minimum values of $FS_{(\text{overturning})} = 3$, $FS_{(\text{sliding})} = 3$, and $FS_{(\text{bearing capacity failure})} = 3$ to 5 are recommended.

Example 13.3

A 10 m-high retaining wall with galvanized steel-strip reinforcement in a granular backfill has to be constructed. Referring to Figure 13.11, given:

Granular backfill: $\phi'_1 = 36°$
  $\gamma_1 = 16.5 \text{ kN/m}^3$

Foundation soil: $\phi'_2 = 28°$
  $\gamma_2 = 17.3 \text{ kN/m}^3$
  $c'_2 = 48 \text{ kN/m}^2$

Galvanized steel reinforcement:

Width of strip, $w = 72 \text{ mm}$
  $S_V = 0.6 \text{ m center-to-center}$
  $S_H = 1 \text{ m center-to-center}$
  $f_y = 242 \text{ MN/m}^2$
  $\phi'_\mu = 20°$

Required $FS_{(B)} = 3$

Required $FS_{(P)} = 3$

Check for the external and internal stability. Assume the corrosion rate of the galvanized steel to be 0.025 mm/year and the life span of the structure to be 50 years.

Solution
Internal Stability Check

Tie thickness: Maximum tie force, $T_{max} = \sigma'_{a(max)} S_V S_H$

$$\sigma_{a(max)} = \gamma_1 H K_a = \gamma H \tan^2\left(45 - \frac{\phi'_1}{2}\right)$$

so

$$T_{max} = \gamma_1 H \tan^2\left(45 - \frac{\phi'_1}{2}\right) S_V S_H$$

From Eq. (13.33), for *tie break*,

$$t = \frac{(\sigma'_a S_V S_H)[FS_{(B)}]}{w f_y} = \frac{\left[\gamma_1 H \tan^2\left(45 - \frac{\phi'_1}{2}\right) S_V S_H\right] FS_{(B)}}{w f_y}$$

or

$$t = \frac{\left[(16.5)(10) \tan^2\left(45 - \frac{36}{2}\right)(0.6)(1)\right](3)}{(0.072 \text{ m}) (242,000 \text{ kN/m}^2)} = 0.00443 \text{ m}$$

$$\approx 4.5 \text{ mm}$$

If the rate of corrosion is 0.025 mm/yr and the life span of the structure is 50 yr, then the actual thickness, t, of the ties will be

$$t = 4.5 + (0.025)(50) = 5.75 \text{ mm}$$

So a **tie thickness of 6 mm** would be enough.

Tie length: Refer to Eq. (13.32). For this case, $\sigma'_a = \gamma_1 z K_a$ and $\sigma'_o = \gamma_1 z$, so

$$L = \frac{(H - z)}{\tan\left(45 + \frac{\phi'_1}{2}\right)} + \frac{FS_{(P)} \gamma_1 z K_a S_V S_H}{2w \gamma_1 z \tan \phi'_\mu}$$

Now the following table can be prepared. (Note: $FS_{(P)} = 3$, $H = 10$ m, $w = 0.006$ m, and $\phi'_\mu = 20°$.)

z(m)	Tie length L (m) [Eq. (13.32)]
2	13.0
4	11.99
6	10.97
8	9.95
10	8.93

So use a **tie length of $L = 13$ m**

External Stability Check

Check for overturning: Refer to Figure 13.13. For this case, using Eq. (13.36)

$$FS_{(overturning)} = \frac{W_1 x_1}{\left[\int_0^H \sigma_a' \, dz\right] z'}$$

$$W_1 = \gamma_1 HL = (16.5)(10)(13) = 2145 \text{ kN}$$

$$x_1 = 6.5\text{m}$$

$$P_a = \int_0^H \sigma_a' \, dz = \tfrac{1}{2}\gamma_1 K_a H^2 = (\tfrac{1}{2})(16.5)(0.26)(10)^2 = 214.5 \text{ kN/m}$$

$$z' = \frac{10}{3} = 3.33 \text{ m}$$

$$FS_{(overturning)} = \frac{(2145)(6.5)}{(214.5)(3.33)} = \mathbf{19.5 > 3—OK}$$

Check for sliding: From Eq. (13.37)

$$FS_{(sliding)} = \frac{W_1 \tan(k\phi_1')}{P_a} = \frac{2145 \tan\left[\left(\dfrac{2}{3}\right)(36)\right]}{214.5} = \mathbf{4.45 > 3—OK}$$

Check for bearing capacity: For $\phi_2' = 28°$, $N_c = 25.8$, $N_\gamma = 16.72$ (Table 12.2). From Eq. (13.38),

$$q_{ult} = c_2' N_c + \tfrac{1}{2}\gamma_2 L' N_\gamma$$

$$e = \frac{L}{2} - \frac{M_R - M_O}{\Sigma V} = \frac{13}{2} - \left[\frac{(2145 \times 6.5) - (214.5 \times 3.33)}{2145}\right] = 0.333 \text{ m}$$

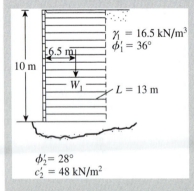

$$\gamma_1 = 16.5 \text{ kN/m}^3$$
$$\phi_1' = 36°$$

6.5 m

10 m

W_1

$L = 13$ m

$\phi_2' = 28°$
$c_2' = 48 \text{ kN/m}^2$

Figure 13.13 Retaining wall with galvanized steel-strip reinforcement in the backfill

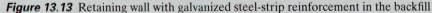

$$L' = 13 - (2 \times 0.333) = 12.334 \text{ m}$$

$$q_{ult} = (48)(25.8) + (\tfrac{1}{2})(17.3)(12.334)(16.72) = 3022 \text{ kN/m}^2$$

From Eq. (13.41),

$$\sigma'_{o(H)} = \gamma_1 H = (16.5)(10) = 165 \text{ kN/m}^2$$

$$\text{FS}_{(bearing\ capacity)} = \frac{q_{ult}}{\sigma'_{o(H)}} = \frac{3022}{165} = \mathbf{18.3 > 5\text{—OK}} \qquad \blacksquare$$

13.12 *Retaining Walls with Geotextile Reinforcement*

Figure 13.14 shows a retaining wall in which layers of geotextile have been used as reinforcement. As in Figure 13.12, the backfill is a granular soil. In this type of retaining wall, the facing of the wall is formed by lapping the sheets as shown with a lap length of l_l. When construction is finished, the exposed face of the wall must be covered; otherwise, the geotextile will deteriorate from exposure to ultraviolet light. *Bitumen emulsion* or *Gunite* is sprayed on the wall face. A wire mesh anchored to the geotextile facing may be necessary to keep the coating on.

The design of this type of retaining wall is similar to that presented in Section 13.11. Following is a step-by-step procedure for design based on the recommendations of Bell et al. (1975) and Koerner (1990):

Internal Stability

Step 1. Determine the active pressure distribution on the wall from the formula

$$\sigma'_a = K_a \sigma'_o = K_a \gamma_1 z \qquad (13.43)$$

where
K_a = Rankine active pressure coefficient = $\tan^2 (45 - \phi'_1/2)$
γ_1 = unit weight of the granular backfill
ϕ'_1 = friction angle of the granular backfill

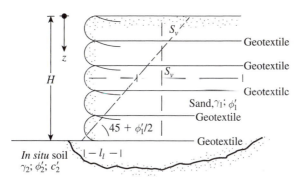

Figure 13.14
Retaining wall with geotextile reinforcement

Step 2.   Select a geotextile fabric with an allowable strength of σ_G (kN/m).

Step 3.   Determine the vertical spacing of the layers at any depth z from the formula

$$S_V = \frac{\sigma_G}{\sigma'_a FS_{(B)}} = \frac{\sigma_G}{(\gamma_1 z K_a)[FS_{(B)}]} \tag{13.44}$$

Note that Eq. (13.44) is similar to Eq. (13.25). The magnitude of $FS_{(B)}$ is generally 1.3 to 1.5.

Step 4.   Determine the length of each layer of geotextile from the formula

$$L = l_r + l_e \tag{13.45}$$

where

$$l_r = \frac{H - z}{\tan\left(45 + \dfrac{\phi'_1}{2}\right)} \tag{13.46}$$

and

$$l_e = \frac{S_V \sigma'_a [FS_{(P)}]}{2\sigma'_o \tan\phi'_F} \tag{13.47}$$

in which

$$\sigma'_a = \gamma_1 z K_a$$

$$\sigma'_o = \gamma_1 z$$

$$FS_{(P)} = 1.3 \text{ to } 1.5$$

$$\phi'_F = \text{friction angle at geotextile-soil interface}$$

$$\approx \tfrac{2}{3}\phi'_1$$

Step 5.   Determine the lap length, l_l, from

$$l_l = \frac{S_V \sigma'_a FS_{(P)}}{4\sigma'_o \tan\phi'_F} \tag{13.48}$$

The minimum lap length should be 1 m.

External Stability

Step 6.   Check the factors of safety against overturning, sliding, and bearing capacity failure as described in Section 13.11 (Steps 9, 10, and 11).

Example 13.4

A geotextile-reinforced retaining wall 5 m high is shown in Figure 13.15. For the granular backfill, $\gamma_1 = 17.3$ kN/m³ and $\phi'_1 = 36°$. For the geotextile, $\sigma_G = 14$ kN/m. For the design of the wall, determine S_V, L, and l_l.

Solution
We have

$$K_a = \tan^2\left(45 - \frac{\phi'_1}{2}\right) = 0.26$$

Determination of S_V

To find S_V, we make a few trials. From Eq. (13.44),

$$S_V = \frac{\sigma_G}{(\gamma_1 z K_a)[FS_{(B)}]}$$

With $FS_{(B)} = 1.5$ at $z = 2.5$ m,

$$S_V = \frac{14}{(17.3)(2.5)(0.26)(1.5)} = 0.83 \text{ m}$$

At $z = 5$ m,

$$S_V = \frac{14}{(17.3)(5)(0.26)(1.5)} = 0.42 \text{ m}$$

So, **use $S_V = 0.83$ m for $z = 0$ to $z = 2.5$ m and $S_V = 0.42$ m for $z > 2.5$ m** (See Figure 13.15.)

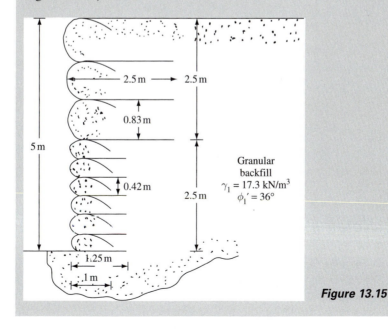

Granular backfill
$\gamma_1 = 17.3$ kN/m³
$\phi_1' = 36°$

Figure 13.15

Determination of L

From Eqs. (13.45), (13.46), and (13.47),

$$L = \frac{(H - z)}{\tan\left(45 + \frac{\phi'_1}{2}\right)} + \frac{S_V K_a [FS_{(P)}]}{2 \tan \phi'_F}$$

For $FS_{(P)} = 1.5$, $\tan \phi'_F = \tan\left[\left(\frac{2}{3}\right)(36)\right] = 0.445$, and it follows that

$$L = (0.51)(H - z) + 0.438 S_V$$

Now the following table can be prepared:

z (m)	S_V (m)	0.51(H − z) (m)	0.438S_V (m)	L (m)
0.83	0.83	2.13	0.364	2.490
1.66	0.83	1.71	0.364	2.074
2.50	0.83	1.28	0.364	1.644
2.92	0.42	1.06	0.184	1.244
3.34	0.42	0.85	0.184	1.034
3.76	0.42	0.63	0.184	0.814
4.18	0.42	0.42	0.184	0.604
4.60	0.42	0.20	0.184	0.384

On the basis of the preceding calculations, **use L = 2.5 m for z ≤ 2.5 m and L = 1.25 m for z > 2.5 m**

Determination of l_l

From Eq. (13.48),

$$l_l = \frac{S_V \sigma'_a [FS_{(P)}]}{4\sigma'_o \tan\phi'_F}$$

With $\sigma'_a = \gamma_1 z K_a$, $FS_{(P)} = 1.5$; with $\sigma'_o = \gamma_1 z$, $\phi'_F = \frac{2}{3}\phi'_1$. So

$$l_l = \frac{S_V K_a [FS_{(P)}]}{4 \tan\phi'_F} = \frac{S_V (0.26)(1.5)}{4 \tan\left[\left(\frac{2}{3}\right)(36)\right]} = 0.219 \, S_V$$

At $z = 0.83$ m,

$$l_l = 0.219 S_V = (0.219)(0.83) = 0.116 \text{m} \le 1\text{m}$$

So, use $l_l = 1$ m   ∎

13.13   *Retaining Walls with Geogrid Reinforcement*

Geogrids can also be used as reinforcement in granular backfill for the construction of retaining walls. Figure 13.16 shows typical schematic diagrams of retaining walls with geogrid reinforcement. The design procedure of a geogrid-reinforced retaining wall is essentially similar to that given in Section 13.12.

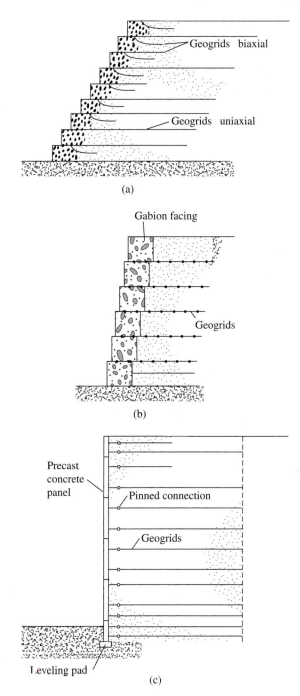

Figure 13.16 Typical schematic diagrams of retaining walls with geogrid reinforcement: (a) geogrid wraparound wall; (b) wall with gabion facing; (c) concrete panel-faced wall

BRACED CUTS

13.14   *Braced Cuts—General*

Figure 13.17 shows two types of braced cuts commonly used in construction work. One type uses the *soldier beam* (Figure 13.17a), which is a vertical steel or timber beam driven into the ground before excavation. *Laggings*, which are horizontal

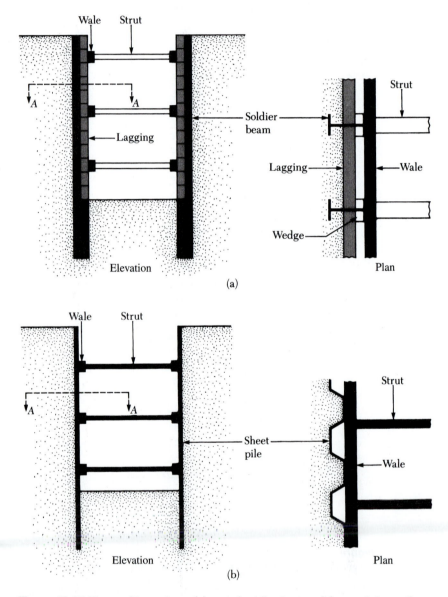

Figure 13.17 Types of braced cut: (a) use of soldier beams; (b) use of sheet piles

timber planks, are placed between soldier beams as the excavation proceeds. When the excavation reaches the desired depth, *wales* and *struts* (horizontal steel beams) are installed. The struts are horizontal compression members. Figure 13.17b shows another type of braced excavation. In this case, interlocking *sheet piles* are driven into the soil before excavation. Wales and struts are inserted immediately after excavation reaches the appropriate depth. A majority of braced cuts use sheet piles.

    Steel sheet piles in the United States are about 10 to 13 mm thick. European sections may be thinner and wider. Sheet pile sections may be *Z, deep arch, low arch*, or *straight web* sections. The interlocks of the sheet pile sections are shaped like a *thumb and finger* or a *ball and socket* for watertight connections. Figure 13.18a shows schematic diagrams of the thumb-and-finger type of interlocking for straight web sections. The ball-and-socket type of interlocking for *Z* section piles is shown in Figure 13.18b. Table 13.2 shows the properties of the sheet pile sections produced by Bethlehem Steel Corporation. The allowable design flexural stress for the steel sheet piles is as follows:

Type of steel	Allowable stress (MN/m^2)
ASTM A-328	170
ASTM A-572	210
ASTM A-690	210

Steel sheet piles are convenient to use because of their resistance to high driving stress developed when being driven into hard soils. They are also lightweight and reusable.

    Figure 13.19 shows the braced cut construction for the Chicago Subway in 1940. Timber lagging, timber struts, and steel wales were used. Figure 13.20 shows a braced cut made during the construction of the Washington, D.C. Metro in 1974. In this cut, timber lagging, steel H-soldier piles, steel wales, and pipe struts were used.

    To design braced excavations (that is, to select wales, struts, sheet piles, and soldier beams), an engineer must estimate the lateral earth pressure to which the braced cuts will be subjected. This topic is discussed in Section 13.15; subsequent sections cover the procedures of analysis and design of braced cuts.

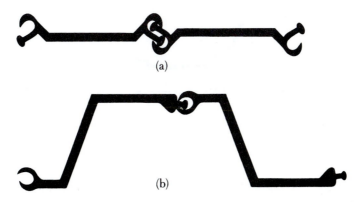

(a)

(b)

Figure 13.18 Nature of sheet pile connections: (a) thumb-and-finger type; (b) ball-and-socket type

Table 13.2 Properties of some sheet pile sections (Produced by Bethlehem Steel Corporation)

Section designation	Sketch of section	Section modulus (m³/m of wall)	Moment of inertia (m⁴/m of wall)
PZ-40		326.4×10^{-5}	670.5×10^{-6}
PZ-35		260.5×10^{-5}	493.4×10^{-6}
PZ-27		162.3×10^{-5}	251.5×10^{-6}
PZ-22		97×10^{-5}	115.2×10^{-6}
PSA-31		10.8×10^{-5}	4.41×10^{-6}
PSA-23		12.8×10^{-5}	5.63×10^{-6}

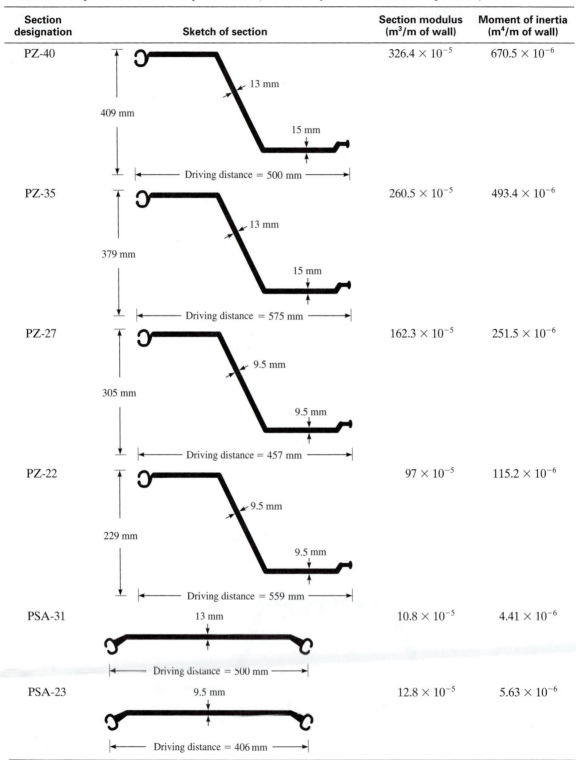

Figure 13.19
Braced cut in Chicago
subway construction
(Courtesy of Ralph B. Peck)

Figure 13.20 Braced cut in the construction of Washington, D.C. Metro
(Courtesy of Ralph B. Peck)

13.15   *Lateral Earth Pressure in Braced Cuts*

Chapter 11 explained that a retaining wall rotates about its bottom (Figure 13.21a). With sufficient yielding of the wall, the lateral earth pressure is approximately equal to that obtained by Rankine's theory or Coulomb's theory. In contrast to retaining walls, braced cuts show a different type of wall yielding (see Figure 13.21b). In this case, deformation of the wall gradually increases with the depth of excavation. The variation of the amount of deformation depends on several factors, such as the type of soil, the depth of excavation, and the workmanship. However, with very little wall yielding at the top of the cut, the lateral earth pressure will be close to the at-rest pressure. At the bottom of the wall, with a much larger degree of yielding, the lateral earth pressure will be substantially lower than the Rankine active earth pressure. As a result, the distribution of lateral earth pressure will vary substantially in comparison to the linear distribution assumed in the case of retaining walls. Also, the lateral earth pressure in a braced cut is dependent on the type of soil, construction method, type of equipment used, and workmanship. For all the uncertainties involved relating to lateral earth pressure distribution, it is a common practice to use an earth pressure envelope for design of braced cuts.

Using the strut loads observed from the Berlin subway cut, the Munich subway cut, and the New York subway cut, Peck (1969) provided the envelope for lateral pressure for design of cuts in *sand*. This is illustrated in Figure 13.22a. Note that in Figure 13.22a

$$\sigma = 0.65 \gamma H K_a \qquad (13.49)$$

where
  γ = unit weight
  H = height of the cut
  K_a = Rankine's active pressure coefficient = $\tan^2(45 - \phi'/2)$

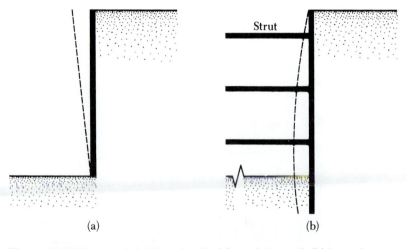

Figure 13.21 Nature of yielding of walls: (a) retaining wall; (b) braced cut

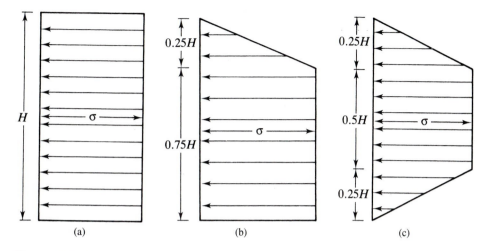

Figure 13.22 Peck's (1969) apparent pressure envelope for (a) cuts in sand; (b) cuts in soft to medium clay; (c) cuts in stiff clay

In a similar manner, Peck (1969) also provided the pressure envelopes in clay. The pressure envelope for soft to medium clay is shown in Figure 13.22b. It is applicable for the condition

$$\frac{\gamma H}{c} > 4$$

where c = undrained cohesion ($\phi = 0$). The pressure, σ, is the larger of

$$\sigma = \gamma H \left[1 - \left(\frac{4c}{\gamma H} \right) \right] \quad \text{or} \quad \sigma = 0.3\gamma H \tag{13.50}$$

where γ = unit weight of clay. The pressure envelope for cuts in stiff clay shown in Figure 13.22c, in which

$$\sigma = 0.2\gamma H \text{ to } 0.4\gamma H \quad \text{(with an average of } 0.3\gamma H) \tag{13.51}$$

is applicable to the condition $\gamma H/c \leq 4$.

Limitations for the Pressure Envelopes

When using the pressure envelopes just described, keep the following points in mind:

1. The pressure envelopes are sometimes referred to as *apparent pressure envelopes*. However, the actual pressure distribution is a function of the construction sequence and the relative flexibility of the wall.

2. They apply to excavations with depths greater than about 6 m.
3. They are based on the assumption that the water table is below the bottom of the cut.
4. Sand is assumed to be drained with 0 pore water pressure.
5. Clay is assumed to be undrained, and pore water pressure is not considered.

13.16   *Soil Parameters for Cuts in Layered Soil*

Sometimes, layers of both sand and clay are encountered when a braced cut is being constructed. In this case, Peck (1943) proposed that an equivalent value of cohesion ($\phi = 0$ concept) should be determined in the following manner (refer to Figure 13.23a):

$$c_{av} = \frac{1}{2H}\left[\gamma_s K_s H_s^2 \tan\phi_s' + (H - H_s)n'q_u\right] \tag{13.52}$$

where

H = total height of the cut
γ_s = unit weight of sand
H_s = height of the sand layer
K_s = a lateral earth pressure coefficient for the sand layer (≈ 1)
ϕ_s' = angle of friction of sand
q_u = unconfined compression strength of clay
n' = a coefficient of progressive failure (ranges from 0.5 to 1.0; average value 0.75)

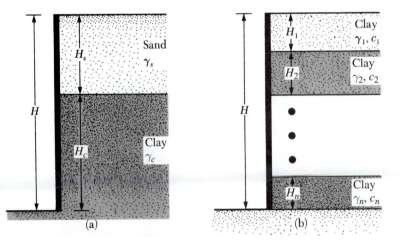

Figure 13.23 Layered soils in braced cuts

The average unit weight, γ_a, of the layers may be expressed as

$$\gamma_a = \frac{1}{H}[\gamma_s H_s + (H - H_s)\gamma_c] \tag{13.53}$$

where γ_c = saturated unit weight of clay layer. Once the average values of cohesion and unit weight are determined, the pressure envelopes in clay can be used to design the cuts.

Similarly, when several clay layers are encountered in the cut (Figure 13.23b), the average undrained cohesion becomes

$$c_{av} = \frac{1}{H}(c_1 H_1 + c_2 H_2 + \cdots + c_n H_n) \tag{13.54}$$

where $c_1, c_2, \cdots, c_n$ = undrained cohesion in layers $1, 2, \cdots, n$
$H_1, H_2, \cdots, H_n$ = thickness of layers $1, 2, \cdots, n$

The average unit weight, γ_a, is

$$\gamma_a = \frac{1}{H}(\gamma_1 H_1 + \gamma_2 H_2 + \gamma_3 H_3 + \cdots + \gamma_n H_n) \tag{13.55}$$

13.17   *Design of Various Components of a Braced Cut*

Struts

In construction work, struts should have a minimum vertical spacing of about 3 m or more. The struts are actually horizontal columns subject to bending. The load-carrying capacity of columns depends on the *slenderness ratio, l/r*. The slenderness ratio can be reduced by providing vertical and horizontal supports at intermediate points. For wide cuts, splicing the struts may be necessary. For braced cuts in clayey soils, the depth of the first strut below the ground surface should be less than the depth of tensile crack, z_o. From Eq. (11.16), we have

$$\sigma'_a = \gamma z K_a - 2c'\sqrt{K_a}$$

where K_a = coefficient of Rankine's active pressure. For determining the depth of tensile crack, we use

$$\sigma'_a = 0 = \gamma z_o K_a - 2c'\sqrt{K_a}$$

or

$$z_o = \frac{2c'}{\sqrt{K_a}\gamma}$$

With $\phi = 0$, $K_a = \tan^2(45 - \phi/2) = 1$. So

$$z_o = \frac{2c}{\gamma} \quad (\textit{Note: } c = c_u)$$

   A simplified conservative procedure may be used to determine the strut loads. Although this procedure will vary depending on the engineers involved in the project, the following is a step-by-step outline of the general procedure (refer to Figure 13.24):

1. Draw the pressure envelope for the braced cut (see Figure 13.22). Also show the proposed strut levels. Figure 13.24a shows a pressure envelope for a sandy soil; however, it could also be for a clay. The strut levels are marked *A, B, C,* and *D.* The sheet piles (or soldier beams) are assumed to be hinged at the strut levels, except for the top and bottom ones. In Figure 13.24a, the hinges are at the level of struts *B* and *C.* (Many designers also assume the sheet piles, or soldier beams, to be hinged at all strut levels, except for the top.)
2. Determine the reactions for the two simple cantilever beams (top and bottom) and all the simple beams between. In Figure 13.24b, these reactions are $A, B_1,$ $B_2, C_1, C_2,$ and $D.$

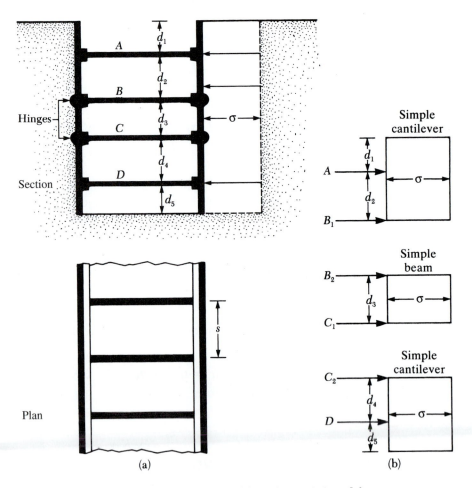

Figure 13.24 Determination of strut loads: (a) section and plan of the cut; (b) method for determining strut loads

3. Calculate the strut loads in Figure 13.24 as follows:

$$P_A = (A)(s)$$
$$P_B = (B_1 + B_2)(s)$$
$$P_C = (C_1 + C_2)(s)$$
$$P_D = (D)(s) \qquad (13.56)$$

where
P_A, P_B, P_C, P_D = loads to be taken by the individual struts at levels A, B, C, and D, respectively
A, B_1, B_2, C_1, C_2, D = reactions calculated in step 2 (note unit: force/unit length of the braced cut)
s = horizontal spacing of the struts (see plan in Figure 13.24a)

4. Knowing the strut loads at each level and the intermediate bracing conditions allows selection of the proper sections from the steel construction manual.

Sheet Piles

The following steps are taken in designing the sheet piles:

1. For each of the sections shown in Figure 13.24b, determine the maximum bending moment.
2. Determine the maximum value of the maximum bending moments (M_{max}) obtained in step 1. Note that the unit of this moment will be, for example, kN-m/m length of the wall.
3. Obtain the required section modulus of the sheet piles:

$$S = \frac{M_{max}}{\sigma_{all}} \qquad (13.57)$$

where σ_{all} = allowable flexural stress of the sheet pile material.
4. Choose a sheet pile that has a section modulus greater than or equal to the required section modulus from a table such as Table 13.2.

Wales

Wales may be treated as continuous horizontal members if they are spliced properly. Conservatively, they may also be treated as though they are pinned at the struts. For the section shown in Figure 13.24a, the maximum moments for the wales (assuming that they are pinned at the struts) are

At level A, $\qquad M_{max} = \dfrac{(A)(s^2)}{8}$

At level B, $\qquad M_{max} = \dfrac{(B_1 + B_2)s^2}{8}$

At level C, $M_{max} = \dfrac{(C_1 + C_2)s^2}{8}$

At level D, $M_{max} = \dfrac{(D)(s^2)}{8}$

where A, B_1, B_2, C_1, C_2, and D are the reactions under the struts per unit length of the wall (step 2 of strut design).

We can determine the section modulus of the wales with

$$S = \frac{M_{max}}{\sigma_{all}}$$

The wales are sometimes fastened to the sheet piles at points that satisfy the lateral support requirements.

Example 13.5

Refer to the braced cut shown in Figure 13.25. Given $\gamma = 17.6 \text{ kN/m}^3$, $\phi' = 32°$, and $c' = 0$. The struts are located at 4 m center-to-center in the plan. Draw the earth pressure envelope and determine the strut loads at levels A, B, and C.

Solution

For this case, the earth pressure envelope shown in Figure 13.22a will apply.

$$K_a = \tan^2\left(45 - \frac{\phi'}{2}\right) = \tan^2\left(45 - \frac{32}{2}\right) = 0.307$$

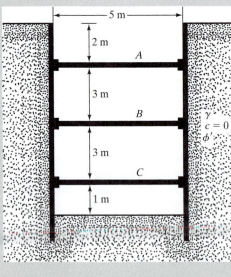

Figure 13.25

From Equation (13.49)

$$\sigma = 0.65\gamma H K_a = (0.65)(17.6)(9)(0.307) = 31.6 \text{ kN/m}^2$$

Figure 13.26a shows the pressure envelope. Now, referring to Figure 13.26b,

$$\sum M_{B_1} = 0$$

$$A = \frac{(31.6)(5)\left(\dfrac{5}{2}\right)}{3} = 131.67 \text{ kN/m}$$

$$B_1 = (31.6)(5) - 131.67 = 26.33 \text{ kN/m}$$

Again, referring to Figure 13.26c,

$$\sum M_{B_2} = 0$$

$$C = \frac{(31.6)(4)\left(\dfrac{4}{2}\right)}{3} = 84.27 \text{ kN/m}$$

$$B_2 = (31.6)(4) - 84.27 = 42.13 \text{ kN/m}$$

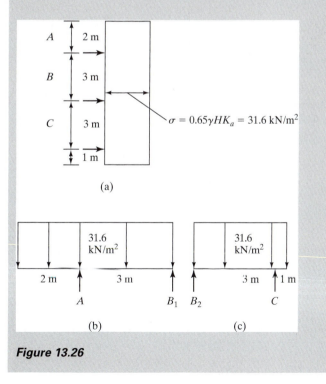

Figure 13.26

Strut load at $A = (131.67)(\text{spacing}) = (131.67)(4)$

$$= \textbf{526.68 kN}$$

Strut load at $B = (B_1 + B_2)(\text{spacing}) = (26.33 + 42.13)(4)$

$$= \textbf{273.84 kN}$$

Strut load at $C = (84.27)(s) = (84.27)(4)$

$$= \textbf{337.08 kN} \qquad \blacksquare$$

Example 13.6

For the braced cut described in Example 13.5, determine the following:
a. The sheet pile section modulus. Use $\sigma_{all} = 170 \times 10^3$ kN/m^2.
b. The required section modulus of the wales at level A. Assume that $\sigma_{all} = 173 \times 10^3$ kN/m^2.

Solution
Part a
Refer to the load diagrams shown in Figures 13.26b and 13.26c. Based on the load diagrams, the shear force diagrams are as given in Figure 13.27.

$$x_1 = \frac{68.47}{31.6} = 2.17$$

$$x_2 = \frac{52.67}{31.6} = 1.67$$

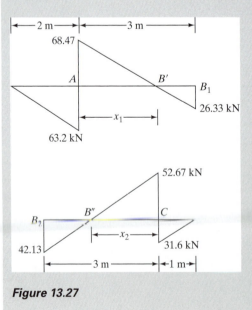

Figure 13.27

$$\text{Moment at } A = \frac{1}{2}(63.2)(2) = 63.2 \text{ kN-m/m}$$

$$\text{Moment at } C = \frac{1}{2}(31.6)(1) = 15.8 \text{ kN-m/m}$$

$$\text{Moment at } B' = \frac{1}{2}(26.33)(0.83) = 10.93 \text{ kN-m/m}$$

$$\text{Moment at } B'' = \frac{1}{2}(42.13)(1.33) = 28.02 \text{ kN-m/m}$$

M_A is maximum.

$$S_x = \frac{M_{max}}{\sigma_{all}} = \frac{63.2 \text{ kN-m/m}}{170 \times 10^3 \text{ kN/m}^2} = 37.2 \times 10^{-5} \text{ m}^3/\text{m}$$

Part b
For the wale at level A,

$$M_{max} = \frac{A(s^2)}{8}$$

$A = 131.67$ kN/m (from Example 13.5). So

$$M_{max} = \frac{(131.67)(4^2)}{8} = 263.34 \text{ kN-m}$$

$$S_x = \frac{M_{max}}{\sigma_{all}} = \frac{263.34}{173 \times 10^3} = \mathbf{1.522 \times 10^{-3} \, m^3/m} \qquad \blacksquare$$

13.18   *Heave of the Bottom of a Cut in Clay*

Braced cuts in clay may become unstable as a result of heaving of the bottom of the excavation. Terzaghi (1943) analyzed the factor of safety of long braced excavations against bottom heave. The failure surface for such a case is shown in Figure 13.28. The vertical load per unit length of the cut at the bottom of the cut along line *bd* and *af* is

$$Q = \gamma H B_1 - cH \qquad (13.58)$$

where
$\quad B_1 = 0.7B$
$\quad c = \text{cohesion } (\phi = 0 \text{ concept})$

This load Q may be treated as a load per unit length on a continuous foundation at the level of *bd* (and *af*) and having a width of $B_1 = 0.7B$. Based on Terzaghi's bearing capacity theory, the net ultimate load-carrying capacity per unit length of this foundation (Chapter 12) is

$$Q_u = cN_cB_1 = 5.7cB_1$$

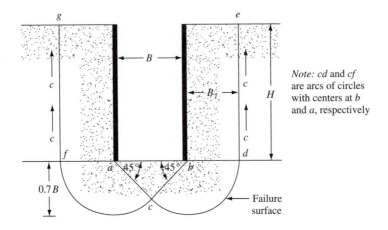

Figure 13.28 Factor of safety against bottom heave

Hence, from Eq. (13.58), the factor of safety against bottom heave is

$$FS = \frac{Q_u}{Q} = \frac{5.7cB_1}{\gamma HB_1 - cH} = \frac{1}{H}\left(\frac{5.7c}{\gamma - \dfrac{c}{0.7B}}\right) \qquad (13.59)$$

This factor of safety is based on the assumption that the clay layer is homogeneous, at least to a depth of $0.7B$ below the bottom of the cut. However, a *hard layer of rock or rocklike material at a depth of $D < 0.7B$* will modify the failure surface to some extent. In such a case, the factor of safety becomes

$$FS = \frac{1}{H}\left(\frac{5.7c}{\gamma - c/D}\right) \qquad (13.60)$$

Bjerrum and Eide (1956) also studied the problem of bottom heave for braced cuts in clay. For the factor of safety, they proposed:

$$FS = \frac{cN_c}{\gamma H} \qquad (13.61)$$

The bearing capacity factor, N_c, varies with the ratios H/B and L/B (where $L =$ length of the cut). For infinitely long cuts ($B/L = 0$), $N_c = 5.14$ at $H/B = 0$ and increases to $N_c = 7.6$ at $H/B = 4$. Beyond that—that is, for $H/B > 4$—the value of N_c

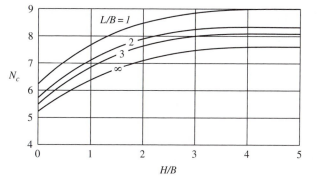

Figure 13.29
Variation of N_c with L/B and H/B [based on Bjerrum and Eide's equation, Eq. (13.62)]

remains constant. For cuts square in plan ($B/L = 1$), $N_c = 6.3$ at $H/B = 0$, and $N_c = 9$ for $H/B \geqslant 4$. In general, for any H/B,

$$N_{c(\text{rectangle})} = N_{c(\text{square})} \left(0.84 + 0.16 \frac{B}{L} \right) \qquad (13.62)$$

Figure 13.29 shows the variation of the value of N_c for $L/B = 1, 2, 3$, and ∞.

When Eqs. (13.61) and (13.62) are combined, the factor of safety against heave becomes

$$FS = \frac{cN_{c(\text{square})} \left(0.84 + 0.16 \dfrac{B}{L} \right)}{\gamma H} \qquad (13.63)$$

Equation (13.63) and the variation of the bearing capacity factor, N_c, as shown in Figure 13.29 are based on the assumptions that the clay layer below the bottom of the cut is homogeneous, and that the magnitude of the undrained cohesion in the soil that contains the failure surface is equal to c (Figure 13.30).

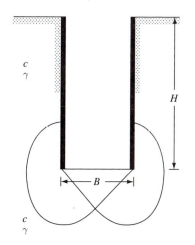

Figure 13.30
Derivation of Eq. (13.63)

13.19 *Lateral Yielding of Sheet Piles and Ground Settlement*

In braced cuts, some lateral movement of sheet pile walls may be expected (Figure 13.31). The amount of lateral yield depends on several factors, the most important of which is the elapsed time between excavation and placement of wales and struts. As discussed before, in several instances the sheet piles (or the soldier piles, as the case may be) are driven to a certain depth below the bottom of the excavation. The reason is to reduce the lateral yielding of the walls during the last stages of excavation. Lateral yielding of the walls will cause the ground surface surrounding the cut to settle (Figure 13.31). The degree of lateral yielding, however, depends mostly on the soil type below the bottom of the cut. If clay below the cut extends to a great depth and $\gamma H/c$ is less than about 6, extension of the sheet piles or soldier piles below the bottom of the cut will help considerably in reducing the lateral yield of the walls.

However, under similar circumstances, if $\gamma H/c$ is about 8, the extension of sheet piles into the clay below the cut does not help greatly. In such circumstances, we may expect a great degree of wall yielding that may result in the total collapse of the bracing systems. If a hard soil layer lies below a clay layer at the bottom of the cut, the piles should be embedded in the stiffer layer. This action will greatly reduce lateral yield.

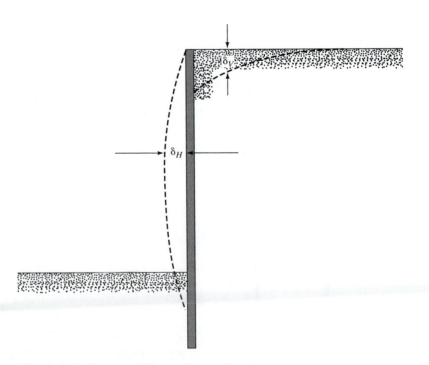

Figure 13.31 Lateral yielding and ground settlement

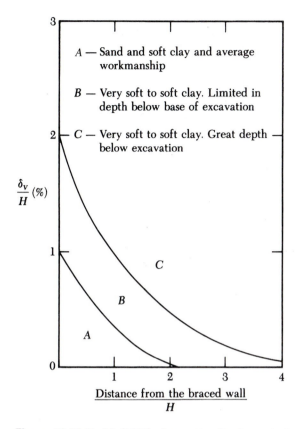

Figure 13.32 Peck's (1969) observation for the variation of ground settlement with distance

The lateral yielding of walls will generally induce ground settlement, δ_V, around a braced cut, which is generally referred to as *ground loss*. Based on several field observations, Peck (1969) provided curves for predicting ground settlement in various types of soil (see Figure 13.32). The magnitude of ground loss varies extensively; however, Figure 13.32 may be used as a general guide.

Problems

For Problems 13.1–13.5, use unit weight of concrete, $\gamma_c = 23.58$ kN/m³. Also assume $k_1 = k_2 = \frac{2}{3}$ in Eq. (13.13).

13.1 For the cantilever retaining wall shown in Figure 13.33, the wall dimensions are $H = 8$ m, $x_1 = 0.4$ m, $x_2 = 0.6$ m, $x_3 = 1.5$ m, $x_4 = 3.5$ m, $x_5 = 0.96$ m, $D = 1.75$ m, and $\alpha = 10°$; and the soil properties are $\gamma_1 = 16.8$ kN/m³, $\phi_1' = 32°$, $\gamma_2 = 17.6$ kN/m³, $\phi_2' = 28°$, and $c_2' = 30$ kN/m². Calculate the factors of safety with respect to overturning, sliding, and bearing capacity.

13.2 Repeat Problem 13.2 for the wall dimensions $H = 6$ m, $x_1 = 0.3$ m, $x_2 = 0.7$ m, $x_3 = 1.4$ m, $x_4 = 2.3$ m, $x_5 = 0.85$ m, $D = 1.25$ m, and $\alpha = 5°$; and the soil

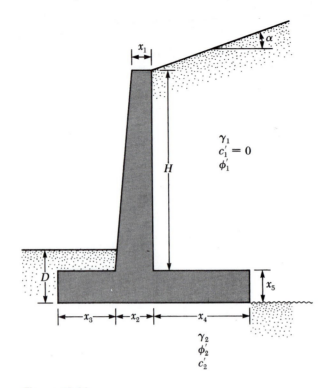

Figure 13.33

properties $\gamma_1 = 18.4$ kN/m³, $\phi'_1 = 34°$, $\gamma_2 = 16.8$ kN/m³, $\phi'_2 = 18°$, and $c'_2 = 50$ kN/m².

13.3 Repeat Problem 13.1 with wall dimensions of $H = 5.49$ m, $x_1 = 0.46$ m, $x_2 = 0.58$ m, $x_3 = 0.92$ m, $x_4 = 1.55$ m, $x_5 = 0.61$ m, $D = 1.22$ m, and $\alpha = 0°$; and soil properties of $\gamma_1 = 18.08$ kN/m³, $\phi'_1 = 36°$, $\gamma_2 = 19.65$ kN/m³, $\phi'_2 = 15°$, and $c'_2 = 44$ kN/m².

13.4 A gravity retaining wall is shown in Figure 13.34. Calculate the factors of safety with respect to overturning and sliding. We have wall dimensions $H = 6$ m, $x_1 = 0.6$ m, $x_2 = 0.2$ m, $x_3 = 2$ m, $x_4 = 0.5$ m, $x_5 = 0.75$ m, $x_6 = 0.8$ m, and $D = 1.5$ m; and soil properties $\gamma_1 = 16.5$ kN/m³, $\phi'_1 = 32°$, $\gamma_2 = 18$ kN/m³, $\phi'_2 = 22°$, and $c'_2 = 40$ kN/m². Use Rankine's active pressure for calculation.

13.5 Repeat Problem 13.4 using Coulomb's active pressure for calculation and $\delta' = \frac{2}{3}\phi'_1$.

13.6 A reinforced earth retaining wall (Figure 13.11) is to be 10 m high. Here,

Backfill: unit weight, $\gamma_1 = 18.7$ kN/m³ and soil friction angle, $\phi'_1 = 34°$

Reinforcement: vertical spacing, $S_V = 1$ m; horizontal spacing, $S_H = 1.3$ m; width of reinforcement = 120 mm, $f_y = 262$ MN/m² $\phi_\mu = 25°$; factor of safety against tie pullout = 3; and factor of safety against tie breaking = 3

Determine:
a. The required thickness of ties
b. The required maximum length of ties

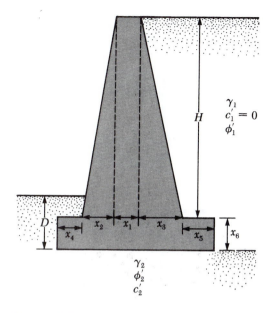

$$\gamma_1$$
$$c'_1 = 0$$
$$\phi'_1$$

$$\gamma_2$$
$$\phi'_2$$
$$c'_2$$

Figure 13.34

13.7 In Problem 13.6 assume that the ties at all depths are the length determined in Part b. For the *in situ* soil, $\phi'_2 = 25°$, $\gamma_2 = 18.2$ kN/m^3, $c'_2 = 31$ kN/m^2. Calculate the factor of safety against (a) overturning, (b) sliding, and (c) bearing capacity failure.

13.8 Redo Problem 13.6 for a retaining wall with a height of 8 m.

13.9 A retaining wall with geotextile reinforcement is 6-m high. For the granular backfill, $\gamma_1 = 15.9$ kN/m^3 and $\phi'_1 = 30°$. For the geotextile, $\sigma_G = 16$ kN/m. For the design of the wall, determine S_V, L, and l_l. Use $FS_{(B)} = FS_{(P)} = 1.5$.

13.10 The S_V, L, and l_l determined in Problem 13.9, check the overall stability (i.e., factor of safety overturning, sliding, and bearing capacity failure) of the wall. For the *in situ* soil, $\gamma_2 = 16.8$ kN/m^3, $\phi'_2 = 20°$, and $c'_2 = 55$ kN/m^2.

13.11 Refer to the braced cut in Figure 13.35, for which $\gamma = 17$ kN/m^3, $\phi' = 30°$, and $c' = 0$. The struts are located at 3 m on center in the plan. Draw the earth pressure envelope and determine the strut loads at levels A, B, and C.

13.12 For the braced cut described in Problem 13.11, assume that $\sigma_{all} = 170$ MN/m^2.
 a. Determine the sheet pile section (section modulus)
 b. What is the section modulus of the wales at level A?

13.13 Refer to Figure 13.36 in which $\gamma = 17.5$ kN/m^3, $c = 60$ kN/m^2, and center-to-center spacing of struts is 5 m. Draw the earth pressure envelope and determine the strut loads at levels A, B, and C.

13.14 Refer to Figure 13.23a. For the braced cut, $H - 6$ m, $H_s = 2$ m, $\gamma_s = 16.2$ kN/m^3, angle of friction of sand, $\phi'_s = 34°$, $H_c = 4$ m, $\gamma_c = 17.5$ kN/m^3, and the unconfined compression strength of the clay layer, $q_u = 68$ kN/m^2.
 a. Estimate the average cohesion, c_{av}, and the average unit weight, γ_{av}, for development of the earth pressure envelope.
 b. Plot the earth pressure envelope.

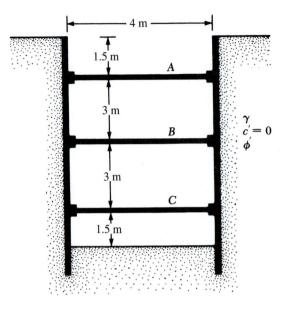

Figure 13.35

13.15 Refer to Figure 13.23b, which shows a braced cut in clay. Here, $H = 7$ m, $H_1 = 2$ m, $c_1 = 102$ kN/m^2, $\gamma_1 = 17.5$ kN/m^3, $H_2 = 2.5$ m, $c_2 = 75$ kN/m^2, $\gamma_2 = 16.8$ kN/m^3, $H_3 = 2.5$ m, $c_3 = 80$ kN/m^2, and $\gamma_3 = 17$ kN/m^3.
   a. Determine the average cohesion, c_{av}, and the average unit weight, γ_{av}, for development of the earth pressure envelope.
   b. Plot the earth pressure envelope.
13.16 Determine the factor of safety against bottom heave for the braced cut described in Problem 13.13. Use Eqs. (13.59) and (13.63). For Eq. (13.63), assume the length of the cut, $L = 18$ m.

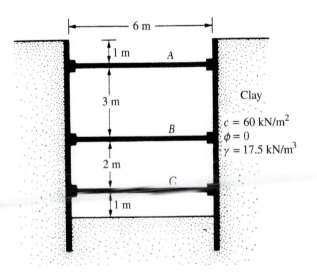

Figure 13.36

References

BELL, J. R., STILLEY, A. N., and VANDRE, B. (1975). "Fabric Retaining Earth Walls," *Proceedings, Thirteenth Engineering Geology and Soils Engineering Symposium*, Moscow, ID.

BINQUET, J., and LEE, K. L. (1975). "Bearing Capacity Analysis of Reinforced Earth Slabs," *Journal of the Geotechnical Engineering Division*, American Society of Civil Engineers, Vol. 101, No. GT12, 1257–1276.

BJERRUM, L, and EIDE, O. (1956). "Stability of Strutted Excavation in Clay," *Geotechnique*, Vol. 6 No. 1, 32–47.

DARBIN, M. (1970). "Reinforced Earth for Construction of Freeways" (in French), *Revue Générale des Routes et Aerodromes*, No. 457, Sept.

KOERNER, R. B. (1990). *Design with Geosynthetics*, 2d ed., Prentice Hall, Englewood Cliffs, NJ.

LEE, K. L., ADAMS, B. D., and VAGNERON, J. J. (1973). "Reinforced Earth Retaining Walls," *Journal of the Soil Mechanics and Foundations Division*, American Society of Civil Engineers, Vol. 99, No. SM10, 745–763.

PECK, R. B. (1943). "Earth Pressure Measurements in Open Cuts, Chicago (Ill.) Subway," *Transactions*, American Society of Civil Engineers, Vol. 108, 1008–1058.

PECK, R. B. (1969). "Deep Excavation and Tunneling in Soft Ground," *Proceedings*, Seventh International Conference on Soil Mechanics and Foundation Engineering, Mexico City, State-of-the-Art Volume, 225–290.

SCHLOSSER, F., and LONG, N. (1974). "Recent Results in French Research on Reinforced Earth," *Journal of the Construction Division*, American Society of Civil Engineers, Vol. 100, No. CO3, 113–237.

SCHLOSSER, F., and VIDAL, H. (1969). "Reinforced Earth" (in French), *Bulletin de Liaison des Laboratoires Routier*, Ponts et Chaussées, Paris, France, Nov., 101–144.

TERZAGHI, K. (1943). *Theoretical Soil Mechanics*, Wiley, New York.

VIDAL, H. (1966). "La terre Armee," *Anales de l'Institut Technique du Bâtiment et des Travaux Publiques*, France, July–August, 888–938.

14

Deep Foundations—Piles and Drilled Shafts

Piles are structural members made of steel, concrete, and/or timber. They are used to build pile foundations, which are deep and more costly than shallow foundations (see Chapter 12). Despite the cost, the use of piles is often necessary to ensure structural safety. Drilled shafts are cast-in-place piles that generally have a diameter greater than 750 mm with or without steel reinforcement and with or without an enlarged bottom. The first part of this chapter considers pile foundations, and the second part presents a detailed discussion on drilled shafts.

PILE FOUNDATIONS

14.1 *Need for Pile Foundations*

Pile foundations are needed in special circumstances. The following are some situations in which piles may be considered for the construction of a foundation.

1. When the upper soil layer(s) is (are) highly compressible and too weak to support the load transmitted by the superstructure, piles are used to transmit the load to underlying bedrock or a stronger soil layer, as shown in Figure 14.1a. When bedrock is not encountered at a reasonable depth below the ground surface, piles are used to transmit the structural load to the soil gradually. The resistance to the applied structural load is derived mainly from the frictional resistance developed at the soil–pile interface (Figure 14.1b).
2. When subjected to horizontal forces (see Figure 14.1c), pile foundations resist by bending while still supporting the vertical load transmitted by the superstructure. This situation is generally encountered in the design and construction of earth-retaining structures and foundations of tall structures that are subjected to strong wind and/or earthquake forces.
3. In many cases, the soils at the site of a proposed structure may be expansive and collapsible. These soils may extend to a great depth below the ground surface. Expansive soils swell and shrink as the moisture content increases and

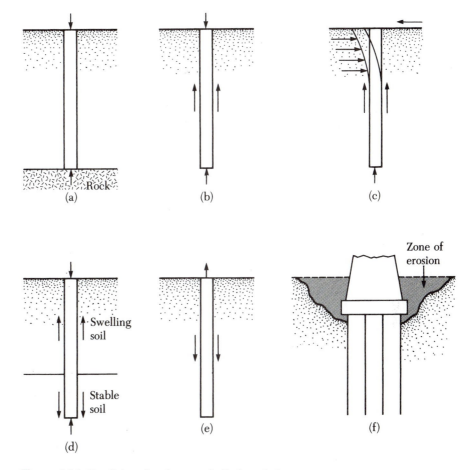

Figure 14.1 Conditions for the use of pile foundations

decreases, and the swelling pressure of such soils can be considerable. If shallow foundations are used, the structure may suffer considerable damage. However, pile foundations may be considered as an alternative when piles are extended beyond the active zone, which swells and shrinks (Figure 14.1d). Soils such as loess are collapsible. When the moisture content of these soils increases, their structures may break down. A sudden decrease in the void ratio of soil induces large settlements of structures supported by shallow foundations. In such cases, pile foundations may be used, in which piles are extended into stable soil layers beyond the zone of possible moisture change.

4. The foundations of some structures, such as transmission towers, offshore platforms, and basement mats below the water table, are subjected to uplifting forces. Piles are sometimes used for these foundations to resist the uplifting force (Figure 14.1e).

5. Bridge abutments and piers are usually constructed over pile foundations to avoid the possible loss of bearing capacity that a shallow foundation might suffer because of soil erosion at the ground surface (Figure 14.1f).

Although numerous investigations, both theoretical and experimental, have been conducted to predict the behavior and the load-bearing capacity of piles in granular and cohesive soils, the mechanisms are not yet entirely understood and may never be clear. The design of pile foundations may be considered somewhat of an "art" as a result of the uncertainties involved in working with some subsoil conditions.

14.2   Types of Piles and Their Structural Characteristics

Different types of piles are used in construction work, depending on the type of load to be carried, the subsoil conditions, and the water table. Piles can be divided into these categories: (a) steel piles, (b) concrete piles, (c) wooden (timber) piles, and (d) composite piles.

Steel Piles

Steel piles generally are either *pipe piles* or *rolled steel* H-*section piles*. Pipe piles can be driven into the ground with their ends open or closed. Wide-flange and I-section steel beams can also be used as piles; however, H-section piles are usually preferred because their web and flange thicknesses are equal. In wide-flange and I-section beams, the web thicknesses are smaller than the thicknesses of the flange. Table 14.1 gives the dimensions of some standard H-section steel piles used in the United States. Table 14.2 shows selected pipe sections frequently used for piling purposes. In many cases, the pipe piles are filled with concrete after they are driven.

When necessary, steel piles are spliced by welding or by riveting. Figure 14.2a shows a typical splicing by welding for an H-pile. A typical splicing by welding for a pipe pile is shown in Figure 14.2b. Figure 14.2c shows a diagram of splicing an H-pile by rivets or bolts.

Table 14.1 Common H-section piles used in the United States

Designation, size (mm) × weight (kN/m)	Depth, d_1 (mm)	Section area (m² × 10⁻³)	Flange and web thickness, w (mm)	Flange width (mm)	Moment of inertia (m⁴ × 10⁻⁶)	
					I_{xx}	I_{yy}
HP 200 × 0.52	204	6.84	11.3	207	49.4	16.8
HP 250 × 0.834	254	10.8	14.4	260	123	42
× 0.608	246	8.0	10.6	256	87.5	24
HP 310 × 1.226	312	15.9	17.5	312	271	89
× 1.079	308	14.1	15.5	310	237	77.5
× 0.912	303	11.9	13.1	308	197	63.7
× 0.775	299	10.0	11.1	306	164	62.9
HP 330 × 1.462	334	19.0	19.5	335	370	123
× 1.264	329	16.5	16.9	333	314	104
× 1.069	324	13.9	14.5	330	263	86
× 0.873	319	11.3	11.7	328	210	69

Table 14.1 (*continued*)

Designation, size (mm) × weight (kN/m)	Depth, d_1 (mm)	Section area (m² × 10⁻³)	Flange and web thickness, w (mm)	Flange width (mm)	Moment of inertia (m⁴ × 10⁻⁶)	
					I_{xx}	I_{yy}
HP 360 × 1.707	361	22.2	20.5	378	508	184
× 1.491	356	19.4	17.9	376	437	158
× 1.295	351	16.8	15.6	373	374	136
× 1.060	346	13.8	12.8	371	303	109

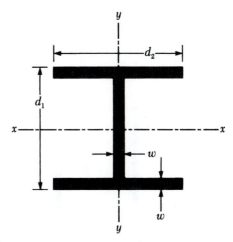

Table 14.2 Selected pipe pile sections

Outside diameter (mm)	Wall thickness (mm)	Area of steel (cm²)	Outside diameter (mm)	Wall thickness (mm)	Area of steel (cm²)
219	3.17	21.5	457	5.56	80
	4.78	32.1		6.35	90
	5.56	37.3		7.92	112
	7.92	52.7	508	5.56	88
254	4.78	37.5		6.35	100
	5.56	43.6		7.92	125
	6.35	49.4	610	6.35	121
305	4.78	44.9		7.92	150
	5.56	52.3		9.53	179
	6.35	59.7		12.70	238
406	4.78	60.3			
	5.56	70.1			
	6.35	79.8			

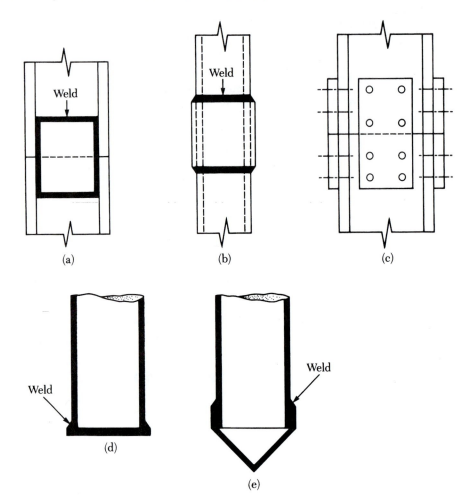

Figure 14.2 Steel piles: (a) splicing of H-pile by welding; (b) splicing of pipe pile by welding; (c) splicing of H-pile by rivets or bolts; (d) flat driving point of pipe pile; (e) conical driving point of pipe pile.

When hard driving conditions are expected, such as driving through dense gravel, shale, and soft rock, steel piles can be fitted with driving points or shoes. Figures 14.2d and e are diagrams of two types of shoe used for pipe piles.

Following are some general facts about steel piles.

Usual length: 15 m–60 m
Usual load: 300 kN–1200 kN
Advantages: a. Easy to handle with respect to cutoff and extension to the desired length
b. Can stand high driving stresses
c. Can penetrate hard layers such as dense gravel, soft rock
d. High load-carrying capacity

Disadvantages: a. Relatively costly material
                    b. High level of noise during pile driving
                    c. Subject to corrosion
                    d. H-piles may be damaged or deflected from the vertical during driving through hard layers or past major obstructions

Concrete Piles

Concrete piles may be divided into two basic types: precast piles and cast-*in-situ* piles. *Precast piles* can be prepared using ordinary reinforcement, and they can be square or octagonal in cross section (Figure 14.3). Reinforcement is provided to enable the pile to resist the bending moment developed during pickup and transportation, the vertical load, and the bending moment caused by lateral load. The piles are cast to desired lengths and cured before being transported to the work sites.

Precast piles can also be prestressed by using high-strength steel prestressing cables. The ultimate strength of these steel cables is about 1800 MN/m^2. During casting of the piles, the cables are pretensioned to 900 to 1300 MN/m^2, and concrete is poured around them. After curing, the cables are cut, thus producing a compressive force on the pile section. Table 14.3 gives additional information about prestressed concrete piles with square and octagonal cross sections.

The general details of the precast concrete piles are as follows:

Usual length: 10 m–15 m
Usual load: 300 kN–3000 kN
Advantages: a. Can be subjected to hard driving
                  b. Corrosion resistant
                  c. Can be easily combined with concrete superstructure
Disadvantages: a. Difficult to achieve proper cutoff
                  b. Difficult to transport

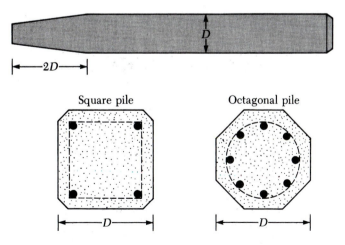

Figure 14.3 Precast piles with ordinary reinforcement

Table 14.3 Typical prestressed concrete piles

| Pile shape* | D (mm) | Area of cross section (cm²) | Perimeter (mm) | Number of strands | | Minimum effective prestress force (kN) | Section modulus (m³ × 10⁻³) | Design bearing capacity (kN) Concrete strength (MN/m²) | |
				12.7-mm diameter	11.1-mm diameter			34.5	41.4
S	254	645	1016	4	4	312	2.737	556	778
O	254	536	838	4	4	258	1.786	462	555
S	305	929	1219	5	6	449	4.719	801	962
O	305	768	1016	4	5	369	3.097	662	795
S	356	1265	1422	6	8	610	7.489	1091	1310
O	356	1045	1168	5	7	503	4.916	901	1082
S	406	1652	1626	8	11	796	11.192	1425	1710
O	406	1368	1346	7	9	658	7.341	1180	1416
S	457	2090	1829	10	13	1010	15.928	1803	2163
O	457	1729	1524	8	11	836	10.455	1491	1790
S	508	2581	2032	12	16	1245	21.844	2226	2672
O	508	2136	1677	10	14	1032	14.355	1842	2239
S	559	3123	2235	15	20	1508	29.087	2694	3232
O	559	2587	1854	12	16	1250	19.107	2231	2678
S	610	3658	2438	18	23	1793	37.756	3155	3786
O	610	3078	2032	15	19	1486	34.794	2655	3186

*S = square section; O = octagonal section

The general details about the precast prestressed piles are as follows:

Usual length: 10 m–45 m
Maximum length: 60 m
Maximum load: 7500 kN–8500 kN

The advantages and disadvantages are the same as in the case of precast piles.

Cast-in-situ, or *cast-in-place, piles* are built by making a hole in the ground and then filling it with concrete. Various types of cast-in-place concrete pile are currently used in construction, and most of them have been patented by their manufacturers. These piles may be divided into two broad categories: cased and uncased. Both types may have a pedestal at the bottom.

Cased piles are made by driving a steel casing into the ground with the help of a mandrel placed inside the casing. When the pile reaches the proper depth, the mandrel is withdrawn and the casing is filled with concrete. Figures 14.4a, b, c, and d

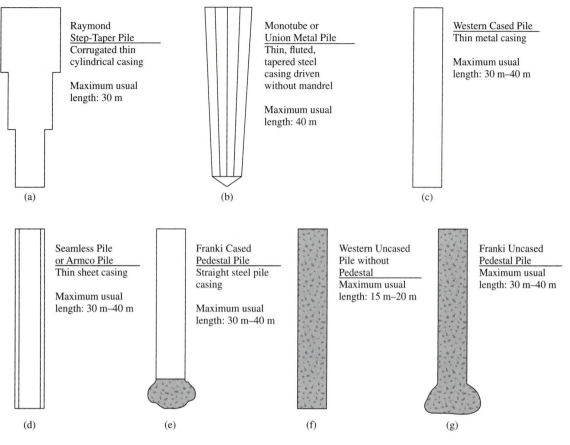

Figure 14.4 Cast-in-place concrete piles

show some examples of cased piles without a pedestal. Figure 14.4e shows a cased pile with a pedestal. The pedestal is an expanded concrete bulb that is formed by dropping a hammer on fresh concrete.

The general details of cased cast-in-place piles are as follows:

Usual length:	5 m–15 m
Maximum length:	30 m–40 m
Usual load:	200 kN–500 kN
Approximate Maximum load:	800 kN
Advantages:	a. Relatively cheap
	b. Possibility of inspection before pouring concrete
	c. Easy to extend
Disadvantages:	a. Difficult to splice after concreting
	b. Thin casings may be damaged during driving
Allowable load:	$Q_{all} = A_s f_s + A_c f_c$ (14.1)

where

A_s = area of cross section of steel
A_c = area of cross section of concrete
f_s = allowable stress of steel
f_c = allowable stress of concrete

Figures 14.4f and 14.4g are two types of *uncased pile*, one without a pedestal, and the other with one. The uncased piles are made by first driving the casing to the desired depth and then filling it with fresh concrete. The casing is then gradually withdrawn.

Following are some general details of *uncased cast-in-place* concrete piles.

Usual length:	5 m–15 m
Maximum length:	30 m–40 m
Usual load:	300 kN–500 kN
Approximate Maximum load:	700 kN
Advantages:	a. Initially economical
	b. Can be finished at any elevation
Disadvantages:	a. Voids may be created if concrete is placed rapidly.
	b. Difficult to splice after concreting.
	c. In soft soils, the sides of the hole may cave in, thus squeezing the concrete.
Allowable load:	$Q_{all} = A_c f_c$ (14.2)

where

A_c = area of cross section of concrete
f_c = allowable stress of concrete

Timber Piles

Timber piles are tree trunks that have had their branches and bark carefully trimmed off. The maximum length of most timber piles is 10 to 20 m. To qualify for use as a pile, the timber should be straight, sound, and without any defects. The American Society of Civil Engineers' *Manual of Practice*, No. 17 (1959), divided timber piles into three classifications:

1. *Class A piles* carry heavy loads. The minimum diameter of the butt should be 356 mm.
2. *Class B piles* are used to carry medium loads. The minimum butt diameter should be 305 to 330 mm.
3. *Class C piles* are used in temporary construction work. They can be used permanently for structures when the entire pile is below the water table. The minimum butt diameter should be 305 mm.

In any case, a pile tip should have a diameter not less than 150 mm.

Timber piles cannot withstand hard driving stress; therefore, the pile capacity is generally limited to about 220 to 270 kN. Steel shoes may be used to avoid damage at the pile tip (bottom). The tops of timber piles may also be damaged during the driving operation. To avoid damage to the pile top, a metal band or cap may be used. The crushing of the wooden fibers caused by the impact of the hammer is referred to as *brooming*.

Splicing of timber piles should be avoided, particularly when they are expected to carry tensile load or lateral load. However, if splicing is necessary, it can be done by using *pipe sleeves* (Figure 14.5a) or *metal straps* and *bolts* (Figure 14.5b). The length of the pipe sleeve should be at least five times the diameter of the pile. The butting ends should be cut square so that full contact can be maintained. The spliced portions should be carefully trimmed so that they fit tightly to the inside of the pipe sleeve. In the case of metal straps and bolts, the butting ends should also be cut square. Also, the sides of the spliced portion should be trimmed plane for putting the straps on.

Timber piles can stay undamaged indefinitely if they are surrounded by saturated soil. However, in a marine environment, timber piles are subject to attack by various organisms and can be damaged extensively in a few months. When located above the water table, the piles are subject to attack by insects. The life of the piles may be increased by treating them with preservatives such as creosote.

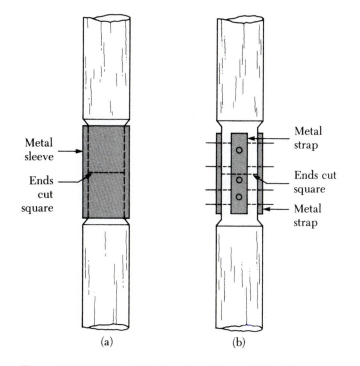

Figure 14.5 Splicing of timber piles: (a) use of pipe sleeves;
(b) use of metal straps and bolts

The usual length of wooden piles is 5 m to 15 m. The maximum length is about 30 m to 40 m. The usual load carried by wooden piles is 300 kN to 500 kN.

Composite Piles

The upper and lower portions of *composite piles* are made of different materials. For example, composite piles may be made of steel and concrete or timber and concrete. Steel and concrete piles consist of a lower portion of steel and an upper portion of cast-in-place concrete. This type of pile is used when the length of the pile required for adequate bearing exceeds the capacity of simple cast-in-place concrete piles. Timber and concrete piles usually consist of a lower portion of timber pile below the permanent water table and an upper portion of concrete. In any case, forming proper joints between two dissimilar materials is difficult, and, for that reason, composite piles are not widely used.

14.3 Estimation of Pile Length

Selecting the type of pile to be used and estimating its necessary length are fairly difficult tasks that require good judgment. In addition to the classifications given in Section 14.2, piles can be divided into two major categories, depending on their lengths and the mechanisms of load transfer to the soil: (a) point bearing piles, and (b) friction piles.

Point Bearing Piles

If soil-boring records establish the presence of bedrock or rocklike material at a site within a reasonable depth, piles can be extended to the rock surface (Figure 14.6a). In this case, the ultimate capacity of the piles depends entirely on the load-bearing capacity of the underlying material; thus, the piles are called *point bearing piles*. In most of these cases, the necessary length of the pile can be fairly well established.

Instead of bedrock, if a fairly compact and hard stratum of soil is encountered at a reasonable depth, piles can be extended a few meters into the hard stratum (Figure 14.6b). Piles with pedestals can be constructed on the bed of the hard stratum, and the ultimate pile load may be expressed as

$$Q_u = Q_p + Q_s \tag{14.3}$$

where
Q_p = load carried at the pile point
Q_s = load carried by skin friction developed at the side of the pile (caused by shearing resistance between the soil and the pile)

If Q_s is very small, then

$$Q_u \approx Q_p \tag{14.4}$$

In this case, the required pile length may be estimated accurately if proper subsoil exploration records are available.

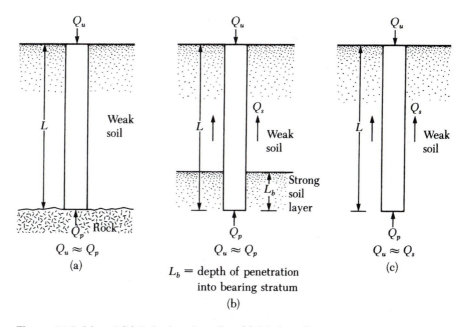

Figure 14.6 (a) and (b) Point bearing piles; (c) friction piles

Friction Piles

When no layer of rock or rocklike material is present at a reasonable depth at a site, point bearing piles become very long and uneconomical. For this type of subsoil condition, piles are driven through the softer material to specified depths (Figure 14.6c). The ultimate load of these piles may be expressed by Eq. (14.3). However, if the value of Q_p is relatively small,

$$Q_u \approx Q_s \tag{14.5}$$

These piles are called *friction piles* because most of the resistance is derived from skin friction. However, the term *friction pile*, although used often in literature, is a misnomer in clayey soils; the resistance to applied load is also caused by *adhesion*.

The length of friction piles depends on the shear strength of the soil, the applied load, and the pile size. To determine the necessary lengths of these piles, an engineer needs a good understanding of soil–pile interaction, good judgment, and experience. Theoretical procedures for calculating the load-bearing capacity of piles are presented in Section 14.6.

14.4   Installation of Piles

Most piles are driven into the ground by *hammers* or *vibratory drivers*. In special circumstances, piles can also be inserted by *jetting* or *partial augering*. The types of hammer used for pile driving include the (a) drop hammer, (b) single-acting air or steam hammer, (c) double-acting and differential air or steam hammer, and (d) diesel hammer. In the driving operation, a cap is attached to the top of the pile. A cushion may be used between the pile and the cap. This cushion has the effect of reducing the impact force and spreading it over a longer time; however, its use is optional. A hammer cushion is placed on the pile cap. The hammer drops on the cushion. Figure 14.7 shows a vibratory pile driver.

In pile driving, when the pile needs to penetrate a thin layer of hard soil (such as sand and gravel) overlying a softer soil layer, a technique called *jetting* is sometimes used. In jetting, water is discharged at the pile point by a pipe 50 to 75 mm in diameter to wash and loosen the sand and gravel.

Based on the nature of their placement, piles may be divided into two categories: *displacement piles* and *nondisplacement piles*. Driven piles are displacement piles because they move some soil laterally; hence, there is a tendency for the densification of soil surrounding them. Concrete piles and closed-ended pipe piles are high-displacement piles. However, steel H-piles displace less soil laterally during driving, and so they are low-displacement piles. In contrast, bored piles are nondisplacement piles because their placement causes very little change in the state of stress in the soil.

Figure 14.7 Vibratory pile driver (Courtesy of Michael W. O'Neill, University of Houston, Texas)

Load Transfer Mechanism

The load transfer mechanism from a pile to the soil is complicated. To understand it, consider a pile of length L, as shown in Figure 14.8a. The load on the pile is gradually increased from 0 to $Q_{(z=0)}$ at the ground surface. Part of this load will be resisted by

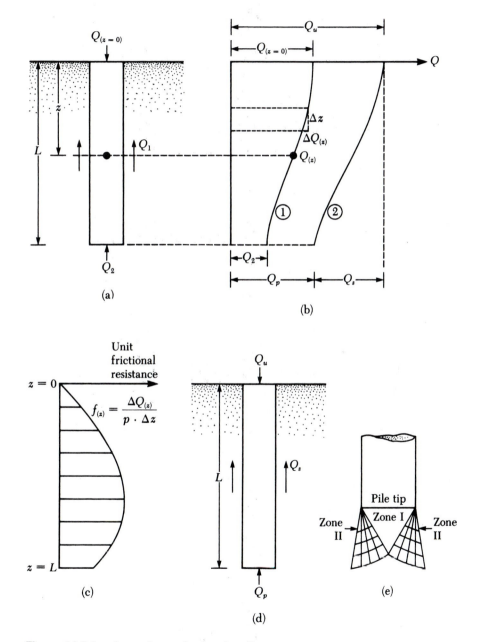

Figure 14.8 Load transfer mechanism for piles

the side friction developed along the shaft, Q_1, and part by the soil below the tip of the pile, Q_2. Now, how are Q_1 and Q_2 related to the total load? If measurements are made to obtain the load carried by the pile shaft, $Q_{(z)}$, at any depth z, the nature of variation will be like curve 1 of Figure 14.8b. The *frictional resistance per unit area, $f_{(z)}$*, at any depth z may be determined as

$$f_{(z)} = \frac{\Delta Q_{(z)}}{(p)(\Delta z)} \tag{14.6}$$

where p = perimeter of the pile cross section. Figure 14.8c shows the variation of $f_{(z)}$ with depth.

   If the load Q at the ground surface is gradually increased, maximum frictional resistance along the pile shaft will be fully mobilized when the relative displacement between the soil and the pile is about 5 to 10 mm, irrespective of pile size and length L. However, the maximum point resistance $Q_2 = Q_p$ will not be mobilized until the pile tip has moved about 10% to 25% of the pile width (or diameter). The lower limit applies to driven piles and the upper limit to bored piles. At ultimate load (Figure 14.8d and curve 2 in Figure 14.8b), $Q_{(z=0)} = Q_u$. Thus,

$$Q_1 = Q_s$$

and

$$Q_2 = Q_p$$

The preceding explanation indicates that Q_s (or the unit skin friction, f, along the pile shaft) is developed at a *much smaller pile displacement compared to the point resistance, Q_p.*

   At ultimate load, the failure surface in the soil at the pile tip (bearing capacity failure caused by Q_p) is like that shown in Figure 14.8e. Note that pile foundations are deep foundations and that the soil fails mostly in a *punching mode*, as illustrated previously in Figures 12.2c and 12.3. That is, a *triangular zone*, I, is developed at the pile tip, which is pushed downward without producing any other visible slip surface. In dense sands and stiff clayey soils, a *radial shear zone*, II, may partially develop. Hence, the load displacement curves of piles will resemble those shown in Figure 12.2c.

14.6   Equations for Estimation of Pile Capacity

The ultimate load-carrying capacity of a pile, Q_u, is given by a simple equation as the load carried at the pile point plus the total frictional resistance (skin friction) derived from the soil–pile interface (Figure 14.9), or

$$Q_u = Q_p + Q_s \tag{14.7}$$

where
   Q_p = load-carrying capacity of the pile point
   Q_s = frictional resistance

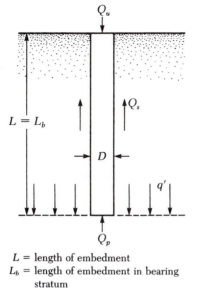

Q_u

Q_s

$L = L_b$

D

q'

Q_p

L = length of embedment
L_b = length of embedment in bearing
stratum

Figure 14.9
Ultimate load-capacity of a pile

Numerous published studies cover the determination of the values of Q_p and Q_s. Excellent reviews of many of these investigations have been provided by Vesic (1977), Meyerhof (1976), and Coyle and Castello (1981). These studies provide insight into the problem of determining the ultimate pile capacity.

Load-Carrying Capacity of the Pile Point, Q_p

The ultimate bearing capacity of shallow foundations was discussed in Chapter 12. The general bearing capacity equation for shallow foundations was given in Chapter 12 (for vertical loading) as

$$q_u = c'N_cF_{cs}F_{cd} + qN_qF_{qs}F_{qd} + \frac{1}{2}\gamma BN_\gamma F_{\gamma s}F_{\gamma d}$$

Hence, in general, the ultimate bearing capacity may be expressed as

$$q_u = cN_c^* + qN_q^* + \gamma BN_\gamma^* \tag{14.8}$$

where N_c^*, N_q^*, and N_γ^* are the bearing capacity factors that include the necessary shape and depth factors.

Pile foundations are deep. However, the ultimate resistance per unit area developed at the pile tip, q_p, may be expressed by an equation similar in form to Eq. 14.8, although the values of N_c^*, N_q^*, and N_γ^* will change. The notation used in this chapter for the width of the pile is D. Hence, substituting D for B in Eq. (14.8) gives

$$q_u = q_p = c'N_c^* + qN_q^* + \gamma DN_\gamma^* \tag{14.9}$$

Because the width, D, of a pile is relatively small, the term γDN_γ^* may be dropped from the right side of the preceding equation without introducing a serious error, or

$$q_p = c'N_c^* + q'N_q^* \tag{14.10}$$

Note that the term q has been replaced by q' in Eq. (14.10) to signify effective vertical stress. Hence, the load-carrying capacity of the pile point is

$$Q_p = A_p q_p = A_p(c'N_c^* + q'N_q^*) \qquad (14.11)$$

where
$$A_p = \text{area of the pile tip}$$
$$c' = \text{cohesion of the soil supporting the pile tip}$$
$$q_p = \text{unit point resistance}$$
$$q' = \text{effective vertical stress at the level of the pile tip}$$
$$N_c^*, N_q^* = \text{bearing capacity factors}$$

There are several methods for calculating the magnitude of q_p. In this text, the method suggested by Meyerhof (1976) will be used.

14.7   *Calculation of q_p—Meyerhof's Method*

In sand, the cohesion c' is equal to 0. Thus, Eq. (14.11) takes the form

$$Q_p = A_p q_p = A_p q' N_q^* \qquad (14.12)$$

The variation of N_q^* with the soil friction angle, ϕ', is shown in Figure 14.10. Meyerhof pointed out that the point bearing capacity, q_p, of a pile in sand generally increases with the depth of embedment in the bearing stratum and reaches a maximum value at an embedment ratio of $L_b/D = (L_b/D)_{cr}$. Note that in a homogeneous soil, L_b is equal to the actual embedment length of the pile, L (see Figure 14.9). However, in Figure 14.6b, where a pile has penetrated into a bearing stratum, $L_b < L$. Beyond the critical embedment ratio, $(L_b/D)_{cr}$, the value of q_p remains constant ($q_p = q_l$). That is, as shown in Figure 14.11 for the case of a homogeneous soil, $L = L_b$. Hence, Q_p should not exceed the limiting value, or $A_p q_l$, so

$$Q_p = A_p q' N_q^* \leq A_p q_l \qquad (14.13)$$

The limiting point resistance is

$$q_l(\text{kN/m}^2) = 50 N_q^* \tan \phi' \qquad (14.14)$$

where $\phi' = $ effective soil friction angle in the bearing stratum.

Based on field observations, Meyerhof (1976) also suggested that the ultimate point resistance, q_p, in a homogeneous granular soil ($L = L_b$) may be obtained from standard penetration numbers as

$$q_p(\text{kN/m}^2) = 40 N_{60} \frac{L}{D} \leq 400 N_{60} \qquad (14.15)$$

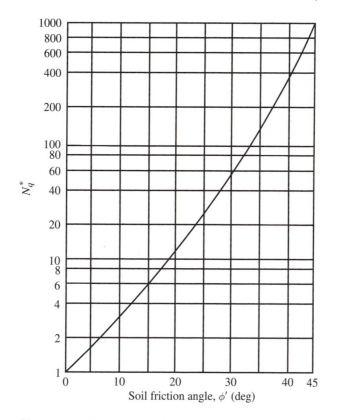

Figure 14.10 Meyerhof's bearing capacity factor, N_q^*

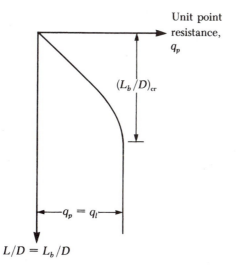

Figure 14.11 Variation of unit point resistance in a homogeneous sand

where N_{60} = average standard penetration number near the pile point (about $10D$ above and $4D$ below the pile point).

For piles in *saturated clays* in undrained conditions ($\phi = 0$),

$$Q_p = N_c^* c_u A_p = 9 c_u A_p \tag{14.16}$$

where c_u = undrained cohesion of the soil below the pile tip.

14.8   *Frictional Resistance, Qs*

The frictional or skin resistance of a pile may be written as

$$Q_s = \sum p \, \Delta L f \tag{14.17}$$

where
   p = perimeter of the pile section
   ΔL = incremental pile length over which p and f are taken constant (Figure 14.12a)
   f = unit friction resistance at any depth z

Frictional Resistance in Sand

The unit frictional resistance at any depth for a pile is

$$f = K \sigma_o' \tan \delta' \tag{14.18}$$

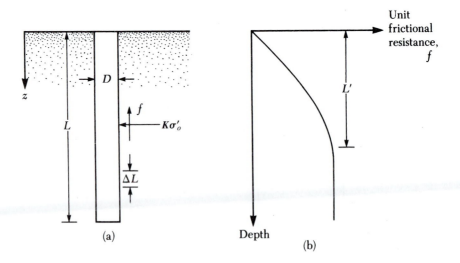

Figure 14.12 Unit frictional resistance for piles in sand

where

K = earth pressure coefficient

σ'_o = effective vertical stress at the depth under consideration

δ' = soil–pile friction angle

In reality, the magnitude of K varies with depth. It is approximately equal to the Rankine passive earth pressure coefficient, K_p, at the top of the pile and may be less than the at-rest earth pressure coefficient, K_o, at the pile tip. It also depends on the nature of the pile installation. Based on presently available results, the following average values of K are recommended for use in Eq. (14.18):

Pile type	K
Bored or jetted	$\approx K_o = 1 - \sin \phi'$
Low-displacement driven	$\approx K_o = 1 - \sin \phi'$ to $1.4K_o = 1.4(1 - \sin \phi')$
High-displacement driven	$\approx K_o = 1 - \sin \phi'$ to $1.8K_o = 1.8(1 - \sin \phi')$

The effective vertical stress, σ'_o, for use in Eq. (14.18) increases with pile depth to a maximum limit at a depth of 15 to 20 pile diameters and remains constant thereafter, as shown in Figure 14.12b. This critical depth, L', depends on several factors, such as the soil friction angle and compressibility and relative density. A conservative estimate is to assume that

$$L' = 15D \tag{14.19}$$

The values of δ' from various investigations appear to be in the range of $0.5\phi'$ to $0.8\phi'$. Judgment must be used in choosing the value of δ'.

Meyerhof (1976) also indicated that the average unit frictional resistance, f_{av}, for high-displacement driven piles may be obtained from average standard penetration resistance values as

$$f_{av} (\text{kN/m}^2) = 2\overline{N}_{60} \tag{14.20}$$

where $\overline{N}_{60}$ = average value of standard penetration resistance. For low-displacement driven piles,

$$f_{av} (\text{kN/m}^2) = \overline{N}_{60} \tag{14.21}$$

Thus,

$$Q_s = pLf_{av} \tag{14.22}$$

The cone penetration test was discussed in Chapter 10. Nottingham and Schmertmann (1975) and Schmertmann (1978) provided correlations for estimating Q_s using the frictional resistance (f_c) obtained during cone penetration tests. According to this method

$$f = \alpha' f_c \tag{14.23}$$

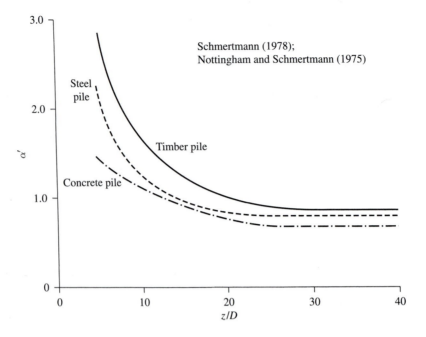

Figure 14.13 Variation of α' with embedment ratio for piles in sand: electric cone penetrometer

The variations of α' with z/D for electric cone and mechanical cone penetrometers are shown in Figures 14.13 and 14.14, respectively. We have

$$Q_s = \Sigma p \, (\Delta L) \, f = \Sigma p \, (\Delta L) \, \alpha' \, f_c \qquad (14.24)$$

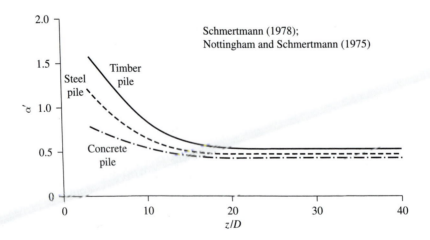

Figure 14.14 Variation of α' with embedment ratio for piles in sand: mechanical cone penetrometer

Frictional (or Skin) Resistance in Clay

Several methods are available for obtaining the unit frictional (or skin) resistance of piles in clay. Three of the presently accepted procedures are described briefly.

1. λ *Method:* This method was proposed by Vijayvergiya and Focht (1972). It is based on the assumption that the displacement of soil caused by pile driving results in a passive lateral pressure at any depth and that the average unit skin resistance is

$$f_{av} = \lambda(\overline{\sigma}_o' + 2c_u) \tag{14.25}$$

where
$\overline{\sigma}_o'$ = mean effective vertical stress for the entire embedment length
c_u = mean undrained shear strength ($\phi = 0$ concept)

The value of λ changes with the depth of pile penetration (see Table 14.4). Thus, the total frictional resistance may be calculated as

$$Q_s = pLf_{av}$$

Care should be taken in obtaining the values of $\overline{\sigma}_o'$ and c_u in layered soil. Figure 14.15 helps explain the reason. According to Figure 14.15b, the mean value of c_u is $(c_{u(1)}L_1 + c_{u(2)}L_2 + \cdots)/L$. Similarly, Figure 14.15c shows the plot of the variation of effective stress with depth. The mean effective stress is

$$\overline{\sigma}_o' = \frac{A_1 + A_2 + A_3 + \cdots}{L} \tag{14.26}$$

where $A_1, A_2, A_3, \ldots$ = areas of the vertical effective stress diagrams.

2. α *Method:* According to the α method, the unit skin resistance in clayey soils can be represented by the equation

$$f = \alpha c_u \tag{14.27}$$

Table 14.4 Variation of λ with L [Eq. (14.25)]

L (m)	λ	L (m)	λ
0	0.5	35	0.136
5	0.318	40	0.127
10	0.255	50	0.123
15	0.205	60	0.118
20	0.177	70	0.117
25	0.155	80	0.117
30	0.145	90	0.117

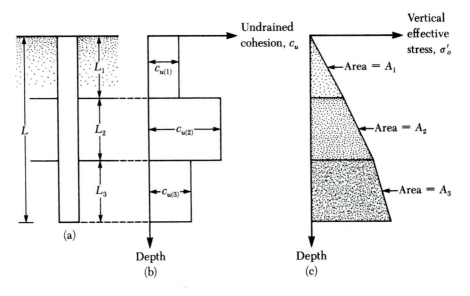

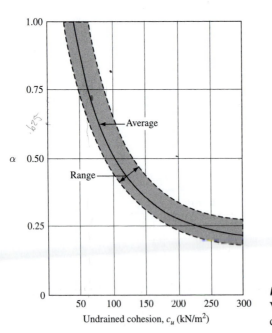

Figure 14.15 Application of λ method in layered soil

where α = empirical adhesion factor. The approximate variation of the value of α is shown in Figure 14.16. Thus,

$$Q_s = \sum fp\,\Delta L = \sum \alpha c_u p\,\Delta L \tag{14.28}$$

Figure 14.16
Variation of α with undrained cohesion of clay

3. *β Method*: When piles are driven into saturated clays, the pore water pressure in the soil around the piles increases. This excess pore water pressure in normally consolidated clays may be 4 to 6 times c_u. However, within a month or so, this pressure gradually dissipates. Hence, the unit frictional resistance for the pile can be determined on the basis of the effective stress parameters of the clay in a remolded state ($c' = 0$). Thus, at any depth,

$$f = \beta \sigma_o' \tag{14.29}$$

where
  $\sigma_o' =$ vertical effective stress
  $\beta = K \tan \phi_R'$ $\tag{14.30}$
  $\phi_R' =$ drained friction angle of remolded clay
  $K =$ earth pressure coefficient

Conservatively, we can calculate the magnitude of K as the earth pressure coefficient at rest, or

$$K = 1 - \sin \phi_R' \quad \text{(for normally consolidated clays)} \tag{14.31}$$

and

$$K = (1 - \sin \phi_R')\sqrt{OCR} \quad \text{(for overconsolidated clays)} \tag{14.32}$$

where $OCR =$ overconsolidation ratio.
Combining Eqs. (14.29), (14.30), (14.31), and (14.32) for normally consolidated clays yields

$$f = (1 - \sin \phi_R')\tan \phi_R' \sigma_o' \tag{14.33}$$

and for overconsolidated clays,

$$f = (1 - \sin \phi_R')\tan \phi_R' \sqrt{OCR}\, \sigma_o' \tag{14.34}$$

With the value of f determined, the total frictional resistance may be evaluated as

$$Q_s = \sum f p \, \Delta L$$

Correlation with Cone Penetration Test Results

Nottingham and Schmertmann (1975) and Schmertmann (1978) found the correlation for unit skin friction in clay (with $\phi = 0$) to be

$$f = \alpha' f_c \tag{14.35}$$

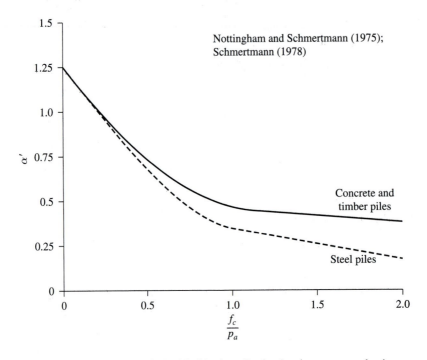

Figure 14.17 Variation of α' with f_c/p_a for piles in clay (p_a = atmospheric pressure $\approx 100 \text{ kN/m}^2$)

The variation of α' with the cone frictional resistance f_c is shown in Figure 14.17. Thus,

$$Q_s = \Sigma f p(\Delta L) = \Sigma \alpha' f_c p(\Delta L) \tag{14.36}$$

14.9   *Allowable Pile Capacity*

After the total ultimate load-carrying capacity of a pile has been determined by summing the point bearing capacity and the frictional (or skin) resistance, a reasonable factor of safety should be used to obtain the total allowable load for each pile, or

$$Q_{\text{all}} = \frac{Q_u}{FS} \tag{14.37}$$

where
Q_{all} = allowable load-carrying capacity for each pile
FS = factor of safety

The factor of safety generally used ranges from 2.5 to 4, depending on the uncertainties of the ultimate load calculation. In large projects involving several piles, generally a specific number of load tests must be conducted to determine the ultimate and allowable bearing capacities. The primary reason for this is the unreliability of prediction methods.

14.10   *Load-Carrying Capacity of Pile Point Resting on Rock*

Sometimes piles are driven to an underlying layer of rock. In such cases, the engineer must evaluate the bearing capacity of the rock. The ultimate unit point resistance in rock (Goodman, 1980) is approximately

$$q_p = q_{u\text{-}R}(N_\phi + 1) \tag{14.38}$$

where
$$N_\phi = \tan^2(45 + \phi'/2)$$
$q_{u\text{-}R}$ = unconfined compression strength of rock
ϕ' = drained angle of friction

The unconfined compression strength of rock can be determined by laboratory tests on rock specimens collected during field investigation. However, extreme caution should be used in obtaining the proper value of $q_{u\text{-}R}$ because laboratory specimens are usually small in diameter. As the diameter of the specimen increases, the unconfined compression strength decreases, which is referred to as the *scale effect*. For specimens larger than about 1 m in diameter, the value of $q_{u\text{-}R}$ remains approximately constant. There appears to be a fourfold to fivefold reduction in the magnitude of $q_{u\text{-}R}$ in this process. The scale effect in rock is primarily caused by randomly distributed large and small fractures and also by progressive ruptures along the slip lines. Hence, we always recommend that

$$q_{u\text{-}R(\text{design})} = \frac{q_{u\text{-}R(\text{lab})}}{5} \tag{14.39}$$

Table 14.5 lists some representative values of (laboratory) unconfined compression strengths of rock along with the rock friction angle, ϕ'.

A factor of safety of *at least* 3 should be used to determine the allowable load-carrying capacity of the pile point. Thus,

$$Q_{p(\text{all})} = \frac{[q_{u\text{-}R}(N_\phi + 1)]A_p}{FS} \tag{14.40}$$

Table 14.5 Typical unconfined compressive strength and angle of friction of rocks

Rock type	$q_{u\text{-}R}$ (MN/m²)	ϕ' (deg.)
Sandstone	70–140	27–45
Limestone	105–210	30–40
Shale	35–70	10–20
Granite	140–210	40–50
Marble	60–70	25–30

Example 14.1

A fully embedded precast concrete pile 12 m long is driven into a homogeneous sand layer ($c' = 0$). The pile is square in cross section with sides measuring 305 mm. The dry unit weight of sand, γ_d, is 16.8 kN/m³, the average soil friction angle is 35°, and the standard penetration resistance near the vicinity of the pile tip is 16. Calculate the ultimate point load on the pile.

    a. Use Meyerhof's method with Eq. (14.13).
    b. Use Meyerhof's method with Eq. (14.15).

Solution
Part a.
This soil is homogeneous, so $L_b = L$. For $\phi' = 35°$, $N_q^* \approx 120$. Thus,

$$q' = \gamma_d L = (16.8)(12) = 201.6 \text{ kN/m}^2$$

$$A_p = \frac{305 \times 305}{1000 \times 1000} = 0.0929 \text{ m}^2$$

$$Q_p = A_p q' N_q^* = (0.0929)(201.6)(120) = 2247.4 \text{ kN}$$

However, from Eq. (14.14), we have

$$q_l = 50 N_q^* \tan \phi' = 50(120)\tan 35° = 4201.25 \text{ kN/m}^2$$

so

$$Q_p = A_p q_l = (0.0929)(4201.25) = 390.3 \text{ kN} < A_p q' N_q^*$$

and

$$Q_p \approx \mathbf{390 \text{ kN}}$$

Part b.
The average standard penetration resistance near the pile tip is 16. So, from Eq. (14.15), we have

$$q_p = 40 N_{60} \frac{L}{D} \le 400 N_{60}$$

$$\frac{L}{D} = \frac{12}{0.305} = 39.34$$

$$Q_p = A_p q_p = (0.0929)(40)(16)39.34 = 2339 \text{ kN}$$

However, the limiting value is

$$Q_p = A_p 400 N_{60} = (0.0929)(400)(16) = 594.6 \text{ kN} \approx \mathbf{595 \text{ kN}} \qquad \blacksquare$$

Example 14.2

Refer to Example Problem 14.1. Determine the total frictional resistance for the pile. Use Eqs. (14.17), (14.18), and (14.19). Also use $K = 1.4$, $\delta' = 0.6\phi'$.

Solution

The unit skin friction at any depth is given by Eq. (14.18) as

$$f = K\sigma'_o \tan \delta'$$

Also from Eq. (14.19), we have

$$L' = 15D$$

So, for depth $z = 0$–$15D$, $\sigma'_o = \gamma z = 16.8z$ (kN/m²), and beyond $z \geq 15D$, $\sigma'_o = \gamma(15D) = (16.8)(15 \times 0.305) = 76.86$ kN/m². This result is shown in Figure 14.18.

The *frictional resistance from z* = 0 to 15D is

$$Q_s = pL'f_{av} = [(4)(0.305)][15D]\left[\frac{(1.4)(76.86)\tan(0.6 \times 35)}{2}\right]$$

$$= (1.22)(4.575)(20.65) = 115.26 \text{ kN}$$

The *frictional resistance from z* = 15D to 12 m is

$$Q_s = p(L - L')f_{z=15D} = [(4)(0.305)][12 - 4.575][(1.4)(76.86)\tan(0.6 \times 35)]$$

$$= (1.22)(7.425)(41.3) = 374.1 \text{ kN}$$

So, the total frictional resistance is

$$115.26 + 374.1 = 489.35 \text{ kN} \approx \textbf{490 kN} \qquad \blacksquare$$

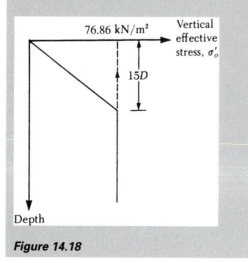

Figure 14.18

Example 14.3

A concrete pile 458 mm $\times$ 458 mm in cross section is embedded in a saturated clay. The length of embedment is 16 m. The undrained cohesion, c_u, of clay is 60 kN/m^2, and the unit weight of clay is 18 kN/m^3. Use a factor of safety of 5 to determine the allowable load the pile can carry.

    a. Use the α method.
    b. Use the λ method.

Solution
Part a.
From Eq. (14.16),

$$Q_p = A_p q_p = A_p c_u N_c^* = (0.458 \times 0.458)(60)(9) = 113.3 \text{ kN}$$

From Eqs. (14.27) and (14.28),

$$Q_s = \alpha c_u p L$$

From the average plot of Figure 14.16 for $c_u = 60$ kN/m^2, $\alpha \approx 0.77$ and

$$Q_s = (0.77)(60)(4 \times 0.458)(16) = 1354 \text{ kN}$$

$$Q_{\text{all}} = \frac{Q_p + Q_s}{FS} = \frac{113.3 + 1354}{5} \approx \mathbf{294 \text{ kN}}$$

Part b.
From Eq. (14.25),

$$f_{av} = \lambda(\overline{\sigma}_o' + 2c_u)$$

We are given $L = 16.0$ m. From Table 14.4 for $L = 16$ m, $\lambda \approx 0.2$, so

$$f_{av} = 0.2\left[\left(\frac{18 \times 16}{2}\right) + 2(60)\right] = 52.8 \text{ kN/m}^2$$

$$Q_s = pLf_{av} = (4 \times 0.458)(16)(52.8) = 1548 \text{ kN}$$

As in part a, $Q_p = 113.3$ kN, so

$$Q_{\text{all}} = \frac{Q_p + Q_s}{FS} = \frac{113.3 + 1548}{5} = \mathbf{332 \text{ kN}}$$

    ■

Example 14.4

A driven pile in clay is shown in Figure 14.19a. The pile has a diameter of 406 mm.

    a. Calculate the net point bearing capacity. Use Eq. (14.16).
    b. Calculate the skin resistance (1) by using Eqs. (14.27) and (14.28) (α method), (2) by using Eq. (14.25) (λ method), and (3) by using

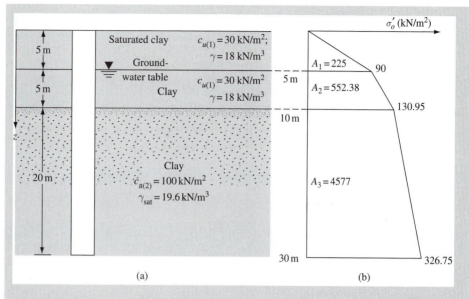

Figure 14.19

Eq. (14.29) (β method). For all clay layers, $\phi_R = 30°$. The top 10 m of clay is normally consolidated. The bottom clay layer has an *OCR* of 2.

c. Estimate the net allowable pile capacity. Use *FS* = 4.

Solution
The area of the cross section of the pile is

$$A_p = \frac{\pi}{4}D^2 = \frac{\pi}{4}(0.406)^2 = 0.1295 \text{ m}^2$$

Part a: Calculation of Net Point Bearing Capacity
From Eq. (14.16), we have

$$Q_p = A_p q_p = A_p N_c^* c_{u(2)} = (0.1295)(9)(100) = \textbf{116.55 kN}$$

Part b: Calculation of Skin Resistance
(1) From Eq. (14.28),

$$Q_s = \Sigma \ \alpha c_u p \ \Delta L$$

For the top soil layer, $c_{u(1)} = 30 \text{ kN/m}^2$. According to the average plot of Figure 14.16, $\alpha_1 = 1.0$. Similarly, for the bottom soil layer, $c_{u(2)} = 100 \text{ kN/m}^2$; $\alpha_2 = 0.5$.
Thus,

$$Q_s = \alpha_1 c_{u(1)}[(\pi)(0.406)]10 + \alpha_2 c_{u(2)}[(\pi)(0.406)]20$$

$$= (1)(30)[(\pi)(0.406)]10 + (0.5)(100)[(\pi)(0.406)]20$$

$$= 382.7 + 1275.5 = \textbf{1658.2 kN}$$

(2) The average value of c_u is

$$\frac{c_{u(1)}(10) + c_{u(2)}(20)}{30} = \frac{(30)(10) + (100)(20)}{30} = 76.7 \text{ kN/m}^2$$

To obtain the average value of $\overline{\sigma}'_o$, the diagram for vertical effective stress variation with depth is plotted in Figure 14.19. From Eq. (14.26),

$$\overline{\sigma}'_o = \frac{A_1 + A_2 + A_3}{L} = \frac{225 + 552.38 + 4577}{30} = 178.48 \text{ kN/m}^2$$

The magnitude of λ from Table 14.4 is 0.145. So

$$f_{av} = 0.145[178.48 + (2)(76.7)] = 48.12 \text{ kN/m}^2$$

Hence,

$$Q_s = pLf_{av} = \pi(0.406)(30)(48.12) = \textbf{1841.3 kN}$$

(3) The top clay layer (10 m) is normally consolidated and $\phi_R = 30°$.
For $z = 0\text{–}5$ m [Eq. (14.33)],

$$f_{av(1)} = (1 - \sin \phi_R)\tan \phi_R \sigma'_{o(av)}$$

$$= (1 - \sin 30°)(\tan 30°)\left(\frac{0 + 90}{2}\right) = 13.0 \text{ kN/m}^2$$

Similarly, for $z = 5\text{–}10$ m,

$$f_{av(2)} = (1 - \sin 30°)(\tan 30°)\left(\frac{90 + 130.95}{2}\right) = 31.9 \text{ kN/m}^2$$

For $z = 10\text{–}30$ m [Eq. (14.34)],

$$f_{av} = (1 - \sin \phi_R)\tan \phi_R \sqrt{OCR} \, \sigma'_{o(av)}$$

For $OCR = 2$,

$$f_{av(3)} = (1 - \sin 30°)(\tan 30°)\sqrt{2}\left(\frac{130.95 + 326.75}{2}\right) = 93.43 \text{ kN/m}^2$$

So

$$Q_s = p[f_{av(1)}(5) + f_{av(2)}(5) + f_{av(3)}(20)]$$

$$= (\pi)(0.406)[(13)(5) + (31.9)(5) + (93.43)(20)] = \textbf{2669.7 kN}$$

Part c: Calculation of Net Ultimate Capacity, Q_u
Comparing the three values of Q_s shows that the α and λ methods give similar results. So we use

$$Q_s = \frac{1658.2 + 1841.3}{2} \approx 1750 \text{ kN}$$

Thus,

$$Q_u = Q_p + Q_s = 116.55 + 1750 = 1866.55 \text{ kN}$$

$$Q_{all} = \frac{Q_u}{FS} = \frac{1866.55}{4} = 466.6 \text{ kN}$$ ∎

14.11   Elastic Settlement of Piles

The elastic settlement of a pile under a vertical working load, Q_w, is determined by three factors:

$$S_e = S_{e(1)} + S_{e(2)} + S_{e(3)} \tag{14.41}$$

where

S_e = total pile settlement
$S_{e(1)}$ = settlement of pile shaft
$S_{e(2)}$ = settlement of pile caused by the load at the pile point
$S_{e(3)}$ = settlement of pile caused by the load transmitted along the pile shaft

Determination of $S_{e(1)}$

If the pile material is assumed to be elastic, the deformation of the pile shaft can be evaluated using the fundamental principles of mechanics of materials:

$$S_{e(1)} = \frac{(Q_{wp} + \xi Q_{ws})L}{A_p E_p} \tag{14.42}$$

where

Q_{wp} = load carried at the pile point under working load condition
Q_{ws} = load carried by frictional (skin) resistance under working load condition
A_p = area of the pile cross section
L = length of the pile
E_p = modulus of elasticity of the pile material

The magnitude of ξ depends on the nature of the unit friction (skin) resistance distribution along the pile shaft. It may vary between 0.5 and 0.67 (Vesic, 1977).

Determination of $S_{e(2)}$

The settlement of a pile caused by the load carried at the pile point may be expressed as

$$S_{e(2)} = \frac{q_{wp}D}{E_s}(1 - \mu_s^2)I_{wp} \tag{14.43}$$

where
D = width or diameter of the pile
q_{wp} = point load per unit area at the pile point = Q_{wp}/A_p
E_s = modulus of elasticity of soil at or below the pile point
μ_s = Poisson's ratio of soil
I_{wp} = influence factor ≈ 0.85

Vesic (1977) also proposed a semiempirical method to obtain the magnitude of the settlement, $S_{e(2)}$:

$$S_{e(2)} = \frac{Q_{wp}C_p}{Dq_p} \tag{14.44}$$

where
q_p = ultimate point resistance of the pile
C_p = an empirical coefficient

Representative values of C_p for various soils are given in Table 14.6.

Determination of $S_{e(3)}$

The settlement of a pile caused by the load carried along the pile shaft is given by a relation similar to Eq. (14.43), or

$$S_{e(3)} = \left(\frac{Q_{ws}}{pL}\right)\frac{D}{E_s}(1 - \mu_s^2)I_{ws} \tag{14.45}$$

where
p = perimeter of the pile
L = embedded length of the pile
I_{ws} = influence factor

Note that the term Q_{ws}/pL in Eq. (14.45) is the average value of f along the pile shaft. The influence factor, I_{ws}, has a simple empirical relation (Vesic, 1977):

$$I_{ws} = 2 + 0.35\sqrt{\frac{L}{D}} \tag{14.46}$$

Table 14.6 Typical values of C_p as recommended by Vesic (1977) [Eq. (14.44)]

Soil type	Driven pile	Bored pile
Sand (dense to loose)	0.02–0.04	0.09–0.18
Clay (stiff to soft)	0.02–0.03	0.03–0.06
Silt (dense to loose)	0.03–0.05	0.09–0.12

Vesic (1977) also proposed a simple empirical relation similar to Eq. (14.44) for obtaining $S_{e(3)}$:

$$S_{e(3)} = \frac{Q_{ws}C_s}{Lq_p}$$ (14.47)

where C_s = an empirical constant = $(0.93 + 0.16\sqrt{L/D})C_p$.   (14.48)
The values of C_p for use in Eq. (14.48) may be estimated from Table 14.6.

Example 14.5

A 12-m-long precast concrete pile is fully embedded in sand. The cross section of the pile measures 0.305 m × 0.305 m. The allowable working load for the pile is 337 kN, of which 240 kN is contributed by skin friction. Determine the elastic settlement of the pile for $E_p = 21 \times 10^6$ kN/m², $E_s = 30,000$ kN/m², and $\mu_s = 0.3$.

Solution
We will use Eq. (14.41):

$$S_e = S_{e(1)} + S_{e(2)} + S_{e(3)}$$

From Eq. (14.42),

$$S_{e(1)} = \frac{(Q_{wp} + \xi Q_{ws})L}{A_p E_p}$$

Let $\xi = 0.6$ and $E_p = 21 \times 10^6$ kN/m². Then

$$S_{e(1)} = \frac{[97 + (0.6)(240)]12}{(0.305)^2(21 \times 10^6)} = 0.00148 \text{ m} = 1.48 \text{ mm}$$

From Eq. (14.43),

$$S_{e(2)} = \frac{q_{wp}D}{E_s}(1 - \mu_s^2)I_{wp}$$

$$I_{wp} = 0.85$$

$$q_{wp} = \frac{Q_{wp}}{A_p} = \frac{97}{(0.305)^2} = 1042.7 \text{ kN/m}^2$$

So

$$S_{e(2)} = \left[\frac{(1042.7)(0.305)}{30,000}\right](1 - 0.3^2)(0.85) = 0.0082 \text{ m} = 8.2 \text{ mm}$$

Again, from Eq. (14.45),

$$S_{e(3)} = \left(\frac{Q_{ws}}{pL}\right)\frac{D}{E_s}(1 - \mu_s^2)I_{ws}$$

$$I_{ws} = 2 + 0.35\sqrt{\frac{L}{D}} = 2 + 0.35\sqrt{\frac{12}{0.305}} = 4.2$$

So

$$S_{e(3)} = \frac{240}{(\pi \times 0.305)(12)}\left(\frac{0.305}{30,000}\right)(1 - 0.3^2)(4.2) = 0.00081\text{ m} = 0.81\text{ mm}$$

Hence, the total settlement is

$$S_e = 1.48 + 8.2 + 0.81 = \textbf{10.49 mm}$$

14.12 *Pile-Driving Formulas*

To develop the desired load-carrying capacity, a point bearing pile must penetrate the dense soil layer sufficiently or have sufficient contact with a layer of rock. This requirement cannot always be satisfied by driving a pile to a predetermined depth because soil profiles vary. For that reason, several equations have been developed to calculate the ultimate capacity of a pile during driving. These dynamic equations are widely used in the field to determine whether the pile has reached a satisfactory bearing value at the predetermined depth. One of the earliest of these dynamic equations—commonly referred to as the *Engineering News Record (ENR) formula*—is derived from the work-energy theory; that is,

energy imparted by the hammer per blow
= (pile resistance)(penetration per hammer blow)

According to the ENR formula, the pile resistance is the ultimate load, Q_u, expressed as

$$Q_u = \frac{W_R h}{S + C} \tag{14.49}$$

where
W_R = weight of the ram
h = height of fall of the ram
S = penetration of the pile per hammer blow
C = a constant

The pile penetration, S, is usually based on the average value obtained from the last few driving blows. In the equation's original form, the following values of C were recommended:

For drop hammers: $C = 2.54$ cm (if the units of S and h are in centimeters)
For steam hammers: $C = 0.254$ cm (if the units of S and h are in centimeters)

Also, a factor of safety of $FS = 6$ was recommended to estimate the allowable pile capacity. Note that, for single- and double-acting hammers, the term $W_R h$ can be replaced by EH_E (where E = hammer efficiency and H_E = rated energy of the hammer). Thus,

$$Q_u = \frac{EH_E}{S + C} \qquad (14.50)$$

The ENR pile-driving formula has been revised several times over the years. A recent form—the *modified ENR formula*—is

$$Q_u = \frac{EW_R h}{S + C} \frac{W_R + n^2 W_p}{W_R + W_p} \qquad (14.51)$$

where
E = hammer efficiency
$C = 0.254$ cm if the units of S and h are in centimeters
W_p = weight of the pile
n = coefficient of restitution between the ram and the pile cap

The efficiencies of various pile-driving hammers, E, are in the following ranges:

Hammer type	Efficiency, E
Single- and double-acting hammers	0.7–0.85
Diesel hammers	0.8–0.9
Drop hammers	0.7–0.9

Representative values of the coefficient of restitution, n, follow:

Pile material	Coefficient of restitution, n
Cast iron hammer and concrete piles (without cap)	0.4–0.5
Wood cushion on steel piles	0.3–0.4
Wooden piles	0.25–0.3

A factor of safety of 4 to 6 may be used in Eq. (14.51) to obtain the allowable load-bearing capacity of a pile.

Another equation, referred to as the *Danish formula*, also yields results as reliable as any other equation's:

$$Q_u = \frac{EH_E}{S + \sqrt{\dfrac{EH_E L}{2A_p E_p}}} \qquad (14.52)$$

where

E = hammer efficiency
H_E = rated hammer energy
E_p = modulus of elasticity of the pile material
L = length of the pile
A_p = area of the pile cross section

Consistent units must be used in Eq. (14.52). A factor of safety varying from 3 to 6 is recommended to estimate the allowable load-bearing capacity of piles.

Example 14.6

A precast concrete pile 305 mm × 305 mm in cross section is driven by a hammer. We have these values:

maximum rated hammer energy = 35 kN-m
weight of ram = 36 kN
total length of pile = 20 m
hammer efficiency = 0.8
coefficient of restitution = 0.45
weight of pile cap = 3.2 kN
number of blows for last 25.4 mm of penetration = 5

Estimate the allowable pile capacity by using each of these equations:

a. Eq. (14.50) (use $FS = 6$)
b. Eq. (14.51) (use $FS = 5$)
c. Eq. (14.52) (use $FS = 4$)

Solution
Part a. Eq. (14.50) is

$$Q_u = \frac{EH_E}{S + C}$$

We have $E = 0.8$, $H_E = 35$ kN-m, and

$$S = \frac{25.4}{5} = 5.08 \text{ mm} = 0.508 \text{ cm}$$

So

$$Q_u = \frac{(0.8)(35)(100)}{0.508 + 0.254} = 3674.5 \text{ kN}$$

Hence,

$$Q_{all} = \frac{Q_u}{FS} = \frac{3674.5}{6} \approx \mathbf{612 \text{ kN}}$$

Part b. Eq. (14.51) is

$$Q_u = \frac{EW_R h}{S + C} \frac{W_R + n^2 W_p}{W_R + W_p}$$

Weight of pile $= LA_p\gamma_c = (20)(0.305)^2(23.58) = 43.87$ kN and

W_p = weight of pile + weight of cap = 43.87 + 3.2 = 47.07 kN

So

$$Q_u = \left[\frac{(0.8)(35)(100)}{0.508 + 0.254}\right]\left[\frac{36 + (0.45)^2(47.07)}{36 + 47.07}\right]$$

$$= (3674)(0.548) \approx 2013 \text{ kN}$$

$$Q_{\text{all}} = \frac{Q_u}{FS} = \frac{2013}{5} = 402.6 \text{ kN} \approx \textbf{403 kN}$$

Part c. Eq. (14.52) is

$$Q_u = \frac{EH_E}{S + \sqrt{\dfrac{EH_E L}{2A_p E_p}}}$$

We have $E_p \approx 20.7 \times 10^6$ kN/m^2. So

$$\sqrt{\frac{EH_E L}{2A_p E_p}} = \sqrt{\frac{(0.8)(35)(20)}{(2)(0.305)^2(20.7 \times 10^6)}} = 0.0121 \text{ m} = 1.21 \text{ cm}$$

Hence,

$$Q_u = \frac{(0.8)(35)(100)}{0.508 + 1.21} = 1630 \text{ kN}$$

$$Q_{\text{all}} = \frac{Q_u}{FS} = \frac{1630}{4} = \textbf{407.5 kN} \qquad \blacksquare$$

14.13 *Negative Skin Friction*

Negative skin friction is a downward drag force exerted on the pile by the soil surrounding it. This action can occur under conditions such as the following:

1. If a fill of clay soil is placed over a granular soil layer into which a pile is driven, the fill will gradually consolidate. This consolidation process will exert a downward drag force on the pile (Figure 14.20a) during the period of consolidation.
2. If a fill of granular soil is placed over a layer of soft clay, as shown in Figure 14.20b, it will induce the process of consolidation in the clay layer and thus exert a downward drag on the pile.

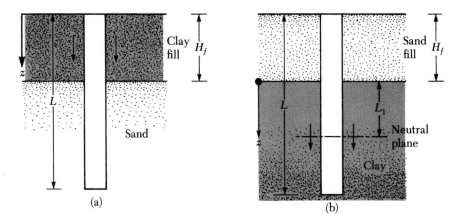

Figure 14.20 Negative skin friction

3. Lowering of the water table will increase the vertical effective stress on the soil at any depth, which will induce consolidation settlement in clay. If a pile is located in the clay layer, it will be subjected to a downward drag force.

In some cases, the downward drag force may be excessive and cause foundation failure. This section outlines two tentative methods for calculating negative skin friction.

Clay Fill over Granular Soil (Figure 14.20a)

Similar to the β method presented in Section 14.8, the negative (downward) skin stress on the pile is

$$f_n = K'\sigma'_o \tan \delta' \tag{14.53}$$

where
   K' = earth pressure coefficient = $K_o = 1 - \sin \phi'$
   σ'_o = vertical effective stress at any depth $z = \gamma'_f z$
   γ'_f = effective unit weight of fill
   δ' = soil–pile friction angle $\approx 0.5\phi'-0.7\phi'$

Hence, the total downward drag force, Q_n, on a pile is

$$Q_n = \int_0^{H_f} (pK'\gamma'_f \tan \delta')z\, dz = \frac{pK'\gamma'_f H_f^2 \tan \delta'}{2} \tag{14.54}$$

where H_f = height of the fill. If the fill is above the water table, the effective unit weight, γ'_f, should be replaced by the moist unit weight.

Granular Soil Fill over Clay (Figure 14.20b)

In this case, the evidence indicates that the negative skin stress on the pile may exist from $z = 0$ to $z = L_1$, which is referred to as the *neutral depth* (see Vesic, 1977, pp. 25–26, for discussion). The neutral depth may be given as (Bowles, 1982)

$$L_1 = \frac{L - H_f}{L_1}\left(\frac{L - H_f}{2} + \frac{\gamma_f' H_f}{\gamma'}\right) - \frac{2\gamma_f' H_f}{\gamma'} \tag{14.55}$$

where γ_f' and γ' = effective unit weights of the fill and the underlying clay layer, respectively.

Once the value of L_1 is determined, the downward drag force is obtained in the following manner: The unit negative skin friction at any depth from $z = 0$ to $z = L_1$ is

$$f_n = K' \sigma_o' \tan \delta' \tag{14.56}$$

where

$K' = K_o = 1 - \sin \phi'$
$\sigma_o' = \gamma_f' H_f + \gamma' z$
$\delta' = 0.5\phi' - 0.7\phi'$

Hence, the total drag force is

$$Q_n = \int_0^{L_1} p f_n \, dz = \int_0^{L_1} pK'(\gamma_f' H_f + \gamma' z)\tan \delta' \, dz$$

$$= (pK'\gamma_f' H_f \tan \delta')L_1 + \frac{L_1^2 pK'\gamma' \tan \delta'}{2} \tag{14.57}$$

For end-bearing piles, the neutral depth may be assumed to be located at the pile tip (i.e., $L_1 = L - H_f$).

If the soil and the fill are above the water table, the effective unit weights should be replaced by moist unit weights. In some cases, the piles can be coated with bitumen in the downdrag zone to avoid this problem. Baligh et al. (1978) summarized the results of several field tests that were conducted to evaluate the effectiveness of bitumen coating in reducing the negative skin friction.

A limited number of case studies of negative skin friction is available in the literature. Bjerrum et al. (1969) reported monitoring the downdrag force on a test pile at Sorenga in the harbor of Oslo, Norway (noted as pile G in the original paper). The study of Bjerrum et al. (1969) was also discussed by Wong and Teh (1995) in terms of the pile being driven to bedrock at 40 m. Figure 14.21a shows the soil profile and the pile. Wong and Teh estimated the following quantities:

- *Fill:* Moist unit weight, $\gamma_f = 16 \text{ kN/m}^3$
  Saturated unit weight, $\gamma_{sat(f)} = 18.5 \text{ kN/m}^3$

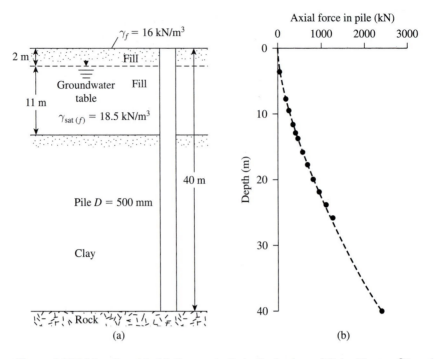

Figure 14.21 Negative skin friction on a pile in the harbor of Oslo, Norway [Based on Bjerrum et al. (1969) and Wong and Teh (1995)]

So

$$\gamma'_f = 18.5 - 9.81 = 8.69 \text{ kN/m}^3$$

and

$$H_f = 13 \text{ m}$$

- *Clay:* $K' \tan \delta' \approx 0.22$
  Saturated effective unit weight, $\gamma' = 19 - 9.81 = 9.19 \text{ kN/m}^3$
- *Pile:* $L = 40$ m
  Diameter, $D = 500$ mm

Thus, the maximum downdrag force on the pile can be estimated from Eq. (14.57). Since in this case the pile is a point bearing pile, the magnitude of $L_1 = 27$ m, and

$$Q_n = (p)(K' \tan \delta')[\gamma_f \times 2 + (13 - 2)\gamma'_f](L_1) + \frac{L_1^2 p \gamma'(K' \tan \delta')}{2}$$

or

$$Q_n = (\pi \times 0.5)(0.22)[(16 \times 2) + (8.69 \times 11)](27) + \frac{(27)^2(\pi \times 0.5)(9.19)(0.22)}{2}$$

$$= 2348 \text{ kN}$$

The measured value of the maximum Q_n was about 2500 kN (Figure 14.21b), which is in good agreement with the calculated value.

Example 14.7

Refer to Figure 14.20a; $H_f = 3$ m. The pile is circular in cross section with a diameter of 0.5 m. For the fill that is above the water table, $\gamma_f = 17.2$ kN/m³ and $\phi' = 36°$. Determine the total drag force. Use $\delta' = 0.7\phi'$.

Solution
From Eq. (14.54),

$$Q_n = \frac{pK'\gamma_f H_f^2 \tan \delta'}{2}$$

$$p = \pi(0.5) = 1.57 \text{ m}$$

$$K' = 1 - \sin \phi' = 1 - \sin 36° = 0.41$$

$$\delta' = (0.7)(36) = 25.2°$$

$$Q_n = \frac{(1.57)(0.41)(17.2)(3)^2 \tan 25.2}{2} = \textbf{23.4 kN}$$

Example 14.8

Refer to Figure 14.20b. Here, $H_f = 2$ m, pile diameter = 0.305 m, $\gamma_f = 16.5$ kN/m³, $\phi'_{clay} = 34°$, $\gamma_{sat(clay)} = 17.2$ kN/m³, and $L = 20$ m. The water table coincides with the top of the clay layer. Determine the downward drag force.

Solution
The depth of the neutral plane is given in Eq. (14.55) as

$$L_1 = \frac{L - H_f}{L_1}\left(\frac{L - H_f}{2} + \frac{\gamma_f H_f}{\gamma'}\right) - \frac{2\gamma_f H_f}{\gamma'}$$

Note that γ'_f in Eq. (14.55) has been replaced by γ_f because the fill is above the water table. So

$$L_1 = \frac{20 - 2}{L_1}\left[\frac{(20 - 2)}{2} + \frac{(16.5)(2)}{(17.2 - 9.81)}\right] - \frac{(2)(16.5)(2)}{(17.2 - 9.81)}$$

$$= \frac{242.4}{L_1} - 8.93$$

$$L = 11.75 \text{ m}$$

Now, referring to Eq. (14.57), we have

$$Q_n = (pK'\gamma_f H_f \tan \delta')L_1 + \frac{L_1^2 pK'\gamma' \tan \delta'}{2}$$

$$p = \pi(0.305) = 0.958 \text{ m}$$

$$K' = 1 - \sin 34° = 0.44$$

$$Q_n = (0.958)(0.44)(16.5)(2)[\tan(0.6 \times 34)](11.75)$$

$$+ \frac{(11.75)^2(0.958)(0.44)(17.2 - 9.81)[\tan(0.6 \times 34)]}{2}$$

$$= 60.78 + 79.97 = \mathbf{140.75\ kN}$$ ∎

14.14   *Group Piles—Efficiency*

In most cases, piles are used in groups to transmit the structural load to the soil (Figure 14.22). A *pile cap* is constructed over *group piles*. (Figure 14.22a). Determination of the load-bearing capacity of group piles is extremely complicated and has not yet been fully resolved. When the piles are placed close to each other, a reasonable assumption is that the stresses transmitted by the piles to the soil will overlap (Figure 14.22b), thus reducing the load-bearing capacity of the piles. Ideally, the piles in a group should be spaced so that the load-bearing capacity of the group is

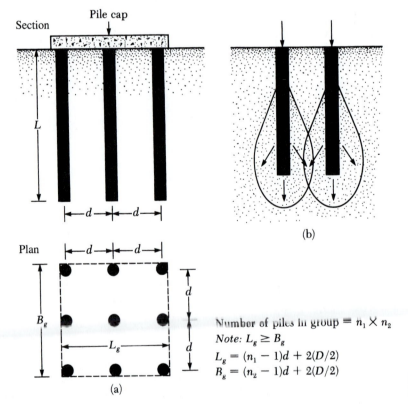

Number of piles in group $= n_1 \times n_2$

Note: $L_g \geq B_g$

$L_g = (n_1 - 1)d + 2(D/2)$

$B_g = (n_2 - 1)d + 2(D/2)$

Figure 14.22
Pile groups

no less than the sum of the bearing capacity of the individual piles. In practice, the minimum center-to-center pile spacing, d, is $2.5D$ and in ordinary situations is actually about $3D$ to $3.5D$.

The efficiency of the load-bearing capacity of a group pile may be defined as

$$\eta = \frac{Q_{g(u)}}{\Sigma\, Q_u} \qquad (14.58)$$

where

η = group efficiency
$Q_{g(u)}$ = ultimate load-bearing capacity of the group pile
Q_u = ultimate load-bearing capacity of each pile without the group effect

Piles in Sand

Feld (1943) suggested a method by which the load capacity of individual piles (friction) in a group embedded in sand could be assigned. According to this method, the ultimate capacity of a pile is reduced by one-sixteenth by each adjacent diagonal or row pile. The technique can be explained by referring to Figure 14.23, which shows the plan of a group pile. For pile type A, there are eight adjacent piles; for pile type B, there are five adjacent piles; and for pile type C, there are three adjacent piles. Now the following table can be prepared:

Pile type	No. of piles	No. of adjacent piles/pile	Reduction factor for each pile	Ultimate capacity[a]
A	1	8	$1 - \dfrac{8}{16}$	$0.5Q_u$
B	4	5	$1 - \dfrac{5}{16}$	$2.75Q_u$
C	4	3	$1 - \dfrac{3}{16}$	$3.25Q_u$
				$\Sigma\, 6.5Q_u = Q_{g(u)}$

[a] (No of piles) (Q_u) (reduction factor)
Q_u = ultimate capacity for an isolated pile

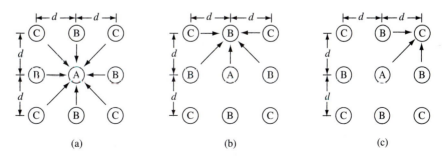

Figure 14.23 Feld's method for estimation of group capacity of friction piles

Hence,

$$\eta = \frac{Q_{g(u)}}{\Sigma Q_u} = \frac{6.5Q_u}{9Q_u} = 72\%$$

Figure 14.24 shows the variation of the group efficiency η for a 3×3 group pile in sand (Kishida and Meyerhof, 1965). It can be seen, that for loose and medium sands, the magnitude of the group efficiency can be larger than unity. This is due primarily to the densification of sand surrounding the pile.

Based on the experimental observations of the behavior of group piles in sand to date, two general conclusions may be drawn:

1. For *driven* group piles in *sand* with $d \geq 3D$, $Q_{g(u)}$ may be taken to be ΣQ_u, which includes the frictional and the point bearing capacities of individual piles.
2. For *bored* group piles in *sand* at conventional spacings ($d \approx 3D$), $Q_{g(u)}$ may be taken to be $\frac{2}{3}$ to $\frac{3}{4}$ times ΣQ_u (frictional and point bearing capacities of individual piles).

Piles in Clay

The ultimate load-bearing capacity of group piles in clay may be estimated with the following procedure:

1. Determine $\Sigma Q_u = n_1 n_2 (Q_p + Q_s)$. From Eq. (14.16),

$$Q_p = A_p[9c_{u(p)}]$$

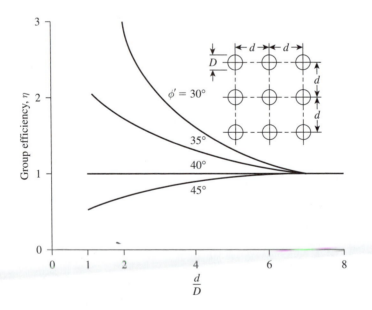

Figure 14.24 Variation of efficiency of pile groups in sand (Based on Kishida and Meyerhof, 1965)

where $c_{u(p)}$ = undrained cohesion of the clay at the pile tip. Also, from Eq. (14.28),

$$Q_s = \sum \alpha p c_u \, \Delta L$$

So

$$\sum Q_u = n_1 n_2 [9 A_p c_{u(p)} + \sum \alpha p c_u \, \Delta L] \tag{14.59}$$

2. Determine the ultimate capacity by assuming that the piles in the group act as a block with dimensions of $L_g \times B_g \times L$. The skin resistance of the block is

$$\sum p_g c_u \, \Delta L = \sum 2(L_g + B_g) c_u \, \Delta L$$

Calculate the point bearing capacity from

$$A_p q_p = A_p c_{u(p)} N_c^* = (L_g B_g) c_{u(p)} N_c^*$$

The variation of N_c^* with L/B_g and L_g/B_g is illustrated in Figure 14.25. Thus, the ultimate load is

$$\sum Q_u = L_g B_g c_{u(p)} N_c^* + \sum 2(L_g + B_g) c_u \, \Delta L \tag{14.60}$$

3. Compare the values obtained from Eqs. (14.59) and (14.60). The *lower* of the two values is $Q_{g(u)}$.

Piles in Rock

For point bearing piles resting on rock, most building codes specify that $Q_{g(u)} = \Sigma Q_u$, provided that the minimum center-to-center spacing of piles is $D + 300$ mm. For H-piles and piles with square cross sections, the magnitude of D is equal to the diagonal dimension of the pile cross section.

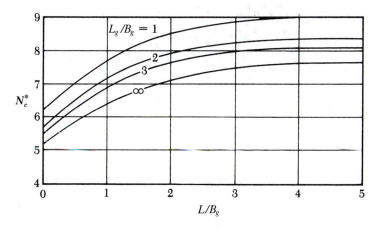

Figure 14.25 Variation of N_c^* with L_g/B_g and L/B_g

General Comments

A pile cap resting on soil, as shown in Figure 14.22a, will contribute to the load-bearing capacity of a pile group. However, this contribution may be neglected for design purposes because the support may be lost as a result of soil erosion or excavation during the life of the project.

Example 14.9

The section of a 3 × 4 group pile in a layered saturated clay is shown in Figure 14.26. The piles are square in cross section (356 mm × 356 mm). The center-to-center spacing, d, of the piles is 890 mm. Determine the allowable load-bearing capacity of the pile group. Use $FS = 4$.

Solution

From Eq. (14.59),

$$\Sigma Q_u = n_1 n_2 [9 A_p c_{u(p)} + \alpha_1 p c_{u(1)} L_1 + \alpha_2 p c_{u(2)} L_2]$$

From Figure 14.16, $c_{u(1)} = 50 \text{ kN/m}^2$; $\alpha_1 = 0.86$ and $c_{u(2)} = 85 \text{ kN/m}^2$; $\alpha_2 = 0.6$.

$$\Sigma Q_u = (3)(4) \left[\begin{array}{l} (9)(0.356)^2(85) + (0.86)(4 \times 0.356)(50)(5) + \\ (0.6)(4 \times 0.356)(85)(15) \end{array} \right] \approx 17{,}910 \text{ kN}$$

For piles acting as a group,

$$L_g = (3)(890) + 356 = 3026 \text{ mm} = 3.026 \text{ m}$$

$$B_g = (2)(890) + 356 = 2136 \text{ mm} = 2.136 \text{ m}$$

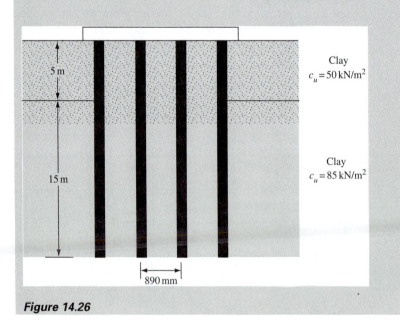

Figure 14.26

$$\frac{L_g}{B_g} = \frac{3.026}{2.136} = 1.42$$

$$\frac{L}{B_g} = \frac{20}{2.136} = 9.36$$

From Figure 14.25, $N_c^* \approx 8.75$. From Eq. (14.60),

$$\Sigma Q_u = L_g B_g c_{u(p)} N_c^* + \Sigma 2(L_g + B_g) c_u \Delta L$$
$$= (3.026)(2.136)(85)(8.75) + (2)(3.026 + 2.136)[(50)(3) + (85)(15)]$$
$$= 19,519 \text{ kN}$$

Hence,

$$\Sigma Q_u = 17,910 \text{ kN}$$

$$\Sigma Q_{\text{all}} = \frac{17,910}{FS} = \frac{17,910}{4} \approx 4478 \text{ kN} \qquad \blacksquare$$

14.15 *Elastic Settlement of Group Piles*

Several investigations relating to the settlement of group piles with widely varying results have been reported in the literature. The simplest relation for the settlement of group piles was given by Vesic (1969) as

$$S_{g(e)} = \sqrt{\frac{B_g}{D}} S_e \qquad (14.61)$$

where
$S_{g(e)}$ = elastic settlement of group piles
B_g = width of pile group section (see Figure 14.21a)
D = width or diameter of each pile in the group
S_e = elastic settlement of each pile at comparable working load (see Section 14.11)

For pile groups in sand and gravel, Meyerhof (1976) suggested the following empirical relation for elastic settlement:

$$S_{g(e)}(\text{mm}) = \frac{0.92q\sqrt{B_g}I}{N_{60}} \qquad (14.62)$$

where

$$q \ (\text{kN/m}^2) = Q_g/(L_g \, B_g) \tag{14.63}$$

L_g and B_g = length and width of the pile group section, respectively (m)
N_{60} = average standard penetration number within seat of settlement ($\approx B_g$ deep below the tip of the piles)
I = influence factor = $1 - L/8B_g \geq 0.5$
L = length of embedment of piles (m) $\tag{14.64}$

Similarly, the pile group settlement is related to the cone penetration resistance as

$$S_{g(e)} = \frac{qB_g I}{2q_c} \tag{14.65}$$

where q_c = average cone penetration resistance within the seat of settlement. In Eq. (14.65), all symbols are in consistent units.

14.16 *Consolidation Settlement of Group Piles*

The consolidation settlement of a pile group can be estimated by assuming an approximate distribution method that is commonly referred to as the 2:1 method. The calculation procedure involves the following steps (Figure 14.27):

1. Let the depth of embedment of the piles be L. The group is subjected to a total load of Q_g. If the pile cap is below the original ground surface, Q_g equals the total load of the superstructure on the piles minus the effective weight of soil above the pile group removed by excavation.
2. Assume that the load Q_g is transmitted to the soil beginning at a depth of $2L/3$ from the top of the pile, as shown in Figure 14.27 ($z = 0$). The load Q_g spreads out along 2 vertical:1 horizontal lines from this depth. Lines aa' and bb' are the two 2:1 lines.
3. Calculate the effective stress increase caused at the middle of each soil layer by the load Q_g:

$$\Delta\sigma_i' = \frac{Q_g}{(B_g + z_i)(L_g + z_i)} \tag{14.66}$$

where
$\Delta\sigma_i'$ = effective stress increase at the middle of layer i
L_g, B_g = length and width of the plan of pile group, respectively
z_i = distance from $z = 0$ to the middle of the clay layer, i

For example, in Figure 14.27 for layer 2, $z_i = L_1/2$; for layer 3, $z_i = L_1 + L_2/2$; and for layer 4, $z_i = L_1 + L_2 + L_3/2$. Note, however, that there will be no stress

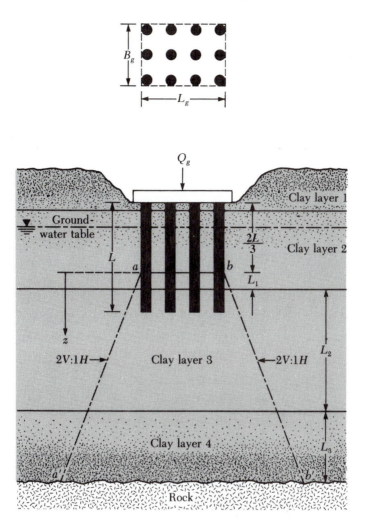

Figure 14.27 Consolidation settlement of group piles

increase in clay layer 1 because it is above the horizontal plane ($z = 0$) from which the stress distribution to the soil starts.

4. Calculate the settlement of each layer caused by the increased stress:

$$\Delta S_{c(i)} = \left[\frac{\Delta e_{(i)}}{1 + e_{0(i)}} \right] H_i \qquad (14.67)$$

where

$\Delta S_{c(i)}$ = consolidation settlement of layer i

$\Delta e_{(i)}$ = change of void ratio caused by the stress increase in layer i

$e_{o(i)}$ = initial void ratio of layer i (before construction)

H_i = thickness of layer i (*Note*: In Figure 14.27, for layer 2, $H_i = L_1$; for layer 3, $H_i = L_2$; and for layer 4, $H_i = L_3$.)

The relations for $\Delta e_{(i)}$ are given in Chapter 7.

5. Calculate the total consolidation settlement of the pile group by

$$\Delta S_{c(g)} = \sum \Delta S_{c(i)} \tag{14.68}$$

Note that the consolidation settlement of piles may be initiated by fills placed nearby, adjacent floor loads, and lowering of water tables.

Example 14.10

A group pile in clay is shown in Figure 14.28. Determine the consolidation settlement of the pile groups. All clays are normally consolidated.

Solution
Because the lengths of the piles are 15 m each, the stress distribution starts at a depth of 10 m below the top of the pile. We have $Q_g = 2000$ kN.

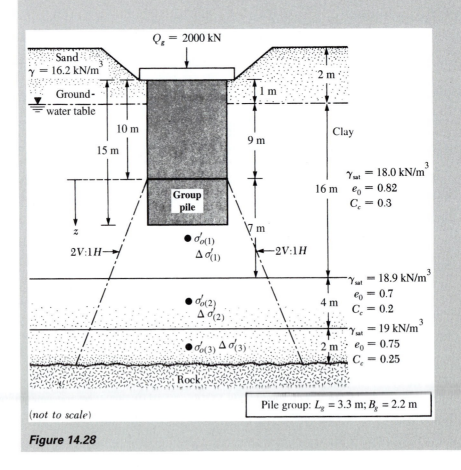

(not to scale)

Figure 14.28

Calculation of Settlement of Clay Layer 1

For normally consolidated clays,

$$\Delta S_{c(1)} = \left[\frac{C_{c(1)}H_1}{1 + e_{0(1)}}\right]\log\left[\frac{\sigma'_{o(1)} + \Delta\sigma'_{(1)}}{\sigma'_{o(1)}}\right]$$

$$\Delta\sigma'_{(1)} = \frac{Q_g}{(L_g + z_1)(B_g + z_1)} = \frac{2000}{(3.3 + 3.5)(2.2 + 3.5)} = 51.6 \text{ kN/m}^2$$

$$\sigma'_{o(1)} = 2(16.2) + 12.5(18.0 - 9.81) = 134.8 \text{ kN/m}^2$$

So

$$\Delta S_{c(1)} = \left[\frac{(0.3)(7)}{1 + 0.82}\right]\log\left[\frac{134.8 + 51.6}{134.8}\right] = 0.1624 \text{ m} = \textbf{162.4 mm}$$

Settlement of Layer 2

$$\Delta S_{c(2)} = \left[\frac{C_{c(2)}H_2}{1 + e_{o(2)}}\right]\log\left[\frac{\sigma'_{o(2)} + \Delta\sigma'_{(2)}}{\sigma'_{o(2)}}\right]$$

$$\sigma'_{o(2)} = 2(16.2) + 16(18.0 - 9.81) + 2(18.9 - 9.81) = 181.62 \text{ kN/m}^2$$

$$\Delta\sigma'_{(2)} = \frac{2000}{(3.3 + 9)(2.2 + 9)} = 14.52 \text{ kN/m}^2$$

Hence,

$$\Delta S_{c(2)} = \left[\frac{(0.2)(4)}{1 + 0.7}\right]\log\left[\frac{181.62 + 14.52}{181.62}\right] = 0.0157 \text{ m} = \textbf{15.7 mm}$$

Settlement of Layer 3

$$\sigma'_{o(3)} = 181.62 + 2(18.9 - 9.81) + 1(19 - 9.81) = 208.99 \text{ kN/m}^2$$

$$\Delta\sigma'_{(3)} = \frac{2000}{(3.3 + 12)(2.2 + 12)} = 9.2 \text{ kN/m}^2$$

$$\Delta S_{c(3)} = \left[\frac{(0.25)(2)}{1 + 0.75}\right]\log\left[\frac{208.99 + 9.2}{208.99}\right] = 0.0054 \text{ m} = \textbf{5.4 mm}$$

Hence, the total settlement is

$$\Delta S_{c(g)} = 162.4 + 15.7 + 5.4 = \textbf{183.5 mm}$$ ∎

DRILLED SHAFTS

As mentioned in the introduction of this chapter, drilled shafts are cast-in-place piles that generally have a diameter of about 750 mm or more. The use of drilled-shaft foundations has many advantages:

1. A single drilled shaft may be used instead of a group of piles and the pile cap.
2. Constructing drilled shafts in deposits of dense sand and gravel is easier than driving piles.
3. Drilled shafts may be constructed before grading operations are completed.
4. When piles are driven by a hammer, the ground vibration may cause damage to nearby structures, which the use of drilled shafts avoids.
5. Piles driven into clay soils may produce ground heaving and cause previously driven piles to move laterally. This does not occur during construction of drilled shafts.
6. There is no hammer noise during the construction of drilled shafts, as there is during pile driving.
7. Because the base of a drilled shaft can be enlarged, it provides great resistance to the uplifting load.
8. The surface over which the base of the drilled shaft is constructed can be visually inspected.
9. Construction of drilled shafts generally utilizes mobile equipment, which, under proper soil conditions, may prove to be more economical than methods of constructing pile foundations.
10. Drilled shafts have high resistance to lateral loads.

There are also several drawbacks to the use of drilled-shaft construction. The concreting operation may be delayed by bad weather and always needs close supervision. Also, as in the case of braced cuts, deep excavations for drilled shafts may cause substantial ground loss and damage to nearby structures.

14.17   *Types of Drilled Shafts*

Drilled shafts are classified according to the ways in which they are designed to transfer the structural load to the substratum. Figure 14.29a shows a drilled shaft that has a *straight shaft*. It extends through the upper layer(s) of poor soil, and its tip rests on a strong load-bearing soil layer or rock. The shaft can be cased with steel shell or pipe when required (as in the case of cased, cast-in-place concrete piles). For such shafts, the resistance to the applied load may develop from end bearing and also from side friction at the shaft perimeter and soil interface.

A *drilled shaft with bell* (Figures 14.29b and c) consists of a straight shaft with a bell at the bottom, which rests on good bearing soil. The bell can be constructed in the shape of a dome (Figure 14.29b), or it can be angled (Figure 14.29c). For angled bells, the underreaming tools commercially available can make 30° to 45° angles with the vertical. For the majority of drilled shafts constructed in the United States, the

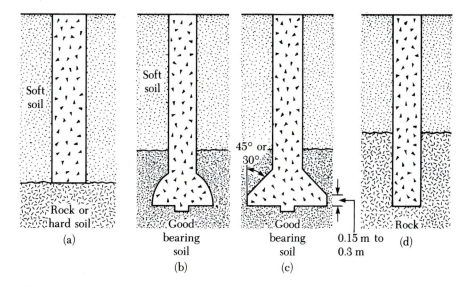

Figure 14.29 Types of drilled shaft: (a) straight shaft; (b) and (c) shaft with bell; (d) straight shafts socketed into rock

entire load-carrying capacity is assigned to the end bearing only. However, under certain circumstances, the end-bearing capacity and the side friction are taken into account. In Europe, both the side frictional resistance and the end-bearing capacity are always taken into account.

Straight shafts can also be extended into an underlying rock layer (Figure 14.29d). In calculating the load-bearing capacity of such drilled shafts, engineers take into account the end bearing and the shear stress developed along the shaft perimeter and rock interface.

14.18 Construction Procedures

The most common construction procedure used in the United States involves rotary drilling. There are three major types of construction methods, and they may be classified as (a) dry method, (b) casing method, and (c) wet method. A brief description of each method follows.

Dry Method of Construction

This method is employed in soils and rocks that are above the water table and will not cave in when the hole is drilled to full depth. The sequence of construction, as shown in Figure 14.30, is as follows:

1. The excavation is completed (and belled if desired) using proper drilling tools, and the spoils from the hole are deposited nearby (Figure 14.30a).
2. Concrete is then poured into the cylindrical hole (Figure 14.30b).

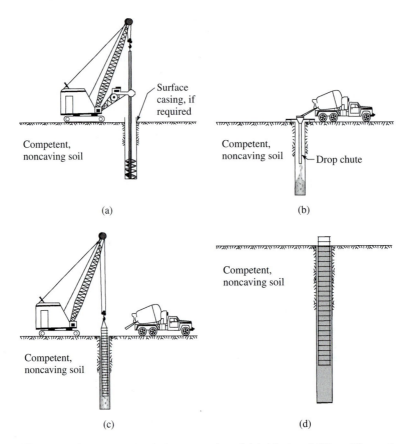

Figure 14.30 Dry method of construction: (a) initiating drilling, (b) starting concrete pour, (c) placing rebar cage, (d) completed shaft (After O'Neill and Reese, 1999)

3. If desired, a rebar cage is placed only in the upper portion of the shaft (Figure 14.30c).
4. The concreting is then completed and the drilled shaft will be as shown in Figure 14.30d.

Casing Method of Construction

This method is used in soils or rocks where caving or excessive deformation is likely to occur when the borehole is excavated. The sequence of construction is shown in Figure 14.31 and may be explained as follows:

1. The excavation procedure is initiated, as in the case of the dry method of construction described earlier (Figure 14.31a).
2. When the caving soil is encountered, bentonite slurry is introduced into the borehole (Figure 14.31b). Drilling is continued until the excavation goes past the caving soil and a layer of impermeable soil or rock is encountered.
3. A casing is then introduced into the hole (Figure 14.31c).

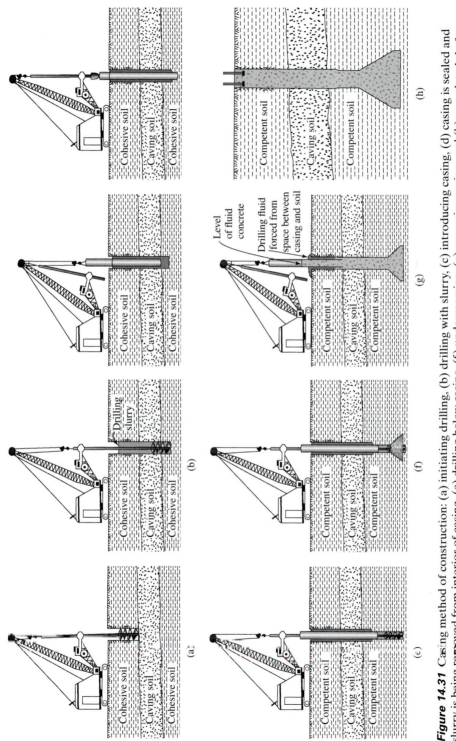

Figure 14.31 Casing method of construction: (a) initiating drilling, (b) drilling with slurry, (c) introducing casing, (d) casing is sealed and slurry is being removed from interior of casing, (e) drilling below casing, (f) underreaming, (g) removing casing, and (h) completed shaft (After O'Neill and Reese, 1999)

4. The slurry is bailed out of the casing using a submersible pump (Figure 14.31d).
5. A smaller drill that can pass through the casing is introduced into the hole and the excavation is continued (Figure 14.31e).
6. If needed, the base of the excavated hole can then be enlarged using an under-reamer (Figure 14.31f).
7. If reinforcing steel is needed, the rebar cage needs to extend the full length of the excavation. Concrete is then poured into the excavation and the casing is gradually pulled out (Figure 14.31g).
8. Figure 14.31h shows the completed drilled shaft.

Wet Method of Construction

This method is sometimes referred to as the slurry displacement method. Slurry is used to keep the borehole open during the entire depth of excavation (Figure 14.32). Following are the steps involved in the wet method of construction.

1. The excavation is continued to full depth with slurry (Figure 14.32a).
2. If reinforcement is required, the rebar cage is placed in the slurry (Figure 14.32b).

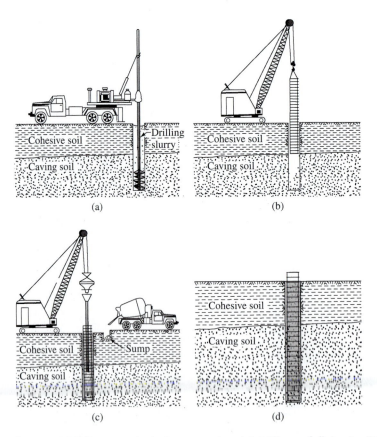

Figure 14.32 Slurry method of construction: (a) drilling to full depth with slurry, (b) placing rebar cage, (c) placing concrete, (d) completed shaft (After O'Neill and Reese, 1999)

3. Concrete that will displace the volume of slurry is then placed in the drill hole (Figure 14.32c).

4. Figure 14.32d shows the completed drilled shaft.

14.19   *Estimation of Load-Bearing Capacity*

The ultimate load of a drilled shaft (Figure 14.33) is

$$Q_u = Q_p + Q_s \tag{14.69}$$

where

Q_u = ultimate load
Q_p = ultimate load-carrying capacity at the base
Q_s = frictional (skin) resistance

The equation for the ultimate base load is similar to that for shallow foundations:

$$Q_p = A_p(c'N_c^* + q'N_q^* + 0.3\gamma D_b N_\gamma^*) \tag{14.70}$$

where

N_c^*, N_q^*, N_γ^* = the bearing capacity factors
q' = vertical effective stress at the level of the bottom of the drilled shaft
D_b = diameter of the base (see Figures 14.33a and b)
A_p = area of the base = $\pi/4 D_b^2$

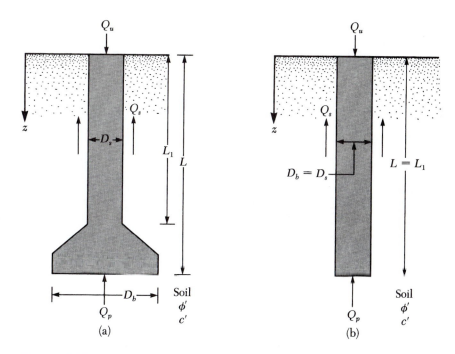

Figure 14.33 Ultimate bearing capacity of drilled shafts: (a) with bell; (b) straight shaft

In most cases, the last term (containing N_γ^*) is neglected except for relatively short shafts, so

$$Q_p = A_p(c'N_c^* + q'N_q^*) \tag{14.71}$$

The net load-carrying capacity at the base (that is, the gross load minus the weight of the drilled shaft) may be approximated as

$$Q_{p(net)} = A_p(c'N_c^* + q'N_q^* - q') = A_p[c'N_c^* + q'(N_q^* - 1)] \tag{14.72}$$

The expression for the frictional, or skin, resistance, Q_s, is similar to that for piles:

$$Q_s = \int_0^{L_1} pf\, dz \tag{14.73}$$

where

p = shaft perimeter = πD_s
f = unit frictional (or skin) resistance

Drilled Shafts in Sand

For shafts in sand, $c' = 0$ and hence, Eq. (14.72) simplifies to

$$Q_{p(net)} = A_p q'(N_q^* - 1) \tag{14.74}$$

The magnitude of $Q_{p(net)}$ can be reasonably estimated from a relationship based on the analysis of Berezantzev et al. (1961), which is a slight modification of Eq. (14.74), or

$$Q_{p(net)} = A_p q'(\omega N_q^* - 1) \tag{14.75}$$

where

$$N_q^* = 0.21 e^{0.17\phi'} \ (\textit{Note: } \phi' \text{ is in degrees}) \tag{14.76}$$

$$\omega = \text{a correction factor} = f\left(\frac{L}{D_b}\right); \text{ see Figure 14.34.}$$

The frictional resistance at ultimate load, Q_s, developed in a drilled shaft may be calculated from the relation given in Eq. (14.73), in which

$$p = \text{shaft perimeter} = \pi D_s$$

$$f = \text{unit frictional (or skin) resistance} = K\sigma_o' \tan \delta' \tag{14.77}$$

where

K = earth pressure coefficient $\approx K_o = 1 - \sin \phi'$
σ_o' = effective vertical stress at any depth z

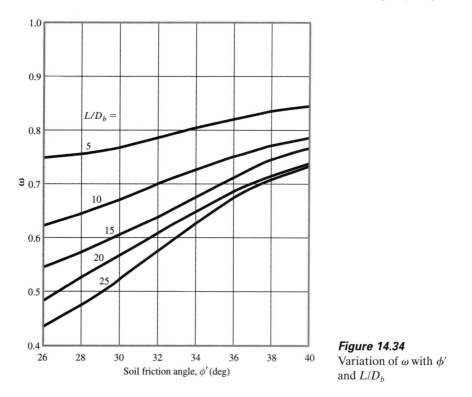

Figure 14.34
Variation of ω with ϕ' and L/D_b

Thus,

$$Q_s = \int_0^{L_1} pf\, dz = \pi D_s (1 - \sin \phi') \int_0^{L_1} \sigma_o' \tan \delta'\, dz \qquad (14.78)$$

The value of σ_o' will increase to a depth of about $15D_s$ and will remain constant thereafter, as shown in Figure 14.12.

An appropriate factor of safety should be applied to the ultimate load to obtain the net allowable load, or

$$Q_{u(net)} = \frac{Q_{p(net)} + Q_s}{FS} \qquad (14.79)$$

Drilled Shafts in Clay

From Eq. (14.72), for saturated clays with $\phi = 0$, $N_q^* = 1$; hence, the net base resistance becomes

$$Q_{p(net)} = A_p c_u N_c^* \qquad (14.80)$$

where

c_u = undrained cohesion
N_c^* = bearing capacity factor = $1.33[(\ln I_r) + 1]$ (for $L \geq 3D_b$) (14.81)
I_r = soil rigidity index

For $\phi = 0$ condition, I_r can be defined as

$$I_r = \frac{E_s}{3c_u}$$ (14.82)

where E_s = modulus of elasticity of soil.

O'Neill and Reese (1999) provided an approximate relationship between c_u and $E_s/3c_u$. Table 14.7 provides the interpolated values of this relationship.

For all practical purposes, if c_u is equal to or greater than 100 kN/m², the magnitude of N_c^* is 9.

The expression for the skin resistance of drilled shafts in clay is similar to Eq. (14.28), or

$$Q_s = \sum_{L=0}^{L=L_1} \alpha^* c_u p \, \Delta L$$ (14.83)

where p = perimeter of the shaft cross section. The value of α^* that can be used in Eq. (14.83) has not been fully established. However, the field test results available at this time indicate that α^* may vary between 1.0 and 0.3.

Kulhawy and Jackson (1989) reported the field test results of 106 drilled shafts without bell: 65 in uplift and 41 in compression. The best correlation for the magnitude of α^* obtained from these results is

$$\alpha^* = 0.21 + 0.25\left(\frac{p_a}{c_u}\right) \leq 1$$ (14.84)

where p_a = atmospheric pressure $\approx$ 100 kN/m² and c_u is in kN/m². So, conservatively, we may assume that

$$\alpha^* = 0.4$$ (14.85)

Table 14.7 Approximate variation of $E_s/3c_u$ with c_u (interpolated from O'Neill and Reese, 1999)

c_u (kN/m²)	$\dfrac{E_s}{3c_u}$	c_u (kN/m²)	$\dfrac{E_s}{3c_u}$
25	25	125	270
50	145	150	285
75	210	175	292
100	250	200	300

Example 14.11

A soil profile is shown in Figure 14.35. A point bearing drilled shaft with a bell is to be placed in the dense sand and gravel layer. The working load, Q_w, is 3000 kN. Use Eq. (14.75) and a factor of safety of 4 to determine the bell diameter, D_b. Ignore the frictional resistance of the shaft.

Solution
From Eq. (14.75),

$$Q_{p(\text{net})} = A_p q'(\omega N_q^* - 1). \quad \phi' = 35°.$$

$$N_q^* = 0.21e^{0.17\phi'} = 0.21e^{(0.17)(35)} = 80.6$$

$$q' = 6(17) + 2(19) = 140 \text{ kN/m}^2$$

$$Q_{p(\text{net})} = (Q_u)(FS) = (3000)(4) = 12,000 \text{ kN}$$

Assume $D_b = 1.5$ m. $L/D_b = 8/1.5 = 5.3$.
From Figure 14.34, for $\phi' = 35°$ and $L/D_b = 5.3$, the value of $\omega \approx 0.82$. So

$$12,000 = \left(\frac{\pi}{4} D_b^2 \right)(140)[(0.82)(80.6) - 1]$$

$D_b = 1.3$ m < 1.5 m (assumed).

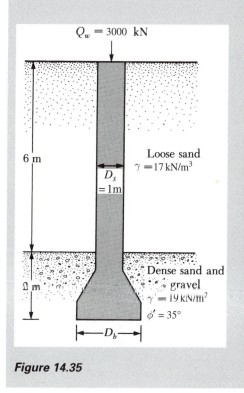

Figure 14.35

Try $D_b = 1.3$. $L/D_b = 8/1.3 = 6.15$.
From Figure 14.34, $\omega \approx 0.8$. So

$$12,000 = \left(\frac{\pi}{4}D_b^2\right)(140)[(0.8)(80.6) - 1]$$

$D_b = 1.31 \text{ m} \approx 1.3$ (assumed).
So

$$D_b \approx \textbf{1.3 m} \qquad \blacksquare$$

Example 14.12

Figure 14.36 shows a drilled shaft without a bell. Here, $L_1 = 8$ m, $L_2 = 3$ m, $D_s = 1.5$ m, $c_{u(1)} = 50$ kN/m^2, and $c_{u(2)} = 105$ kN/m^2. Determine

    a. The net ultimate point bearing capacity
    b. The ultimate skin resistance
    c. The working load, Q_w ($FS = 3$)

Use Eqs. (14.80), (14.83), and (14.85).

Solution
Part a
From Eq. (14.80),

$$Q_{p(\text{net})} = A_p c_u N_c^* = A_p c_{u(2)} N_c^* = \left[\left(\frac{\pi}{4}\right)(1.5)^2\right](105)(9) \approx \textbf{1670 kN}$$

(*Note:* Since $c_u/p_a > 1$, $N_c^* \approx 9$.)

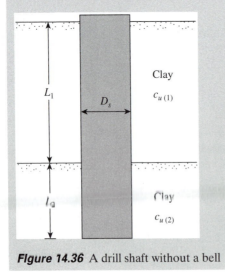

Figure 14.36 A drill shaft without a bell

Part b
From Eq. (14.83),

$$Q_s = \Sigma \alpha * c_u p \Delta L$$

From Eq. (14.85),

$$\alpha * = 0.4$$

$$p = \pi D_s = (3.14)(1.5) = 4.71 \text{ m}$$

and

$$Q_s = (0.4)(4.71)[(50 \times 8) + (105 \times 3)] \approx \textbf{1347 kN}$$

Part c

$$Q_w = \frac{Q_{p(\text{net})} + Q_s}{\text{FS}} = \frac{1670 + 1347}{3} = \textbf{1005.7 kN} \qquad \blacksquare$$

14.20 *Settlement of Drilled Shafts at Working Load*

The settlement of drilled shafts at working load is calculated in a manner similar to the one outlined in Section 14.11. In many cases, the load carried by shaft resistance is small compared to the load carried at the base. In such cases, the contribution of $S_{e(3)}$ may be ignored. Note that, in Eqs. (14.43) and (14.44), the term D should be replaced by D_b for shafts.

14.21 *Load-Bearing Capacity Based on Settlement*

Based on a data base of 41 loading tests, Reese and O'Neill (1989) proposed a method to calculate the load-bearing capacity of drilled shafts. The method is applicable to the following ranges:

1. Shaft diameter: $D_s = 0.52$ to 1.2 m
2. Bell depth: $L = 4.7$ to 30.5 m
3. $c_u = 29$ to 287 kN/m^2
4. Standard field penetration resistance: $N_{60} = 5$ to 60
5. Overconsolidation ratio: 2 to 15
6. Concrete slump: 100 to 225 mm

Reese and O'Neill's procedure, with reference to Figure 14.37, gives

$$Q_u = \sum_{i=1}^{N} f_i p \, \Delta L_i + q_p A_p \qquad (14.86)$$

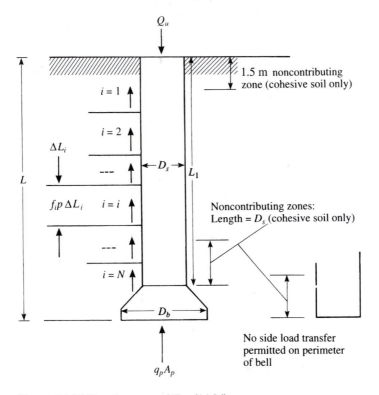

Figure 14.37 Development of Eq. (14.86)

where

f_i = ultimate unit shearing resistance in layer i
p = perimeter of the shaft = πD_s
q_p = unit point resistance
A_p = area of the base = $(\pi/4)D_b^2$

Following are the relationships for determining Q_u in cohesive and granular soils.

Cohesive Soil

Based on Eq. (14.86), we have

$$f_i = \alpha_i^* c_{u(i)} \tag{14.87}$$

The following values are recommended for α_i^*:

α_i^* = 0 for the top 1.5 m and bottom 1 diameter, D_s, of the drilled shaft.
  (*Note:* If $D_b > D_s$, then $\alpha^* = 0$ for 1 diameter above the top of the bell and for the peripheral area of the bell itself.)
α_i^* = 0.55 elsewhere

and

$$q_p \, (\text{kN/m}^2) = 6c_{ub}\left(1 + 0.2\frac{L}{D_b}\right) \le 9c_{ub} \le 3.83 \text{ MN/m}^2 \qquad (14.88)$$

where c_{ub} = average undrained cohesion within $2D_b$ below the base (kN/m^2).

If D_b is large, excessive settlement will occur at the ultimate load per unit area, q_p, as given by Eq. (14.88). Thus, for $D_b > 1.9$ m, q_p may be replaced by q_{pr}, or

$$q_{pr} = F_r q_p \qquad (14.89)$$

where

$$F_r = \frac{2.5}{0.0254\psi_1 D_b \,(\text{m}) + \psi_2} \le 1 \qquad (14.90)$$

$$\psi_1 = 0.0071 + 0.0021\left(\frac{L}{D_b}\right) \le 0.015 \qquad (14.91)$$

$$\psi_2 = 7.787(c_{ub})^{0.5} \qquad (0.5 \le \psi_2 \le 1.5) \qquad (14.92)$$
$$\uparrow$$
$$\text{kN/m}^2$$

If the load-bearing capacity at a limited level of settlement is required, then Tables 14.8 and 14.9 may be used for the procedure described next. The values provided in these tables are based on the average curve from field observations by Reese and O'Neill (1989).

1. Select a value of settlement, S_e.
2. Calculate $\sum_{i=1}^{N} f_i p \, \Delta L_i$ and $q_p A_p$, as given in Eq. (14.86).
3. Using Tables 14.8 and 14.9 and the calculated values in step 2, determine the *side load* and the *end bearing load*.
4. The sum of the side load and the end bearing load is the total applied load.

Table 14.8 Normalized side load transfer with settlement for cohesive soils (based on average curve)

Settlement $\dfrac{}{D_s}$ (%)	Side load transfer $\Sigma f_i p \, \Delta L_i$	Settlement $\dfrac{}{D_s}$ (%)	Side load transfer $\Sigma f_i p \, \Delta L_i$
0	0	0.8	0.95
0.1	0.48	1.0	0.94
0.2	0.74	1.2	0.92
0.3	0.86	1.4	0.91
0.4	0.91	1.6	0.89
0.6	0.95	1.8	0.85
0.7	0.955	2.0	0.82

Table 14.9 Normalized base load transfer with settlement for cohesive soils (based on average curve)

$\dfrac{\text{Settlement}}{D_b}$ (%)	$\dfrac{\text{End bearing}}{q_p A_p}$	$\dfrac{\text{Settlement}}{D_b}$ (%)	$\dfrac{\text{End bearing}}{q_p A_p}$
0	0	4.0	0.951
0.5	0.363	5.0	0.971
1.0	0.578	6.0	0.971
1.5	0.721	7.0	0.971
2.0	0.804	8.0	0.971
2.5	0.863	9.0	0.971
3.0	0.902	10.0	0.971

Cohesionless Soil

Based on Eq. (14.86), we have

$$f_i = \beta \sigma'_{ozi} \tag{14.93}$$

where

σ'_{ozi} = vertical effective stress at the middle of layer i

$$\beta = 1.5 - 0.244 z_i^{0.5} \qquad (0.25 \le \beta \le 1.2) \tag{14.94}$$

z_i = depth of the middle of layer i (m)

More recently, Rollins et al. (2005) have modified Eq. (14.94) for gravelly sands as follows:

For sand with 25 to 50% gravel:

$$\beta = 2.0 - 0.15 z_i^{0.75} \quad (0.25 \le \beta \le 1.8) \tag{14.95}$$

For sand with more than 50% gravel:

$$\beta = 3.4 e^{-0.085 z_i} \quad (0.25 \le \beta \le 3.0) \tag{14.96}$$

In Eqs. (14.95) and (14.96), z is in meters (m).

The point bearing capacity is

$$q_p \, (\text{kN/m}^2) = 57.5 N_{60} \le 4.3 \text{ MN/m}^2 \tag{14.97}$$

where N_{60} = mean *uncorrected* standard penetration number within a distance of $2D_b$ below the base of the drilled shaft.

As in Eq. (14.89), to control excessive settlement, the magnitude of q_p may be modified as follows:

$$q_{pr} = \frac{1.27}{D_b (\text{m})} q_p \qquad (\text{for } D_b \ge 1.27 \text{ m}) \tag{14.98}$$

Tables 14.10 and 14.11 may be used to calculate settlement based load-bearing capacity. They are similar to Tables 14.8 and 14.9 for clay.

Table 14.10 Normalized side load transfer with settlement for cohesionless soils (based on average curve)

Settlement $\dfrac{}{D_s}$ (%)	Side load transfer $\Sigma f_i p \, \Delta L_i$	Settlement $\dfrac{}{D_s}$ (%)	Side load transfer $\Sigma f_i p \, \Delta L_i$
0	0	0.8	0.974
0.1	0.371	1.0	0.987
0.2	0.590	1.2	0.974
0.3	0.744	1.4	0.968
0.4	0.846	1.6	0.960
0.5	0.910	1.8	0.940
0.6	0.936	2.0	0.920

Table 14.11 Normalized base load transfer with settlement for cohesionless soils (based on average curve)

Settlement $\dfrac{}{D_b}$ (%)	End bearing $q_p A_p$	Settlement $\dfrac{}{D_b}$ (%)	End bearing $q_p A_p$
0	0	6	1.10
1	0.32	7	1.20
2	0.56	8	1.29
3	0.73	9	1.38
4	0.87	10	1.44
5	0.98		

Example 14.13

A drilled shaft in a cohesive soil is shown in Figure 14.38. Use the procedure outlined in this section to determine these values:

a. The ultimate load-carrying capacity
b. The load-carrying capacity for an allowable settlement of 12.7 mm

Solution
Part a
From Eq. (14.87), we have

$$f_i = \alpha_i^* c_{u(i)}$$

From Figure 14.38,

$$\Delta L_1 = 3.66 - 1.5 = 2.16 \text{ m}$$

$$\Delta L_2 = (6.1 - 3.66) - D_s = 2.44 - 0.76 = 1.68 \text{ m}$$

$$c_{u(1)} = 38 \text{ kN/m}^2$$

$$c_{u(2)} = 57.5 \text{ kN/m}^2$$

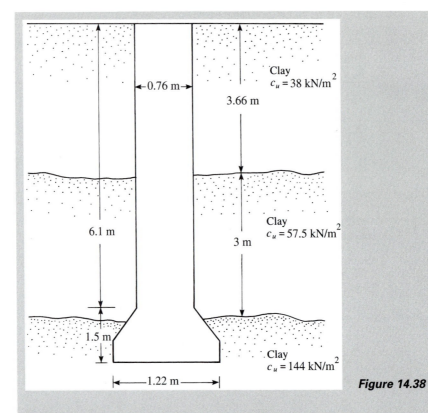

Clay
$c_u = 38$ kN/m^2

0.76 m

3.66 m

6.1 m

Clay
$c_u = 57.5$ kN/m^2

3 m

1.5 m

Clay
$c_u = 144$ kN/m^2

1.22 m

Figure 14.38

Hence,

$$\sum f_i p \, \Delta L_i = \sum \alpha_i^* c_{u(i)} p \, \Delta L_i$$

$$= (0.55)(38)(\pi \times 0.76)(2.16) + (0.55)(57.5)(\pi \times 0.76)(1.68)$$

$$= 234.6 \text{ kN}$$

From Eq. (14.88), we have

$$q_p = 6c_{ub}\left(1 + 0.2\frac{L}{D_b}\right) = (6)(144)\left[1 + 0.2\left(\frac{6.1 + 1.5}{1.22}\right)\right] = 1940 \text{ kN/m}^2$$

Check:

$$q_p = 9c_{ub} = (9)(144) = 1296 \text{ kN/m}^2 < 1940 \text{ kN/m}^2$$

So, we use $q_p = 1296$ kN/m^2:

$$q_p A_p = q_p\left(\frac{\pi}{4}D_b^2\right) = (1296)\left[\left(\frac{\pi}{4}\right)(1.22)^2\right] \approx 1515 \text{ kN}$$

Hence,

$$Q_u = \sum \alpha_i^* c_{u(i)} p \, \Delta L_i + q_p A_p = 234.6 + 1515 = \mathbf{1749.6 \text{ kN}}$$

Part b
We have

$$\frac{\text{allowable settlement}}{D_s} = \frac{12.7}{(0.76)(1000)} = 0.0167 = 1.67\%$$

From Table 14.8, for a normalized settlement of 1.67%, the normalized side load is about 0.87. Thus, the side load is

$$(0.87)\left(\sum f_{ip}\, \Delta L_i\right) = (0.87)(234.6) = 204.1 \text{ kN}$$

Again,

$$\frac{\text{allowable settlement}}{D_b} = \frac{12.7}{(1.22)(1000)} = 0.0104 = 1.04\%$$

From Table 14.9, for a normalized settlement of 1.04%, the normalized end bearing is about 0.58. So the base load is

$$(0.58)(q_p A_p) = (0.58)(1515) = 878.7 \text{ kN}$$

Thus, the total load is

$$Q = 204.1 + 878.7 = \textbf{1082.8 kN}$$ ∎

Example 14.14

A drilled shaft is shown in Figure 14.39. The uncorrected average standard penetration number (N_{60}) within a distance of $2D_b$ below the base of the shaft is about 30. Determine

    a. The ultimate load-carrying capacity
    b. The load-carrying capacity for a settlement of 12 mm. Use Eq. (14.95).

Solution
Part a
From Eqs. (14.93) and (14.95),

$$f_i = \beta \sigma'_{ozi}$$

and

$$\beta = 2.0 - 0.15z^{0.75}$$

For this problem, $z_i = 6/2 = 3$ m, so

$$\beta = 2 - (0.15)(3)^{0.75} = 1.658$$

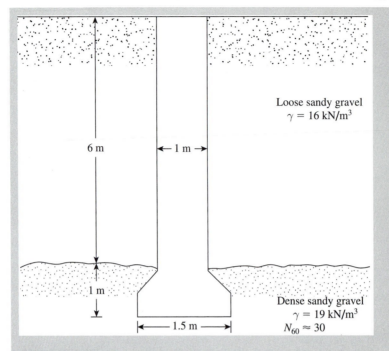

Figure 14.39 Drilled shaft supported by a dense layer of sandy gravel

and

$$\sigma'_{ozi} = \gamma z_i = (16)(3) = 48 \text{ kN/m}^2$$

Thus,

$$f_i = (48)(1.658) = 79.58 \text{ kN/m}^2$$

and

$$\Sigma f_i p \Delta L_i = (79.58)(\pi \times 1)(6) = 1500 \text{ kN}$$

From Eq. (14.97),

$$q_p = 57.5 N_{60} = (57.5)(30) = 1725 \text{ kN/m}^2$$

Note that D_b is greater than 1.27 m. So we will use Eq. (14.98).

$$q_{pr} = \left(\frac{1.27}{D_b}\right) q_p = \left(\frac{1.27}{1.5}\right)(1725) \approx 1461 \text{ kN/m}^2$$

Now,

$$q_{pr} A_p = (1461)\left(\frac{\pi}{4} \times 1.5^2\right) \approx 2582 \text{ kN}$$

Hence,

$$Q_{u(\text{net})} = q_{pr}A_p + \Sigma f_i p \Delta L_i = 2582 + 1500 = \textbf{4082 kN}$$

Part b
We have

$$\frac{\text{Allowable settlement}}{D_s} = \frac{12}{(1.0)(1000)} = 0.12 = 1.2\%$$

Table 14.10 shows, that for a normalized settlement of 1.2%, the normalized load is about 0.974. Thus, the side load transfer is (0.974) (1500) ≈ 1461 kN. Similarly,

$$\frac{\text{Allowable settlement}}{D_b} = \frac{12}{(1.5)(1000)} = 0.008 = 0.8\%$$

Table 14.11 indicates, that for a normalized settlement of 0.8%, the normalized base load is about 0.25. So the base load is (0.25)(2582) = 645.5 kN. Hence, the total load is

$$Q = 1461 + 645.5 \approx \textbf{2102 kN}$$ ∎

Problems

14.1 A concrete pile is 15 m long and 406 mm × 406 mm in cross section. The pile is fully embedded in sand, for which $\gamma = 17.3$ kN/m³ and $\phi' = 30°$.
   a. Calculate the ultimate point load, Q_p [Use Eq. (14.13)]
   b. Determine the total frictional resistance for $K = 1.3$ and $\delta' = 0.8\phi'$. [Use Eqs. (14.17), (14.18), and (14.19)]
14.2 Redo Problem 14.1 for $\gamma = 18.4$ kN/m³ and $\phi' = 37°$.
14.3 A driven closed-ended pipe pile is shown in Figure 14.40.
   a. Find the ultimate point load.
   b. Determine the ultimate frictional resistance, Q_s; use $K = 1.4$ and $\delta' = 0.6\phi'$.
   c. Calculate the allowable load of the pile; use $FS = 4$.
14.4 A concrete pile 20 m long with a cross section of 381 mm × 381 mm is fully embedded in a saturated clay layer. For the clay, $\gamma_{\text{sat}} = 18.5$ kN/m³, $\phi = 0$, and $c_u = 70$ kN/m². Assume that the water table lies below the tip of the pile. Determine the allowable load that the pile can carry $(FS = 3)$. Use the α method to estimate the skin resistance.
14.5 Redo Problem 14.4 using the λ method for estimating the skin resistance.
14.6 A concrete pile 381 mm × 381 mm in cross section is shown in Figure 14.41. Calculate the ultimate skin resistance using each of the following methods:
   a. α method
   b. λ method
   c. β method
Use $\phi'_R = 25°$ for all clays, which are normally consolidated.

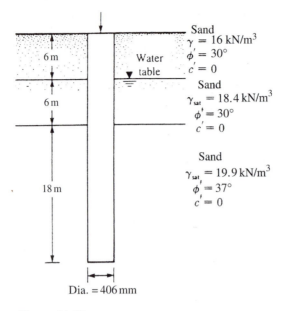

Figure 14.40

14.7 A steel pile (H-section; HP 310 × 1.226; see Table 14.1) is driven into a layer of sandstone. The length of the pile is 20 m. Following are the properties of the sandstone:

Unconfined compression strength $= q_{u(\text{lab})} = 73.5$ MN/m^2

Angle of friction $= 37°$

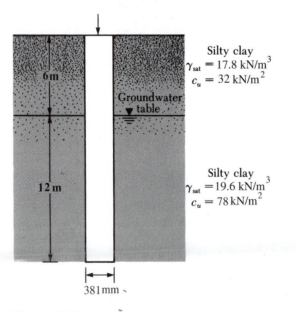

Figure 14.41

Using a factor of safety of 5, estimate the allowable point load that can be carried by the pile.

14.8 A concrete pile is 18 m long and has a cross section of 405 mm × 405 mm. The pile is embedded in sand having $\gamma = 17.5$ kN/m^3 and $\phi' = 36°$. The allowable working load is 650 kN. If 450 kN are contributed by the frictional resistance and 200 kN are from the point load, determine the elastic settlement of the pile. Here, $E_p = 21 \times 10^6$ kN/m^2, $E_s = 28 \times 10^3$ kN/m^2, $\mu_s = 0.4$, and $\zeta = 0.6$.

14.9 A steel pile (H-section; HP 330 × 1.462; see Table 14.1) is driven by a hammer. The maximum rated hammer energy is 50 kN-m, the weight of the ram is 58 kN, and the length of the pile is 25 m. Also given are the following:

- Coefficient of restitution = 0.3
- Weight of the pile cap = 4.3 kN
- Hammer efficiency = 0.8
- Number of blows for the last 25.4 mm of penetration = 12
- $E_p = 207 \times 10^6$ kN/m^2

Estimate the pile capacity using Eq. (14.51). Use $FS = 4$.

14.10 Solve Problem 14.9 using the Danish formula [Eq. (14.52)]. Use $FS = 3$.

14.11 Figure 14.20a shows a pile. Let $L = 18$ m, $D = 356$ mm, $H_f = 4$ m, $\gamma_f = 18$ kN/m^3, $\phi'_{fill} = 28°$. Determine the total downward drag force on the pile. Assume that the fill is located above the water table and that $\delta' = 0.6\phi'_{fill}$.

14.12 Refer to Figure 14.20b. Let $L = 19$ m, $\gamma_{fill} = 15.2$ kN/m^3, $\gamma_{sat(clay)} = 19.5$ kN/m^3, $\phi'_{clay} = 30°$, $H_f = 3.2$ m, and $D = 0.46$ m. The water table coincides with the top of the clay layer. Determine the total downward drag on the pile. Assume that $\delta' = 0.5\phi'_{clay}$.

14.13 The plan of a group pile is shown in Figure 14.42. Assume that the piles are embedded in a saturated homogeneous clay having $c_u = 80$ kN/m^2. For the piles, $D = 356$ mm, center-to-center spacing = 850 mm, and $L = 22$ m. Find the allowable load-carrying capacity of the pile group. Use $FS = 3$.

14.14 Redo Problem 14.13 for $d = 762$ mm, $L = 15$ m, $D = 381$ mm, and $c_u = 50$ kN/m^2.

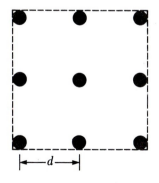

Figure 14.42

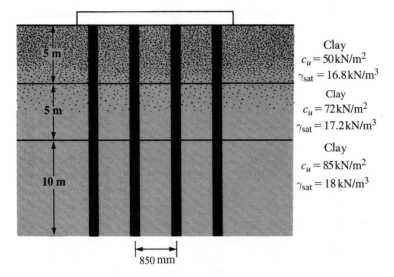

Clay
$c_u = 50\,\text{kN/m}^2$
$\gamma_{\text{sat}} = 16.8\,\text{kN/m}^3$

Clay
$c_u = 72\,\text{kN/m}^2$
$\gamma_{\text{sat}} = 17.2\,\text{kN/m}^3$

Clay
$c_u = 85\,\text{kN/m}^2$
$\gamma_{\text{sat}} = 18\,\text{kN/m}^3$

5 m

5 m

10 m

850 mm

Figure 14.43

14.15 The section of a 4×4 group pile in a layered saturated clay is shown in Figure 14.43. The piles are square in cross section (356 mm $\times$ 356 mm). The center-to-center spacing of the piles, d, is 850 mm. Assuming that the groundwater table is located 3 m below the pile tip, determine the allowable load-bearing capacity of the pile group. Use $FS = 4$.

14.16 Figure 14.44 shows a group pile in clay. Determine the consolidation settlement of the group.

14.17 A drilled shaft is shown in Figure 14.45. For the shaft, $L_1 = 5$ m, $L_2 = 2$ m, $D_s = 1$ m, and $D_b = 2$ m. For the soil, $\gamma_c = 17$ kN/m³, $c_u = 35$ kN/m², $\gamma_s = 19$ kN/m³, and $\phi' = 38°$. Determine the net allowable point bearing capacity ($FS = 3$).

14.18 For the drilled shaft described in Problem 14.17, what skin resistance would develop for the top 5 m, which is in clay?

14.19 Figure 14.46 shows a drilled shaft without a bell. Here, $L_1 = 9$ m, $L_2 = 2.8$ m, $D_s = 1.1$ m, $c_{u(1)} = 50$ kN/m², and $c_{u(2)} = 105$ kN/m². Find these values:
a. The net ultimate point bearing capacity
b. The ultimate skin resistance
c. The working load, Q_w ($FS = 3$)

14.20 For the drilled shaft described in Problem 14.19, estimate the total elastic settlement at working load. Use Eqs. (14.42), (14.44), and (14.45). Assume that $E_p = 21 \times 10^6$ kN/m², $\mu_s = 0.3$, $E_s = 14 \times 10^3$ kN/m², $\xi = 0.65$. Assume 50% mobilization of skin resistance at working load.

14.21 For the drilled shaft described in Problem 14.19, determine these values:
a. The ultimate load-carrying capacity
b. The load-carrying capacity for a settlement of 12.7 mm
Use the procedure outlined in Section 14.21.

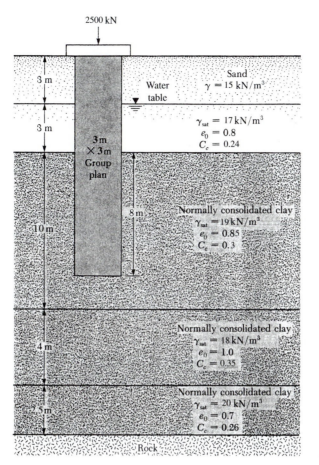

Sand
$\gamma = 15 \, \text{kN/m}^3$

Water table

$\gamma_{sat} = 17 \, \text{kN/m}^3$
$e_0 = 0.8$
$C_c = 0.24$

3 m

3 m

3 m × 3 m Group plan

8 m

10 m

Normally consolidated clay
$\gamma_{sat} = 19 \, \text{kN/m}^3$
$e_0 = 0.85$
$C_c = 0.3$

Normally consolidated clay
$\gamma_{sat} = 18 \, \text{kN/m}^3$
$e_0 = 1.0$
$C_c = 0.35$

4 m

Normally consolidated clay
$\gamma_{sat} = 20 \, \text{kN/m}^3$
$e_0 = 0.7$
$C_c = 0.26$

2.5 m

2500 kN

Rock

Figure 14.44

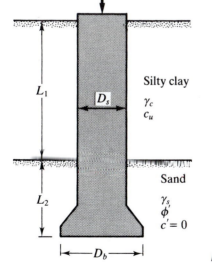

Silty clay
γ_c
c_u

L_1

D_s

Sand
γ_s
ϕ'
$c' = 0$

L_2

D_b

Figure 14.45

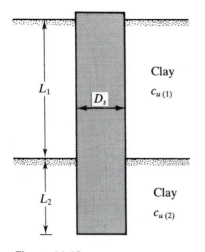

Figure 14.46

14.22 Refer to Figure 14.47, for which $L = 6$ m, $L_1 = 5$ m, $D_s = 1.2$ m, $D_b = 1.7$ m, $\gamma = 15.7$ kN/m³, and $\phi' = 33°$. The average uncorrected standard penetration number within $2D_b$ below the base is 32. Determine these values:
a. The ultimate load-carrying capacity
b. The load-carrying capacity for a settlement of 12.7 mm
Use the procedure outlined in Section 14.21.

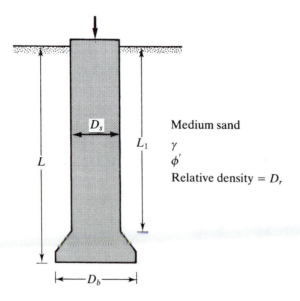

Figure 14.47

References

AMERICAN SOCIETY OF CIVIL ENGINEERS (1959). "Timber Piles and Construction Timbers," *Manual of Practice*, No. 17, American Society of Civil Engineers, New York.

BALIGH, M. M., VIVATRAT, V., and PIGI, H. (1978). "Downdrag on Bitumen-Coated Piles," *Journal of the Geotechnical Engineering Division*, American Society of Civil Engineers, Vol. 104, No. GT11, 1355–1370.

BEREZARTZEV, V. G., KHRISTOFOROV, V. S., and GOLUBKOV, V. N. (1961). "Load Bearing Capacity and Deformation of Piled Foundations," *Proceedings*, Fifth International Conference on Soil Mechanics and Foundation Engineering, Paris, Vol. 2, 11–15.

BJERRUM, L., JOHANNESSEN, I. J., and EIDE, O. (1969). "Reduction of Skin Friction on Steel Piles to Rock," *Proceedings*, Seventh International Conference on Soil Mechanics and Foundation Engineering, Mexico City, Vol. 2, 27–34.

BOWLES, J. E. (1982). *Foundation Design and Analysis*, McGraw-Hill, New York.

FELD, J. (1943). "Friction Pile Foundations," *Discussion, Transactions*, American Society of Civil Engineers, Vol. 108.

GOODMAN, R. E. (1980). *Introduction to Rock Mechanics*, Wiley, New York.

KISHIDA, H., and MEYERHOF, G. G. (1965). "Bearing Capacity of Pile Groups under Eccentric Loads in Sand," *Proceedings, Sixth International Conference on Soil Mechanics and Foundation Engineering*, Montreal, Vol. 2, 270–274.

KULHAWY, F. H., and JACKSON, C. S. (1989). "Some Observations on Undrained Side Resistance of Drilled Shafts," *Proceedings*, Foundation Engineering: Current Principles and Practices, American Society of Civil Engineers, Vol. 2, 1011–1025.

MCCLELLAND, B. (1974). "Design of Deep Penetration Piles for Ocean Structures," *Journal of the Geotechnical Engineering Division*, American Society of Civil Engineers, Vol. 100, No. GT7, 709–747.

MEYERHOF, G. G. (1976). "Bearing Capacity and Settlement of Pile Foundations," *Journal of the Geotechnical Engineering Division*, American Society of Civil Engineers, Vol. 102, No. GT3, 197–228.

NOTTINGHAM, L. C., and SCHMERTMANN, J. H. (1975). *An Investigation of Pile Capacity Design Procedures*, Research Report No. D629, Department of Civil Engineering, University of Florida, Gainesville, FL.

O'NEILL, M. W., and REESE, L. C. (1999). *Drilled Shafts: Construction Procedure and Design Methods*, FHWA Report No. IF-99–025.

REESE, L. C., and O'NEILL, M. W. (1989). "New Design Method for Drilled Shafts from Common Soil and Rock Tests," *Proceedings*, Foundation Engineering: Current Principles and Practices, American Society of Civil Engineers, Vol. 2, 1026–1039.

VESIC, A. S. (1977). *Design of Pile Foundations*, National Cooperative Highway Research Program Synthesis of Practice No. 42, Transportation Research Board, Washington, D.C.

VIJAYVERGIYA, V. N., and FOCHT, J. A., JR. (1972). *A New Way to Predict Capacity of Piles in Clay*, Offshore Technology Conference Paper 1718, Fourth Offshore Technology Conference, Houston.

WONG, K. S., and TEH, C. I. (1995). "Negative Skin Friction on Piles in Layered Soil Deposit," *Journal of Geotechnical and Geoenvironmental Engineering*, American Society of Civil Engineers, Vol. 121, No. 6, 457–465.

Answers to Selected Problems

Chapter 2

2.1   **b.** $D_{60} = 0.41$ mm, $D_{30} = 0.185$ mm, and $D_{10} = 0.09$ mm
    c. 4.56   **d.** 0.929

2.3 $C_u = 6.22$; $C_c = 2.01$

2.5   **b.** $D_{60} = 0.48$ mm, $D_{30} = 0.33$ mm, and $D_{10} = 0.23$ mm
    c. 2.09   **d.** 0.99

2.7   Sand: 46%, Silt: 31%, Clay: 23%

2.9   Sand: 70%, Silt: 16%, Clay: 14%

2.11   Sand: 66%, Silt: 20%, Clay: 14%

2.13   0.0059 mm

Chapter 3

3.3   **a.** 1975 kg/m^3 **b.** 1795.5 kg/m^3
    c. 0.515   **d.** 0.34   **e.** 52.8%
    f. 0.503 kg

3.5   **a.** 16.98 kN/m^3 **b.** 0.58
    c. 0.367   **d.** 78.4%

3.7   19.55%

3.9   **a.** 21.08 kN/m^3
    b. 17.86 kN/m^3
    c. 20.12 kN/m^3

3.11   11.7%

3.13 $e = 0.566$, $\gamma_d = 16.91$ kN/m^3

3.15   17.28%

Chapter 4

4.1 $\rho_{d(max)} = 1885$ kg/m^3, $w_{opt} = 11.5\%$

4.3 $\gamma_{d(max)} = 18.3$ kN/m^3, $w_{opt} = 15.5\%$, and $w = 13\%$ at 0.95 $\gamma_{d(max)}$

3.17

Soil	Group symbol	Group name
1	SC	Clayey sand with gravel
2	GC	Clayey gravel with sand
3	CH	Sandy fat clay
4	CL	Lean clay with sand
5	CH	Fat clay with sand
6	SP	Poorly graded sand with gravel
7	CH	Sandy fat clay
8	SP–SC	Poorly graded sand with clay and gravel
9	SW	Well graded sand
10	SP–SM	Poorly graded sand with silt

4.5 B

4.7 **a.** 14.9 kN/m^3 **b.** 20.4%
   c. 16.39 kN/m^3

4.9 **a.** 18.6 kN/m^3 **b.** 97.9%

4.11 6.71 m

Chapter 5

5.1 0.0754 m^3/hr/m

5.3 0.0288 m^3/hr/m

5.5 2.15 × 10^{-2} cm/sec

5.7 376.4 mm

5.9 0.015 cm/sec

5.11 5.67 × 10^{-2} cm/sec

5.13 0.709 × 10^{-6} cm/sec

5.15 0.0108 cm/sec

5.17 $k_{H(eq)}$ = 0.0000375 cm/sec,
   $k_{V(eq)}/k_{H(eq)}$ = 0.0467

5.19 17.06 × 10^{-6} m^3/m/sec

5.21 2.42 × 10^{-5} m^3/m/sec

Chapter 6

6.1

Point	kN/m^2		
	σ	u	σ'
A	0	0	0
B	30	0	30
C	83.4	29.43	53.97
D	213.60	98.1	115.5

6.3 −26.98 kN/m^2

6.5 1.014 × 10^{-2} m^3/min

6.7 6.04 m

6.9 0.042 kN/m^2

6.11 0.84 kN/m^2

6.13 16.52 kN/m^2

6.15 8 kN/m^2

6.17 143.5 kN/m^2

6.19 163.99 kN/m^2

6.21 106.24 kN/m^2

Chapter 7

7.1 **b.** 47 kN/m^2 **c.** 0.133

7.3 1.33

7.5 152 mm

7.7 172 mm

7.9 5.08 × 10^{-4} m^2/kN

7.11 600.6 days

7.13 648 sec

7.15 1.622 × 10^{-7} m/min

7.17 232 mm

7.19 98 kN/m^2

7.21 24%

7.23

t (yrs)	$U_{r, v}$
0.2	0.615
0.4	0.829
0.8	0.964
1.0	0.984

Chapter 8

8.1 ϕ' = 34°, shear force = 142 N

8.3 0.164 kN

8.5 23.5°

8.7 **a.** 61.55°
   b. σ' = 294.5 kN/m^2, τ = 109.4
      kN/m^2

8.9 **a.** 24.5°
   b. σ' = 236.76 kN/m^2,
      τ = 188.17 kN/m^2

8.11 105.2 kN/m^2

8.13 **a.** 414 kN/m^2
   b. Shear force on plane with
      θ = 45° is 138 kN/m^2 < τ_f =
      146.2 kN/m^2

8.15 94 kN/m^2

8.17 ϕ = 15°, ϕ' = 23.3°

8.19 185.8 kN/m^2

8.21 91 kN/m^2

8.23 −83 kN/m^2

Chapter 9

9.1 **a.** 5.58 m   **b.** 1.207
   c. 0.77 m

9.3 1.26

9.5 5.76

9.7 39.4 m

9.9 1.8

9.11 **a.** 8.21 m   **b.** 14.1 m
   c. 6.98 m

9.13 4.4 m

9.15 1.27

9.17 **a.** 43.2 m   **b.** 31.7 m
         c. 35.9 m   **d.** 21.8 m
9.19 **a.** 1.77   **b.** 2.1
9.21 1.83
9.23 1.0

Chapter 10

10.1 **a.** 13.9%   **b.** 48.44 mm
10.3 50.4 kN/m^2
10.5

Depth (m)	$(N_1)_{60}$
1.5	14
3	12
4.5	13
6	11
7.5	13

10.7 $\phi' \approx 35°$ (average)
10.9 81.4%
10.11 **a.** 35 kN/m^2 **b.** 30.32 kN/m^2
10.13 **a.** 30 kN/m^2 **b.** 1.84
10.15 **a.** 0.65   **b.** 1.37

Chapter 11

11.1 **a.** $P_o = 139.86$ kN/m, $\bar{z} = 1.67$ m
         b. $P_o = 68.79$ kN/m, $\bar{z} = 1.33$ m
11.3 **a.** $P_p = 169.6$ kN/m, $\sigma'_p = 138.5$ kN/m^2
         b. $P_p = 593.3$ kN/m, $\sigma'_p = 296.8$ kN/m^2
11.5 **a.** σ_a (top) $= -33.6$ kN/m^2,
            σ_a (bottom) $= 80.4$ kN/m^2
         b. 1.77 m   **c.** 140.4 kN/m
11.7 **a.** σ_a (top) $= -29.4$ kN/m^2,
            σ_a (bottom) $= 89.4$ kN/m^2
         b. 1.48 m   **c.** 180 kN/m
         d. 201.83 kN/m
11.9 1096 kN/m
11.11 **a.** 1426 kN/m   **b.** 3222 kN/m
         c. 4082 kN/m

Chapter 12

12.1 89.7 kN/m^2
12.3 2400 kN
12.5 1.65 m
12.7 2 m

12.9 1450 kN
12.11 14.42 mm
12.13 65 mm
12.15 2.1 m
12.17 25.2 mm
12.19 792.35 kN/m^2
12.21 11.7 m

Chapter 13

13.1 $FS_{(overturning)} = 3.41$, $FS_{(sliding)} = 1.5$, and $FS_{(bearing)} = 5.49$
13.3 $FS_{(overturning)} = 2.81$, $FS_{(sliding)} = 1.56$, and $FS_{(bearing)} = 3.22$
13.5 $FS_{(overturning)} = 2.79$, $FS_{(sliding)} = 1.66$
13.7 $FS_{(overturning)} = 24.42$, $FS_{(sliding)} = 4.48$, and $FS_{(bearing)} = 11.14$
13.9 $S_V = 0.336$ m, $L = 3.7$ m, and $l_l = 1$ m (minimum)
13.11 A→335.64 kN
         B→223.8 kN
         C→335.64 kN
13.13 A→306.5 kN
         B→439.1 kN
         C→219.15 kN
13.15 **a.** $c_{av} = 84.6$ kN/m^2, $\gamma_{av} = 17.07$ kN/m^3
         b. Use Figure 13.22c with $\sigma = 35.85$ kN/m^2

Chapter 14

14.1 **a.** 261.7 kN
         b. 1184 kN
14.3 **a.** 868.3 kN
         b. 1760.5 kN
         c. 657 kN
14.5 615 kN
14.7 234.8 kN
14.9 1339 kN
14.11 25.81 kN
14.13 3640 kN
14.15 5830 kN
14.17 14,557 kN
14.19 **a.** 898 kN   **b.** 1028 kN
         c. 642 kN
14.21 **a.** 1950 kN   **b.** 1506 kN

INDEX

A

A line, 62, 63
A parameter, triaxial, 265
AASHTO classification system, 63–66
Absolute permeability, 116
Active earth pressure:
  Coulomb, 407–413
  Rankine, 377–381
Activity, 60–61
Adsorbed water, 23
Aeolian soil, 13, 17–18
Allowable bearing capacity, shallow
    foundation:
  based on settlement, 462
  definition of, 431–432
Alluvial soil, 13, 14–17
Alumina octahedron, 20
Angle of friction:
  consolidated, undrained, 267
  correlation, standard penetration
    number, 342
  definition of, 243, 246
  drained, 259
  typical values for, 245, 261
Angle of repose, 286
Area ratio, 339
At-rest earth pressure, 373–375
Auger:
  continuous flight, 333
  helical, 333
  hollow stem, 334
  post hole, 333

Average degree of consolidation:
  radial drainage, 232
  vertical drainage, 211
Average pressure increase, foundation,
    220

B

B parameter, pore water pressure:
  definition of, 256
  typical values for, 218
Backswamp deposit, 15
Bernoulli's equation, 111
Bishop's simplified method, slope
    stability, 314–317
Blasting, compaction, 103
Boiling, 153
Boring, soil exploration:
  auger, 333
  depth, 332
  percussion, 335
  rotary, 335
  spacing, 333
  wash, 335
Boring log, 365–366
Boussinesq's equation, 161–162
Braced cut:
  design, 517–520
  general, 510–511
  ground settlement, 526–527
  heave, 523–525
  lagging, 510
  lateral earth pressure, 514–516

Braced cut: (*continued*)
  lateral yielding, 526
  layered soil, 518–519
  pressure envelope, 514–515
  steel sheet pile, 511–512
  strut, 511
  wale, 511
Braided stream, 15
Brooming, pile, 541

C

Classification, 63–72
Clay:
  activity, 60–61
  definition of, 19
  mineral, 20–23
Coefficient:
  compressibility, 209
  consolidation, radial drainage, 232
  consolidation, vertical
    drainage, 209
  earth pressure at rest, 373–374
  gradation, 32
  uniformity, 32
  volume compressibility, 209
Cohesion, 243, 245
Compacted soil, structure, 104–105
Compaction:
  bell-shaped curve, 84
  double-peak curve, 84
  effect of energy, 83–85
  effect of soil type, 83
  general principles, 78–79
  odd-shaped curve, 84
  one and one-half peak curve, 84
Compression index:
  definition of, 198
  empirical relation for, 198–199
Cone penetration resistance,
    correlation:
  friction angle, 356
  overconsolidation ratio, 358
  preconsolidation pressure, 358
  undrained shear strength, 358
Cone penetration test:
  cone resistance, 351
  electric friction cone, 352

friction ratio, 353
frictional resistance, 351
mechanical friction cone, 352
Confined aquifer, hydraulic
    conductivity, 133
Consistency, clay 53–54
Consolidated drained test, triaxial,
    256–260
Consolidated undrained friction angle,
    267
Consolidated undrained test, triaxial,
    256–268
Consolidation:
  average degree of, 211
  compression index, 198–199
  excess pore water pressure, 210
  fundamentals of, 186–188
  laboratory test for, 188–190
  secondary consolidation, settlement,
    203–205
  settlement, foundation, 220–221
  settlement, primary, 196–198
  settlement, Skempton-Bjerrum
    modifications, 223–226
  swell index, 199
  time-deformation plot, 190
  void ratio-pressure plot, 190–192
Consolidation coefficient:
  logarithm-of-time method, 213
  square-root-of-time
    method, 214
Consolidometer, 188
Constant head test, hydraulic
    conductivity, 116–117
Coring, rock:
  coring bit, 363
  double-tube core barrel, 364
  recovery ratio, 365
  rock quality designation, 365
  single-tube core barrel, 364
Coulomb's earth pressure:
  active, 407–413
  passive, 413–418
Creep, 203
Critical hydraulic gradient, 153
Culman's method, slope stability,
    287–289

D

Darcy's law, 113
Degree of saturation, 39
Density, 41
Depth of boring, 332
Diffuse double layer, 23
Dilatometer test:
    dilatometer modulus, 361
    flat plate, dimensions, 360
    horizontal stress index, 361
    material index, 361
Dipole, 23
Direct shear test:
    saturated clay, 261
    strain controlled, 248
    stress controlled, 248
Dispersing agent, 29
Double layer water, 23
Drift, glacier, 17
Drilled shaft:
    casing method, 586–588
    construction procedure, 585–589
    dry method, 585–586
    load-bearing capacity, 589–592
    settlement, 595–599
    types, 584–585
    wet method, 588–589
Dry density, 41
Dry unit weight, 40
Dune, 17
Dynamic compaction, 103

E

Earth pressure at rest:
    coefficient, 373–374
    coefficient correlation, friction
        angle, 374
    coefficient correlation, plasticity
        index, 374
    partially submerged soil, 375–377
Economical compaction, 95–96
Effective size, 32
Effective stress:
    downward seepage, 149
    partially saturated soil, 156–157
    upward seepage, 151–153
    without seepage, 147–150

Effective stress concept, 147–150
Elevation head, 111
Empirical relations, hydraulic
        conductivity, 122–126
Equipotential line, 136
Exploration report, 367

F

Factor of safety, slope:
    cohesion, 283
    friction, 283
    strength, 283, 284
Failure plane inclination, shear, 246
Falling head test, hydraulic
        conductivity, 117–118
Field compaction, 91–94
Field unit weight:
    nuclear method, 99
    rubber balloon method, 98
    sand cone method, 96–98
Finite slope, definition of, 287
Flow net:
    boundary condition, 136, 137
    definition of, 136
    equipotential line, 136
    flow channel, 138
    flow line, 136
    potential drop, 139
    seepage calculations, 138–140
Fluvial soil, 13
Friction circle, slope stability, 302

G

Gap graded soil, 33
Gibbsite sheet, 20
Glacial soil, 17
Grain-size distribution curve, 27
Gradation, coefficient of, 32
Gravel, 19
Gravity transported soil, 14
Group index, 65–66
Group name, 70–72
Group symbol, 69

H

Hammer, pile-driving:
    double-acting, 544

Hammer (*continued*)
  drop hammer, 543
  single-acting, 543
  vibratory, 543, 544
Head loss, 112
Heave, braced cut, 523–525
Hydraulic conductivity:
  constant head test, 116–117
  definition of, 113–114
  effect of compaction, 105, 106
  empirical relations for, 122–126
  falling head test, 117–118
  pumping from wells, 131–133
  stratified soil, 129–131
  typical values for, 115
Hydraulic gradient, 112–113
Hydrogen bonding, 23
Hydrometer analysis, 27–31

I

Illite, 20
Immediate settlement, shallow
  foundation, 447–455
Infinite slope, stability:
  with seepage, 286
  without seepage, 284–286
Isomorphous substitution, 20

K

Kaolinite, 20
Kozeny-Carman equation, 123–125

L

Lagging, braced cut, 510
Laminary flow zone, 113
Laplace's equation of continuity,
  134–136
Line load, stress, 165–168
Liquid limit, 54–57
Liquidity index, 62
Load transfer mechanism, pile,
  545–546
Loess, 18
Logarithm-of-time method, coefficient
  of consolidation, 213–214

M

Mat foundation:
  bearing capacity, 464–466
  compensated, 457–468
  definition of, 463
  types, 463–464
Meander belt deposit, 15–16
Mechanical analysis, 24–31
Mechanically stabilized earth
    retaining wall:
  external stability, 496
  internal stability, 496
  geogrid reinforcement, 508–509
  geotextile reinforcement, 505–506
  metallic strip reinforcement,
    496–502
Method of slices, slope, 310–317
Mid-plane degree of consolidation,
    228, 230
Modified Proctor test, 86–87
Mohr-Coulomb failure criteria, 243
Mohr's theory, rupture of
    material, 243
Moist density, 41
Moist unit weight, 40
Moisture content, 40
Montmorillonite, 22
Moraine, 17

N

Natural levee, 15
Negative skin friction, pile, 569–572
Neutral stress, 149
Normally consolidated clay, 192–193
Nuclear method, field unit
    weight, 99

O

Octahedral sheet, 20
Oedometer, 188
Optimum moisture content, 79
Organic soil, 18
Overconsolidated clay, 192–193
Overconsolidation ratio, 193
Oxbow lake, 15

P

Partially saturated soil, effective stress, 156–157
Particle-size distribution curve, 27
Passive pressure:
   Coulomb, 413–418
   Rankine, 381–384
Peak shear strength, 250
Percent finer, 27
Percussion drilling, 335
Piezometer, 344
Piezometric level, 112
Pile:
   brooming, 541
   cased, 538, 539
   cast-in-place, 538
   composite, 542
   concrete, 537–538
   displacement, 544
   friction, 543
   nondisplacement, 544
   pipe sleeves, 541
   point bearing, 542
   precast, 537–538
   steel, 534–537
   timber, 540–542
   uncased, 540
Pile capacity:
   allowable, 556
   friction, 550–556
   group, 574–578
   point, 547–550
   rock, 557
   α method, 553
   β method, 555
   λ method, 553
Pile driving:
   Danish formula, 567
   ENR formula, 566
   modified ENR, 567
Pile group:
   consolidation, 580–582
   elastic settlement, 579–580
Plastic limit, 57
Plasticity chart, 62–63

Plasticity index, 57
Pneumatic rubber-tired roller, 93
Point bar deposit, 15
Point load, stress, 161–165
Poisson's ratio, 163
Poorly graded soil, 33
Pore air pressure, 157
Pore water pressure, 149
Porosity, 39
Potential drop, 139
Precompression:
   general considerations, 227–228
   general equation, 228–230
Preconsolidation pressure:
   definition of, 193
   determination of, 193–194
Pressure envelope, braced cut:
   sand, 514, 515
   soft and medium clay, 515
   stiff clay, 515
Pressure head, 111
Pressuremeter test:
   bore-hole diameter, 359, 360
   guard cell, 359
   limit pressure, 360
   measuring cell, 359
   modulus, 360
Proctor compaction test, 79–83
Pumping from well, hydraulic conductivity, 131–133

Q

Quick condition, 153

R

Rankine active pressure:
   coefficient, 381
   depth of tensile crack, 389, 392
   Rankine active state, 377–381
   slip plane, 381
Rankine passive pressure:
   coefficient, 384
   Rankine passive state, 381–384
Reconnaissance, exploration, 331
Recovery ratio, 365

Rectangular loaded area, stress, 174–178
Reinforcement, soil:
    geogrid, 495
    geotextile, 494
    metallic strip, 493
Relative compaction, 94
Relative density, 51
Residual soil, 14
Retaining wall:
    cantilever, 476
    counterfort, 476
    gravity, 475
    proportioning, 477
Rock, friction angle, 557
Rock-quality designation, 365
Rotary drilling, 338
Rubber balloon method, field unit weight, 98

S

Sampling:
    spring core catcher, 339
    standard split spoon, 338
    thin-wall tube, 243
Sand, 19
Sand cone method, 96–98
Sand drain, 231–235
Saturated density, 43
Secondary compression index, 204
Secondary consolidation, 203–205
Seepage calculation, flow net, 138–140
Sensitivity, 274–275
Settlement, consolidation, 196–198
Settlement, elastic:
    flexible, 447, 448
    material parameters, 457
    profile, 448
    rigid, 447, 448
    strain influence factor, 458–460
Settlement, pile, elastic, 448
Shallow foundations:
    bearing capacity factors, 426, 427, 428, 429
    bearing capacity theory, 425–430
    depth factor, 428, 430
    eccentric load, 438–444

effect of water table, bearing capacity, 430–431
effective area, 436
factor of safety, 431–432
general bearing capacity equation, 428
general shear failure, 423
inclination factor, 428, 430
local shear failure, 423
punching shear failure, 423
shape factor, 428, 429
two-way eccentricity, bearing capacity, 438–444
ultimate bearing capacity, 423
Sheepsfoot roller, 93
Sheet pile:
    allowable design, flexural stress, 511
    interlock, 511
    section, 512
Shrinkage limit, 58–60
Sieve analysis, 24–27
Sieve size, 25
Silica sheet, 20
Silica tetrahedron, 20
Silt, 19
Sleeve, pile, 541
Slope stability:
    base failure, 290
    Bishop's simplified method, 314–317
    $c-\phi$ soil, 300–307
    critical height, Culman, 289
    Culman's method, 287–289
    eqrthquake foraces, 322–325
    friction circle, 302
    mass procedure, saturated clay, 292–300
    method of slices, 310–313
    midpoint circle, 290
    rotational collapse mechanism, 305, 306, 307
    slope failure, 282
    stability number, 293
    steady-state seepage, 312–313, 317, 318–320
    toe circle, 290
Smooth-wheel roller, 91
Soil-separate size limits, 19

Specific gravity, 23–24
Specific surface, 20
Specification:
    field compaction, 94–95
    modified Proctor test, 88
    standard Proctor test, 88
Split-spoon sampler, 338
Spring core catcher, 338, 339
Square-root-of-time method, coefficient
        of consolidation, 214
Stability, retaining wall:
    bearing capacity failure, 484–487
    overturning, 480–484
    sliding, 482–484
Stability number, vibroflotation, 102
Standard penetration number:
    correction factor, sand, 339–340
    definition of, 338
    relative density, correlation, 342
    undrained shear strength,
        correlation, 341
Standard Proctor:
    hammer, 79, 80
    mold, 79, 80
    test, 79–83
Stokes' law, 27
Stress:
    circular area, 172–174
    line load, 165–168
    point load, 161–165
    rectangular area, 174–178
    strip, 170–171
Structure, compacted soil, 104
Strut, 511
Swell index:
    definition of, 197
    empirical relations for, 199

T
Thin-wall tube, 343
Thixotropy, 274–276
Time factor, 210
Time rate of consolidation,
        206–212
Total stress, 147
Transient flow zone, 113
Transported soil, 13

Triaxial test:
    consolidated drained, 256–261
    consolidated undrained, 265–268
    deviator stress, 256
    general, 255–256
    Skempton's parameters, 256, 265
    unconsolidated undrained, 270–272
Turbulent flow zone, 113

U
U line, 62, 63
Ultimate strength, shear, 248
Unconfined compression strength:
    definition of, 272
    effect of compaction, 106–107
    typical values for, 273
Unconfined compression test, 272–274
Unconsolidated undrained test, triaxial,
        270–272
Undrained shear strength:
    definition of, 270
Unified classification system:
    group name, 70–72
    group symbol, 68–69
Uniformity coefficient, 32
Uniformly loaded circular area, stress,
        172–174
Unit weight, 40

V
Vane shear test:
    Bjerrum's correction, 348
    field vane dimensions, 347
    vane shear, 345–347
Vibratory plate, 93
Vibratory roller, 93
Vibroflotation, 99–103
Virgin compression curve, 194
Viscosity, 116
Void ratio, 39
Void ratio pressure plot, consolidation:
    effect of disturbance, 194–196
    from laboratory tests, 190–192

W
Wash boring, 335
Water content, 40

Water table, observation of, 343–345
Weathering:
  chemical, 13
  mechanical, 13
Well-graded soil, 33
Westergaard material, stress:
  circularly loaded area, 180
  line load, 179
  point load, 163–165
  rectangular area, 179

Y

Yielding of wall, earth pressure, 384–385

Z

Zero-air-void unit weight, 81